DESIGN OF MODERN
CONCRETE HIGHWAY BRIDGES

DESIGN OF MODERN CONCRETE HIGHWAY BRIDGES

CONRAD P. HEINS

Institute for Physical Science and Technology
University of Maryland

RICHARD A. LAWRIE

T.Y. Lin International
Alexandria, Virginia

A Wiley-Interscience Publication

JOHN WILEY & SONS

New York • Chichester • Brisbane • Toronto • Singapore

Library of Congress Cataloging in Publication Data:

Heins, Conrad P.
   Design of modern concrete highway bridges.

   "A Wiley-interscience-publication."
   Bibliography: p.
   Includes index.
   1. Bridges, Concrete—Design and construction.
I. Lawrie, Richard A.   II. Title.
TG335.H46   1983   624$'$.2   83-6748
ISBN 0-471-87544-9

To our parents

Conrad and Olga Heins
Joseph and Frances Lawrie

IN MEMORIAM

This book was written and in the process of being published when Dr. Heins was tragically killed in an airplane accident on December 24, 1982. Certainly, Dr. Heins' death is a great loss to the profession of structural engineering. I personally will miss his professional wisdom and friendly spirit and I am sure many students and practicing engineers share this sentiment. It is my hope that this text may be a reminder of a man who had a desire to contribute his best to the field of structural engineering. His record shows that he did.

RICHARD A. LAWRIE

Preface

Various factors in recent decades have caused concrete bridges to become the most prevalent type of highway structure throughout the world. In addition to such developments as higher strength concrete and higher yield strength reinforcing, economy and aesthetics are two major contributions that have produced this abundance. Also, prestressed concrete has become more competitive in many span ranges because of the many innovations in construction methods, materials, and equipment.

A study of the design procedures used by different consultants, highway departments, and government agencies would indicate that bridge designers encounter many areas in which the proper method of design is not well defined. These areas often require the designer to make judgments and reasonable approximations of complicated problems. One of our prime objectives in writing this text is to provide guidance to those who make these judgments and approximations. Many example problems and solutions are presented in this regard because the designer can learn best how to deal with them by working with similar problems. In each case the examples are preceded by a detailed explanation of techniques and procedures and a narrative of the application of today's AASHTO Code.

Many books provide thorough solutions to analytical problems in structural design, but considerable effort is required to implement these solutions in bridge design. Therefore, a second prime objective of this text is to present designers with explanations and examples that can be applied directly to the design problems they may encounter.

Often, attention is focused on the significant bridge structures of the present day and the multitude of not so significant bridges is forgotten. Because guidance is just as necessary in the design of routine structures, every effort has been made to provide examples of all types. Chapters are included on short-span slab, Tee-beam, long-span posttensioned, and segmental designs.

The AASHTO Standard Specifications for Highway Bridges (12th ed., 1977) and *Interim Specifications* (1978–1980) are used throughout this text as the design code, and reasonable guidance to the application of these design specifications is given. This sometimes involves modification of the code when practical, but an effort has always been made to follow its intent. The text also utilizes English units because the AASHTO specifications and the bridge building industry in the United States currently use them. A table of Conversion Factors for English and Metric (SI) Units is provided for the convenience of those accustomed to the metric system.

Finally, it should be said that we have attempted to be practical in trying to describe design procedures in a usable manner on the basis of experience and knowledge in the design of these types of bridges.

CONRAD P. HEINS
RICHARD A. LAWRIE

College Park, Maryland
Alexandria, Virginia
August 1983

Acknowledgments

I wish to acknowledge the instruction received from my professors during my collegiate days at Drexel Institute and Lehigh University. Also, appreciation must be extended to the AASHTO bridge committee and FHWA staff who encouraged me to pursue this writing venture.

C.P.H.

I wish to express appreciation to former FHWA supervisors whose guidance and instruction in the field of bridge design has offered a rewarding career. In particular, David M. Goodall, chief of FHWA's Western Bridge Design Office for 15 years is to be thanked for his diligent effort to impart knowledge of practical design procedures to his staff. Much of the information presented in this book is the result of his instruction and insight.

R.A.L.

Contents

Conversion Factors

English to SI	SI to English

Length

1 in. = 25.40 mm	1 mm = 0.0394 in.
1 ft = 0.3048 m	1 m = 3.281 ft
1 in. = 0.0254 m	1 m = 39.37 in.
1 chain = 20.117 m	1 m = 0.0497 chains
1 mile = 1.609 km	1 km = 0.622 mile

Area

$1 \text{ in.}^2 = 645.2 \text{ mm}^2$	$1 \text{ mm}^2 = 0.00155 \text{ in.}^2$
$1 \text{ ft}^2 = 0.0929 \text{ m}^2$	$1 \text{ m}^2 = 10.76 \text{ ft}^2$
$1 \text{ yd}^2 = 0.835 \text{ m}^2$	$1 \text{ m}^2 = 1.196 \text{ yd}^2$

Volume

$1 \text{ in.}^3 = 16.387 \text{ mm}^3$	$1 \text{ cm}^3 = 0.0610 \text{ in.}^3$
$1 \text{ ft}^3 = 0.0283 \text{ m}^3$	$1 \text{ m}^3 = 35.31 \text{ ft}^3$
$1 \text{ yd}^3 = 0.765 \text{ m}^3$	$1 \text{ m}^3 = 1.564 \text{ yd}^3$

Moment of Inertia

$1 \text{ in.}^4 = 41.62 \text{ cm}^4 = 416{,}200 \text{ mm}^4$	$1 \text{ mm}^4 = 2.40 \times 10^{-6} \text{ in.}^4$
$\quad = 0.4162 \times 10^{-6} \text{ m}^4$	$1 \text{ m}^4 = 2.40 \times 10^6 \text{ in.}^4$

Mass

1 lb = 0.454 kg	1 kg = 2.205 lb
1 ton (2000 lb) = 907.2 kg	1 Mg = 1.102 ton (2000 lb)
1 tonne (metric) = 1.102 ton (2000 lb)	

Force

$1 \text{ lb} = 4.448 \ N$	$1 \ N = 0.2248 \text{ lb}$
1 kip (k) = 4.448 kN	

Stress

$1 \text{ psi} = 6.895 \text{ kPa (kN/m}^2)$	1 MPa = 145.0 psi
$1 \text{ ksi} = 6.895 \text{ MN/m}^2$	$1 \text{ N/mm}^2 = 145.0 \text{ ksi}$
	$1 \text{ MN/m}^2 = 145.0 \text{ ksi}$

DESIGN OF MODERN
CONCRETE HIGHWAY BRIDGES

ONE

Introduction

1.1 HISTORICAL DEVELOPMENT

For a few thousand years the classical form in bridge design has been the vault or arch. This structure, because of its inherent contour, utilized masonry as its material.

The use of concrete as a building material, however, was not considered until late in the nineteenth century. In general, the first practical application of reinforced concrete is credited to Monier in 1867. In 1866 Wayss and Koenen, in Germany, conducted a series of tests on reinforced concrete beams, and in the years between 1891 and 1894 extensive research in this field was done by Moeller of Germany, Wunsch of Hungary, Melan of Austria, and Hennebique of France.

The first application of reinforced concrete to bridge structures was pioneered by Hennebique. One notable project, completed in 1905 by Hennebique, is a 55-m, single-span bridge over the Ourthe at Liège, Belgium.

Robert Maillart of Switzerland was successful also in the design and construction of reinforced concrete bridges. In 1901 Maillart built a three-hinged, box-section arch with solid spandrels over the Rhine at Tavanas, Switzerland, which became the prototype for similar structures in the next 40 years.

In addition to the work by Hennebique and Maillart, Freyssinet of France constructed numerous large-span, reinforced concrete arch bridges, among which the Plougasted over the Elom in 1930 was significant and which consisted of arch spans 178 m in length.

In the same period prestressed concrete concepts[1,2] were being formulated by Jackson and Doehring (1886 and 1888). Their application

1

Table 1.1 Span Lengths for Bridge Structures

Type	Material	Span Range (m)	Span Range (ft)	Maximum Constructed (m)	Projected (m)
Slab	Concrete	0–12	0–40		
Beam	Concrete	12–210	40–700	Bendorf 208	
	Steel	30–300	100–860	Rio-Niteroi 300	
Truss	Steel	90–550	300–1800	Quebec 549	
Arch rib	Concrete	90–130	300–1000	Gladesville 305	
	Steel	120–370	400–1200	Fremont 382	
Arch truss	Steel	240–520	800–1700	Bayonne 504	
Cable-stayed	Concrete	90–270	300–800	Waal 267	
	Steel	90–350	300–1100	Duisburg 350	Great Belt 600
Suspension	Steel	300–1400	1000–4500	Verrazano 1298	Humber 1400

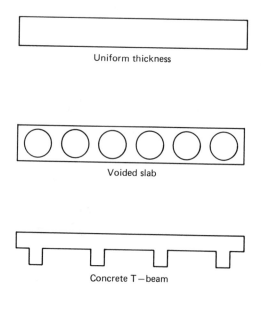

Uniform thickness

Voided slab

Concrete T–beam

Concrete box girder

Figure 1.1 Types of concrete bridges.

2

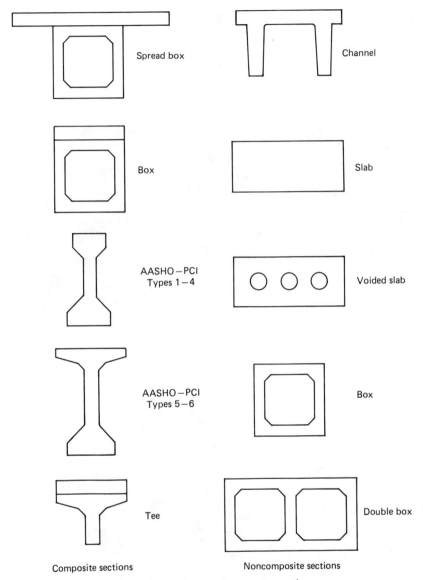

Figure 1.2 Types of concrete section.

was not successful because of the high losses in prestress caused by shrinkage and creep of the concrete. It was not until 1926–1928 when Freyssinet was able to control these losses with high-strength steel that prestressing was considered feasible.

Between 1930 and 1940 reinforced concrete bridges consisted of solid-state spans (40-ft max.) or combinations of slab and girders (T-beam) in a d/L ratio equal to 12 and spans of 100 ft.

Figure 1.3　Johnston Street Bridge—Liberia.

Figure 1.4　Rio Colorado Bridge—Costa Rica.

Because of the steel shortage in the postwar years, prestressed concrete became the more prevalent choice for long-span structures.

Today prestressed, post-tensioned reinforced concrete structures have been combined successfully in major bridges that incorporate the integration of precasting and erection (i.e., segmental). The design concepts of this technique, as well as other conventional methods, are described in detail in this text.

1.2 BRIDGE TYPES

The major advantage in the use of concrete for bridges is the wide variation that can be achieved in form. This flexibility, however, does not limit its exclusive use for all major structures. Because of factors like ratio of dead to live load, depth constraints, availability of material, and labor costs, steel structures may be cost effective and must be considered as a possible alternate. Many major bridge projects today (1983) include alternate designs (steel and concrete) with appropriate working drawings for use in bidding.

The range in span of various bridge types is given in Table 1.1.[1,2] As illustrated in this table, steel and concrete can be utilized for many bridge types.

General reinforced concrete bridges can consist of decks, T-beams, or cells (Figure 1.1). Combinations of these types and precasting these

Figure 1.5   Rio Higriamio—Dominican Republic.

elements can enhance their versatility. Concrete section types may consist of the following (Figure 1.2):

Composite spread box
Composite box
Composite AASHO-PCI standard sections
Type I 30–45 ft span
Type II 40–60 ft span
Type III 55–80 ft span

Figure 1.6   I-205 Bridge over Columbia River, Portland, Oregon.

Figure 1.7   Hegenberger Overpass, Oakland, California.

Type IV 70–160 ft span
Type V 90–120 ft span
Type VI 110–140 ft span
Composite/Tee 40–60 ft
Channel 20–40 ft
Solid slab
Voided slab
Single box
Double box

Although these elements are standardized, the engineer can modify the geometry to accommodate the bridge under design. Examples of modern bridges of various configurations are shown in Figures 1.3 through 1.7 (courtesy of T. Y. Lin International).

BIBLIOGRAPHY

1. C. O'Connor, *Design of Bridge Superstructures*, Wiley, New York, 1971.
2. A. R. Cusens and R. P. Paina, *Bridge Deck Analysis*, Wiley, New York, 1975.

TWO

Materials and Properties

2.1 COMPRESSIVE STRENGTH

The compressive strength of concrete is controlled by the amount of cement, aggregates, water, and various admixtures contained in the mix. The principal factor, however, in the determination of concrete strength is the water-to-cement ratio (see Figure 2.1).

The strength of concrete is designated by the term f_c' and represents the 28-day compressive strength determined from a 6-in. round by 12-in. cylinder.

The compressive strength, stress-strain response of these cylinders is dependent primarily on the concrete strength f_c' (see Figure 2.2). In general, the maximum usable compressive strain is 0.003 in. in.$^{-1}$ The specified concrete strengths can vary from 3 to 11 ksi and are selected according to the design requirements. Reinforced concrete structures usually specify 3–4 ksi, prestressed concrete, 5–6 ksi, and special structures, 6–11 ksi.

2.2 TENSILE STRENGTH

The strength of concrete in tension is important in that it affects the amount and frequency of cracking. The tensile strength of concrete is determined by split cylinder tests in which the cylinders are compressed when positioned on their sides.

According to the AASHTO code, the allowable tensile stress of concrete is computed from

$$f_t \leqslant 0.21 f_r \tag{2.1}$$

8

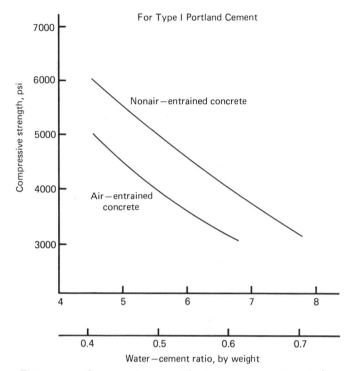

Figure 2.1 Compressive strength versus water cement ratio.

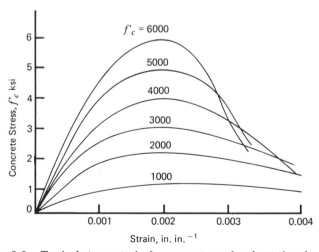

Figure 2.2 Typical stress–strain for concrete under short-time loading.

9

where normal weight concrete $f_r = 7.5\sqrt{f_c'}$,

sand-light weight concrete $f_r = 6.3\sqrt{f_c'}$,

all-light weight concrete $f_r = 5.5\sqrt{f_c'}$.

2.3 MODULUS OF ELASTICITY

The modulus of elasticity of concrete varies with its compressive strength f_c'. A plot of a typical stress (f_c')-strain curve is shown in Figure 2.3 which contains three tangent lines to the curve:

1. Initial modulus (slope at origin).
2. Tangent modulus (slope at stress = $0.5 f_c'$).
3. Secant modulus (slope from origin to $0.5 f_c'$).

These lines can be used to approximate the modulus of elasticity. A study of many data,[2] has resulted in the direct determination of E_c and incorporates the weight of the concrete. According to the AASHTO code, the modulus E_c (psi) is computed from

$$E_c = W^{1.5} \times 33\sqrt{f_c'} \qquad (2.2)$$

for the concrete weight W between 90 and 155 lb/pcf and f_c' in psi.

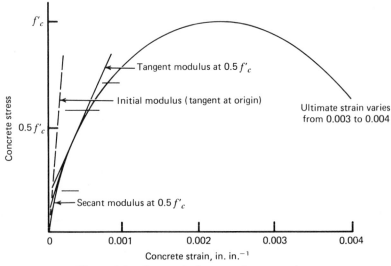

Figure 2.3 Stress–strain curve for concrete.

For normal weight concrete (W = 145 pcf)

$$E_c = 57,000\sqrt{f_c'} \qquad (2.3)$$

We apply eq. (2.3) for normal weight concrete and vary f_c' between 3 and 5 ksi to obtain the following modulus E_c:

f_c' (psi)	E_c (ksi)
3000	3140
3500	3390
4000	3620
4500	3850
5000	4050

The modulus of elasticity of nonprestressed or reinforcement steel is assumed to be

$$E_s = 29 \times 10^3 \text{ ksi} \qquad (2.4)$$

2.4   CREEP AND SHRINKAGE

Creep and shrinkage of concrete are time-dependent deformations and must be included in the design of bridge structures. Short-term loading (live loads) on a concrete bridge induces elastic deformations. Dead loads or superimposed dead loads however are long-term effects that must be considered.

Creep of concrete is the phenomenon in which the deformation continues with time under constant load. This response can be related to the initial elastic deformation or strain by the following equation:

$$C_t = \frac{\text{creep strain}}{\text{initial elastic strain}}$$

$$C_t = \left(\frac{t^{0.60}}{10 + t^{0.60}}\right) C_u \qquad (2.5)$$

which is represented by Figure 2.4; C_u = 2.35 and t = time.

In general, the AASHTO code incorporates the effects of creep by multiplying the factor

$$\left[2 - 1.2\left(\frac{A_s'}{A_s}\right)\right] \geqslant 0.6 \qquad (2.6)$$

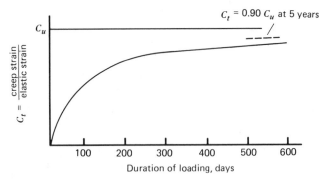

Figure 2.4   Standard creep coefficient curve.

where A_s' = area of compressive reinforcement,
 A_s = area of tension reinforcement.

Shrinkage is defined as the volume change in the concrete with respect to time. The associated shrinkage strain can be computed from

$$E_{sh} = \left(\frac{t}{35 + t}\right)(E_{sh})_u \qquad (2.7)$$

(see Figure 2.5); $(E_{sh})_u = 800 \times 10^{-6}$ in. in.$^{-1}$
   According to the AASHTO code the shrinkage strain is 200×10^{-6} in. in.$^{-1}$

2.5   STEEL REINFORCEMENT

Steel reinforcement can consist of bars, welded wire fabric, or wires. In general, the main reinforcement has formed deformations to provide

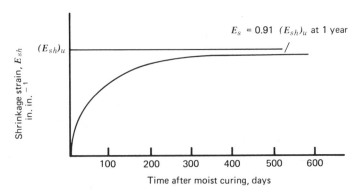

Figure 2.5   Standard shrinkage strain curve.

Table 2.1   Reinforcing Bar, Weight and Dimensions

Bar Number	Weight (lb ft^{-1})	Nominal Dimensions		
		Diameter	Area	Perimeter
3	0.376	0.375	0.11	1.178
4	0.668	0.500	0.20	1.571
5	1.043	0.625	0.31	1.963
6	1.502	0.750	0.44	2.356
7	2.044	0.875	0.60	2.749
8	2.670	1.000	0.79	3.142
9	3.400	1.128	1.00	3.544
10	4.303	1.270	1.27	3.990
11	5.313	1.410	1.56	4.430
14	7.65	1.693	2.25	5.32
18	13.60	2.257	4.00	7.09

better bonding to the concrete. These steel bars are rolled in diameters of $\frac{3}{8}$-$2\frac{1}{4}$ in. and are designated by bar sizes 3–18 (Table 2.1).

Reinforcing bar steel has various ASTM designations:

(Billet steel) ASTM A 615, Grade 40 F_y = 40 ksi
Grade 60 F_y = 60 ksi
Grade 75 F_y = 75 ksi

where F_y is the minimum yield point.

For service load design grade 40 or 60 reinforcement is specified by AASHTO.

2.6   ALLOWABLE STRESSES

2.6.1   Service Load Design

According to the AASHTO code, the following stresses for concrete and reinforcement are required:

Concrete

For service load design the stresses in concrete shall not exceed the following:

1.   **Flexure.**   Extreme fiber stress in compression, f_c, $0.40f_c'$; extreme fiber stress in tension for plain concrete, f_t, $0.21f_r$; modu-

lus of rupture, f_r, from tests or if data are not available:

(a) normal weight concrete $7.5\sqrt{f_c'}$ or $(0.623\sqrt{f_c'})$

(b) "sand-lightweight" concrete $6.3\sqrt{f_c'}$ or $(0.523\sqrt{f_c'})$

(c) "all-lightweight" concrete $5.5\sqrt{f_c'}$ or $(0.456\sqrt{f_c'})$

2. **Shear.** Beams and one-way slabs and footings, shear, carried by concrete v_c, $0.9\sqrt{f_c'}$ or $(0.79\sqrt{f_c'})$; maximum shear carried by concrete plus shear reinforcement v, $v_c + 4\sqrt{f_c'}$ or $(v_c + 0.332\sqrt{f_c'})$; two-way slabs and footings on shear, carried by concrete, v_c, $1.8\sqrt{f_c'}$ or $(0.149\sqrt{f_c'})$; maximum shear carried by concrete plus shear reinforcement v, $3\sqrt{f_c'}$ or $(0.249\sqrt{f_c'})$.

3. Bearing on loaded area, f_b, $0.30 f_c'$:

(a) When the supporting surface is wider on all sides than the loaded area the allowable bearing stress on the loaded area may be increased by $\sqrt{A_2/A_1}$ but not more than 2.

(b) When the supporting surface is sloped or stepped A_2 may be taken as the area of the lower base of the largest frustrum of a right pyramid or cone contained wholly within the support and having for its upper base the loaded area and side slopes of one vertical to two horizontal.

(c) When the loaded area is subjected to high edge stresses due to deflection or eccentric loading the allowable bearing stress on the loaded area shall be multiplied by a factor of 0.75. The requirements of (a) and (b) shall also apply.

Reinforcement

For service load design the tensile stress in the reinforcement f_s shall not exceed the following:

Grade 40 reinforcement, 20,000 psi (137.895 MPa)
Grade 60 reinforcement, 24,000 psi (165.474 MPa)

Fatigue Stress Limit. The range between a maximum tension stress and minimum stress in straight reinforcement caused by live load plus impact shall not exceed $21 - 0.33 f_{min} + 8(r/h)$ where r/h is normally taken as 0.3. Bends in primary reinforcement shall be avoided in regions of high stress.

Design Theory

For investigation of stresses at service loads the straight-line theory of stress and strain in flexure shall be used with the following assumptions:

1. Strains vary linearly as the distance from the neutral axis except for deep flexural members with overall depth span ratios greater than $\frac{2}{5}$ for continuous spans and $\frac{4}{5}$ for simple spans; a nonlinear distribution of strain should be considered.

2. The stress-strain relationship of concrete is a straight line under service loads within the allowable service-load stresses.

3. In reinforced concrete members concrete resists no tension.

4. The modular ratio $n = E_s/E_c$ may be taken as the nearest whole number (but not less than 6). Except in calculations for deflections, the value of n for lightweight concrete shall be the same as that for normal weight concrete of the same strength.

5. In doubly reinforced flexural members an effective modular ratio of $2E_s/E_c$ shall be used to transform the compression reinforcement for stress computations. The compressive stress in such reinforcement shall not be greater than the allowable tensile stress.

2.6.2   Load Factor Design

The basic criteria for load factor design involve factoring the design loads and then comparing the results with the ultimate capacity. In general, the relationship takes the form of

$$\gamma[\beta_D(DL) + \beta_{LL}(L + I)] \leqslant P_u$$

where γ = load factor,
         β = coefficient,
$DL, L + I$ = induced dead- and live-load plus impact forces (M, V, P),
         P_u = ultimate resistance of section.

The γ and β factors are specified according to the loading combinations group given in Table 3.1. The ultimate resistance of the section is presented in subsequent chapters.

BIBLIOGRAPHY

1. *Recommended Practice for Selecting Proportions for Normal Weight Concrete* (ACI Std. 211.1-70), American Concrete Institute, Detroit, 1970.

2. Adrian Pauw, Static Modulus of Elasticity of Concrete as Affected by Density, *Proc. Am. Concr. Inst.*, 57, 679–687 (December 1960).

3. C. K. Wang and C. G. Salmon, Reinforced Concrete Design, *Intext*, New York, 1973.

THREE

Loads on Bridges

3.1 DEAD LOADS

The dead load of a highway bridge consists of the weight of the structure plus any equipment attached. Some bridges carry water or utility lines that may add appreciable weight. It is necessary to make a preliminary estimate of the dead load on which to base the initial design. The weight of the structure can then be calculated and compared with the estimated weight. Most likely the two weights will not agree and a second cycle of design based on the new dead load must be produced. If changes have been made in size of members at the completion of this second cycle, the dead load must be recalculated. This process of design refinement is repeated until the designer is sure that the calculations contain the "as-built" weight of the bridge. An experienced design engineer usually arrives at a convergence in one or two cycles. If a computer program is used in the design process, several cycles can be performed in a short time.

It is possible to arrive at a final value of the dead load for one part of the structure before proceeding with the design of a supporting part; for example, the floor system can be designed before the main girders or trusses and the final weight of the entire superstructure can be determined before beginning the design of the substructure. It is obvious therefore that it is not difficult to select a preliminary dead load for any part of the structure. Some published data include empirical equations for determining the estimated weights of bridge structures, most of which, however, are limited in application or are obsolete as a result of changes in design specifications. A study of similar bridges is a good method of obtaining preliminary dead-load estimates.

The following unit weights of materials in pounds per cubic foot are commonly used in highway bridges:

Concrete	150
Steel	490
Cast iron	450
Asphalt paving	140
Timber	35–60
Stone masonry	150–175
Aluminum	170

The distribution of the dead load is another significant factor. In most instances it is assumed to be uniformly distributed along the length of a structural element, such as a slab, beam, or truss. Continuous bridges may have main load-carrying elements of varying depth. To consider the weight as uniformly distributed for nonprismatic bridges may lead to some, though usually negligible, error. Any error would normally be on the side of safety. Other cases of dead-load distribution are discussed in the appropriate sections of this book.

Adequate allowance should be made in the design for additions to the structure such as future wearing surface and utility lines.

3.2   LIVE LOADS

A highway bridge should be designed to support safely all vehicles that might pass over it in its lifetime. It is not possible for the designer to know what vehicles will use the structure or what the required life of the bridge will be. To ensure the safety of the structure some form of control must be maintained and provision must be made for sufficient strength to carry present and predicted loads. The regulation of vehicles must be such that those of excessive weight are prohibited from crossing. Design control is provided in the United States by AASHTO, which specifies acceptable live loads; traffic regulations covered by state laws specify the legal weights of motor vehicles.

The present design vehicles contained in the AASHTO specifications[1] were adopted in 1944. The loadings consist of five weight classes, namely, H 10, H 15, H 20, HS 15, and HS 20. The vehicles for each of the five classes shown in Figure 3.1 are hypothetical and were not selected to resemble any particular existing design. Any actual vehicle that would be permitted to cross a bridge should not produce stresses greater than those allowed to the hypothetical vehicle.

The lighter loads, H 10 and H 15, are used in the design of lightly traveled state roads, whereas the H 20 and HS 20 are adopted for national highways, the HS 20 in particular for the Interstate Highway Sys-

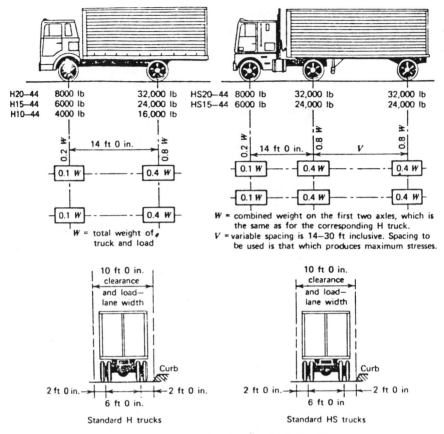

Figure 3.1   AASHTO design vehicle loading.

tem for which an additional alternate loading has been allotted. This
loading consists of two axles spaced at 4 ft and weighing 24 k each.

The HS truck loadings show a variable spacing of the two rear axles
of 14–30 ft. The correct spacing is the length that produces the maximum effect. For stresses in simple span bridges this spacing has the
minimum value of 14 ft. For continuous spans, however, a spacing
greater than 14 ft may produce the maximum effect. The influence
diagram indicates the proper spacing of the axles for maximum stresses.

In addition, the specifications contain equivalent loadings (Figure
3.2) to replace truck loadings when they produce greater stresses than
the truck. Before the introduction of the 1944 specifications the design live load consisted of a basic H truck, preceded and followed by a
train of trucks that totaled three-quarters of the weight of the basic
truck. In 1944 the HS truck was developed and equivalent lane loading
took the place of the train of trucks. Presently only one truck is used
per lane. For longer spans the equivalent loading produces greater

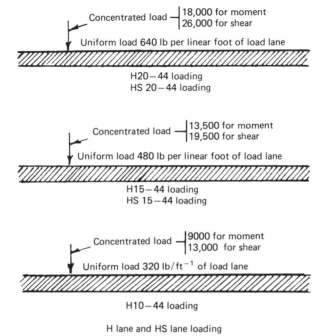

Figure 3.2   AASHTO design land loading.

stresses than the single truck; for example, the H 20 truck produces the greater bending moment in simple beam spans for spans up to 56 ft. For spans longer than 56 ft the equivalent lane loading produces the greater bending moment. The span length at which the loading changes in shear calculations is different from that in moment calculations. The equivalent lane loading approximates the shear and moment produced on long spans by a train of trucks specified before 1944.

The concentrated load in equivalent lane loadings is different for moment than for shear. Only one concentrated load is used in a simple span or for a positive moment in continuous spans. Two concentrated loads are used for a negative moment. The equivalent lane load is placed to produce the maximum stresses, but the uniformly distributed load can be divided into segments when applied to continuous spans. Both concentrated and uniform loads are distributed over a 10-ft lane width on a line normal to the centerline of the lane.

3.3   DYNAMIC EFFECT OF VEHICLES

It is well-known that a vehicle moving across a bridge at a normal rate of speed produces greater stresses than a vehicle that remains in a static position on the structure. This increment in stress can be called the *dy-*

namic effect. The terminology for the dynamic effect, among bridge designers and in bridge design specifications, is *impact.* The latter term, however, is not scientifically correct because it denotes one body striking another, which takes place only when a wheel falls into a "chuck hole." The total dynamic effect is felt not only when the wheels of a vehicle strike deck imperfections but also when the live load is applied to the structure for a very short time. It can be proved by simple calculations of the theory of dynamics that a load instantly applied to a beam causes stresses of twice the magnitude obtained when the same load is static on the beam. In actual load applications to highway bridges the total live loading is never instantaneous but is applied to the structure in a finite period. The dynamic effect due to sudden loading is variable for all structural elements of a bridge.

In addition to the true impact and sudden loading effects, a third effect is caused by the vehicle vibrating on its springs. Uneven roadway surfaces contribute to this effect. The vibration of the vehicle induces vibrations in the structure. The magnitude of stresses is dependent on the relative masses of the vehicle and the bridge, the natural frequency of the structure, and the damping characteristics of the bridge.

An attempt to make an analytical study of the dynamic effect of a particular vehicle crossing a specific bridge would lead to many complications because of the quantities that could be evaluated only approximately. Because it is difficult to make a confident analysis, the procedure has been to assign an approximate quantity for the dynamic effect. This quantity is determined by a simple equation given in the AASHTO specifications for determining the "impact factor":

$$I = \frac{50}{125 + L} \tag{3.1}$$

The stresses due to impact are calculated by multiplying the live load stress by the value of I obtained from eq. 3.1. They are then added directly to the live load stresses to obtain the total stresses due to vehicle loads. The quantity L is defined as the loaded length of the structure, a value that is more specifically the length of structure that would be loaded to produce maximum stress if the live load consisted of a uniformly distributed load. Further explanations of L for continuous structures are given in the appropriate chapters.

The maximum value of the impact factor suggested by AASHTO is 0.30. This corresponds to a simple span of 41.7 ft. All shorter spans use the value 0.30 for impact.

Most bridge designers would prefer to use a more scientific approach than that allowed by eq. 3.1; the amount of research on this subject has, however, been limited and until more studies are forthcoming eq. 3.1 will have to suffice. We believe that except in unusual cases the value obtained from eq. 3.1 is conservative. The percentage of stress due to the dynamic effect is relatively small for long-span bridges.

3.4 LONGITUDINAL FORCES

When a vehicle brakes or accelerates longitudinal forces are transmitted from its wheels to the deck of the bridge. The magnitude of the longitudinal force depends on the amount of acceleration or deceleration. The maximum longitudinal force results from sudden braking of the vehicle, the magnitude of which is dependent on its weight, its velocity at the instant of braking, and the time it takes to come to a complete stop. This force is given by

$$F = \frac{W}{g} \left(\frac{\Delta V}{\Delta t} \right) \tag{3.2}$$

where W = weight of vehicle (k),
g = acceleration of gravity—32.2 ft s^{-2},
ΔV = change in velocity (ft/s) in the interval Δt (s).

If a truck of weight W were traveling at 60 mph (88 ft s^{-1}) and the brakes were suddenly applied, the longitudinal force would be dependent on the time it took the truck to change from a velocity of 60 mph to a velocity of zero. If we assume that the vehicle had come to a complete stop in 6 s and the deceleration was uniform on this interval, the longitudinal force would be

$$F = \frac{W}{32.2} \times \frac{88}{6} = 0.46W$$

Because the coefficient of friction of the rubber tires on a dry roadway is approximately 0.75, this force can be transmitted to the roadway.

The AASHTO specifications provide for a longitudinal force of 5% of the live load in all lanes carrying traffic headed in the same direction. The live load as specified is the equivalent uniform lane load plus the concentrated load for moment. For spans shorter than 84 ft the total equivalent lane load weight is less than the weight of the HS 20 truck. It appears that the AASHTO provision is low for short-span, two-lane structures. For long-span, multiple-lane structures the AASHTO provision is more realistic; for example, for a 400-ft span with two lanes in in the Interstate System (traffic in both lanes in the same direction) the longitudinal force from AASHTO would be

$$F = 0.05(0.64 \times 2 \times 400 + 2 \times 18) = 27.4 \text{ k}$$

The comparative force from one HS 20 truck suddenly braking on the structure (deceleration of 14.7 ft/s^2) would be

$$F = 0.46 \times 72 = 33 \text{ k}$$

There is the possibility that more than one vehicle will brake simultaneously on a bridge with a span of 400 ft.

The specifications provide for the center of gravity of the longitudinal force to be applied 6 ft above the surface of the road. This longitudinal force adds little stress to members of the superstructure but is important in the design of the bearings and substructure (see Chapter 10). An additional longitudinal force due to friction on expansion bearings should be considered in the substructure design. This frictional force is selected after consideration of the type of bearing and possible service maintenance. The use of Teflon sliding bearings, which have a low coefficient of friction, minimizes the longitudinal force due to bearing friction.

3.5   CENTRIFUGAL FORCE

When a body travels on a curvilinear path it produces a force perpendicular to the tangent of the path. This force is given by

$$F = \frac{W}{g} \frac{v^2}{r} \tag{3.3}$$

where W = weight of the body,
  g = acceleration of gravity = 32.2 ft s^{-2},
  v = velocity of the body,
  r = radius of the path of the body.

If the velocity v is expressed in miles per hour, eq. 3.3 can be written as

$$F = 0.0668 \frac{Wv^2}{r} \tag{3.4}$$

Equation 3.4 is given in the AASHTO specifications in a slightly different form. This centrifugal force is applied 6 ft above the surface of the road. When the reinforced concrete slab or steel grid deck is keyed or attached to its supporting members the deck will resist this centrifugal force within its plane. It can then be transmitted to the substructure by bracing at the end of the span. The application of centrifugal force 6 ft above the deck can produce vertical forces in the main supporting girders or trusses.

3.6 WIND LOADS

Wind loads have been the concern of bridge design engineers for many years, but the determination of their effect is complex. Wind loads contained in the specifications are approximate for any particular structure. This is a complex problem because of the many variables, such as size and shape of the bridge, probable angles of attack of the wind, shielding effects of the terrain, and the velocity-time relationship of the wind, that affect the wind force.

Studies of wind loads on bridges[2,3] have yielded some information on this subject and summaries of the latest studies are contained in refs. 4 and 5.

The wind load is a dynamic force. A peak wind velocity may be reached in a short period and may remain for a time or decay rapidly (gust). If the time to reach peak pressure is equal to or greater than the natural frequency of the structure, the wind load can be treated for all practical purposes as a static load equal to the peak pressure. This is usually the condition for most bridges.

The wind pressure on a solid object can be expressed by the equation

$$p = C_D \rho \frac{v^2}{2}$$

where C_D = drag coefficient,
ρ = density of the air,
v = wind velocity.

The total force on a bridge is the pressure multiplied by the effective area, which is the area seen in elevation multiplied by C_A, the area coefficient. The value of C_A depends on the angle of attack of the wind, the type of bridge and floor structure, and the distance between beams or girders. Another factor, expressed as coefficient C_ψ, which affects the total horizontal force, is the yaw angle (angle of wind direction in the horizontal plane measured from a perpendicular to the plane of the superstructure). The total horizontal force can be expressed as

$$H = C_D C_A C_\psi A_\tau \rho \frac{v^2}{2}$$

For girder bridges ref. 3 gives the horizontal force acting transversely on the bridge as

$$H = C_H A_H \rho \frac{v^2}{2} \tag{3.5}$$

where A_H is the total projected area seen in elevation and C_H varies from 1.36 to 1.87 for the bridges tested. This corresponds to pressures of 34.8 (two-girder bridge)–47.8 psf (three- and four-girder bridge).

The values given are for a maximum angle of attack of $\pm15°$, which is considered maximum for most bridges.

Reference 3 also discusses uplift forces and the overturning moments that result. These factors are usually neglected in most design specifications.

The AASHTO specifications give basic wind forces for a velocity of 100 mph as

$$\text{girders and beams} = 50 \text{ psf}$$

A minimum force of $300/\text{lb ft}^{-1}$ should be used on girder and beam spans.

The unit pressure is applied to the total area of the structure seen in elevation at an angle of $90°$ to the longitudinal axis of the structure.

An additional wind force contained in the specifications works against a live load equal to 100 lb ft^{-1} applied 6 ft above the roadway. When this additional load is applied, only 30% of the wind load given above is to be applied to the structure itself.

These loadings are those applied only to the superstructure. The values of wind force are included in the substructure design when the wind is perpendicular to the bridge, but winds at other angles are considered for skew (yaw) angles up to $60°$. When the wind is at a skew longitudinal forces on the substructure should also be considered. The reader is referred to the AASHTO specifications for details of the perpendicular and longitudinal forces for various skew angles and the wind on the substructure itself.

The AASHTO specifications also provide for a vertical overturning force of 20 psf on the deck and sidewalk areas at the windward quarter points of these areas.

The wind forces discussed are treated as static, a treatment that is usually satisfactory when the structure is rigid. For flexible structures, however, this procedure is not sufficient. Suspended structures require a more investigative procedure. A steady-state wind may set up an effect called "vortex shedding" if the structure has a long period of natural frequency and a minimum of natural damping. The vortex shedding refers to the production of vortices as the wind passes around the structure with the result that pressures and suctions alternate on each side of the obstruction to the wind. This alternating direction of forces at a right angle to the wind causes the structure to vibrate in a direction perpendicular to the wind. If the period of shedding of vortices is the same as the natural period of the structure, amplitudes of vibration that are large enough to damage the structure take place. These vibrations occurred on the Tacoma Narrows suspension bridge in 1940.

For stiff structures the wind velocity would have to be greater than normal for resonance to develop. For flexible structures, however, resonance can occur at low wind velocities. This condition in a tower-type structure at a velocity of 10 mph is known to one of us.

The bridge designer should be aware of the vortex-shedding phenomenon even though it is unlikely to appear in most bridges. Several members of truss bridges have vibrated in the wind as a result of vortex-shedding effects and repairs have been necessary.

3.7 EARTHQUAKE FORCES

Until recently the effects of earthquakes on highway bridges were largely ignored or given consideration only with regard to the design of substructures. Superstructures most likely have adequate strength to resist any inertia force in the vertical or horizontal direction, yet recently in Alaska, southern California, and Guatamala several bridges were destroyed. The failure was not caused by the collapse of any element of the superstructure but rather by (1) the superstructure shaking off the bearings and falling to the ground, and (2) the structural failure of the loss of strength of the soil under the substructure as a result of the vibrations induced in the ground.

In times of earthquakes it is important that bridges remain in place to allow critical disaster relief vehicles such as ambulances and fire trucks to function.

The effect of an earthquake on a structure depends on the bridge's elastic characteristics and the distribution of weight. A rigorous analysis is complex and involves the application of structural dynamics. In addition, it is necessary to establish the ground motion expected under the substructure. The usual procedure is to simplify the problem by considering that the earthquake produces lateral forces acting in any direction at the center of gravity of the structure and having a magnitude equal to a percentage of the weight of the structure or any part of the structure under consideration. These lateral loads are then treated as static loads according to the following criteria:

For structures with supporting members of approximately equal stiffness an equivalent horizontal force, EQ, computed by the following formulas, may be applied to the structure. The force EQ is applied as a distributed load or as a series of concentrated loads acting at the center of gravity of the structure. EQ is computed as

$$EQ = C \cdot F \cdot W$$

where $F = 1.0$ for structures in which single columns or piers resist the horizontal forces, $F = 0.8$ for structures in which a continuous frame resists horizontal forces, W = the total dead load of the structure (lb), and

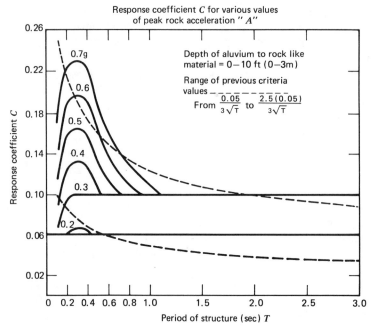

Figure 3.3   Response coefficient curve (d_m, 10 ft).

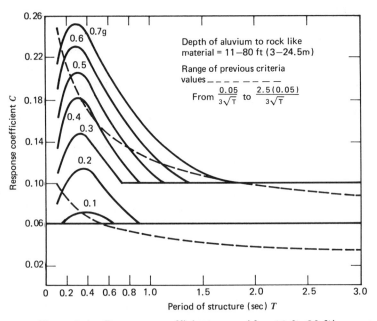

Figure 3.4   Response coefficient curve (d_m, 11 ft, 80 ft).

26

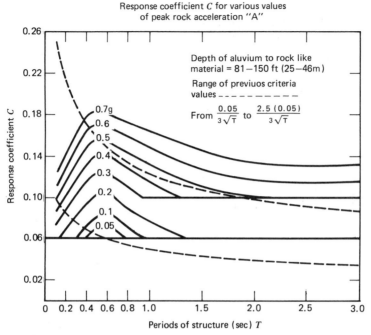

Figure 3.5 Response coefficient curve (d_m, 81 ft, 150 ft).

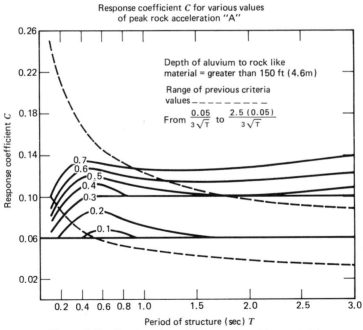

Figure 3.6 Response coefficient curve (d_m, 150 ft).

27

$C = ARS/z$, which is defined as a combined response coefficient and is determined from Figures 3.3–3.6. The evaluation of C requires the computation of a natural period of vibration T of the structure and maximum expected rock acceleration A. These factors are determined as follows:

$$T = 0.32 \left(\frac{W}{p} \right)^{1/2}$$

where W = *total DL* of structure,
   p = total uniform force (lb) to cause a 1-in. maximum horizontal
       deflection of the entire structure,
   A = 0.09 g for zone I,
   A = 0.22 g for zone II,
   A = 0.50 g for zone III.

(The zone classifications are obtained from Figures 3.7 and 3.8.)

3.8   STREAM FLOW PRESSURE

Substructures constructed in a region of flowing water should be designed to withstand water pressure which could cause the pier to slide or overturn. Of considerable concern is the scour around the bottom of the piers.

   The design for every bridge over a stream should involve the careful study of possible stream velocities. Almost all major streams in developed countries have flow records from which velocities can be ob-

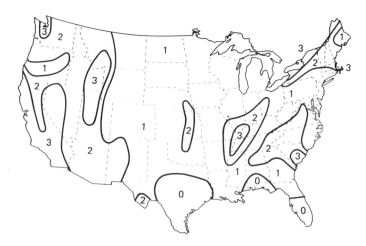

Figure 3.7   Seismic Risk Map USA.

Figure 3.8   Seismic Risk Map Hawaii.

tained. If this information is lacking, the engineer must make the best estimate possible with whatever data are available.

The pressure of flowing water against a body can be predicted by the equation

$$P = KV^2 \qquad\qquad (3.6)$$

where P = unit pressure in lb ft^{-2},

  V = the maximum possible velocity of the water in ft s^{-1},

  K = constant depending on the weight of the water and the shape of the pier.

The AASHTO specifications give the following values for K:

  $K = 1\frac{3}{8}$ piers with square ends

  $K = \frac{1}{2}$ pier with ends angled at $30°$ and less

  $K = \frac{2}{3}$ pier with circular ends.

Although it is known that the velocity of the water varies with depth, it is satisfactory to use a constant pressure for the full depth of the water.

3.9   FLOATING ICE PRESSURE

In cold climates floating ice can cause high forces to work against piers. Bridges have been completely demolished by the pressure of floating ice.

The AASHTO specifications suggest a pressure of 400 lb in.$^{-2}$ of contact area of ice and substructure. The thickness of the ice and its point of application on the piers should be determined by the engineer in an investigation as complete as possible.

The value of 400 psi is based on an estimated crushing pressure of ice of this value. Higher values for the crushing strength of ice have been determined for low temperatures; more than 800 lb in.$^{-2}$ has been recorded for ice at $2°$F. A study of ice pressures against dams[6] due to a rise in temperature—the condition producing the greatest pressure—has led to the conclusion that 400 psi is an overly severe loading for dams. An expanding ice sheet behind a dam is quite different, however, than floating ice forced against a bridge pier by flow of water and possible wind drag. Little is known of these factors and until more information is forthcoming the engineer should use the generally accepted value of 400 psi.

3.10 MISCELLANEOUS

Other forces applied to some types of structures under particular conditions are caused by temperature changes, shrinkage, elastic shortening, and earth pressures. For quantitative values of these items the engineer should consult the specifications or make a judgment based on a study of the particular conditions.

3.11 LOADING COMBINATIONS

The whole or parts of a bridge may be subjected to several loads simultaneously. The engineer has to decide what combination of loads and magnitudes will most likely be applied at one time.

The AASHTO specifications have set up combinations of loads that can act together on a structure. For some combinations the normal allowable stresses can be increased, which is an indirect way of saying that only a percentage of the sum of the maximum effects is considered because the possibility that the peak value of these events will occur simultaneously is remote; for example, when dead load, earth pressure, effects of buoyancy, stream flow, and wind loads are applied simultaneously the allowable stresses can be 125% of the normal allowable stresses. This is the same as saying that only 80% of the peak values of these loads would be considered as acting together.

In the twelfth (1977) edition of the AASHTO Bridge Design Specifications a concept of ultimate strength design of bridges is provided. This procedure does not replace the working design method; it supplements it. The specifications give the following general equation for the combination of loads:

$$\text{group } N = \gamma[\beta_D \cdot D + \beta_L(L + I) + \beta_C \cdot CF + \beta_E \cdot E$$
$$+ \beta_B \cdot B + \beta_S \cdot SF + \beta_W \cdot W + \beta_{WL} \cdot WL$$
$$+ \beta_{LF} \cdot LF + \beta_F \cdot F + \beta_R(R + S + T)$$
$$+ \beta_{EQ} \cdot EQ + \beta_{ICE} \cdot ICE] \tag{3.7}$$

where
D = dead load,
L = live load,
I = impact due to live load,
E = earth pressure,
B = buoyancy,
W = wind force on the structure,
WL = wind load on the live load,
LF = longitudinal force from the live load,

    CF = centrifugal force,
     F = longitudinal force due to bearing friction or shear,
     R = rib shortening (arches or frames),
     S = shrinkage,
     T = force due to temperature change,
    EQ = earthquake,
    SF = stream flow pressure,
   ICE = ice pressure,
     N = loading group number,
     γ = load factor,
     β = coefficient.

The values of the load factor (γ) and coefficients (β) (Table 3.1) depend on the group loading and whether working stress (allowable stress) or load factor design is used.

The specifications designate nine different loading groups that are applicable to bridge design. Groups I–III apply to bridge superstructures and substructures. Groups IV–VI are primarily suited to arches and frames. The last three group loadings are for the design of substructures.

The concept of group loadings is based on which loads might act simultaneously. In working stress design an increase in allowable stress is permitted for some group loadings. The rationale for this increase is that when the peak values occur simultaneously (few times in the life of a structure) the factor of safety can be reduced.

In the application of load factor design the limit of stress is generally the yield point stress for steel and the limit of strength is usually a per-

Table 3.1   Table of Coefficients γ and β

Col. No.	1	2	3	4	5	6	7	8	9	10	11	12	13	14	
								β Factors							
Group	γ	D	$L+I$	CF	E	B	SF	W	WL	LF	$R+S+T$	EQ	ICE	$\%$	
Service Load															
I	1.0	1	1	1	β_E	1	1	0	0	0	0	0	0	100	
II	1.0	1	0	0	1	1	1	1	0	0	0	0	0	125	
III	1.0	1	1	1	β_E	1	1	0.3	1	1	0	0	0	125	
IV	1.0	1	1	1	β_E	1	1	0	0	0	1	0	0	125	
V	1.0	1	0	0	1	1	1	1	0	0	1	0	0	140	
VI	1.0	1	1	1	β_E	1	1	0.3	1	1	1	0	0	140	
VII	1.0	1	0	0	1	1	1	0	0	0	0	1	0	133	
VIII	1.0	1	1	1	1	1	1	0	0	0	0	0	1	140	
IX	1.0	1	0	0	1	1	1	1	0	0	0	0	1	150	
X	1.0	1	1	0	β_E	0	0	0	0	0	0	0	0	100	Culvert

Table 3.1 (*Continued*)

Col. No.	1	2	3	4	5	6	7	8	9	10	11	12	13	14
		β Factors												
Group	γ	D	$L+I$	CF	E	B	SF	W	WL	LF	$R+S+T$	EQ	ICE	%
Load Factor Design														
I	1.3	β_D	1.67	1.0	β_E	1	1	0	0	0	0	0	0	Not Applicable
IA	1.3		2.20	0	0	0	0	0	0	0	0	0	0	
II	1.3		0	0	β_E	1	1	1	0	0	0	0	0	
III	1.3		1	1		1	1	0.3	1	1	0	0	0	
IV	1.3		1	1		1	1	0	0	0	1	0	0	
V	1.25		0	0		1	1	1	0	0	1	0	0	
VI	1.25		1	1		1	1	0.3	1	1	1	0	0	
VII	1.3		0	0		1	1	0	0	0	0	1	0	
VIII	1.3		1	1		1	1	0	0	0	0	0	1	
IX	1.20	β_D	0	0	β_E	1	1	1	0	0	0	0	1	
X	1.50	1	1.67	0	β_E	0	0	0	0	0	0	0	0	Culvert

For service load design:

% (column 14) percentage of basic unit stress.

No increase in allowable unit stresses shall be permitted for members or connections carrying wind loads only.

β_E = 0.70 for reinforced concrete boxes; 0.83 for all other culverts.

β_E = 1.0 and 0.5 for lateral loads on rigid frames (check both loadings to see which one governs).

β_E = 1.3 for lateral earth pressure and 0.5 for checking positive moments in rigid frames.

β_E = 1.0 for vertical earth pressure.

β_D = 0.75 when checking members for minimum axial load and maximum moment or or maximum eccentricity for column design.

β_D = 1.0 when checking member for maximum axial load and minimum moment for column design.

β_D = 1.0 for flexural and tension members.

β_E = 1.0 rigid culverts.

β_E = 1.67 flexible culverts.

centage of the 28-day compressive strength for concrete. The bridge designer should consult the appropriate sections of the specifications for the proper limit of stress. Later chapters in this book show the use of working stress and load-factor design for concrete bridges.

3.12 DESIGN EXAMPLES

Because superstructure design is routinely controlled by a Group I combination of $DL + LL + I$, examples of its application are given in Chapters 4–9 to cover the various types.

Application of loadings and their combinations becomes most significant in substructure design. Therefore two examples of load computations applicable to the actual substructure designs cited in Chapter 10
are given here.

Example 3.1. The bridge shown in elevation and plan in Figure 3.9
and in section in Figures 3.10 and 3.11 details the first example for
load computation. The superstructure for this bridge was designed by
service load design in Example 5.1 and by load factor in Example 5.2.
The frame analysis is used to compute moments and shears on the substructure that result from these loadings. The pier for this bridge is designed in Example 10.2 by load factor design and in Example 10.3 by
service load design. The load tabulations used in Example 10.2 and
10.3 for design of the pier shaft and footing are based on the load computations in this example.

The live loading on this structure is HS 20 with an overload vehicle
(Figure 3.12) which should be used with $\beta_L = 1.0$ when computing capacity and compared with $\beta_L = 1.67$ used with HS 20 when computing
capacity.

The dead load reaction is taken from Example 5.1 and computed for
the pier with and without buoyancy.

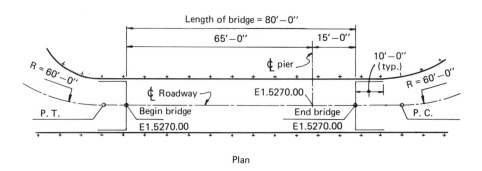

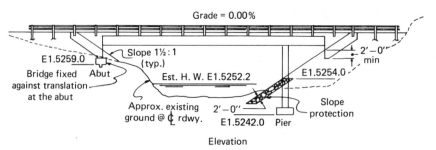

Figure 3.9   Bridge plan and elevation details—Example 3.1.

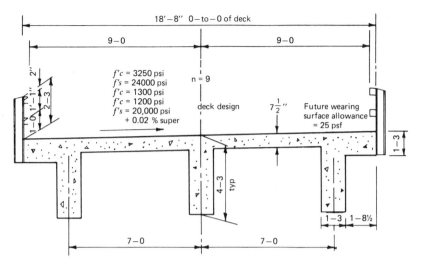

Figure 3.10   Bridge cross section—Example 3.1.

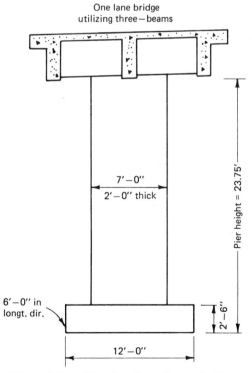

One lane bridge
utilizing three—beams

Figure 3.11   Pier elevation—Example 3.1.

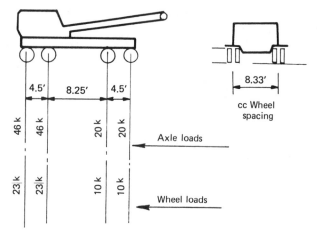

Figure 3.12   Overload vehicle—Example 3.1.

Dead Load (Reaction) w/o Buoyancy (kips)

Exterior girder, 2 × 100.7	201.4
Interior girder, 91.0	91.0
Cap, 2 × 4.7	9.4
Subtotal @ *BD*	301.8 k
Shaft, 2.0 × 7.0 × 21.25 × 0.150	44.6
Subtotal @ *DB* Col	346.4 k
Footing, 6.0 × 12.0 × 2.5 × 0.150	27.0
Earth, (72.0 - 14.0) × 6.5 × 0.120	45.2
Total @ *DB* footing	418.6 k

Dead Load (Reaction w/Buoyancy) (kips)

Subtotal @ *BD*	301.8
Shaft, 2.0 × 7.0 × 13.75 × 0.150	28.9
Shaft, 2.0 × 7.0 × 7.5 × 0.088	9.2
Subtotal @ *DB* Col.	339.9 k
Footing, 6.0 × 12.0 × 2.5 × 0.088	15.8
Earth, (72.0 - 14.0) × 6.5 × 0.058	21.9
Total @ *DB* footing	377.6 k

The dead load moments on the pier are also taken from Example 5.1.

Dead Load Moments (k-ft)

@ *BD*
   Exterior girder, 2 × (+129)   +258 ⎫
   Interior girder   +172 ⎬ +430 k-ft

@ *DB* Column
  Exterior girder, 2 × (+45)   + 90 ⎫
                                      ⎬ +149 k-ft
  Interior girder   + 59 ⎭
@ *DB* Footing
  Exterior girder, 2 × (+65)   +130 ⎫
                                      ⎬ +216 k-ft
  Interior girder   + 86 ⎭

Example 5.1 shows the computation of the loads and moments for a number of live loads for HS 20 and overload vehicles. To analyze the pier shaft we examine the maximum positive and negative bending moments produced by these loadings. For the pier footing we examine the maximum induced reaction. The transverse pier moment is produced by an eccentric live load (Figure 3.13) for HS 20 and overload vehicles. The loads and moments are taken from Example 5.1 and summarized in Table 3.2. The values taken from Example 5.1 are the reactions and moments due to one line of wheels. The transverse moment is computed from the list and multiplied by the impact coefficient and by 2 to expand the values to one full lane. The impact coefficient is 1.263 for span *BC* and 1.30 for cantilever span *AB*.

These loads and moments are summarized in Table 3.2.

The structure is braced longitudinally against sidesway at the abutment; therefore, except for those induced by temperature plus shrinkage, the longitudinal loads on the pier are neglected. Transverse wind on the superstructure is computed; any wind on the substructure is ignored because of its insignificance. These transverse loadings are applied by assuming a free-standing cantilever as shown in Figure 3.14.

Wind on Superstructure. Using alternate AASHTO loading w/span lengths less than 125 ft, $W = 50$ psf transverse at top of pier:

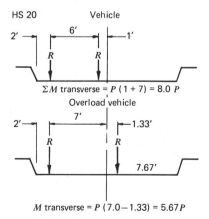

HS 20   Vehicle
ΣM transverse = $P(1 + 7) = 8.0P$
Overload vehicle
M transverse = $P(7.0 - 1.33) = 5.67P$ **Figure 3.13**   Eccentric live load.

Table 3.2

| @ *BD* | One-Line Wheels | | One Truck | LL + I | | |
Case	R_B	M_{BD}	$M_{transverse}$	R_B	M_{BD}	$M_{transverse}$
IB	23.1	+137	185	58.4	+346	234
II	34.6	− 84	277	90.0	−218	360
VB[a]	40.6	+258	230	102.6	+652	290
VI[a]	52.1	−192	295	135.5	−499	384

| @ *DB* Column | One-Line Wheels | | One Truck | LL + I | | |
Case	R_B	M_{DB}	$M_{transverse}$	R_B	M_{DB}	$M_{transverse}$
IB	23.1	+ 46	185	58.4	+116	234
II	34.6	− 29	277	90.0	− 75	360
VB[a]	40.6	+ 88	230	102.6	+222	290
VI[a]	52.1	− 66	295	135.5	−172	384

| @ *DB* Footing | One-Line Wheels | | One Truck | LL Only | | |
Case	R_B	M_{DB}	$M_{transverse}$	R_B	M_{DB}	$M_{transverse}$
IB	23.1	+ 68	185	46.2	+136	185
II	34.6	− 42	277	69.2	− 84	277
IIA	38.0	− 36	304	76.0	− 72	304
VB[a]	40.6	+129	230	81.2	+258	230
VI[a]	52.1	− 96	295	104.2	−192	295
VIA[a]	72.1	− 96	409	144.2	−192	409

[a]Overload vehicle; otherwise HS 20 vehicle.

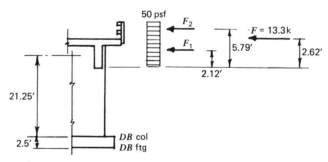

Figure 3.14   Wind loading.

Depth of girder,

$$F_1 = 4.25 \times 50 = \frac{F}{212} \times \frac{\bar{Y}}{2.12} = \frac{F\bar{Y}}{450}$$

Rail,

$$F_2 = 0.67 \times 50 = \frac{34}{246 \text{ lb ft}^{-1}} \times 5.79 = \frac{194}{644} \qquad \therefore \bar{Y} = \frac{644}{246} = 2.62 \text{ ft}$$

The resultant force $F = (246)(0.6 \times 65 \text{ ft} + 15.0) = 13.3$ k, using 0.6 of the span plus the cantilever,

$$M_{DB \text{ col}} = 13.3(21.25 + 2.62) = 317 \text{ k-ft}$$

$$M_{DB \text{ ftg}} = 13.3(23.75 + 2.62) = 351 \text{ k-ft}$$

These moments are combined with those created by wind on the superstructure of 20 psf of uplift, applied at the quarter point of the bridge width.

Overturning

$$W = 20(18.67 \text{ ft bridge width}) = 0.373 \text{ k ft.}^{-1}$$

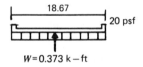

$$F = 0.373(0.6 \times 65 + 15.0) = 20.1 \text{ k}$$

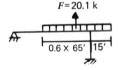

$$M_{OT} = 20.1 \left(\frac{18.67}{4}\right) = 94 \text{ k-ft}$$

Wind on Live Load.   The transverse wind is applied to the live load according to the AASHTO criteria in Article 1.2.14B(1) by using the alternate loading for span lengths less than 125 ft:

$$W = 100 \text{ plf}$$

$$F = (0.10)(0.6 \times 65 + 15.0) = 5.4 \text{ k}$$

$$M_{DB \text{ col}} = (21.25 + 4.25 + 6.0)(5.4) = 170 \text{ k-ft}$$

$$M_{DB \text{ ftg}} = (23.75 + 4.25 + 6.0)(5.4) = 184 \text{ k-ft}$$

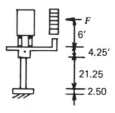

Thermal forces are computed in accordance with AASHTO Article 1.2.15 and combined with shrinkage when temperature drop is considered.   The change in length due to this combination is computed as follows:

Temperature Drop + Shrinkage

$$\Delta = (\alpha_t)(L)(\Delta t) \quad \text{(temperature change)}$$

$$\Delta = (\alpha_s)(L) \quad\quad \text{(shrinkage)}$$

Moderate Climate

$$\left. \begin{array}{l} \Delta = 0.000006 \times 40° \times 65.0 = 0.0156 \\ 0.0002 \times 65.0 \quad\quad\quad = 0.0130 \end{array} \right\} \; 0.0286 \text{ ft}$$

The moment distribution method is used to determine the indeterminate moments in the bridge.   Details of this method are given in Chapter 4.   The fixed end moments from this change in length are

$$M^c_{BD} = M^c_{BD} = \frac{6EI\Delta}{h^2} = \frac{6(468{,}000)(5.80)(0.0286)}{23.75^2}$$

$$= 8.26 \text{ k-ft}$$

Table 3.3[a]

				CO = 0.5		
			B		**D**	
	BA	BC	BD		DB	
	0.673		0.327			Comment
FEM			-826		-826	
Bal	+556		+270		0	$DF \times FEM$
Co					+135	$\frac{1}{2} \times M$
Σ	+556		-556		-691	

[a]All moments are in kip-feet.

The moment distribution of these fixed end moments is given in Table 3.3:

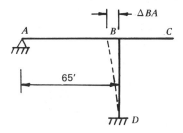

The moment of -691 k-ft at DB is assumed at the bottom of the footing. The moment at the top of the footing is then computed:

$$M_{DB \, col} = -691 - \left(\frac{2.50}{28.75}\right)(-1247) = -560 \text{ k-ft}$$

This is based on a straight-line relationship between the top and bottom moments. The temperature rise movement according to AASHTO article 1.2.15 is the following:

Temperature Rise

$$\Delta = 0.000006 \times 30° \times 65 = 0.0117 \text{ ft}$$

The moments for temperature rise are proportional and of opposite sign to the temperature drop plus shrinkage moments:

$$M_{BD} = \left(-\frac{0.0117}{0.0286} \right) \times (-556) = +227 \text{ k-ft}$$

$$M_{DB \text{ ftg}} = \left(-\frac{0.0117}{0.0286} \right) \times (-691) = +283 \text{ k-ft}$$

$$M_{DB \text{ col}} = \left(-\frac{0.0117}{0.0286} \right) \times (-560) = +229 \text{ k-ft}$$

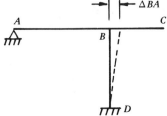

Next, the loads and moments due to earthquake action transverse to the structure are computed. The bridge is in a Zone 3 earthquake area which is the most severe case. The assumed column section is shown in Figure 3.15.

The reinforcement thickness, using a transformed area, is

$$t = \frac{26\,As \times n}{144 \times 6.48 \text{ ft}} \qquad t = \frac{33.02(9)}{144(6.48)} = 0.3185 \text{ ft}$$

The transformed moment of inertia is computed as

$$I_{C \text{ transv}} = \frac{2.0(7.0)^3}{12} = \qquad 57.2$$

$$I_{R/F \text{ transv}} = \frac{0.3185(6.48)^3}{12} = \frac{7.2}{64.4 \text{ ft}^4}$$

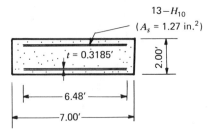

Figure 3.15   Pier section.

The structure weight tributary to the pier is

$$W = 301.8 \text{ k}$$

The natural frequency or period of the pier is $T = 0.32\sqrt{W/P}$, where P is the lateral load required to produce a 1-in. (0.0833-ft) lateral deflection. Therefore for a free standing cantilever

$$P = \frac{3EI\Delta}{h^3} = \frac{3(468,000)(64.4)(.0833)}{(23.75)^3} = 562.0 \text{ k}$$

The period is then computed:

$$T = 0.32 \sqrt{\frac{W}{P}} = 0.32 \sqrt{\frac{301.8}{562}} = 0.23 \text{ s}$$

Assuming that rock is close to the ground surface and using the chart in AASHTO article 1.2.20(A), we see that Figure 3.3, for a depth to rock-like material of 0–10 ft, gives a response coefficient $C = 0.16$ with a Zone III value of acceleration equal to 0.50 g and a period $T = 0.23$ s. The earthquake force then becomes

$$C = 0.16 \qquad F = 1.0$$
$$EQ = CFW$$
$$EQ = 0.16(1.0)(301.8) = 48.3 \text{ k}$$

and the earthquake moments are

$$M_{DB\,col} = 48.3(21.25 + 2.79) = 1161 \text{ k-ft}$$
$$M_{DB\,ftg} = 48.3(23.75 + 2.79) = 1282 \text{ k-ft}$$

acting transverse to the structure.

All moments and load combinations are tabulated in Examples 10.2 and 10.3.

Example 3.2.   The bridge for our next example of load calculation is shown in plan and elevation in Figure 3.16 and in section in Figure 3.17. The longitudinally reinforced slab superstructure for this bridge was designed by service load procedures in Example 4.1 and by load factor procedures in Example 4.2. The frame analysis from these examples is used to compute shears and moments that result from the various loadings considered in this example. The round columns for this bridge were designed by load factor design in Example 10.4 and by service

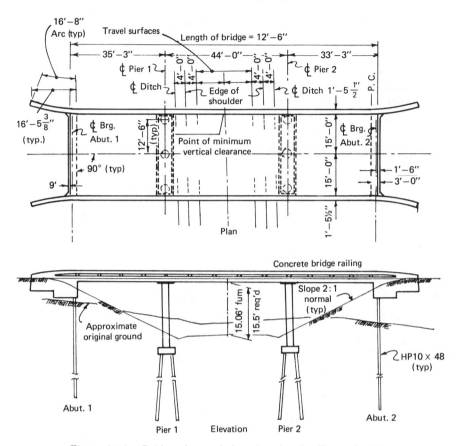

Figure 3.16 Bridge plan and elevation details—Example 3.2.

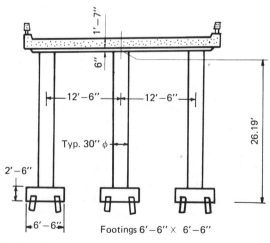

Figure 3.17 Bridge cross section—Example 3.2.

44

load design in Example 10.5. The load tabulations used in Examples 10.4 and 10.5 for the design of the substructure to this bridge apply the results of the load computations in this example.

The live load on this structure is the H 15, as indicated in AASHTO. It is obvious that the interior column of this three-column bent will carry the most load and a larger bending moment from the super-structure loading. Therefore only the loadings on this column are considered.

The dead load and dead load moments and the live load plus impact reactions and moments are given in Example 4.1.

Initially, the dead load reactions (per foot in the transverse direction) and the additional dead load reaction from each traffic railing are computed. Also included in the transverse dead load per foot is the increase in the thickening at the bent cap, given as follows:

Reaction @ Bent 1

$$R_B = 11.55 + 0.5 \times 5.5 \times 0.150 = 11.96 \text{ k-ft}^{-1}$$

$$\text{Rail } R_B = \frac{0.331}{0.263} \times (11.55) = 14.5 \text{ k per rail}$$

In Example 4.1 for the cap design, these loads give a reaction at the interior column of 152.6 k. In the same example the 11.55 k ft^{-1} uniform load gives a M_{BE} - 2.0 k-ft and M_{EB} - 1.2 k-ft, respectively. Therefore we proportion the moments to get the interior column moments:

Interior Column

$$M_{BE} = \frac{152.6}{11.55} (-2.0) = -26.0 \text{ k}$$

$$M_{EB} = \frac{152.6}{11.55} (-1.2) = -16.0 \text{ k}$$

$$M_{EB \text{ col}} = -12.0 \text{ k}$$

The dead load reactions at the interior column are computed as follows:

Interior Column Reactions

Reaction from cap	152.6
Column DL $\pi(1.25)^2$ (26.19 - 2.50) $\times$ 0.150 =	17.4
subtotal	170.0
Footing 6.5 $\times$ 6.5 $\times$ 2.5 $\times$ 0.150 =	15.8
Earth (42.25 - 4.91) $\times$ 3.0 $\times$ 0.120 =	13.4
	199.2 k

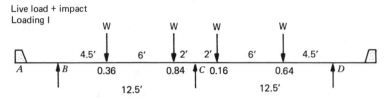

Figure 3.18   Live load interior column.

For the live load a loading similar to that shown as loading I in Example
4.1 is used to produce the maximum response at the interior column
(Figure 3.18).

For this loading the reaction at the column is computed as

$$R_c = 2.9608W$$

For cases III, IV, and VI loadings, shown in Example 4.1, the maximum
negative bending moment and the maximum reaction are produced.
The reactions and moments from these loadings are listed in Table 3.4.

A low shallow structure like this will not be affected by wind; therefore, group loadings II, III, V, and VI are neglected. Now examine the
effects of the temperature drop plus shrinkage. The tributary movement to each pier, due to temperature drop plus shrinkage is computed
as follows:

Temperature Drop + Shrinkage

$$\Delta = (\alpha_t)(L)(\Delta t), \quad \text{temperature change}$$

$$\Delta = (\alpha_s)(L), \qquad \text{shrinkage}$$

$$\Delta = 0.000006 \times 40 \times 22.0 = 0.0053$$
$$0.0002 \times 22.0 \qquad = \underline{0.0044}$$
$$0.0097 \text{ in.}$$

Table 3.4[a]

| Case | LL Only | | | | LL + I | | | |
	R_C	M_{BE}	M_{EB}	Column M_{EB}	I	R_B	M_{BE}	Column M_{EB}
III	31.1 k	+65	+15	+ 7	0.300	40.4 k	+ 84	+ 9
IV	26.4 k	-92	-44	-31	0.296	34.2 k	-119	-40
VI	61.3 k	-20	-17	-13	0.293	79.6 k	- 26	-17

[a]Moments are in-kip feet.

The fixed end moments in each column due to this movement are

$$E_c = 57,000\sqrt{4000} = 3,605,000 \text{ psi} = 519,100 \text{ ksf}$$

$$M_{BE}^F = M_{EB}^F = \frac{6EI\Delta}{h^2} = \frac{6(519,100)(2.12)(0.0097)}{(26.19)^2} = 93 \text{ k-ft}$$

$$M_{CF}^F = M_{FC}^F = \frac{6(519,100)(2.12)(0.0097)}{(25.52)^2} = 98 \text{ k-ft}$$

The moment distribution of these fixed end moments is shown in Table 3.5. Corrections for sidesway, using the sidesway correction given in Example 4.1, are included.

The temperature rise movements and moments are proportional and of opposite sign to the temperature drop plus shrinkage moments just computed,

Table 3.5

| | B | | | E | F | | C | |
	BA	BC	BE	EB	FC	CF	CB	CD
	0.326	0.340	0.334			0.333	0.331	0.336
FEM			+93	+93	-98	-98		
B	-30	-32	-31			+33	+32	+33
C		+16		-16	+16		-16	
B	- 5	- 6	- 5			+ 5	+ 5	+ 6
Σ	-35	-22	+57	+77	-82	-60	+21	+39
SS	- 1	- 2	+ 3	+ 3	+ 3	+ 3	- 2	- 1
M	-36	-24	+60	+80	-79	-57	+19	+38

$$M_{EB\,col} = +67 \text{ k-ft}$$

$$M_{FC\,col} = -66 \text{ k-ft}$$

$$V_{EB} = \frac{+57 + 77}{26.19} = +5.12 \text{ k}$$

$$V_{FC} = \frac{-82 - 60}{25.52} = -5.56 \text{ k}$$

unbal V = 0.45 k

$$\text{coefficient for sidesway} = \frac{0.45}{34.8} = 0.0129 \times \text{mom}$$

Temperature Rise

$$\Delta = 0.000006 \times 30 \times 22.0 = 0.0040$$

$$M_{BE} = -\frac{0.0040}{0.0097} \times (+60) = -24 \text{ k-ft}$$

$$M_{EB} = -\frac{0.0040}{0.0097} \times (+80) = -33 \text{ k-ft}$$

$$M_{EB\,col} = -\frac{0.0040}{0.0097} \times (+67) = -27 \text{ k-ft}$$

$$M_{CF} = -\frac{0.0040}{0.0097} \times (-57) = +23 \text{ k-ft}$$

$$M_{FC} = -\frac{0.0040}{0.0097} \times (-79) = +32 \text{ k-ft}$$

$$M_{FC\,col} = -\frac{0.0040}{0.0097} \times (-66) = +27 \text{ k-ft}$$

M_{EB} would be the footing moment and $M_{EB\,col}$ would be the moment at the top of the footing.

For the earthquake design the response of the structure acting in the longitudinal direction is examined first.

The superstructure weight for such longitudinal earthquakes involves the entire bridge length; therefore W equals the sum of all the superstructure reactions:

Earthquake—longitudinal

$$W = 2(245.9 + 422.7) = 1337 \text{ k}$$

The fixed-end column moments for a 1-in. deflection are computed by using $M^F = 6EI\Delta/h^2$:

For $\Delta = 1$ in. (0.0833 ft)

$$M_{BE}^F = M_{EB}^F = \frac{6(519,100)(2.12)(0.0833)}{(26.19)^2} = 802 \text{ k-ft per column}$$

$$M_{CF}^F = M_{FC}^F = \frac{6(519,100)(2.12)(0.0833)}{(25.52)^2} = 845 \text{ k-ft per column}$$

These fixed end moments can be prorated by using the sidesway computation given in Example 4.1 to compute P which is used in the computation of the natural frequency.

$$P = \frac{3(802)}{278} (34.8) = 303 \text{ k}$$

or

$$\frac{3(845)}{292} (34.8) = 303 \text{ k}$$

The period is then computed as

$$T = 0.32 \sqrt{\frac{W}{P}} = 0.32 \sqrt{\frac{1337}{303}} = 0.67 \text{ s}$$

Assuming a depth of alluvium to rocklike material, dm = 81–150 ft, and using AASHTO Article 1.2.20(a), Figure 3.5 gives C = 0.08 (assuming Zone I and A = 0.09 g) and F = 0.8. The earthquake force on the entire structure is computed as

$$EQ = 0.08(0.8)(1337) = 85.6 \text{ k}$$

Using the sidesway computation in Example 4.1, we evaluate the moments as

$$M_{BE} = \frac{85.6}{34.8} \times 199 = 489 \times \frac{1}{3} = 163 \text{ k-ft per col}$$

$$M_{EB} = \frac{85.6}{34.8} \times 240 = 590 \times \frac{1}{3} = 197 \text{ k-ft per col}$$

$$M_{EB \text{ col}} = \qquad\qquad 487 \times \frac{1}{3} = 162 \text{ k-ft per col}$$

$$M_{CF} = \frac{85.6}{34.8} \times 208 = 511 \times \frac{1}{3} = 170 \text{ k-ft per col}$$

$$M_{FC} = \frac{85.6}{34.8} \times 252 = 620 \times \frac{1}{3} = 207 \text{ k-ft per col}$$

$$M_{FC \text{ col}} = \qquad\qquad 509 \times \frac{1}{3} = 170 \text{ k-ft per col}$$

Again M_{EB} is the footing moment and $M_{EB \text{ col}}$ is the moment corrected to the top of the footing.

The next consideration is an earthquake movement that acts transverse to the structure, for which the tributary weight to each (three-column) pier is the following:

Earthquake—transverse

W = 422.7 k for three columns

The columns are fixed top and bottom. Therefore P for a 1-in. deflection equals the fixed end moment divided by one-half of column height:

$$P = \frac{802}{13.10} = 61.2 \text{ k per col}$$

$$P = 3(61.2) = 183.7 \text{ k for 3 col}$$

The period is then computed by using $T = 0.32\sqrt{W/P}$ which gives

$$T = 0.32 \sqrt{\frac{422.7}{183.7}} = 0.049 \text{ s}$$

Assuming the same depth of alluvium to rocklike material (81–150 ft) and using Figure 3.5, C = 0.08 (assuming Zone I, A = 0.09 g) and F = 0.8. The earthquake force at the pier therefore is

$$EQ = 0.8(0.08)(422.7) = 27.1 \text{ k per 3 col}$$

The moment then equals the fixed end moment, which is the column force times one-half the column height or

$$M = 27.1(13.10)(\tfrac{1}{3}) = 118 \text{ k-ft per col}$$

These loads and moments are used in the substructure designs in Examples 10.4 and 10.5.

BIBLIOGRAPHY

1. American Association of State Highway and Transportation Officials, *Standard Specifications for Highway Bridges*, 1977, with the 1978–1981 interim specifications.
2. J. M. Biggs, Wind Loads on Truss Bridges, *Proc. Am. Soc. Civ. Eng.*, Separate No. 201 (July 1953).
3. J. M. Biggs, S. Namyet, and J. Adachi, Wind Loads on Girder Bridges, *Proc. Am. Soc. Civ. Eng.*, Separate No. 587 (January 1955).
4. W. W. Pagon, Wind Forces on Structures, Plate Girders and Trusses, *J. Struc. Div. Am. Soc. Civ. Eng.*, **84** (ST4) (July 1958).
5. Wind Forces on Structures, final report, Task Committee on Wind Forces, *Trans. Am. Soc. Civ. Eng.*, 1124–1198 (1961).

6. E. Rose, Thrust Exerted by Expanding Ice Sheet, *Trans. Am. Soc. Civ. Eng.* (1947).

7. C. Y. Liaw and A. K. Chopra, Earthquake Response of Axisymmetric Tower Structures Surrounded by Water, University of California, Berkeley, Report No. EERC 73-25, October 1973.

8. C. P. Heins and D. A. Firmage, *Design of Modern Steel Highway Bridges*, Wiley, New York, 1980.

FOUR

Design of Bridge Slabs

Bridge slabs are solid or voided sections that span between supports in the longitudinal or transverse direction. As supporting members they are also riding surfaces for vehicular traffic, which means that they must provide durability against abrasive deterioration and repetitive cycles of loading in flexure and shear.

Durability of deck slabs is a subject of universal concern. Various protective systems, special dense concretes, and epoxy-coated reinforcing bars have been proposed to protect decks against repeated contact with wheel loadings, deicing chemicals, and moisture.

It is a common opinion that the major portion of deck deterioration is due to corrosion of the reinforcing steel which causes spalling of the concrete cover. These protective systems are designed to keep the corrosive elements away from the reinforcing. Another means of protecting against penetration of corrosive elements is to design deck slabs with particular attention to serviceability requirements for crack control. This is achieved primarily by using low values of stress in the deck reinforcing at service load and is best accomplished by using deeper slabs, not by larger reinforcing.

The deck slabs discussed in this chapter are of two types:

1. Longitudinally reinforced.
2. Transversely reinforced.

4.1 LONGITUDINALLY REINFORCED DECK SLABS

4.1.1 General

Longitudinally reinforced deck slabs are practical for shorter spans, for example, between transverse floor beams, or for spans up to 45 ft be-

52

tween substructure units. They also offer a minimal structure depth for which structure opening and vertical clearance are significant.

For this type of concrete deck slab it is suggested that the clearance for the bottom reinforcing be 1.5 in. clear. This is more than required by AASHTO Art. 1.5.6(a) in which 1 in. clear is indicated for bottom reinforcing on bridge deck slabs. The AASHTO value of 2 in. for top reinforcing clearance is satisfactory.

Haunching of continuous or framed longitudinal deck slabs is a reasonable way to match the differential in requirements between positive and negative moments. The charts in Appendixes J, K, L, and M (1–4) provide valuable design assistance for analyzing haunched slabs by moment distribution.

AASHTO Art. 1.3.2(d) requires that an edge beam be designed for longitudinally reinforced slabs. The edge beam can be one of three sections:

1. A slab section additionally reinforced.
2. A beam integral with the slab and deeper than the slab.
3. An integral reinforced section of slab and curb.

The edge beam is designed for a "live load moment of 0.10 PS." This is a simple span moment:

$$P = \text{wheel load, in pounds}$$

$$S = \text{span length, in feet}$$

For continuous spans this moment is reduced by 20%. Our interpretation of this section of AASHTO is that 10% of the live load moment caused by one line of wheels should be added to the other computed moments $(DL + LL + I) + (RDL)$ for the width selected for the edge beam. The width of edge beam selected should not exceed 3 ft.

Example 4.1.   The bridge shown in elevation and plan in Figure 4.1 is also shown in section in Figure 4.2. The superstructure for this bridge is designed as a longitudinally reinforced slab by service load design procedures. The substructure for this bridge is designed by service load in Example 10.4.

The allowable stresses for design are

$$f_c' = 4000 \text{ psi}$$

$$f_c = 1600 \text{ psi}, n = 8$$

$$f_y = 60,000 \text{ psi}$$

$$f_s = 24,000 \text{ psi}$$

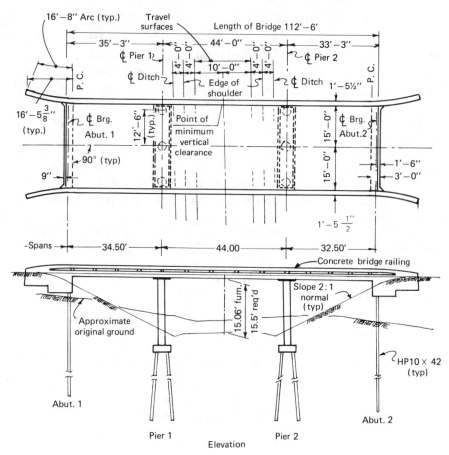

Figure 4.1 Longitudinally reinforced slab bridge: plan and elevation.

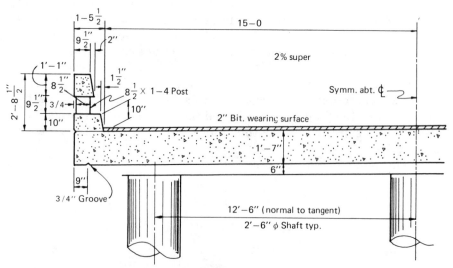

Figure 4.2 Typical section: longitudinally reinforced slab bridge.

54

This structure is a simple rigid frame that is analyzed by moment distribution for the development of influence lines and dead load effects. Frame constants must therefore be computed for this moment distribution. For moment of inertia computations the gross concrete section is used for the superstructure, the transformed gross section, for the substructure, which is in agreement with AASHTO specifications. The current AASHTO specifications state only that the assumptions are to be reasonable and consistent.

The moment of inertia of the superstructure is

$$I = \frac{bh^3}{12}$$

$$I = \frac{32.92(1.5833)^3}{12} = 10.89 \text{ ft}^4 \text{ (superstructure)}$$

If we assume 1% reinforcing in the columns, the substructure moment of inertia is (Figure 4.3)

$$A_c = \pi(1.25)^2 = 4.91 \text{ ft}^2$$

$$A_s = 0.01 A_c = 0.49 \text{ ft}^2$$

$$nA_s = 8(0.49) = 0.39 \text{ ft}^2$$

$$I_c = 0.049087(d)^4$$

$$I_c = 0.049087(2.50)^4 = 192$$

$$I_s = nA_s(d)^2\left(\tfrac{1}{8}\right) \qquad \Big\} \text{ sum of 2.12 ft}^4$$

$$I_s = 0.39(2.00)^2\left(\tfrac{1}{8}\right) = 0.20$$

$$I_{\text{substr}} = 3(2.12) = 6.36 \text{ ft}^4 \text{ for 3 col}$$

The bridge layout is shown in Figure 4.4 and members CD and BA are considered hinged at one end.

At joint B the relative stiffness for each member, along with the dis-

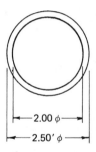

—2.00 ϕ—

—2.50' ϕ—   **Figure 4.3** Column section.

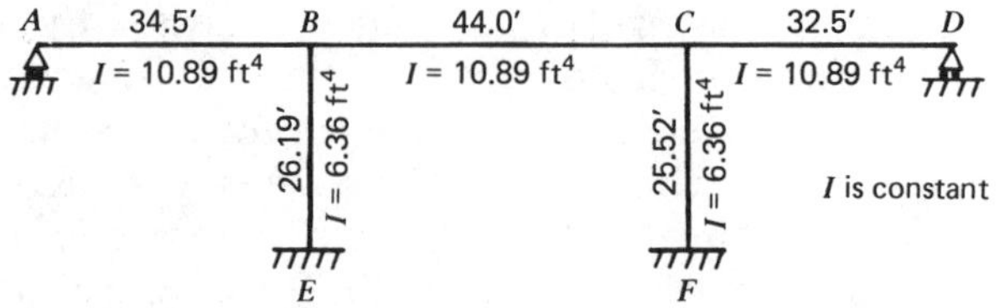

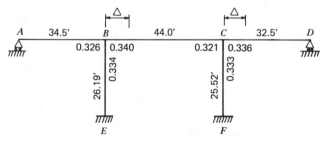

Figure 4.4   Bridge layout.

tribution factors, is computed based on $K = 3I/L$ (for one end hinged), $K = 4I/L$ (both ends fixed), and $D = K/\Sigma K$:

$$K_{BA} = \frac{3(10.89)}{34.5} = 0.947 \qquad D_{BA} = \frac{0.947}{2.908} = 0.326$$

$$K_{BC} = \frac{4(10.89)}{44.0} = 0.990 \qquad D_{BC} = \frac{0.990}{2.908} = 0.340$$

$$K_{BE} = \frac{4(6.36)}{26.19} = 0.971 \qquad D_{BE} = \frac{0.971}{2.908} = 0.334$$

At joint C the same relationships are

$$K_{CB} = \qquad\qquad 0.990 \qquad D_{CB} = \frac{0.990}{2.992} = 0.331$$

$$K_{CD} = \frac{3(10.89)}{32.5} = 1.005 \qquad D_{CD} = \frac{1.005}{2.992} = 0.336$$

$$K_{CF} = \frac{4(6.36)}{25.52} = 0.997 \qquad D_{CF} = \frac{0.997}{2.992} = 0.333$$

The distribution factors on the members at the joints are shown in Figure 4.5.

If a vertical load is applied to the structure and moment distribution is performed, examination of the internal column shears V_{EB} and V_{FC}

Figure 4.5   Bridge analysis details.

will indicate that no external shear is available to balance these actions. Therefore an external, equal, and opposite force must be introduced into the system that will then induce lateral deformations and balance the internal forces. To account for these actions the response of the basic system, due to a lateral displacement, is examined. These results are then used to modify the unit fixed end moment solutions applied at each joint.

For sidesway computations an arbitrary longitudinal deflection of the frame is assumed. If $M = 6EI\Delta/h^2$ and $E = 5000$ k-ft^2 and $\Delta = 1$ ft, then

$$M_{BE}^F = \frac{6(6.36)(5000)}{(26.19)^2} = 278 \text{ k-ft}$$

$$M_{CF}^F = \frac{6(6.36)(5000)}{(25.52)^2} = 292 \text{ k-ft}$$

The moment distribution given in Table 4.1 is performed for the sidesway fixed end moments (counterclockwise on the member is positive).

The column shears at E and F can now be computed:

$$V_{EB} = \frac{+199 + 240}{26.19} = +16.8 \text{ k}$$

$$V_{FC} = \frac{+252 + 208}{25.52} = \frac{+18.0 \text{ k}}{\sum 34.8 \text{ k}}$$

Table 4.1[a]

CO = 0.5 (B ↔ C)

CO = 0.5 (BE ↔ EB)   CO = 0.5 (FC ↔ CF)

Joint		*B*		*E*	*F*		*C*	
Member	*BA*	*BC*	*BE*	*EB*	*FC*	*CF*	*CB*	*CD*
Distribution factor	0.326	0.340	0.334			0.333	0.331	0.336
FEM			+278	+278	+292	+292		
BAL	-91	-95	-92			-97	-97	-98
CO		-48		-46	-48		-48	
BAL	+16	+16	+16			+16	+16	+16
CO		+8		+8	+8		+8	
BAL	-2	-3	-3			-3	-2	-3
ΣM_Δ	-77	-122	+199	+240	+252	+208	-123	-85

[a]Moments are in kip-feet.

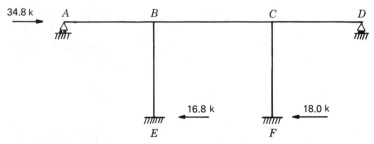

Figure 4.6   Balanced shears.

The total external shear to be applied to the structure must equal the unbalance of the column shears at the base (Figure 4.6). This shear force is used to correct the moment distribution solution caused by vertical load effects.

The induced moments in the structure due to vertical loads are evaluated by first applying a unit fixed end moment at each superstructure joint. All other distributed moments are taken from these unit distributions. The fixed end moment is assumed to be equal to 1000 k-ft to eliminate the use of decimals. Again for moment distribution counterclockwise on the member is considered positive.

First a unit fixed end moment of -1000 k-ft is applied at BA, as shown in Table 4.2, and the moments are distributed and summed. As described earlier, the resulting unbalance between the column shears at E and F are then computed:

$$V_{unbal} = \frac{+343 + 167}{26.19} + \frac{-28 - 57}{25.52} = +16.2 \text{ k}$$

This unbalanced shear is then proportioned to the sidesway shear (34.8 k) to give sidesway moments, that is, $(16.2/34.8) \times M_\Delta$, which are added to the distributed moments as shown in the last column of Table 4.2. These modified moments should give a balanced column shear condition between points E and F [i.e., $(250 + 55)/26.19 = (154 + 145)/25.52$].

The unit distribution for a fixed end moment of +1000 at BC along with the sidesway modification is similarly computed as shown in Table 4.3.

The shear unbalance before sidesway is the same as that given in the preceding distribution. Similarly, the distributed moments for a unit fixed end moment of -1000 k-ft @ CB are computed in Table 4.4.

The shear unbalance before sidesway is

$$V_{unbal} = \frac{-55 - 28}{26.19} + \frac{166 + 342}{25.52} = -3.2 + 19.9 = +16.7 \text{ k}$$

CO = 0.5

CO = 0.5   CO = 0.5

Joint	B			E	F		C	
Member	BA	BC	BE	EB	FC	CF	CB	CD
Distribution factor	0.326	0.340	0.334			0.333	0.331	0.336
FEM	-1000							
BAL	+326	+340	+334					
CO				+167			+170	
BAL						-57	-56	-57
CO		-28			-28			
BAL	+9	+10	+9					
ΣM	-665	+322	+343	+167	-28	-57	+114	-57
Sidesway	+36	+57	-93	-112	-117	-97	+57	+40
$\left(\dfrac{16.2}{34.8}\right) \times M_\Delta$								
M	-629	+379	+250	+55	-145	-154	+171	-17

^{a}Moments are in kip-feet.

CO = 0.5

CO = 0.5   CO = 0.5

Joint	B			E	F		C	
Member	BA	BC	BE	EB	FC	CF	CB	CD
Distribution factor	0.326	0.340	0.334			0.325	0.335	0.340
FEM		+1000						
BAL	-326	-340	-334					
CO				-167			-170	
BAL						+57	+56	+57
CO		+28			+28			
BAL	-9	-10	-9					
ΣM	-335	+678	-343	-167	+28	+57	-114	+57
Sidesway	-36	-57	+93	+112	+117	+97	-57	-40
$\left(\dfrac{16.2}{34.8}\right) \times M_\Delta$								
M	-371	+621	-250	-55	+145	+154	-171	+17

^{a}Moments are in kip-ft.

Table 4.4[a]

			CO = 0.5					
		CO = 0.5		CO = 0.5			
Joint		B		E	F		C	
Member	BA	BC	BE	EB	FC	CF	CB	CD
Distribution factor	0.326	0.340	0.334			0.333	0.331	0.336
FEM							-1000	
BAL						+333	+331	+336
CO		+166			+166			
BAL	-54	-56	-55					
CO				-28			-28	
BAL						+9	+9	+10
ΣM	-54	+110	-55	-28	+166	+342	-688	+346
Sidesway $\left(\dfrac{16.2}{34.8}\right) \times M_\Delta$	+37	+59	-95	-115	-121	-100	+59	+41
M	-17	+169	-150	-143	+45	+242	-629	+387

[a]Moments are in kip-ft.

Table 4.5[a]

			CO = 0.5					
		CO = 0.5		CO = 0.5			
Joint		B		E	F		C	
Member	BA	BC	BE	EB	FC	CF	CB	CD
Distribution factor	0.326	0.340	0.334			0.333	0.331	0.336
FEM								+1000
BAL						-333	-331	-336
CO		-166			-166			
BAL	+54	+56	+55					
CO				+28			+28	
BAL						-9	-9	-10
Σ	+54	-110	+55	+28	-166	-342	-312	+654
Sidesway $\left(\dfrac{16.2}{34.8}\right) \times M_\Delta$	-37	-59	+95	+115	+121	+100	-59	-41
M	+17	-169	+150	+143	-45	-242	-371	+613

[a]Moments are in kip-ft.

Table 4.6   Moment Distribution Summary

	M_{BA}	M_{BC}	M_{BE}	M_{EB}	M_{FC}	M_{CF}	M_{CB}	M_{CD}
M_{BA}^F	-0.629	+0.379	+0.250	+0.055	-0.145	-0.154	+0.171	-0.017
M_{BC}^F	-0.371	+0.621	-0.250	-0.055	+0.145	+0.154	-0.171	+0.017
M_{CB}^F	-0.017	+0.169	-0.150	-0.143	+0.045	+0.242	-0.629	+0.387
M_{CD}^F	+0.017	-0.169	+0.150	+0.143	-0.045	-0.242	-0.371	+0.613

The distributed moments for a unit fixed end moment of +1000 k-ft @ CD are computed in Table 4.5. The shear unbalance is the same as that for the distribution at CB but of opposite sign.

The resulting moment at each joint, due to a unit (k-ft) moment, is then determined by summing the resultants at each joint and dividing by 1000, as shown in Table 4.6. The signs are maintained in the system to be consistent with vertical gravity loads applied to the superstructure.

These moments are used to determine the joint actions due to dead and unit loads for the development of influence lines.

Dead Load.   The railing dead load is computed by taking the dimensions from Figure 4.2 (assume concrete weight = 0.150 k ft^{-3}):

$$0.83 \times 1.40 \times 0.150 = 0.175$$

$$0.708 \times 1.33 \times 0.792 \times 0.150 \div 8.0 = 0.014$$

$$0.875 \times 1.08 \times 0.150 = \underline{0.142}$$
$$0.331 \text{ k-ft}^{-1}$$

Length of railing is to be applied to a 2.0-ft edge section only. The slab and wearing surface dead load is

Unif. *DL*

w = slab $1.5833 \times 0.150 = 0.238$

WS $= \underline{0.025}$
$$0.263 \text{ ksf}$$

For the slab and wearing surface the fixed end moments for an interior section are computed and the unit moment distribution summary is used to compute the final dead load distributed moments.

The critical moments to be examined in addition to the end shears are $M_{0.4AB}$ and $M_{0.5BC}$.

The fixed end moments are computed as

$$M_{BA}^F = \frac{0.263(34.5)^2}{8} = 39.1 \text{ k-ft}$$

$$M_{BC}^F = \frac{0.263(44.0)^2}{12} = 42.4 \text{ k-ft}$$

$$M_{CB}^F = \qquad\qquad = 42.4 \text{ k-ft}$$

$$M_{CD}^F = \frac{0.263(32.5)^2}{8} = 34.7 \text{ k-ft}$$

The moment summary is shown in Table 4.7:

Table 4.7[a]

	M_{BA}	M_{BC}	M_{BE}	M_{EB}	M_{FC}	M_{CF}	M_{CB}	M_{CD}
$M_{BA}^F = 39.1$	-24.6	$+14.8$	$+9.8$	$+2.2$	-5.7	-6.0	$+6.7$	-0.7
$M_{BC}^F = 42.4$	-15.7	$+26.3$	-10.6	-2.3	$+6.1$	$+6.5$	-7.2	$+0.7$
$M_{CB}^F = 42.4$	-0.7	$+7.2$	-6.4	-6.1	$+1.9$	$+10.3$	-26.7	$+16.4$
$M_{CD}^F = 34.7$	$+0.6$	-5.9	$+5.2$	$+5.0$	-1.6	-8.4	-12.9	$+21.3$
Total DL moments	-40.4	$+42.4$	-2.0	-1.2	$+0.7$	$+2.4$	-40.1	$+37.7$

[a]Moments are in kip-feet.

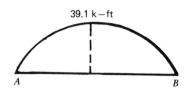

39.1 k–ft

A   B

$$M_{0.4AB} \cong 0.263(34.5)^2 \left(\tfrac{1}{8}\right) = +39.1$$
$$-0.4(40.4) = \underline{-16.2}$$
$$+22.9 \text{ k-ft}$$

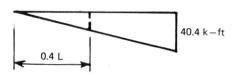

40.4 k–ft

0.4 L

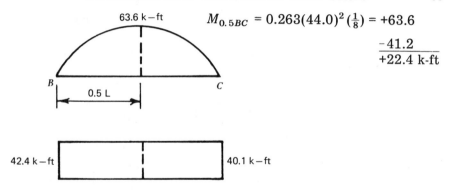

$$M_{0.5BC} = 0.263(44.0)^2\left(\tfrac{1}{8}\right) = +63.6$$

$$\frac{-41.2}{+22.4 \text{ k-ft}}$$

63.6 k−ft

B   0.5 L   C

42.4 k−ft   40.1 k−ft

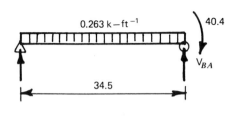

0.263 k−ft^{-1} 40.4

34.5

V_{BA}

$$V_{AB} = 0.263(17.25) = +4.54$$

$$\frac{-40.4}{34.5} = \frac{-1.17}{+3.37}$$

$$V_{BA} = -0.263(17.25) = -4.54$$

$$\frac{-40.4}{34.5} = \frac{-1.17}{-5.71 \text{ k}}$$

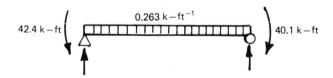

42.4 k−ft   0.263 k−ft^{-1} 40.1 k−ft

$$V_{BC} = 0.263(22.0) = +5.79$$

$$\frac{42.4-40.1}{44.0} = \frac{+0.05}{+5.84 \text{ k}}$$

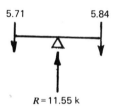

5.71   5.84

$R = 11.55$ k

Reaction per transverse width is therefore $R = 11.55$ k.

Next, the critical live load plus impact moments are computed to verify or revise the initial superstructure sections. An early check on the critical section size is essential to avoid changing the frame constants and the entire design at a later stage.

The impact on the end spans is 30% because the formula for impact $[I = 50/(L + 125)]$, AASHTO Article 1.2.12(c), is not less than the allowed maximum of 30% until the span length L exceeds 41.67 ft. The distribution of live load plus impact per foot is computed by using the distribution value E taken from AASHTO Article 1.3.2(c), Case B. These values for live load distribution and impact along with the coefficient, which represent live load plus impact, are computed as follows:

Span AB $I = 0.300$

$E = 4.0 + 0.06(34.5) = 6.07$ ft

$D = 1/6.07 = 0.165$ per foot of width

Coefficient $= 1.3(0.165) = 0.214$

Span BC $I = \dfrac{50}{44 + 125} = 0.296$

$E = 4.0 + 0.06(44.0) = 6.64$ ft

$\dfrac{0.204}{\text{av}}$

$D = 1/6.64 = 0.151$ per foot of width

Coefficient $= 1.296(0.151) = 0.195$

Span CD $I = 0.300$

$E = 4.0 + 0.06(32.5) = 5.95$ ft

$\dfrac{0.206}{\text{av}}$

$D = 1/5.95 = 0.168$ per foot of width

Coefficient $= 1.3(0.168) = 0.218$

The first loading case is intended to produce maximum positive moment in the end span. These loadings can be properly placed by visualizing the general form of the influence line for a given point (Muller-Breslau principle).

Case I (Figure 4.7). $M_{BA}^{F} = 12[0.1680 + (0.1393)/4]\ (34.5) = 84.0$ k-ft. (coefficients from Appendix O where $M^F = CPL$). From the moment distribution summary, Table 4.6,

$$M_{BA} = 84.0(-0.629) = -52.8 \text{ k-ft}$$

$$\text{SBV}_{AB} = 12\left(0.6 + \frac{0.19}{4}\right) = 7.77 \text{ k (simple beam shear)}$$

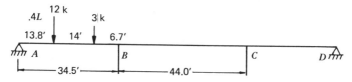

Figure 4.7 End span maximum live load moment loading (positive).

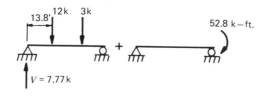

$$M_{0.4} = 7.77(13.8) = +107$$

$$0.4(-52.8) = -\ 21$$

$$+\ 86 \text{ k-ft per line wheels}$$

$$M_{0.4}^{LL+I} = +86 \times 0.214 = 18.4 \text{ k-ft per foot of width}$$

The second loading case is intended to produce maximum positive moment in the interior span.

Case II (Figure 4.8)

$$M_{BC}^{F} = 12(44.0)\left(0.1250 + \frac{0.0266}{4}\right) = 70 \text{ k-ft}$$

$$\text{(Appendix O, where } M^{F} = \text{CPL)}$$

$$M_{CB}^{F} = 12(44.0)\left(0.1250 + \frac{0.1210}{4}\right) = 82 \text{ k-ft}$$

$$SBV_{BC} = 12\left(0.50 + \frac{0.18}{4}\right) = 6.54 \text{ k (simple beam shear)}$$

$$M_{BC} = 70(-0.621) + 82(-0.169) = -57 \text{ k-ft}$$

$$M_{CB} = 70(-0.171) + 82(-0.629) = 64 \text{ k-ft}$$

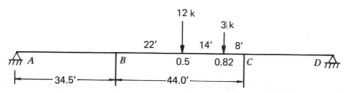

Figure 4.8 Interior span maximum live load moment loading (positive).

$$M_{0.5BC} = 6.54(22.0) = \left.\begin{array}{r} +144 \\ -\ 60 \end{array}\right\} \Sigma = +84 \text{ k-ft per line wheels}$$

$$M_{0.5BC}^{LL+I} = 0.195(+84) = +16.4 \text{ k-ft per foot of width}$$

The Case III loading (Figure 4.9) is intended to produce the maximum negative moment at *BA* which results from the H 15 truck loading. This should be compared with a lane loading to obtain the actual maximum live load moment. This loading should also produce the maximum positive (rotational signs) moment in column *BE*. These loads and moments are included in the tabulations given in Chapters 3 and 10, in which the substructure design is discussed.

Case III (Figure 4.9)

$$M_{BA}^F = 12(34.5)\left(\frac{0.0916}{4} + 0.1920\right) = 89 \text{ k-ft}$$

$$(\text{Appendix O, } M^F = \text{CPL})$$

$$M_{BA} = 89(-0.629) = -56 \text{ k-ft per line wheels}$$

$$M_{BC} = 89(-0.379) = -34 \text{ k-ft per line wheels}$$

$$M_{CB} = 89(+0.171) = +15 \text{ k-ft per line wheels}$$

$$M_{BE} = 89(+0.250) = +22 \text{ k-ft per line wheels}$$

$$M_{EB} = 89(+0.055) = +5 \text{ k-ft per line wheels}$$

$$M_{BA}^{LL+I} = 0.214(-56) = -12.0 \text{ k-ft ft}^{-1}$$

$$M_{BC}^{LL+I} = 0.214(-34) = -7.3 \text{ k-ft ft}^{-1}$$

$$R_B = 12\left(\frac{0.19}{4} + 0.6\right) + \frac{56}{34.5} + \frac{34+15}{44.0}$$

$$= 7.77 + 1.62 + 1.11 = 10.5 \text{ k per line wheels}$$

The Case IV (Figure 4.10) loading is intended to produce the maximum negative moment at *BC* which results from the H 15 truck loading. This is to be compared with a lane loading to obtain the actual

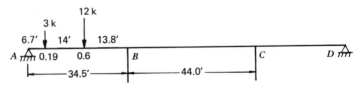

Figure 4.9 End span maximum live load moment loading (negative).

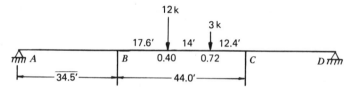

Figure 4.10   Interior span maximum live load moment loading (negative).

maximum live load moment.   This loading should also produce the maximum negative (rotational signs) moment in column BE.   The column loads and moments are included in the tabulations in Example 10.5, in which substructure design is discussed.

Case IV (Figure 4.10)

$$M_{BC}^F = 12(44.0)\left(0.1440 + \frac{0.0564}{4}\right) = 83 \text{ k-ft (Appendix O)}$$

$$M_{CB}^F = 12(44.0)\left(0.0960 + \frac{0.1452}{4}\right) = 70 \text{ k-ft}$$

$M_{BA} = 83(-0.371) + 70(-0.017) = -27$ k-ft per line wheels

$M_{BC} = 83(-0.621) + 70(-0.169) = -63$ k-ft per line wheels

$M_{CB} = 83(-0.171) + 70(-0.629) = -58$ k-ft per line wheels

$M_{BE} = 83(-0.250) + 70(-0.150) = -31$ k-ft per line wheels

$M_{EB} = 83(-0.055) + 70(-0.143) = -15$ k-ft per line wheels

$M_{BA}^{LL+I} = 0.195(-27) = -5.3$ k-ft ft^{-1}

$M_{BC}^{LL+I} = 0.195(-63) = -12.3$ k-ft ft^{-1}

$$R_B = 12\left(0.60 + \frac{0.28}{4}\right) + \frac{27}{34.5} + \frac{63 - 58}{44.0}$$

$$= 8.04 + 0.78 + 0.11 = 8.93 \text{ k per line wheels}$$

Cases V and VI (Figures 4.11 and 4.12) are truck loading and lane loading, respectively, for maximum reaction at B.   The maximum of

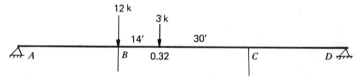

Figure 4.11   Truck loading for maximum reaction at B.

these two loadings is to be used in the footing design, Examples 10.3 and 10.4.

Case V (Figure 4.11)

$$M_{BC}^F = 3(44)(0.1480) = 19.5 \text{ k-ft (Appendix O)}$$

$$M_{CB}^F = 3(44)(0.0696) = 9.2 \text{ k-ft}$$

$$M_{BA} = 19.5(-0.371) + 9.2(-0.017) = -7.4 \text{ k-ft per line wheels}$$

$$M_{BC} = 19.5(-0.621) + 9.2(-0.169) = -13.7 \text{ k-ft per line wheels}$$

$$M_{CB} = 19.5(-0.171) + 9.2(-0.629) = -9.1 \text{ k-ft per line wheels}$$

$$M_{BE} = 19.5(-0.250) + 9.2(-0.150) = -6.3 \text{ k-ft per line wheels}$$

$$M_{EB} = 19.5(-0.055) + 9.2(-0.143) = -2.4 \text{ k-ft per line wheels}$$

$$R_B = 12.0 \left(1.0 + \frac{0.32}{4}\right) + \frac{7.4}{34.5} + \frac{13.7 - 9.1}{44.0}$$

$$= 12.96 + 0.21 + 0.10 = 13.27 \text{ k per line wheels}$$

Case VI (Figure 4.12)

$$M_{BA}^F = \frac{0.240(34.5)^2}{8} = 35.7 \text{ k-ft}$$

$$M_{BC}^F = \frac{0.240(44.0)^2}{12} = 38.7 \text{ k-ft}$$

$$M_{CB}^F = \qquad\qquad = 38.7 \text{ k-ft}$$

$$M_{BA} = 35.7(-0.629) + 38.7(-0.371 - 0.017)$$
$$= -22.5 - 15.0 = -37.5 \text{ k-ft per line of wheels}$$

$$M_{BC} = 35.7(-0.379) + 38.7(-0.621 - 0.169)$$
$$= -13.5 - 30.6 = -44.1 \text{ k-ft per line of wheels}$$

$$M_{CB} = 35.7(+0.171) + 38.7(-0.171 - 0.629)$$
$$= +6.1 - 31.0 = -24.9 \text{ k-ft per line of wheels}$$

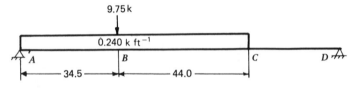

Figure 4.12   Lane loading for maximum reaction at B.

$$M_{BE} = 35.7(+0.250) + 38.7(-0.250 - 0.150)$$

$$= +8.9 - 15.5 = -6.6 \text{ k-ft per line of wheels}$$

$$M_{EB} = 35.7(+0.055) + 38.7(-0.055 - 0.143)$$

$$= +2.0 - 7.7 = -5.7 \text{ k-ft per line of wheels}$$

$$R_B = 9.75 + 0.240 \left(\frac{34.5 + 44.0}{2} \right) + \frac{37.5}{34.5} + \frac{44.1 - 24.9}{44.0}$$

$$= 9.75 + 9.42 + 1.09 + 0.44 = 20.70 \text{ k per line wheels}$$

The Case VII loading (Figure 4.13) is the lane loading to produce the maximum negative moment at BA and BC. These moments are to be compared with truck moments computed in cases III and IV to obtain the actual maximum negative moments:

$$M_{BA}^F = 35.7 + 6.75(34.5)(0.1920) = 80.4 \text{ k-ft}$$

$$M_{BC}^F = 38.7 + 6.75(44.0)(0.1440) = 81.5 \text{ k-ft}$$

$$M_{CB}^F = 38.7 + 6.75(44.0)(0.0960) = 67.2 \text{ k-ft}$$

$$M_{BA} = 80.4(-0.629) + 81.5(-0.371) + 67.2(-0.17)$$

$$= -82.0 \text{ k-ft per line of wheels}$$

$$M_{BC} = 80.4(-0.379) + 81.5(-0.621) + 67.2(-0.169)$$

$$= -92.4 \text{ k-ft per line of wheels}$$

$$M_{BA}^{LL+I} = 82.0(0.204) = 16.7 \text{ k-ft per foot of width}$$

$$M_{BC}^{LL+I} = 92.4(0.204) = 18.8 \text{ k-ft per foot of width}$$

From experience the positive moment will always be controlled by truck loadings for this span range.

The maximum moments at the critical points can now be tabulated. Note that with the H 15 loading the overload provision of AASHTO Article 1.2.4 must be checked. The specification states that this provision need not apply to the roadway deck; to be conservative, however, a check should be made for longitudinally reinforced decks and the ex-

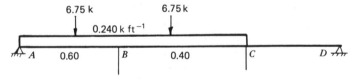

Figure 4.13 Maximum live load moment loading (negative) at BA, BC.

ception should be applied to transversely reinforced decks. At these critical locations the deck is approximately proportioned and checked for stress. At $0.4AB$, using maximum positive moment, the approximate required depth and reinforcing are computed by the equations $d = \sqrt{M/Kb}$ and $A_s = M/f_s jd$, where K and j are conservatively assumed to be equal to $K = 246$ and $j = 0.87$, respectively.

@ 0.4AB

$$M = +41.3 \text{ k-ft ft}^{-1}$$

$$d \text{ required} = \sqrt{\frac{41.3(12)}{0.246(12)}} = 12.96 \text{ in.} < (19.0 - 2.0)$$

$$= 17.0 \text{ in. available}$$

$$A_s \text{ (required)} = \frac{41.3(12)}{24(0.87)(17.0)} = 1.40 \text{ in.}^2 \text{ ft}^{-1}$$

#8 @ $6\frac{1}{2}$ is selected which gives $A_s = 1.46$ in.^{2} ft^{-1}. The effective depth is therefore the depth (19.0 in.) minus the cover (1.5 in.), minus $\frac{1}{2}$ bar thickness, which gives $d = 19.0 - 1.5 - 0.5 = 17.00$ in.
The stresses are checked by referring to Appendix A and the formulas

$$f_s = \frac{M}{A_s jd} \quad f_c = \frac{f_s k}{n(1 - k)} \quad \text{and} \quad p = \frac{A_s}{bd}$$

where $b = 12.0$ in.

Table 4.8[a]

	DL	LL + I	Σ	Overload Provision $\dfrac{DL + 2(LL + I)}{1.50}$
0.4AB	+22.9	+18.4	+41.3	+39.8
0.5BC	+22.4	+16.4	+38.8	+36.8
BA	-40.4	-16.7	-57.1	-49.2
BC	-42.4	-18.8	-61.2	-53.3

[a]Moments are in kip-ft ft^{-1}.

Computing $p = 1.46/[12(17.0)] = 0.0071$,

$$k = 0.285 \quad \text{and} \quad j = 0.905, n = 8$$

from Appendix A. Therefore

$$f_s = \frac{41.3(12)}{1.46(0.905)(17.00)} = 22.1 \text{ ksi} < 24.0 \text{ ksi}$$

$$f_c = \frac{22.1(0.285)}{8(0.715)} = 1099 \text{ psi} < 1600 \text{ psi}$$

In the same manner the positive moment section at $0.5BC$ is proportioned and stresses are checked.

@ 0.5BC

$$M = 38.8 \text{ k-ft}$$

$$A_s \text{ required} = \frac{38.8(12)}{24(0.87)(17.0)} = 1.31 \text{ in.}^2 ; \#8 @ 7 \text{ in. } A_s = 1.35 \text{ in.}^2$$

$$d = 19 - 1.5 - 0.50 = 17.00 \text{ in.}$$

$$p = \frac{1.35}{12(17.00)} = 0.0066; \text{ therefore } k = 0.276, j = 0.908$$

$$f_s = \frac{38.8(12)}{1.35(0.908)(17.00)} = 22.3 \text{ ksi} < 24.0 \text{ ksi}$$

$$f_c = \frac{22.3(0.276)}{8(0.724)} = 1065 \text{ psi} < 1600 \text{ psi}$$

In the same manner the negative moment section at B is proportioned and stresses are checked.

@ B

$$M = -61.2 \text{ k-ft}$$

$$d \text{ required} = \sqrt{\frac{61.2(12)}{0.246(12)}} = 15.8 \text{ in.} < 16.5 \text{ in. } \pm \text{ available}$$

$$A_s \text{ required} = \frac{61.2(12)}{24(0.87)(16.5)} = 2.13 \text{ in.}^2 ; \#10 @ 7 \text{ in.}; A_s = 2.18 \text{ in.}^2$$

$$d = 19.0 - 2.0 - 0.64 = 16.36 \text{ in.}$$

$$p = \frac{2.18}{12(16.36)} = 0.0111; \text{ therefore } k = 0.342, j = 0.886$$

$$f_s = \frac{61.2(12)}{2.18(0.886)(16.36)} = 23.2 \text{ ksi} < 24.0 \text{ ksi}$$

$$f_c = \frac{23.2(0.342)}{8(0.658)} = 1510 \text{ psi} < 1600 \text{ psi}$$

With the critical deck sections designed we examine the deck-column details. In general, for a longitudinally reinforced concrete slab bridge that is constructed monolithically with the substructure it is appropriate to make any necessary transverse cap beam integral with the slab. The cap for this particular bridge is shown in Figure 4.14. Its design requires evaluation of the dead and live load shears, reactions, and moments.

The dead load reaction per foot is

$$R_B = 11.55 \text{ k} + 0.5(5.5)(0.150) = 11.96 \text{ k ft}^{-1}$$

11.55 k is computed on page 63. Proportioning the rail loading to the slab and ws load gives the concentrated load applied at the edge of the slab:

$$\text{rail } R_B = \frac{0.331}{0.263} \times 11.55 \text{ k} = 14.5 \text{ k at each edge}$$

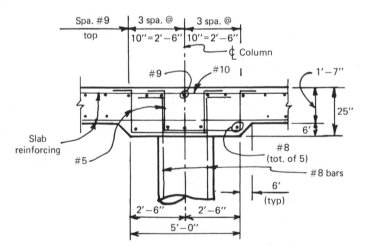

Figure 4.14   Pier cap details.

The cap can be approximated as a two-structure span with cantilever ends as shown in Figure 4.15 with the dead load.

Dead Load.   The dead load moments at the supports are by use of statics

$$M_B = 14.5 \text{ k } (3.38 \text{ ft})\qquad\qquad = 49.0 \text{ k-ft}$$
$$\left.\begin{array}{l}11.96 \text{ k } (3.96 \text{ ft})(1.98 \text{ ft}) = 93.8 \text{ k-ft}\end{array}\right\}\ \Sigma = -143 \text{ k-ft}$$

$$M_C = \frac{11.96(12.5)^2}{8} - \frac{143}{2} = -163 \text{ k-ft}$$

The moment at midspan between the columns is

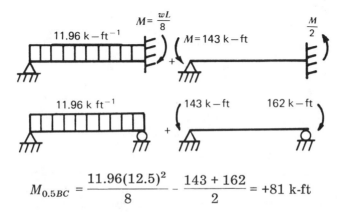

$$M_{0.5BC} = \frac{11.96(12.5)^2}{8} - \frac{143 + 162}{2} = +81 \text{ k-ft}$$

The dead load shears and reactions are

$$V_{BA} = 3.96(11.96) + 14.5 = 61.9 \text{ k}$$
$$V_{BC} = 11.96(6.25) = +74.8 \text{ k}$$
$$\frac{-162 - 143}{12.5} = \underline{-1.5 \text{ k}}$$
$$+73.3 \text{ k}$$
$$V_{CB} = +74.8 + 1.5 = +76.3 \text{ k}$$

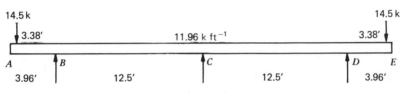

Figure 4.15   Pier cap span.

Reaction R_B is therefore

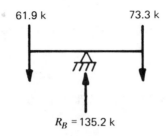

and reaction R_c is

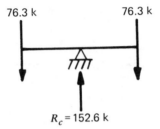

The influence line ordinates for the moment at C are tabulated and plotted in Figure 4.16 for $L = 12.5$ ft. These ordinates can be obtained from a number of sources such as the AISC publication, *Moments, Shears, and Reactions for Continuous Highway Bridges.*[6]

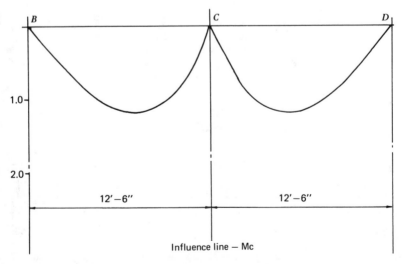

Figure 4.16   Influence line for M_c of cap.

Load @	M_c
B	0
0.1	−0.31
0.2	−0.60
0.3	−0.85
0.4	−1.05
0.5	−1.17
0.6	−1.20
0.7	−1.12
0.8	−0.90
0.9	−0.54
C	0

For live load plus impact on the cap four different loadings are evaluated (Figures 4.17–4.20):

Live Load + Impact

Loading I

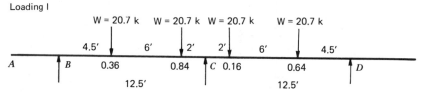

Figure 4.17   Pier cap live loading—I.

Loading II

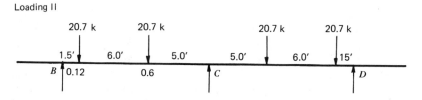

Figure 4.18   Pier cap live loading—II.

Loading III

Figure 4.19   Pier cap live loading—III.

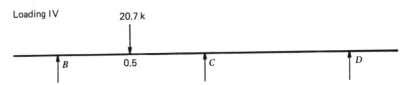

Figure 4.20   Pier cap live loading—IV.

The impact factor is an average of the factors for loading spans AB and BC or $I_{Avg} = (0.300 + 0.296)/2 = 0.298$. Loading I or II gives the maximum negative moment at C.

For Loading I

$$Ld \times (IL)_{ord} \times 2Lds \times (1 + I)$$

$$M_C = (20.7)(1.02 + 0.75)(2)(1.298) = -95 \text{ k-ft}$$

$$V_{CB} = [20.7(0.36 + 0.84)] (1.298) + \frac{95}{12.5} = 39.7 \text{ k}$$

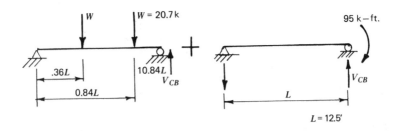

For Loading II

$$M_C = [20.7(1.20 + 0.35)] (2)(1.298) = -83.3 \text{ k-ft}$$

Loading III yields the maximum negative moment at B:

$$M_B = 20.7(1.5)(1.298) = -40.3 \text{ k-ft}$$

Loading IV can be used to give the maximum midspan positive moment

$$M_C = 20.7(-1.17) = -24.2 \text{ k-ft}$$

$$M_{0.5BC} = \left[20.7(12.5) \left(\frac{1}{4}\right) - \frac{24.2}{2} \right] (1.298) = +68.3 \text{ k-ft}$$

Table 4.9 summarizes the critical moments and shears.
Analysis and design of the maximum negative moment are given for

Table 4.9[a]

Force	DL	LL + I	ΣM	Overload $\dfrac{DL + 2(LL + I)}{1.50}$
M_B	-143	-40	-183	-149
M_C	-162	-95	-257	-235
$M_{0.5BC}$	+81	+68	+149	+145
V_{CB}	76.3	39.7	116.0	103.8

[a]Moments are in kip-ft; shears are in kips.

the section at C, where $M_C = -257$ k-ft. The approximate section depth is computed from $d = \sqrt{(M \times 12)/Kb}$; assuming $K = 246$ and $j = 0.87$ for a 5.0-ft wide section,

$$d_{\text{required}} = \sqrt{\frac{257 \times (12)}{246 \times (60)}} = 14.5 \text{ in.};$$

$$d = 24.0 - 2.0 - 1.27 - 0.56 = 21.17 \text{ in.}$$

Using $A_s = (M \times 12)/f_s jd$, we get the approximate $A_{s(\text{required})} = 257(12)/24(0.87)(21.17) = 6.98 \text{ in.}^2$; use 7-#9; $A_s = 7.00 \text{ in.}^2$

Minimum reinforcing requirements (AASHTO Article 1.5.7), based on 1.2 times the cracking moment, are computed as $f_{cr} = 7.5\sqrt{f_c'} = 7.5\sqrt{4000} = 474$ psi; the section modulus is $S = bd^2/6$, or $S = 60(25)^2/6 = 6250 \text{ in.}^3$ The cracking moment is computed by using $474 = [M_{cr}(12)]/6250$ and solving for M_{cr} gives 247 k-ft and $1.2M_{cr} = 296$ k-ft. Considering the section at ultimate moment and summing forces, we obtain $0.85(4)(60) \times a = 7.00 \times 60$, and solving for a we get $a = 2.06$ in.

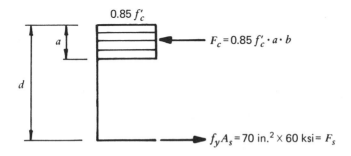

The ultimate moment is $\phi M_n = \phi[A_s f_y(d - a/2)] \left(\frac{1}{12}\right)$ or $\phi M_n = 0.90 [(7.00)(60)(21.17 - 1.03)] \left(\frac{1}{12}\right) = 634$ k-ft > 296 k-ft. Therefore the section is adequate for minimum reinforcing requirements.

Now, using Appendix A to obtain k and j values, we compute the allowable stresses and compare them with the allowables.

$$p = \frac{A_s}{bd} = \frac{7.00}{60(21.17)} = 0.0055$$

From Appendix A

$$k = 0.256, j = 0.914, \text{ using } n = 8$$

$$f_s = \frac{257(12)}{7.00(0.914)(21.17)} = 22.8 \text{ ksi} < 24.0 \text{ ksi}$$

$$f_c = \frac{22.8(0.256)}{8(0.744)} = 979 \text{ psi} < 1600 \text{ psi}$$

For positive moment

$$M_{0.5BC} = +149 \text{ k-ft}$$

$$\text{approx } A_{s \text{ required}} = \frac{149(12)}{24(0.87)(22.3)} = 3.84 \text{ in.}^2$$

Try 5-#8 $A_s = 3.95$ in.^{2}, $d = 25.0 - 1.5 - 0.62 - 0.50 = 22.38$ in.

$$p = \frac{3.95}{60(22.38)} = 0.0029$$

From Appendix A $k = 0.204$, $j = 0.932$, $n = 8$; therefore

$$f_s = \frac{149(12)}{3.95(0.932)(22.38)} = 21.7 \text{ ksi} < 24.0 \quad \text{OK}$$

$$f_c = \frac{21.7(0.204)}{8(0.796)} = 695 \text{ psi} < 1600 \text{ psi} \quad \text{OK}$$

The stirrup requirements for shear are as follows:

$$v = \frac{V}{bd}$$

$$v = \frac{116.0}{60(21.17)} = 91 \text{ psi}$$

Use $v_c = 66$ psi.

The spacing of the stirrups is therefore

$$S_{required} = \frac{A_v f_y}{b(v - v_c)}$$

$$s = \frac{4(0.31)(24,000)}{60(91 - 55)} = 13.8 \text{ in.}$$

$$S_{max} = \frac{d}{2} \quad \text{or} \quad \frac{A_v f_s}{b \times 50}$$

$$S_{max} = \frac{21.17}{2} = 10.5 \text{ in.} \quad \text{or}$$

$$S_{max} = \frac{4(0.31)(60,000)}{60(50)} = 24.8 \text{ in.}$$

Therefore 10.5 in. governs; use 4 legs of #5 @ 10.5 in.

Final Superstructure Design

An initial design of the critical superstructure sections has already been made. It is generally advisable not to complete the superstructure design in a frame analysis until the substructure analysis is complete to ensure that no change in frame constants is necessary. Therefore, assuming that the substructure analysis is complete as shown in Example 10.4, we perform the complete analysis of the superstructure. In general, for normal spans, moment curves are established at each tenth point across the span, but because this bridge has such short spans values at 0.2, 0.4, 0.5, 0.6, and 0.8 of the span along with the support locations are used to establish the moment curves.

The dead load moments are computed at these points. First the dead load moments are computed by the moment distribution method at all points and tabulated as negative moments. The simple span moments (positive moments) are then tabulated and added algebraically to the negative moments (see Table 4.6) to produce the actual dead load moments (M_{DL}).

For uniform loads (w), such as the dead load, the simple span moments are a function of the value KwL^2; K is listed in the table in Appendix N. The resulting dead load moments are given in Table 4.10.

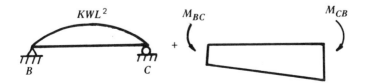

Table 4.10

Point	Negative Moment	Positive Moment	M_{DL} (kip-ft)
AB			0
0.2	-8.1	+25.0	+16.9
0.4	-16.2	+37.6	+21.4
0.5	-20.2	+39.1	+18.9
0.6	-24.2	+37.6	+13.4
0.8	-32.3	+25.0	-7.3
BA	-40.4		-40.4
BC	-42.4		-42.4
0.2	-41.9	+40.7	-1.2
0.4	-41.5	+61.1	+19.6
0.5	-41.2	+63.6	+22.4
0.6	-41.0	+61.1	+20.1
0.8	-40.6	+40.7	+0.1
CB	-40.1		-40.1
CD	-37.7		-37.7
0.2	-30.2	+22.2	-8.0
0.4	-22.6	+33.4	+10.8
0.5	-18.8	+34.8	+16.0
0.6	-15.1	+33.4	+18.3
0.8	-7.5	+22.2	+14.7
DC			0

$$\text{Positive moment} \begin{cases} \text{span } AB & wl^2 = 0.263(34.5)^2 = 313 \times K \\ \text{span } BC & wl^2 = 0.263(44.0)^2 = 509 \times K \\ \text{span } CD & wl^2 = 0.263(37.5)^2 = 278 \times K \end{cases}$$

where K = coefficient given in Appendix N.

In preparation for the computation of live load moments influence lines are computed. Their ordinates and areas are detailed in the following computations.

To obtain the distributed moments at the support points the fixed end moments for a 1-k load at each point along the bridge must be calculated. The fixed end moments are obtained by using the information given in Appendix O. The distributed moments are then computed by using these fixed end moments and Table 4.6. The resulting fixed end moments, the distributed moments, and the areas under the influence lines for the distributed moments are tabulated in Table 4.11.

The areas under the influence line diagrams were determined by calculating areas of rectangles and triangles.

The influence lines for moment at points in spans other than the span on which the 1-k load is located are spaced equally between the influence lines for the end moments for that span. Therefore there is no

Table 4.11

Load @	M_{BA}^F	M_{BC}^F	M_{CB}^F	M_{CD}^F	M_{BA}	M_{BC}	M_{CB}	M_{CD}
AB								
0.1								
0.2	3.31				-2.08	-1.25	$+0.57$	$+0.06$
0.3								
0.4	5.80				-3.65	-2.20	$+0.99$	$+0.10$
0.5	6.47				-4.07	-2.45	$+1.11$	$+0.11$
0.6	6.62				-4.16	-2.51	$+1.13$	$+0.11$
0.7								
0.8	4.97				-3.13	-1.88	$+0.85$	$+0.08$
0.9								
B								
0.1								
0.2		5.63	1.41		-2.11	-3.73	-1.85	-0.64
0.3								
0.4		6.34	4.22		-2.42	-4.65	-3.74	-1.74
0.5		5.50	5.50		-2.13	-4.34	-4.40	-2.22
0.6		4.22	6.34		-1.67	-3.69	-4.71	-2.53
0.7								
0.8		1.41	5.63		-0.62	-1.83	-3.78	-2.20
0.9								
C								
0.1								
0.2				4.68	$+0.08$	$+0.79$	-1.74	-2.87
0.3								
0.4				6.24	$+0.11$	$+1.05$	-2.32	-3.83
0.5				6.09	$+0.10$	$+1.03$	-2.26	-3.73
0.6				5.46	$+0.09$	$+0.92$	-2.03	-3.35
0.7								
0.8				3.12	$+0.05$	$+0.53$	-1.16	-1.91
0.9								
D								
			Area span *AB*		-93	-56	$+25$	$+2$
			Area span *BC*		-62	-126	-128	-65
			Area span *CD*		$+2$	$+22$	-48	-80
			Total + area		$+2$	$+22$	$+25$	$+2$
			Total − area		-155	-182	-176	-145

need to tabulate these influence line ordinates for intermediate points in spans other than those that are loaded.

For the loaded span the tabulation shown is a simple method of computing the influence line ordinates and areas. The influence line ordinates for moment at the ends of the support are tabulated as previously computed. At the point at which the load is located compute the positive simple span moment and the interpolated negative moment. Add the two algebraically; one will have the ordinate of the peak of the moment curve for the given loading. The moment will then vary in a straight line from the peak to each end moment. Intermediate points are picked from this straight line. A programmable desk calculator can work this procedure rapidly. If we observe Table 4.12 closely, it will be obvious that the horizontal ordinates form the moment diagrams and the vertical ordinates form the influence lines. The areas and ordinates of each influence line are shown individually for the three spans in Tables 4.12, 4.13, and 4.14.

The resulting influence lines are plotted as shown in Figure 4.21. Using the influence lines, tabulate the induced maximum live load moments (Table 4.15).

The maximum positive and negative moments due to a truck loading can be obtained by using a straight edge, with the H 15 truck wheels at a scaled spacing, and scaling the ordinates for each point. Note that for positive moment one large wheel always falls at the peak of the curve.

Table 4.12   Influence Lines: Span AB

Load @	AB	0.2	0.4	0.5	0.6	0.8	BA
0.1	0						−1.08
		+5.52					
0.2	0	−0.42					−2.08
		+5.10	+3.30	+2.41	+1.51	−0.28	
0.3	0						−2.96
			+8.28				
0.4	0		−1.46				−3.65
		+3.41	+6.82	+5.08	+3.33	−0.16	
				+8.62			
0.5	0			−2.03			−4.07
		+2.64	+5.27	+6.59	+4.46	+0.19	
					+8.28		
0.6	0				−2.50		−4.16
		+1.93	+3.85	+4.82	+5.78	+0.81	
0.7	0						−3.87
						+5.52	
0.8	0					−2.50	−3.13
		+0.76	+1.51	+1.89	+2.26	+3.02	
0.9	0						−1.86
Positive area		+77	+107	+104	+89	+26	
Negative area						−3	

Table 4.13 Influence Lines: Span *BC*

Load @	BC	0.2	0.4	0.5	0.6	0.8	CB
0.1	-2.28						-0.86
		+7.04					
0.2	-3.73	-3.35					-1.85
		+3.69	+2.30	+1.61	+0.92	-0.46	
0.3	-4.49						-2.85
			+10.56				
0.4	-4.65		-4.29				-3.74
		+0.81	+6.27	+4.60	+2.93	-0.40	
				+11.00			
0.5	-4.34			-4.37			-4.40
		+0.05	+4.44	+6.63	+4.42	+0.01	
				+10.56			
0.6	-3.69			-4.30			-4.71
		-0.37	+2.94	+4.60	+6.26	+0.78	
0.7	-2.81						-4.54
						+7.04	
0.8	-1.83					-3.39	-3.78
		-0.46	+0.91	+1.60	-2.28	+3.65	
0.9	-0.85						-2.31
Positive area		+40	+109	+118	+109	+39	
Negative area		-7				-8	

Therefore the maximum positive ordinates, already computed, are listed and the ordinates for other wheels are scaled. Note also that the maximum negative moments for the intermediate points result from the same truck loading as the end moments and therefore can be prorated from the end moments. It is usual practice to compute moments and shears for one line of wheels and then include the effects of lateral distribution and impact. This is done here, and the moments for one line of wheels with a truck loading are compared with those for one line of wheels with a lane loading. A check mark next to the number indicates the controlling value, which is then magnified for $(LL + I)$ and distribution.

The same procedure is applied to lane loadings, by using areas for uniform loading and maximum ordinates for concentrated loads. Note that AASHTO Article 1.2.8(e) allows two concentrated loads in different spans for computing maximum negative moment. For maximum positive moment only one concentrated load is permitted. Uniform loading can be made discontinuous to cover just the positive or negative

Table 4.14 Influence Lines: Span CD

Load @	CD	0.2	0.4	0.5	0.6	0.8	DC
0.1	-1.70						0
		+5.20					
0.2	-2.87	-2.30					0
		+2.90	+2.18	+1.81	+1.45	+0.72	
0.3	-3.56						0
			+7.80				
0.4	-3.83		-2.30				0
		+0.83	+5.50	+4.58	+3.67	+1.83	
				+8.12			
0.5	-3.73			-1.86			0
		+0.26	+4.26	+6.26	+5.01	+2.50	
					+7.80		
0.6	-3.35				-1.34		0
		-0.08	+3.19	+4.82	+6.46	+3.23	
0.7	-2.72						0
						+5.20	
0.8	-1.91					-0.38	0
		-0.23	+1.46	+2.30	+3.14	+4.82	
0.9	-0.99						0
Positive area		+24	+80	+93	+96	+69	
Negative area		-2					

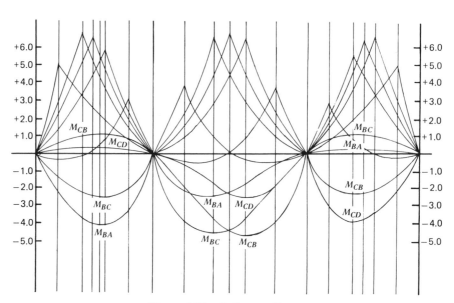

Figure 4.21 Influence lines.

84

Table 4.15   Live Load + Impact Moments

	Truck: Positive Moment				Truck: Negative Moment				
									Load
			M				M		
			One Line				One Line		
			Wheels	M_{LL+I}			Wheels	M_{LL+I}	Position
Point	12 k	3 k	(k-ft)	(k-ft)	12 k	3 k	(k-ft)	(k-ft)	Span
AB									
0.2	+5.10	+1.9	+67 √	+14.3			-7 √	-1.4	
0.4	+6.82	+1.4	+86 √	+18.4			-14 √	-2.7	
0.5	+6.59	+1.2	+83 √	+17.8			-17 √	-3.3	
0.6	+5.78	+1.5	+74 √	+15.8			-20 √	-3.9	
0.8	+3.02		+36 √	+7.7			-27		
BA					-2.5	-1.4	-34		*BC*
					-4.2	-1.8	-56[a]		*BA*
					-4.5	-2.8	-62[a]		*BC*
BC					-2.5	-1.3	-34		*BA*
					+1.0	+0.5	+13		*CD*
0.2	+3.69		+44 √	+8.6			-24		
0.4	+6.27	+1.5	+80 √	+15.6			-14		
0.5	+6.63	+1.5	+84 √	+16.4			-10		
0.6	+6.26	+1.6	+80 √	+15.4			-12		
0.8	+3.65		+44 √	+8.6			-21		
					+1.1	+0.6	+15		*BA*
CB					-2.3	-1.0	-31		*CD*
					-4.6	-2.8	-64[a]		*BC*
CD					-3.8	-1.7	-51[a]		*CD*
					-2.6	-1.5	-36		*BC*
0.2	+2.90		+35 √	+7.6			-29 √	-5.7	
0.4	+5.50	+1.1	+69 √	+15.0			-22 √	-4.3	
0.5	+6.26	+0.8	+78 √	+17.0			-18 √	-3.5	
0.6	+6.46	+1.1	+81 √	+17.7			-14 √	-2.7	
0.8	+4.82	+1.6	+63 √	+13.7			-7 √	-1.4	
DC									

[a]Maximum values.

areas of the influence line. Maximum positive moment with lane loading has been computed for purposes of illustration. For short spans such as these, however, lane loading does not control and need not be computed for positive moment. Note that it does control the negative moment condition.

The resulting induced moments (positive and negative) due to the lane loading are listed in Tables 4.16 and 4.17.

The maximum positive and negative $(DL + LL + I)$ moments are summed and tabulated in Table 4.18.

Table 4.16 Live Load + Impact Moments.
Lane: Positive Moments

Point	0.240 (kip ft^{-1})		6.75 (k)	M One Line Wheels (kip-ft)	M_{LL+I} (kip-ft)
AB					
0.2	+77		+5.10	+53	
0.4	+107	+1	+6.82	+72	
0.5	+104	+1	+6.59	+70	
0.6	+89	+1	+5.78	+61	
0.8	+26	+2	+3.02	+27	
BA					
BC					
0.2	+40	+8	+3.69	+36	
0.4	+109		+6.27	+68	
0.5	+118		+6.63	+73	
0.6	+109		+6.26	+68	
0.8	+39	+9	+3.65	+36	
CB					
CD					
0.2	+24	+2	+2.90	+26	
0.4	+80	+1	+5.50	+57	
0.5	+93	+1	+6.26	+65	
0.6	+96	+1	+6.46	+67	
0.8	+69		+4.82	+49	
DC					

Table 4.17 Live Load + Impact Moments.
Lane: Negative Moments

Point	0.240 (kip ft^{-1})		6.75 (k)		M One Line Wheels (kip-ft)	M_{LL+I} (kip-ft)
AB	0					
0.2	-12			-0.49	-6	
0.4	-25			-0.98	-13	
0.5	-31			-1.22	-16	
0.6	-37			-1.47	-19	
0.8	-3	-50		-1.96	-28 ✓	-5.7
BA	-93	-62	-0.28	-2.45	-82 ✓	-16.7
BC	-56	-126	-4.16	-4.65	-92 ✓	-18.8
0.2	-40	-7	-2.51	-0.46	-26 ✓	-5.3
0.4	-24		-6	-1.78	-0.30	-16 ✓	-3.3

Table 4.17   (Continued)

Point	0.240 (kip ft^{-1})		6.75 (k)		M One Line Wheels (kip-ft)	M_{LL+I} (kip-ft)
0.5	-16	-13	-0.69	-0.64	-16 √	-3.2
0.6	-8	-20	-0.33	-0.97	-15 √	-3.1
0.8		-8	-34	-0.46	-1.65	-24 √	-4.9
CB		-128	-48	-4.71	-2.32	-90 √	-18.5
CD		-65	-80	-2.55	-3.83	-78 √	-16.1
0.2		-52	-2	-2.04	-0.23	-28	
0.4		-39		-1.53		-20	
0.5		-32		-1.28		-16	
0.6		-26		-1.02		-13	
0.8		-13		-0.51		-7	
DC							

Table 4.18   Maximum Moments[a]

Point	M_{DL}	$+M_{L+I}$	$+M_T$	M_{DL}	$-M_{L+I}$	$-M_T$
AB	0			0		
0.2	+16.9	+14.3	+31.2	+16.9	-1.4	
0.4	+21.4	+18.4	+39.8	+21.4	-2.7	
0.5	+18.9	+17.8	+36.7	+18.9	-3.3	
0.6	+13.4	+15.8	+29.2	+13.4	-3.9	+9.5
0.8	-7.3	+7.7	+0.4	-7.3	-5.7	-13.0
BA	-40.4			-40.4	-16.7	-57.1
BC	-42.4			-42.4	-18.8	-61.2
0.2	-1.2	+8.6	+7.4	-1.2	-5.3	-6.5
0.4	+19.6	+15.6	+35.2	+19.6	-3.3	+16.3
0.5	+22.4	+16.4	+38.8	+22.4	-3.2	
0.6	+20.1	+15.4	+35.5	+20.1	-3.1	+17.0
0.8	+0.1	+8.6	+8.7	+0.1	-4.9	-4.8
CB	-40.1			-40.1	-18.5	-58.6
CD	-37.7			-37.7	-16.1	-53.8
0.2	-8.0	+7.6	-0.4	-8.0	-5.7	-13.7
0.4	+10.8	+15.0	+25.8	+10.8	-4.3	+6.5
0.5	+16.0	+17.0	+33.0	+16.0	-3.5	
0.6	+18.3	+17.7	+36.0	+18.3	-2.7	
0.8	+14.7	+13.7	+28.4	+14.7	-1.4	
DC	0			0		

[a]kip-ft.

By using these moments we can design each section for the required reinforcement. Computation of the resisting moments (M_R) is based initially on the preliminary section designs for positive and negative moments.

The resisting moments M_R are computed by using the reinforcement (A_s) and j values, as computed previously. The reinforcement computed for the interior span is also used for the end spans because the increase in required reinforcement is slight and uniformity makes construction easier. Cutoffs reduce the reinforcing area by $\frac{2}{3}$ and $\frac{1}{3}$ and, therefore, the resisting moment is reduced by $\frac{2}{3}$ and $\frac{1}{3}$, which ignores a slight increase in the j values, which would be insignificant.

Longitudinal Resisting Moments

Positive Moment: All Spans, #8 @ 7 in.

$$M_R = \frac{(A_s \times jd \times f_s)}{12}$$

$$M_R = 1.35(0.908)(17.00)(24.0)(\tfrac{1}{12}) = 41.6 \text{ k-ft}$$

Negative Moment #10 @ 7 in.

$$M_R = 2.18(0.886)(16.36)(24.0)(\tfrac{1}{12}) = 63.2 \text{ k-ft}$$

The required lateral distribution reinforcement, according to AASHTO Article 1.3.2(e) is computed as follows:

Transverse	Distribution Reinforcement $[A_s = 100/S$ when $S = $ span $(f_f)]$	
Span AB		
#8 @ 7 $A_s = 1.35$ in.^{2}	$A_s = \dfrac{100}{\sqrt{34.5}}(1.35) = 0.23$ in.^{2} ;	#4 @ 10" $A_s = 0.24$ in.^{2}
Span BC		
#8 @ 7 $A_s = 1.35$ in.^{2}	$A_s = \dfrac{100}{\sqrt{44}}(1.35) = 0.20$ in.^{2} ;	#4 @ 1'-0" $A_s = 0.20$ in.^{2}
Span CD		
#8 @ 7 $A_s = 1.35$ in.^{2}	$A_s = \dfrac{100}{\sqrt{32.5}}(1.35) = 0.24$ in.^{2} ;	#4 @ 10" $A_s = 0.24$ in.^{2}

The required bar extensions, according to AASHTO Article 1.5.13(a), are as follows:

$$\#10\text{--}15\phi = 1.6 \text{ ft}$$

$$\#8\text{--}15\phi = 1.2 \text{ ft}$$

or

$$\tfrac{1}{20} \text{ span } AB = 1.8 \text{ ft}$$

$$\tfrac{1}{20} \text{ span } BC = 2.2 \text{ ft}$$

$$\tfrac{1}{20} \text{ span } CD = 1.8 \text{ ft}$$

Next, the maximum moments are plotted with the resisting moments (Figure 4.22). The cutoffs are selected by using equal or greater bar extensions than previously given. The effect of development length is shown by sloping lines. It is obvious from this figure that the development length need not be considered in normal circumstances. For a #8 bar the development length applicable is $0.8l_d = 26$ in.

The edge beam is designed in accordance with AASHTO Article 1.3.2(d), as explained in Section 4.2.1. A 2 ft-0 in. width of slab is selected as the edge beam.

The dead load of the slab and wearing surface is tabulated for a 2 ft-0 in. width. The dead load of the rail is prorated with the slab and wearing surface dead load and tabulated at all the required points. The live load and impact moments are computed and tabulated for a 2 ft-0 in. width and 0.10 times the live load moment for one line of wheels. The total dead and live load maximum moments are then tabulated (Table 4.19).

The maximum negative moment at BC is checked by the method described in ACI publication SP-3 *Reinforced Concrete Design Handbook—Working Stress Method*,[7] which accounts for compressive reinforcement. When this information is required, it is recommended that the following formulas be used for computing j and k:

$$m = \frac{nA_s}{bd} + \frac{(2n-1)\,A_s'}{bd}$$

$$q = \frac{nA_s}{bd} + \frac{(2n-1)\,A_s'}{bd} \times \frac{d'}{d}$$

$$k = \sqrt{m^2 + 2q} - m$$

$$z = \frac{0.167 + \{[(2n-1)\,A_s']/kbd\}\,(d'/kd)(1 - d'/kd)}{0.500 + \{[(2n-1)\,A_s']/kbd\}\,(1 - d'/kd)}$$

$$j = 1 - zk$$

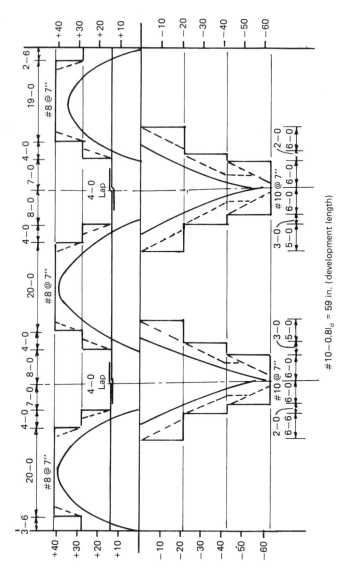

Figure 4.22 Maximum moments—Slab.

#10—0.8l_d = 59 in. (development length)

Table 4.19   Edge Beam

Point	2'-0" Slab + WS M_{DL}	Rail M_{DL}	2'-0" M_{LL+I}	0.10 × One Line Wheels M_{LL}	ΣM
Positive Moment (k-ft)					
AB	0				
0.2	+33.8	+21.3	+28.6	+6.7	+90.4
0.4	+42.8	+26.9	+36.8	+8.6	+115.1
0.5	+37.8	+23.8	+35.6	+8.3	+105.5
0.6	+26.8	+16.9	+31.6	+7.4	+82.7
0.8	−14.6	−9.2	+15.4	+3.6	−4.8
BA	−80.8				
BC	−84.8				
0.2	−2.4	−1.5	+17.2	+4.4	+17.7
0.4	+39.2	+24.7	+31.2	+8.0	+103.1
0.5	+44.8	+28.2	+32.8	+8.4	+114.2
0.6	+40.2	+25.3	+30.8	+8.0	+104.3
0.8	+0.2	+0.1	+17.2	+4.4	+21.9
CB	−80.2				
CD	−75.4				
0.2	−16.0	−10.1	+15.2	+3.5	+2.5
0.4	+21.6	+13.6	+30.0	+6.9	+72.1
0.5	+32.0	+20.1	+34.0	+7.8	+93.9
0.6	+36.6	+23.0	+35.4	+8.1	+103.1
0.8	+29.4	+18.5	+27.4	+6.3	+81.6
DC	0				
Negative Moment (k-ft)					
0.8	−14.6	−9.2	−11.4	−2.8	−38.0 check,
BA	−80.8	−50.8	−33.4	−8.2	−173.2 most
BC	−84.8	−53.3	−37.6	−9.2	−184.9 ← critical
0.2	−2.4	−1.5	−10.6	−2.6	−17.1
0.8	+0.2	+0.1	−9.8	−2.4	−11.9
CB	−80.2	−50.5	−37.0	−9.0	−176.7
CD	−75.4	−47.4	−32.2	−7.8	−162.8
0.2	−16.0	−10.1	−11.4	−2.9	−40.4

Edge Beam Negative Moment Section @ BC

$$M = -184.9 \text{ k-ft}, \, b = 24 \text{ in.}, \, K = 246, \, j = 0.87$$

The depth required is

$$d = \sqrt{\frac{184.9(12)}{0.246(24)}} = 19.4 \text{ in.} > 16.30 \text{ available}$$

∴ Compression reinforcing is required.

The area of the tension steel required is therefore

$$A_s = \frac{184.9(12)}{24(0.87)(16.30)} = 6.52 \text{ in.}^2 \text{; use } 4\text{-#}11 \, A_s = 6.24 \text{ in.}^2$$

Try 3-#9 compressive reinforcement: $A_s' = 3.00 \text{ in.}^2$:

$$d = 16.30 \text{ in.; } d' = 2.06 \text{ in.; } d'/d = 0.126; n = 8$$

Solving for m yields

$$m = \frac{8(6.24)}{24(16.30)} + \frac{15(3.00)}{24(16.30)} = 0.128 + 0.115 = 0.243$$

q can also be computed as

$$q = 0.128 + (0.115)(0.126) = 0.128 + 0.014 = 0.142$$

k is therefore

$$k = \sqrt{(0.243)^2 + 2(0.142)} - 0.243 = 0.343$$

The terms associated with j are

$$\frac{1}{k} \times \frac{(2n - 1) A_s'}{bd} = \frac{0.115}{0.343} = 0.335$$

and

$$\frac{1}{k} \times \frac{d'}{d} = \frac{0.126}{0.343} = 0.367$$

Therefore z equals

$$z = \frac{0.167 + (0.335)(0.367)(0.633)}{0.500 + (0.335)(0.633)} = 0.344;$$

$$j = 1 - zk = 1 - 0.344(0.343) = 0.882$$

Knowing j and k, we can compute the stress in the steel and concrete as

$$f_s = \frac{184.9(12)}{6.24(0.882)(16.30)} = 24.7 \text{ ksi} > 24.0 \text{ ksi; } 3.0\% \text{ overstress} \quad \text{OK}$$

$$f_c = \frac{24.7(0.343)}{8(0.657)} = 1614 \text{ psi} > 1600 \text{ psi}, 0.8\% \text{ overstress} \quad \text{OK}$$

$$f_s' = 2(24.7)\left(\frac{0.343 - 0.126}{0.657}\right) = 16.3 \text{ ksi} < 24.0 \text{ ksi} \quad \text{OK}$$

It is reasonable to use the same reinforcing on the edge beam for positive moments in all spans. Therefore, only the analysis for the positive moment section at $0.4AB$ is evaluated. First an approximation for the required reinforcement area and required d value is obtained by assuming that $j = 0.87$ and $k = 246$:

Positive Moment @ 0.4AB

$$M = 115.1 \text{ k-ft}$$

$$\text{Approx. } A_{s(\text{required})} = \frac{115.1(12)}{24.0(0.87)(16.94)}$$

$$= 3.91 \text{ in.}^2 \text{; use } 4\text{-}\#9 \ A_s = 4.00 \text{ in.}^2$$

$$\text{Approx. } d_{\text{required}} = \sqrt{\frac{115.1(12)}{0.246(24)}} = 15.3 \text{ in.} < 16.94 \text{ in.}$$

By using 4–#9 and $d = 16.94$ in. and the constants given the actual stresses in the edge beam can be computed as follows:

$$p = \frac{4.00}{24(16.94)} = 0.0098$$

From Appendix A

$$k = 0.325, j = 0.892$$

$$f_s = \frac{115.1(12)}{4.00(0.892)(16.94)} = 22.9 \text{ ksi} < 24.0 \text{ ksi}$$

$$f_c = \frac{22.9(0.325)}{8(0.675)} = 1375 \text{ psi} < 1600 \text{ psi}$$

The resisting moments based on these computations follow:

Negative Moment (4-#11)

$$M_R = 6.24(0.882)(16.30)(24.0)(\tfrac{1}{12}) = 179.4 \text{ k-ft}$$

Positive Moment (4-#9)

$$M_R = 4.00(0.892)(16.94)(24.0)(\tfrac{1}{12}) = 120.9 \text{ k-ft}$$

Half the 4-#11 used for the negative moment steel is cut off and conservatively half the resisting moment computed for 4-#11 is used for 2-#11.

Cutoffs for the positive moment steel are not used for reasons of simplicity. Full-length reinforcement is more cost effective than a minor savings in reinforcement cutoffs.

Figure 4.23 is a plot of the maximum edge beam moments and resisting moments. Again development length is indicated by the sloping lines and it is evident that it does not have to be examined.

Computation of deflection due to dead load is done to provide information for setting proper deck grades. The conjugate beam method is used where the conjugate beam is loaded with the dead load moment diagram divided by *EI*. The shears on the conjugate beam indicate slope or rotation; the moments on the conjugate beam indicate deflection. For this problem *EI* is constant and therefore is divided after completion of the problem. The moment diagram for each span is divided into standard shapes, as listed in Appendix N, to obtain simple span moments or deflections.

The total weight of the superstructure is used for dead load and the moments are computed from a proration with the per foot dead-load moments. These moments are then applied to the gross section, which gives reasonable results, in lieu of the *I* value given in AASHTO Article 1.5.23.

DL Deflection

Total w = deck + *WS*, $30.0 \times 0.263 = 7.89$

Deck @ rail, $2(1.46 \times 0.238) = 0.70$

Rail, $2(0.331) = \underline{0.66}$

9.25 k-ft; ratio $= \dfrac{9.25}{0.263} = 35.2$

$E = 519{,}000$ ksf $\times \tfrac{1}{3} = 173{,}000$ ksf

$I = 10.89$ ft^4

The span *AB* moment diagram can be broken into a triangle and parabola as shown in Figure 4.24, where M_{BA} is computed by using the per foot moment of -40.4 k-ft taken from Table 4.10.

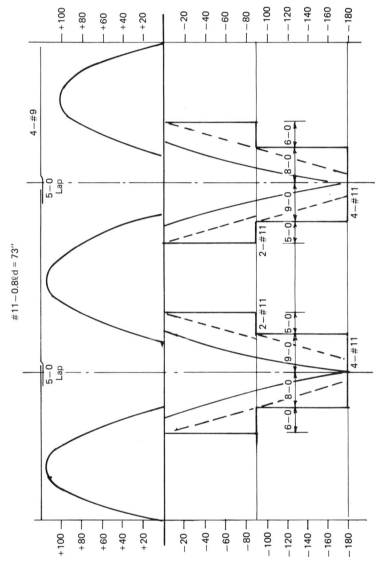

Figure 4.23 Maximum moments—edge beam.

95

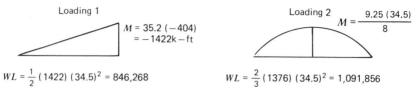

Loading 1

$M = 35.2 \, (-404)$
$= -1422 \text{k-ft}$

$WL = \dfrac{1}{2}(1422)(34.5)^2 = 846,268$

Loading 2

$M = \dfrac{9.25\,(34.5)}{8}$

$WL = \dfrac{2}{3}(1376)(34.5)^2 = 1,091,856$

Figure 4.24   Span AB dead load moments.

Point	Loading 1 CM_1	Loading 2 CM_2	$\sum CM$	$\Delta = \dfrac{\sum CM}{EI}$	
A					
0.2	$-54,161$	$+101,324$	$+47,163$	0.0250	$\frac{1}{4}$ in.
0.4	$-94,782$	$+162,468$	$+67,686$	0.0359	$\frac{3}{8}$ in.
0.6	$-108,322$	$+162,468$	$+54,146$	0.0287	$\frac{3}{8}$ in.
0.8	$-81,242$	$+101,324$	$+20,082$	0.0107	$\frac{1}{8}$ in.
B					

CM uses the coefficient from Appendix N. Similarly, Appendix N can be used to compute the deflections in span BC (Figure 4.25) by using a rectangle, a triangle, and parabola for loading shapes:

Point	CM_1	CM_2	CM_3	$\sum CM$	$\Delta = \dfrac{\sum CM}{EI}$	
B						
0.2	-218.691	$-7,527$	$+268,054$	$+41,836$	0.0222	$\frac{1}{4}$ in.
0.4	$-328,036$	$-10,036$	$+429,811$	$+91,739$	0.0487	$\frac{5}{8}$ in.
0.6	$-328,036$	$-8,782$	$+429,811$	$+92,993$	0.0494	$\frac{5}{8}$ in.
0.8	$-218,691$	$-5,018$	$+268,054$	$+44,345$	0.0235	$\frac{1}{4}$ in.
C						

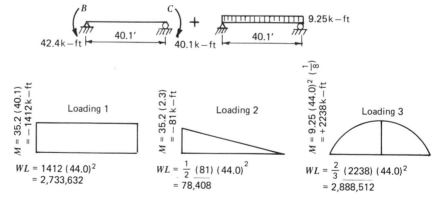

B C $+$ 9.25k-ft

42.4k-ft $\leftarrow$ 40.1' $\rightarrow$ 40.1k-ft $\leftarrow$ 40.1' $\rightarrow$

Loading 1

$M = 35.2\,(40.1)$
$= -1412\text{k-ft}$

$WL = 1412\,(44.0)^2$
$= 2,733,632$

Loading 2

$M = 35.2\,(2.3)$
$= -81\text{k-ft}$

$WL = \dfrac{1}{2}(81)(44.0)^2$
$= 78,408$

Loading 3

$M = 9.25\,(44.0)^2\,\left(\dfrac{1}{8}\right)$
$= +2238\text{k-ft}$

$WL = \dfrac{2}{3}(2238)(44.0)^2$
$= 2,888,512$

Figure 4.25   Span BC dead load moments.

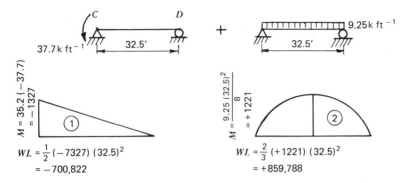

Figure 4.26   Span *CD* dead load moments.

Span *CD* deflections (Figure 4.26) are computed like span *AB*:

Point	CM_1	CM_2	$\sum CM$	$\Delta = \dfrac{\sum CM}{EI}$	
C					
0.2	−67,279	+79,788	+12,509	+0.0066	$\frac{1}{8}$ in.
0.4	−89,705	+127,936	+38,231	+0.0203	$\frac{1}{4}$ in.
0.6	−78,492	+127,936	+49,444	+0.0262	$\frac{3}{8}$ in.
0.8	−44,853	+79,788	+34,935	+0.0185	$\frac{1}{4}$ in.
D					

Example 4.2.   The bridge superstructure in Example 4.1 is designed by the load factor method. The substructure for this bridge is also designed by load factor as described in Chapter 10, Example 10.3.

The frame constants, moment distribution, and computation of moments are discussed in Example 4.1. The required moment capacity is summarized for the critical locations in deck design;

Deck Design

	1.3 DL (1)	$\gamma[\beta_L](LL + I)$ 1.3(2.20)(LL + I) 2.86(LL + I) (2)	$M_u = M_1 + M_2$
0.4 AB	+29.8	+52.6	+82.4 k-ft
0.5 BC	+29.1	+46.9	+76.0 k-ft
BA	−52.5	−47.8	−100.3 k-ft
BC	−55.1	−53.8	−108.9 k-ft

Group IA loading is used because the live load is H 15.

The service load moment summary for these critical points, computed in Example 4.1, follows:

	DL (1)	LL + I (2)	$M = M_1 + M_2$	$\dfrac{DL + 2(LL + I)}{1.50}$
0.4 AB	+22.9	+18.4	+41.3	+39.8 k-ft
0.5 BC	+22.4	+16.4	+38.8	+36.8 k-ft
BA	-40.4	-16.7	-57.1	-49.2 k-ft
BC	-42.4	-18.8	-61.2	-53.3 k-ft

The last column is the overload check in accordance with AASHTO Article 1.2.4. The critical end span section at 0.4 AB is then checked for capacity and serviceability.

An approximate reinforcement requirement is computed by assuming that $(d - a/2) = \pm 16.0$ in., based on the moment capacity given as; $\phi M_n = \phi[A_s f_y(d - (a/2))]\frac{1}{12}$, where $\phi = 0.90$; therefore

$$A_{s(\text{required})} = \frac{82.4(12)}{60(16.0\pm)(0.90)} = 1.14 \text{ in.}^2$$

Use #7 @ 6 in.; $A_s = 1.20$ in.^{2} as the approximate required reinforcement. The effective depth d is

$$d = 19.0 - 1.50 - 0.44 = 17.06 \text{ in.}$$

With #7 @ 6 in. and $d = 17.06$ in. the computed capacity is

$$a = \frac{A_s f_y}{0.85 f'_c b} = \frac{1.20(60.0)}{0.85(4.0)(12)} = 1.76 \text{ in.}$$

$$\phi M_n = \phi \left[A_s f_y \left(d - \frac{a}{2} \right) \right]$$

$$= 0.90[(1.20)(60)(17.06-0.88)] \left(\tfrac{1}{12}\right)$$

$$= 87.4 \text{ k-ft} > 82.4 \text{ k-ft required} \quad \text{OK}$$

Crack control and fatigue are checked for seviceability; these stresses are examined at the service load level and k and j are computed according to Appendix A.

$$p = \frac{1.20}{12(17.06)} = 0.0059$$

Therefore from Appendix A, $k = 0.264$, $j = 0.912$. The maximum service load level moment at 0.4 AB is 41.3 k-ft and the maximum steel stress is

$$\text{max } f_s = \frac{41.3(12)}{(1.20)(0.912)(17.06)} = 26.5 \text{ ksi}$$

The minimum moment and stress in the reinforcing steel are

$$\text{min } M = 0.4(-6.6) + 22.9 = 20.2 \text{ k-ft}$$

$$\text{min } f_s = \frac{20.2(12)}{(1.20)(0.912)(17.06)} = 13.0 \text{ ksi}$$

The maximum steel stress range is

$$f_t = 26.5 - 13.0 = 13.5 \text{ ksi}$$

According to AASHTO Article 1.5.38, the allowable stress range for fatigue requirement is

$$f_t = 21 - 0.33 f_{\min} + 8 \left(\frac{r}{h}\right)$$

where $r/h = 0.30$; therefore

$$f_t = 21.0 - 0.33(13.0) + 8(0.3) = 19.0 \text{ ksi} > 13.0 \text{ ksi} \quad \text{OK}$$

The crack control criteria of AASHTO Article 1.5.39 are checked by using $z = 130$ k in.$^{-1}$ for severe exposure conditions; d_c = cover to the center of reinforcing and A = area of concrete affected by reinforcing bar in tension or $2d_c \times$ bar spacing.

The maximum stress allowed by AASHTO eq. 6.30 is computed by using

$$d_c = 1.50 + 0.44 = 1.94 \text{ in. and } A = z(1.94) \times 6.0 = 23.48 \text{ in.}$$

$$f_s = \frac{z}{\sqrt[3]{d_c A}}$$

Therefore

$$f_s = \frac{130}{\sqrt[3]{1.94(23.28)}} = 36.5 \text{ ksi}$$

f_s may not exceed $0.6 f_y = 36 \text{ ksi} > 26.5 \text{ ksi} \quad \text{OK}$

The maximum positive moment in span BC is less than the maximum positive moment span AB. Therefore the section at 0.5 BC is analyzed for a moment of +76.0 k-ft.

The approximate reinforcement required, based on capacity and assuming that $d = -a/2 = \pm16.0$ in., is

$$A_{s(required)} = \frac{76.0(12)}{60(16.0\pm)(0.90)} = 1.06 \text{ in.}^2$$

Use #7 @ $6\frac{1}{2}$ in. $A_s = 1.11$ in.^{2}

Using #7 @ $6\frac{1}{2}$ in. and $d = 17.06$ in., we obtain the computed capacity:

$$a = \frac{A_s f_y}{0.85 f_c' b} = \frac{1.11(60)}{0.85(4.0)(12)} = 1.63 \text{ in.}$$

$$\phi M_n = \phi \left[A_s f_y \left(d - \frac{a}{2} \right) \right]$$

$$= 0.90 \left[(1.11)(60) \left(17.06 - \frac{1.63}{2} \right) \right] \left(\frac{1}{12} \right)$$

$$= 81.1 \text{ k-ft} > 76.0 \text{ k-ft} \quad \text{OK}$$

For serviceability stress computations

$$p = \frac{1.11}{12(17.06)} = 0.0054$$

Using Appendix A, $k = 0.254, j = 0.915$; the maximum reinforcing steel stress

$$\max f_s = \frac{36.9(12)}{1.11(0.915)(17.06)} = 25.5 \text{ ksi}$$

for minimum moment assume equal end spans and lane loading with concentrated loads at each end span:

$$M_{LL+I} = 80.4(-0.629)(0.204) = -10.3 \text{ k-ft}$$

$$M_{min} = +22.4 - 10.3 = +12.1 \text{ k-ft}$$

$$\min f_s = \frac{12.1(12)}{1.11(0.915)(17.06)} = 8.4 \text{ ksi}$$

The stress range in the reinforcing is $f_f = 25.5 - 8.4 = 17.1$ ksi. The allowable fatigue stress range is

$$f_f = 21 - 0.33f_{min} + 8\left(\frac{r}{h}\right)$$

$$= 21.0 - 0.33(8.4) + 8(0.3) = 20.6 \text{ ksi} > 17.1 \text{ ksi} \quad \text{OK}$$

The allowable stress for crack control is computed as

$$d_c = 1.94 \text{ in.}$$
$$A = (1.94)(2)(6.5) = 25.22$$
$$f_s = \frac{z}{\sqrt[3]{d_c A}}$$
$$f_s = \frac{130}{\sqrt[3]{1.94(25.22)}} = 35.5 \text{ ksi} > 25.5 \text{ ksi} \quad \text{OK}$$

The same capacity analysis is applied to the maximum negative moment of -108.9 k-ft at BC.

Capacity Analysis

$$\text{Approx. } A_{s(required)} = \frac{108.9(12)}{60.0(15.5\pm)(0.90)} = 1.56 \text{ in.}^2$$

Use #8 @ 6 in.; $A_s = 1.58$ in.^{2}

$$d = 19.0 - 2.0 - 0.50 = 16.50 \text{ in.}$$
$$a = \frac{A_s f_y}{0.85f'_c b} - \frac{1.58(60)}{0.85(4.0)(12)} = 2.32 \text{ in.}$$
$$\phi M_n = \phi\left[A_s f_y\left(d - \frac{a}{2}\right)\right]$$
$$= 0.90\left[(1.58)(60)\left(16.50 - \frac{2.32}{2}\right)\right]\left(\frac{1}{12}\right)$$
$$= 109.1 \text{ k-ft} > 108.9 \text{ k-ft}$$

Serviceability

$$\rho = \frac{1.58}{12(16.50)} = 0.0080$$

Therefore from Appendix A $k = 0.299$, $j = 0.900$, and

$$f_s = \frac{53.3(12)}{1.58(0.900)(16.50)} = 27.3 \text{ ksi max}$$

min moment $= -42.4 + (+15)(0.214) = -39.2$ k-ft

$$f_s = \frac{39.2(12)}{1.58(0.900)(16.50)} = 20.0 \text{ ksi}$$

$$f_f = 27.3 - 20.0 = 7.3 \text{ ksi}$$

allowable $f_f = 21.0 - 0.33 f_{min} + 8 \left(\dfrac{r}{h} \right)$

$$= 21.0 - 0.33(20.0) + 8(0.3)$$

$$= 16.8 \text{ ksi} > 7.3 \text{ psi} \quad \text{OK}$$

Crack Control

$$dc = 2.0 + 0.5 = 2.50 \text{ in.}$$

$$A = (2.50)(2)(6.0) = 30.0 \text{ in.}^2$$

$$f_s = \frac{130}{\sqrt[3]{(2.50)(30.0)}} = 30.8 \text{ ksi} > 27.3 \text{ ksi} \quad \text{OK}$$

The minimum reinforcing requirements in AASHTO Article 1.5.7 must be examined. This article states that reinforcement must be provided to give a capacity greater than 1.2 times the cracking moment. Application of this requirement at the critical locations yields the following:

Minimum Reinforcement

$$f_{cr} = 7.5\sqrt{4000} = 474 \text{ psi}$$

$$S = \frac{12(19)^2}{6} = 722 \text{ in.}^3$$

$$474 = \frac{M_{cr}(12)}{722}$$

$$M_{cr} = 28.5 \text{ k-ft}$$

$$1.2M_{cr} = 34.2 \text{ k-ft}$$

Reinforcing must be provided to give capacity equal to a minimum moment of 34.2 k-ft:

$$M_{0.4AB} = +82.4 \text{ k-ft} > 34.2 \text{ k-ft}$$

$$M_{0.5BC} = +76.0 \text{ k-ft} > 34.2 \text{ k-ft}$$

$$M_{BA} = -100.3 \text{ k-ft} > 34.2 \text{ k-ft}$$

$$M_{BC} = -108.9 \text{ k-ft} > 34.2 \text{ k-ft}$$

Therefore the required reinforcement for capacity will be greater than that required for crack control.

The maximum reinforcement requirements of AASHTO Article 1.5.32(A) (1) which sets a maximum reinforcement ratio at 75% of the balanced ratio are now checked.

Maximum Reinforcement

$$\rho_b = \frac{0.85\beta_1 f_c'}{f_y} \left(\frac{87}{87 + f_y} \right) = \frac{0.85(0.85)(4.0)}{60} \left(\frac{87}{87 + 60} \right) = 0.0285$$

$$0.75\rho_b = 0.75(0.0285) = 0.0214$$

Positive moment #7 @ 6 in. $A_s = 1.20$ in.^{2}, $d = 17.06$ in.

$$\rho = \frac{1.20}{12(17.06)} = 0.0059 < 0.0214 \quad \text{OK}$$

#7 @ $6\frac{1}{2}$ in., $A_s = 1.11$ in.^{2}, $d = 17.06$ in.

$$\rho = \frac{1.11}{12(17.06)} = 0.0054 < 0.214 \quad \text{OK}$$

Negative moment #8 @ 6 in., $A_s = 1.58$ in.^{2}, $d = 16.50$ in.

$$\rho = \frac{1.53}{12(16.50)} = 0.0080 < 0.0214 \quad \text{OK}$$

Cap Design

The loads and moments for cap design are computed in Example 4.1. The maximum required ultimate moments and shears are tabulated as follows:

	1.3DL	$\gamma(\beta_L)(LL + I)$ $1.3(2.20)(LL + I)$ $2.86(LL + I)$	M_u or V_u
M_B	-186	-114	-300 k-ft
M_C	-211	-272	-483 k-ft
$M_{0.5BC}$	+105	+194	+299 k-ft
V_{CB}	99.2	113.5	212.7 k-ft

At service load level these moments are (including the check for overload according to AASHTO Article 1.2.4):

	DL (1)	LL + I (2)	$M = M_1 + M_2$	$\dfrac{DL + 2(LL + I)}{1.50}$
M_B	-143	-40	-183 k-ft	-149 k-ft
M_C	-162	-95	-257 k-ft	-235 k-ft
$M_{0.5BC}$	-81	+68	+149 k-ft	+145 k-ft
V_{CB}	76.3	39.7	116.0 k-ft	103.8 k-ft

The negative moment serviceability requirement is often critical because of the large cover over the main reinforcement steel. This cover is the result of placing the cap negative moment steel under the top mat of the deck reinforcement. It results in a low allowable steel stress at service load and a low allowable stress due to the cracking criteria of AASHTO Article 1.5.39. In this case it is advisable to check serviceability crack control before checking the capacity:

d_c = cover to center of reinforcing

A = area of concrete affected by reinforcing bars in tension

Negative Moment

Serviceability: assume 7-#9

$$d = 25.0 - 2.9 - 1.0 - 0.56 = 21.44 \text{ in.}$$

$$d_c = 2.0 + 1.0 + 0.56 = 3.56 \text{ in.}$$

$$A = 3.56(2)(60) \div 7 = 61.0 \text{ in.}^2$$

$$f_s = \frac{z}{\sqrt[3]{d_c A}} = \frac{130}{\sqrt[3]{(3.56)(61.0)}}$$

= 21.6 ksi $<$ 24.0 ksi allowable for service load design

It is unreasonable to use an allowable $f_s <$ 24.0 ksi; therefore an allowable f_s = 24.0 ksi is used compared with the actual stress:

$$7\text{-}\#9, \; A_s = 7.00 \text{ in.}^2$$

$$\rho = \frac{7.00}{60(21.44)} = 0.0054,$$

From Appendix A k = 0.254 j = 0.915; therefore

$$f_s = \frac{257(12)}{7.00(0.915)(21.44)} = 22.5 \text{ ksi} < 24.0 \text{ ksi} \quad \text{OK}$$

The capacity is then checked:

$$a = \frac{A_s f_y}{0.85 f_c' b} = \frac{7.00(60)}{0.85(4.0)(60)} = 2.06 \text{ in.}$$

$$\phi M_n = \phi \left[A_s f_y \left(d - \frac{a}{2} \right) \right]$$

$$= 0.90[(7.00)(60)(21.44 - 1.03)](\tfrac{1}{12})$$

$$= 643 \text{ k-ft} > 483 \text{ k-ft required} \quad \text{OK}$$

Fatigue is not a critical condition; however, according to AASHTO Article 1.5.7, minimum reinforcement requirements must be checked.

Minimum Reinforcement

$$f_{cr} = 7.5\sqrt{4000} = 474 \text{ psi}$$

$$S = \frac{60(25)^2}{2} = 6250 \text{ in.}^3$$

$$474 = \frac{M_{cr}(12)}{6250}$$

$$M_{cr} = 247 \text{ k-ft}$$

$$1.2 M_{cr} = 296 \text{ k-ft}$$

Provide reinforcement for minimum moment = 296 k-ft. Compare with critical moments:

$$M_B = -300 \text{ k-ft} > 296 \text{ k-ft}$$
$$M_C = -483 \text{ k-ft} > 296 \text{ k-ft}$$
$$M_{0.5BC} = +299 \text{ k-ft} > 296 \text{ k-ft}$$

OK for min. reinforcement

The maximum reinforcement requirements are checked according to AASHTO Article 1.5.32(A)(1) for the negative moment:

Maximum Reinforcement

$$\rho_b = \frac{0.85\beta_1 f_c'}{f_y}\left(\frac{87}{87+f_y}\right) = \frac{0.85(0.85)(4.0)}{60}\left(\frac{87}{87+60}\right) = 0.0285$$

$0.75\rho_b = 0.0214$

$7\text{-}\#9,\ \rho = 0.0054$ OK

Preliminary calculations show that 7-#7 is a reasonable reinforcement for the cap positive moment section. The capacity is checked:

Positive Moment

$$7\text{-}\#7,\ A_s = 4.20 \text{ in.}^2,\ d = 25 - 1.5 - 0.62 - 0.44 = 22.44 \text{ in.};$$

the capacity is therefore

$$a = \frac{A_s f_y}{0.85 f_c' b} = \frac{4.20(60)}{0.85(4.0)(60)} = 1.24 \text{ in.}$$

$$\phi M_n = \phi\left[A_s f_y\left(d - \frac{a}{2}\right)\right]$$

$$= 0.90[(4.20)(60)(22.44 - 0.62)]\left(\tfrac{1}{12}\right)$$

$$= 412 \text{ k-ft} > 299 \text{ k-ft}   \text{OK}$$

As previously explained; the serviceability crack control criteria are checked for the positive moment section of the cap.

Serviceability

$$d = 22.44 \text{ in.},\ b = 60 \text{ in.}$$

$$d_c = 1.5 + 0.625 + 0.44 = 2.565 \text{ in.}$$

$$A = (2.565)(2)(60)/7 = 43.97 \text{ in.}$$

$$f_s = \frac{z}{\sqrt[3]{d_c A}} = \frac{130}{\sqrt[3]{(2.565)(43.97)}} = 26.9 \text{ ksi}$$

$$\rho = \frac{4.20}{60(22.44)} = 0.0031$$

From Appendix A $k = 0.199$, $j = 0.934$.

$$f_s = \frac{149(12)}{4.20(0.934)(22.34)} = 20.4 \text{ ksi} > 26.9 \text{ ksi} \quad \text{OK}$$

The shear and stirrup requirements for the cap are checked. The ultimate unit shear is

$$v_\mu = \frac{V_c}{\phi bd} = \frac{212.7}{0.85(60)(21.44)} = 194 \text{ psi}$$

The allowable concrete shear strength, according to AASHTO Article 1.5.35(B)(1), is $v_c = 2\sqrt{4000} = 126$ psi. Then $(v_\mu - v_c) = 194 - 126 = 68$ psi and the required spacing is

$$s = \frac{A_v f_y}{(v_\mu - v_c)(b)} = \frac{4(0.31)(60,000)}{68(60)} = 18.2 \text{ in.}$$

The minimum spacing according to AASHTO Article 1.5.10(A)(2) is

$$\min s = \frac{4(0.31)(60,000)}{50(60)} = 24.8 \text{ in.}$$

According to AASHTO Article 1.5.10(c), the reinforcing spacing should not exceed $\frac{1}{2}d = 10.5$ in. Therefore use #5 stirrups @ $10\frac{1}{2}$ in.

The influence lines and the dead and live load moments for the entire span, are computed in Example 4.1. These moments are factored and tabulated in Table 4.20 to give the maximum positive ultimate moments over the full length of the bridge.

Table 4.21 lists the factored maximum negative ultimate moments.

The positive moment resisting capacities computed for span AB are

DESIGN OF BRIDGE SLABS

Table 4.20[a]

Point	$1.3M_{DL}$	$2.86\,(+M_{LL+I})$	$+M_u$
AB	0	0	0
0.2	+22.0	+40.9	+62.9
0.4	+27.8	+52.6	+80.4
0.5	+24.6	+50.9	+75.5
0.6	+17.4	+45.2	+62.6
0.8	-9.5	+22.0	+12.5
BA	-52.5		
BC	-55.1		
0.2	-1.6	+24.6	+23.0
0.4	+25.5	+44.6	+70.1
0.5	+29.1	+46.9	+76.0
0.6	+26.1	+44.0	+70.1
0.8	+0.1	+24.6	+24.7
CB	-52.1		
CD	-49.0		
0.2	-10.4	+21.7	+11.3
0.4	+14.0	+42.9	+56.9
0.5	+20.8	+48.6	+69.4
0.6	+23.8	+50.6	+74.4
0.8	+19.1	+39.2	+58.3
DC	0		

[a]Moments in kip-ft.

based on the previous analysis for the moment at $0.4AB$. The reinforcing is cut off in three equal increments:

Positive Moment

Section 1. #7 @ 6 in., A_s = 1.20 in.^{2}; d = 17.06 in.

$$a = \frac{A_s f_y}{0.85 f'_c b} = \frac{1.20(60)}{0.85(4.0)(12)} = 1.76 \text{ in.,} \quad \frac{a}{2} = 0.88 \text{ in.}$$

$$\phi M_n = \phi \left[A_s f_y \left(d - \frac{a}{2} \right) \right] \left(\frac{1}{12} \right)$$

$$= 0.90[(1.20)(60)(17.06 - 0.88)] \left(\frac{1}{12} \right) = 87.4 \text{ k-ft}$$

Section 2. Cutoff $\frac{1}{3}$ reinforcement, A_s = 0.80 in.^{2}, d = 17.06 in.

$$a = \frac{0.80(60)}{0.85(4.0)(12)} = 1.18 \text{ in.,} \quad \frac{a}{2} = 0.59 \text{ in.}$$

$$\phi M_n = 0.90[(0.80)(60)(17.06 - 0.59)](\tfrac{1}{12}) = 59.3 \text{ k-ft}$$

Table 4.21[a]

Point	$1.3M_{DL}$	$2.86\ (-M_{LL+I})$	$-M_u$
AB	0		
0.2	+22.0	-4.0	
0.4	+27.8	-7.7	
0.5	+24.6	-9.4	
0.6	+17.4	-11.2	+6.2
0.8	-9.5	-16.3	-25.8
BA	-52.5	-47.8	-100.3
BC	-55.1	-53.8	-108.9
0.2	-1.6	-15.2	-16.8
0.4	+25.5	-9.4	+16.1
0.5	+29.1	-9.2	
0.6	+26.1	-8.9	-17.2
0.8	+0.1	-14.0	-13.9
CB	-52.1	-52.9	-105.0
CD	-49.0	-46.0	-95.0
0.2	-10.4	-16.3	-26.7
0.4	+14.0	-12.3	-1.7
0.5	+20.8	-10.0	+10.8
0.6	+23.8	-7.7	
0.8	+19.1	-4.0	
DC	0		

[a]Moments are in kip-ft.

Section 3.   Cutoff $\frac{2}{3}$ reinforcement, $A_s = 0.40$ in.^{2}, $d = 17.06$ in.

$$a = \frac{0.40(60)}{0.85(4.0)(12)} = 0.59 \text{ in.}, \frac{a}{2} = 0.29 \text{ in.}$$

$$\phi M_n = 0.90[(0.40)(60)(17.06 - 0.29)](\tfrac{1}{12}) = 30.2 \text{ k-ft}$$

The positive resisting moment capacities computed for spans BC and CD are based on the previous analysis for the moment at $0.5BC$. Again the cutoffs are in three equal increments.

Positive Moment

Section 1.   #7 @ $6\frac{1}{2}$ in., $A_s = 1.11$ in.^{2}, $d = 17.06$ in.

$$a = \frac{1.11(60)}{0.85(4.0)(12)} = 1.63 \text{ in.}, \frac{a}{2} = 0.82 \text{ in.}$$

$$\phi M_n = 0.90[(1.11)(60)(17.06 - 0.82)](\tfrac{1}{12}) = 81.1 \text{ k-ft}$$

Section 2.   Cutoff $\frac{1}{3}$ reinforcement, $A_s = 0.74$ in.^{2}, $d = 17.06$ in.

$$a = \frac{0.74(60)}{0.85(4.0)(12)} = 1.09 \text{ in.}^2, \frac{a}{2} = 0.54 \text{ in.}^2$$

$\phi M_n = 0.90[(0.74)(60)(17.06 - 0.54)](\frac{1}{12}) = 55.0$ k-ft.

Section 3.   Cutoff $\frac{2}{3}$ reinforcement, $A_s = 0.37$ in.^{2}, $d = 17.06$ in.

$$a = \frac{0.37(60)}{0.85(4.0)(12)} = 0.54 \text{ in.}, \frac{a}{2} = 0.27 \text{ in.}$$

$\phi M_n = 0.90[(0.37)(60)(17.06 - 0.27)](\frac{1}{12}) = 28.0$ k-ft.

The negative resisting moment capacities are based on an analysis of the section at *BC*. Also for negative moment the cutoffs are in three equal increments.

Negative Moment (@ B and C)

Section 1.   8 @ 6 in., $A_s = 1.58$ in.^{2}, $d = 16.50$ in.

$$a = \frac{1.58(60)}{0.85(4.0)(12)} = 2.32 \text{ in.}, \frac{a}{2} = 1.16 \text{ in.}$$

$\phi M_n = 0.90[(1.58)(60)(16.50 - 1.16)](\frac{1}{12}) = 109.1$ k-ft

Section 2. $\frac{1}{3}$ reinforcement cutoff, $A_s = 1.05$ in.^{2}, $d = 16.50$ in.

$$a = \frac{1.05(60)}{0.85(4.0)(12)} = 1.54 \text{ in.}, \frac{a}{2} = 0.77 \text{ in.}$$

$\phi M_n = 0.90[(1.05)(60)(16.50 - 0.77)](\frac{1}{12}) = 74.3$ k-ft

Section 3. $\frac{2}{3}$ reinforcement cutoff, $A_s = 0.53$ in.^{2}, $d = 16.50$ in.

$$a = \frac{0.53(60)}{0.85(4.0)(12)} = 0.78 \text{ in.}, \frac{a}{2} = 0.39 \text{ in.}$$

$\phi M_n = 0.90[(0.53)(60)(16.50 - 0.39)](\frac{1}{12}) = 38.4$ k-ft

The distribution reinforcement required by AASHTO article 1.3.2(E) is as follows:

Distribution Reinforcement

Span AB.   #7 @ 6 in., $A_s = 1.20$ in.^{2}

$$A_{s(\text{required})} = \frac{100}{\sqrt{34.5}} (1.20) = 0.20 \text{ in.}^2$$

#4 @ 1 ft 0 in., $A_s = 0.20$ in.^{2}

Span BC.   #7 @ $6\frac{1}{2}$ in., A_s = 1.11 in.^{2}

$$A_{s(\text{required})} = \frac{100}{\sqrt{44.0}} (1.11) = 0.17 \text{ in.}^2$$

#4 @ 1 ft 0 in., A_s = 0.20 in.^{2}

Span CD.   #7 @ $6\frac{1}{2}$ in., A_s = 1.11 in.^{2}

$$A_{s(\text{required})} = \frac{100}{\sqrt{32.5}} (1.11) = 0.19 \text{ in.}^2$$

#4 @ 1 ft 0 in., A_s = 0.20 in.^{2}

The required bar extensions according to AASHTO Article 1.5.13(A) are as follows:

Bar Extension

$$
\left.
\begin{array}{l}
\#8 \\[4pt]
15\phi = 1.25 \text{ ft} \\[4pt]
\#7 \\[4pt]
15\phi = 1.1 \text{ ft}
\end{array}
\right\} \quad 15 \text{ diameters}
$$

$$
\left.
\begin{array}{l}
\dfrac{s}{20} = \dfrac{34.5}{20} = 1.7 \text{ ft} \\[10pt]
\dfrac{s}{20} = \dfrac{44.0}{20} = 2.2 \text{ ft} \\[10pt]
\dfrac{s}{20} = \dfrac{325}{20} = 1.6 \text{ ft}
\end{array}
\right\} \quad \dfrac{1}{20} \times \text{span}
$$

The maximum ultimate moments are plotted in Figure 4.27. The cutoffs of the reinforcement are selected with equal or greater bar extensions than required. Development length is ignored because it is apparent from Example 4.1 that it is not important to the location of the reinforcement cutoffs. Development length requirements should be considered in the selection of a minimum length bar to be used in the negative moment regions.

The edge beam moments are the same as those computed in Example 4.1. These moments are tabulated in Table 4.22 and the maximum required ultimate moments are listed for both positive and negative moments when the edge beam width is 2 ft–0 in.

The positive and negative moment sections must be analyzed for the edge beam. For positive moment the maximum ultimate moment is +220.5 k-ft at $0.4AB$. Based on capacity, an approximate reinforcement requirement is computed, assuming that $(d - a/2) = \pm 17.0$ in.

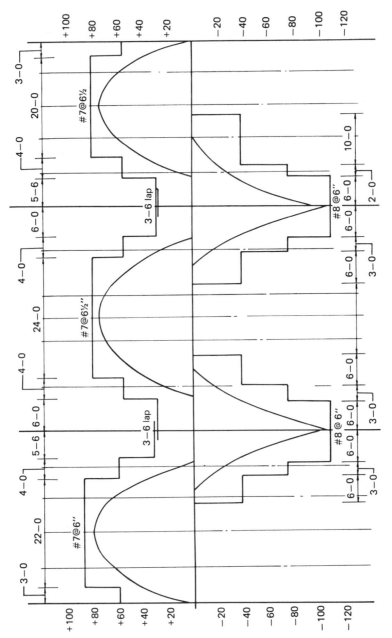

Figure 4.27 Maximum ultimate moments: slab.

Table 4.22   Edge Beam[a]

Point	Slab + WS M_{DL}	Railing M_{DL}	ΣM_{DL}	2 ft-0 in. M_{LL+I}	0.10 × One Line Wheels M_{LL}	ΣM_{LL+I}	$M_u = 1.3 M_{DL} + 2.86 M_{LL+I}$
Positive Moment[b]							
AB	0						
0.2	+33.8	+21.3	+55.1	+28.6	+6.7	+35.3	+172.6
0.4	+42.8	+26.9	+69.7	+36.8	+8.6	+45.4	+220.5
0.5	+37.8	+23.8	+61.6	+35.6	+8.3	+43.9	+205.6
0.6	+26.8	+16.9	+43.7	+31.6	+7.4	+39.0	+168.4
0.8	−14.6	−9.2	−23.8	+15.4	+3.6	+19.0	+23.4
BA	−80.8						
BC	−84.8						
0.2	−2.4	−1.5	−3.9	+17.2	+4.4	+21.6	+56.7
0.4	+39.2	+24.7	+63.9	+31.2	+8.0	+39.2	+195.2
0.5	+44.8	+28.2	+73.0	+32.8	+8.4	+41.2	+212.7
0.6	+40.2	+25.3	+65.5	+30.8	+8.0	+38.8	+196.1
0.8	+0.2	+0.1	+0.3	+17.2	+4.4	+21.6	+62.2
CB	−80.2						
CD	−75.4						
0.2	−16.0	−10.1	−26.1	+15.2	+3.5	+18.7	+19.6
0.4	+21.6	+13.6	+35.2	+30.0	+6.9	+36.9	+151.3
0.5	+32.0	+20.1	+52.1	+34.0	+7.8	+41.8	+187.3
0.6	+36.6	+23.0	+59.6	+35.4	+8.1	+43.5	+201.9
0.8	+29.4	+18.5	+47.9	+27.4	+6.3	+33.7	+158.7
DC	0						
Negative Moment[b]							
0.8	−14.6	−9.2	−23.8	−11.4	−2.8	−14.2	−71.6
BA	−80.8	−50.8	−131.6	−33.4	−8.2	−41.6	−290.1
BC	−84.8	−53.3	−138.1	−37.6	−9.2	−46.8	−313.4
0.2	−2.4	−1.5	−3.9	−10.6	−2.6	−13.2	−42.8
0.8	+0.2	+0.1	+0.3	−9.8	−2.4	−12.2	−34.5
CB	−80.2	−50.5	−130.7	−37.0	−9.0	−46.0	−301.5
CD	−75.4	−47.4	−122.8	−32.2	−7.8	−40.0	−274.0
0.2	−16.0	−10.1	−26.1	−11.4	−2.9	−14.3	−74.8

[a]2-0 width.
[b]Moments are in kip-ft.

Maximum Moment

$$M_{0.4AB} = +220.5 \text{ k-ft}$$

$$\text{approx } A_{s(\text{required})} = \frac{220.5(12)}{60(17.0\pm)(0.90)} = 2.88 \text{ in.}^2$$

4-#8, A_s = 3.16 in.^{2} are selected for investigation and the capacity is checked:

$$d = 19.0 - 1.50 - 0.50 = 17.00 \text{ in.}$$

$$a = \frac{A_s f_y}{0.85 f'_c b} = \frac{3.16(60.0)}{0.85(4.0)(24)} = 2.32 \text{ in.}, \quad \frac{a}{2} = 1.16 \text{ in.}$$

$$\phi M_n = \phi \left[A_s f_y \left(d - \frac{a}{2} \right) \right] = 0.90[(3.16)(60)(17.00 - 1.16)] \left(\frac{1}{12} \right)$$

$$= 225.2 \text{ k-ft} > 220.5 \text{ k-ft} \quad \text{OK}$$

For serviceability the maximum service load level stress is computed in the reinforcing steel.

Serviceability

$$\rho = \frac{3.16}{24(17.00)} = 0.0077$$

From Appendix A, $k = 0.294, j = 0.901$

$$M^{max}_{0.4AB} = \overbrace{(21.4 + 18.4)(2.0)}^{DL + LL + I} + \overbrace{0.10(+86)}^{0.10 \, LL} + \overbrace{\frac{0.331}{0.263}(+21.4)}^{RDL}$$

$$= +79.6 + 8.6 + 26.9 = 115.1 \text{ k-ft}$$

$$f^{max}_s = \frac{115.1(12)}{3.16(0.901)(17.00)} = 28.5 \text{ ksi}$$

The miminum service load level stress is computed in the reinforcing steel.

$$M^{min}_{0.4AB} = \overbrace{[21.4 + 0.4(-5.3)]2.0}^{DL + LL + I} + \overbrace{0.10(0.4)(-5.3)}^{0.10 \, LL}$$

$$+ \overbrace{\frac{0.331}{0.263}(+21.4)}^{RDL} = +65.3 \text{ k-ft}$$

$$f^{min}_s = \frac{65.3(12)}{3.16(0.901)(17.00)} = 16.2 \text{ ksi}$$

The stress range is

$$f_f = 28.5 - 16.2 = 12.3 \text{ ksi}$$

The allowable stress range is

$$f_f = 21 - 0.33f_{min} + 8\left(\frac{r}{h}\right)$$

$$= 21.0 - 0.33(16.2) + 8(0.3) = 18.0 \text{ ksi} > 12.3 \text{ ksi} \quad \text{OK}$$

The allowable stress for crack control, as previously explained, is

$$d_c = 1.50 + 0.50 = 2.00 \text{ in.}$$

$$A = 2(2.00)(6.0) = 24.0 \text{ in.}^2$$

$$f_s = \frac{130}{\sqrt[3]{(2.00)(24.0)}} = 35.8 \text{ ksi} > 28.5 \text{ ksi} \quad \text{OK}$$

The same approach is applied to an examination of the negative moment section for the maximum ultimate moment at BC of -313.4 k-ft. The approximate required reinforcement based on capacity is computed:

$$M_{BC}^{max} = -313.4 \text{ k-ft}$$

$$\text{approx. } A_{s(required)} = \frac{313.4(12)}{60(16.5\pm)(0.90)} = 4.22 \text{ in.}^2$$

4-#10, $A_s = 5.08$ in.^{2}, is selected for investigation. The capacity is checked.

$$d = 19.0 - 2.00 - 0.64 = 16.36 \text{ in.}$$

$$a = \frac{A_s f_y}{0.85 f_c' b} = \frac{5.08(60)}{0.85(4.0)(24)} = 3.74 \text{ in.}, \frac{a}{2} = 1.87 \text{ in.}$$

$$\phi M_n = (0.90)[(5.08)(60)(16.36 - 1.87)]\left(\tfrac{1}{12}\right)$$

$$= 331.2 \text{ k-ft} > 313.4 \text{ k-ft} \quad \text{OK}$$

For serviceability we compute the maximum and minimum stresses in the reinforcement.

Serviceability

$$\rho = \frac{5.08}{24(16.36)} = 0.0129,$$

From Appendix A, $k = 0.363$, $j = 0.879$;

$$M_{BC}^{max} = \overbrace{(-42.4 - 18.8)(2.0)}^{DL+LL+I} + \overbrace{0.10(-92)}^{0.10\ LL} + \overbrace{\left(\frac{0.331}{0.263}\right)(-42.4)}^{RDL}$$

$$= -122.4 - 9.2 - 53.4 = -185.0 \text{ k-ft}$$

$$f_s^{max} = \frac{185.0(12)}{5.08(0.879)(16.36)} = 30.4 \text{ ksi}$$

$$M_{BC}^{min} = \overbrace{(-42.4 + 3.2)(2.0)}^{DL+LL+I} + \overbrace{0.10(+15.0)}^{0.10\ LL} + \overbrace{\left(\frac{0.331}{0.263}\right)(-42.4)}^{RDL}$$

$$= -78.4 + 1.5 - 53.4 = -130.3 \text{ k-ft}$$

$$f_s^{min} = \frac{130.3(12)}{5.08(0.879)(16.36)} = 21.4 \text{ ksi}$$

The stress range is

$$f_f = 30.4 - 21.4 = 9.0 \text{ ksi}$$

The stress range allowable for fatigue is

$$f_f = 21 - 0.33 f_{min} + 8 \left(\frac{r}{h}\right)$$

$$= 21 - 0.33(21.4) + 8(0.3)$$

$$= 16.3 \text{ ksi} > 9.0 \text{ ksi} \quad \text{OK}$$

The serviceability allowable stress in the reinforcement for crack control is computed:

$$d_c = 2.0 + 0.64 = 2.64 \text{ in.}$$

$$A = 2(2.64)(6.0) = 31.7 \text{ in.}^2$$

$$f_s = \frac{130}{\sqrt[3]{(2.64)(31.7)}} = 29.7 \text{ ksi} > 30.4 \text{ ksi}$$

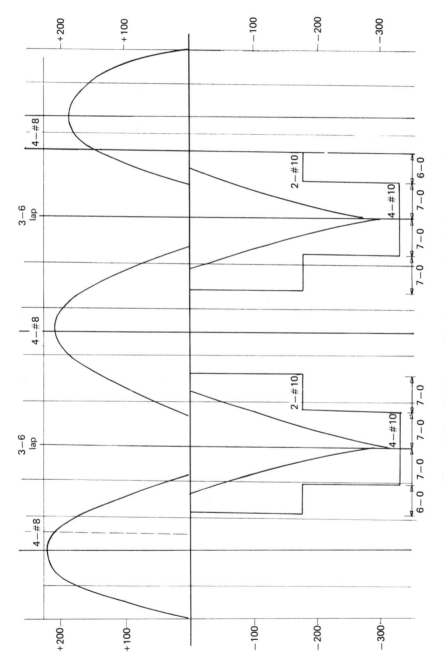

Figure 4.28 Maximum ultimate moments: edge beam.

Overstress at 2.3% is close enough. 2–#10 are cutoff in equal increments. The resisting capacity for 2–#10 is negative moment reinforcement, 2–#10, A_s = 2.54 in.^{2}, d = 16.36 in.

$$a = \frac{2.54(60)}{0.85(4.0)(24)} = 1.87 \text{ in.}, \frac{a}{2} = 0.93 \text{ in.}$$

$$\phi M_n = 0.90[(2.54)(60)(16.36 - 0.93)](\tfrac{1}{12}) = 176.4 \text{ k-ft}$$

The maximum ultimate moments on the edge beam are plotted against
the resisting moments and cutoffs are selected for the negative moment
steel by using the appropriate bar extensions shown in Figure 4.28.

The dead load deflection is computed in Example 4.1.

4.2   TRANSVERSELY REINFORCED BRIDGE SLABS

4.2.1   General

A transversely reinforced concrete slab is the most commonly used
bridge deck throughout the world and one of the most significant portions of bridge design in terms of dollar investment. Unfortunately it is
the one portion of the bridge structure in which significant deterioration has been a real problem in recent years. Heavy wheel loads, excessive salt usage, studded tires, and poor construction control have been
contributing factors.

Considerable research has been done but complete solutions have not
been attained. It would be reasonable to say that most of these research efforts have been directed toward a "materials type solution."
Such things as membranes, waterproofing, special concretes, wearing
surfaces, and epoxy-coated reinforcement have been recommended in
recent studies. It appears that it would also be appropriate to conduct
studies that would make recommendations in the areas of construction
control and design. From the design viewpoint further research in the
area of concrete cracking and microcracking would be beneficial. Such
research should develop better crack-control criteria for serviceability
analysis.

Most data indicate that deck deterioration is due to corrosion of the
reinforcement which results in spalling of the surrounding concrete.
Therefore use of very heavy deck reinforcement in thin sections probably should be avoided. Research in Ontario[8] has indicated that with
appropriate details providing sufficient restraint against horizontal
thrust the deck will perform as an arch and only a nominal (0.3%)
amount of reinforcement need be provided. It is believed that by re-

ducing the required reinforcement to these minimal amounts deck deterioration can be reduced significantly.

Edge support for transversely reinforced slabs is normally provided by cast-in-place end diaphragms. These diaphragms are often placed only between girders. Caution should be exercised to provide an edge support on slab overhangs where a substantial length of overhang might exist and where moments due to wheel loads might be a major portion of the total moment requirement. Cast-in-place decks on structural steel superstructures is another place where edge support might not naturally be provided. Edge support should be designed for each condition to be capable of carrying a wheel load.

Placement of transverse slab reinforcing on skewed bridges is a subject of some debate. A reasonable rule used by many designers, however, places the reinforcement on the skew up to 20° and for 20° or greater places the reinforcing normal to the roadway with variable length bars at the skewed ends. For reinforcement placed on the skew the span should be increased to the skewed length and the area of reinforcement increased for the spacing normal to the skew.

The AASHTO specifications require a 2-in. cover over the top reinforcing steel and a 1-in. cover over the bottom reinforcement in deck slabs. This means that the effective depth d for negative moment is less than that for positive moment, and because transverse slab spans are designed for the same moment at midspan and at the support the negative moment top reinforcement is more critical. Therefore it is common practice to design the top reinforcement and to make the bottom reinforcement the same to avert confusion during construction.

Conservative judgment suggests a capacity reduction factor of $\phi = 0.80$, in lieu of the 0.90 provided by AASHTO, because slight deviations in thickness and bar location could have a greater effect on the capacity of a thin slab than on a deep beam.

Example 4.3. The transversely reinforced deck shown in Figure 4.29 was designed by service load design procedures. This is the deck slab for the bridge designed as a T-beam in Example 5.1.

Note that the slab design has lower allowable stresses than the normal requirements in AASHTO. The value $f_c = 1200$ psi is in agreement with a 1967 memorandum issued for the design of bridge decks by the FHWA. The intent is to control flexibility and to lessen fatigue. The steel stress allowable, $f_s = 20,000$ psi, is intended to control cracking. These two serviceability requirements will often affect design and should be applied to service load design and load factor design.

The design overload vehicle requirements appear in Figure 4.30. The deck is designed for service load stresses using an HS20 design vehicle and checked for capacity with an overload vehicle.

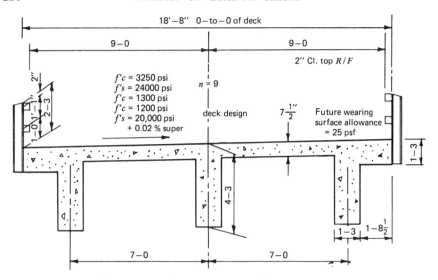

Figure 4.29 Roadway section: T-beam bridge.

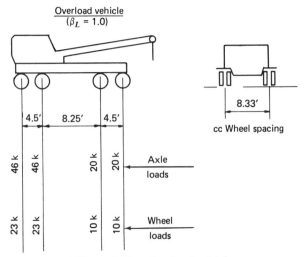

Figure 4.30 Overload vehicle.

The overhangs in this example are short and thickened; therefore it is not necessary to consider them because it is obvious that the span between the girders will control the design.

Span Between the Girders

The span according to AASHTO Article 1.3.2(a) is $S = 7.00 - 1.25 = 5.75$ ft. The dead load and dead load moment are computed by considering continuous spans and assuming $M_{DL} = wl^2/10$, which incorporates an 0.8 factor for continuity similar to the specification provision for live load.

The uniform dead load and resulting moment are

$$w = t_s \times 150 + ws$$

$$w = 0.625 \times 150 + 25 = 119 \text{ psf}$$

$$M_{DL} = \frac{119(5.75)^2}{10} = 393 \text{ lb-ft}$$

For the HS20 service load design the live load moment is computed by using the AASHTO equation in Article 1.3.2(c) and the 0.8 continuity factor. The impact is an assumed 30%; hence the 1.3 coefficient:

$$M_{LL} + I = 1.3(0.8) \left(\frac{7.75}{32}\right) (16.0) = 4030 \text{ lb-ft}$$

The total design moment is

$$M_T = 393 + 4030 = 4423 \text{ lb-ft}$$

Next some estimate should be made of the required reinforcement; therefore we try #5 @ 6 in. reinforcement and then check stresses with the normal reinforced concrete design formulas for stress $f_s = M/A_s jd$ and $f_c = f_s k/n(1 - k)$:

$$\text{#5 @ 6 in. } A_s = 0.62 \text{ in.}^2, d = 7.5 - 2.0 - 0.31 = 5.19 \text{ in.}$$

$$p = \frac{0.62}{12(5.19)} = 0.0100$$

From Appendix A, $k = 0.344, j = 0.885$; therefore

$$f_s = \frac{4423(12)}{0.62(0.885)(5.19)} = 18.6 \text{ ksi} < 20.0 \text{ ksi}$$

$$f_c = \frac{18.6(0.344)}{9(0.656)} = 1086 \text{ psi} < 1200 \text{ psi} \quad \text{OK}$$

Next, using the vehicle shown in Figure 4.30, it is necessary to check the slab for overload. This is a tandem-axle vehicle and because the current AASHTO specifications do not provide for wheel distribution and moments with tandem axles the 1957 AASHTO requirements for tan-

	Main Reinforcement Perpendicular to Traffic	
	Formulas for Moments per Foot Width of Slab	
Distribution of wheel loads: 　single axle	Freely supported spans	Continuous spans
Spans 2–7 ft, $E = 0.6S + 2.5$	$M = +0.25\,\dfrac{P_1}{E}\,S$	$M = \pm 0.2\,\dfrac{P_1}{E}\,S$
Spans of more than 7 ft, 　$E = 0.4S + 3.75$	$M = +0.25\,\dfrac{P_1}{E}\,S$	$M = \pm 0.2\,\dfrac{P_1}{E}\,S$
Tandem axles spans 2–7 ft 　$E = 0.36S + 2.58$	$M = +0.25\,\dfrac{P_2}{E}\,S$	$M = +0.2\,\dfrac{P_2}{E}\,S$
Spans over 7 ft, 　$E = .063S + 4.65$	$M = +0.25\,\dfrac{P_2}{E}\,S$	$M = \pm 0.2\,\dfrac{P_2}{E}\,S$

Figure 4.31　1957 AASHTO requirements: slab live load distribution and moments.

dem axles are used (Figure 4.31). The definitions of the terms are as follows:

S = effective span length defined under "Span Lengths" (Articles 1.3.2(a) and 1.7.2)
E = width of slab over which a wheel load is distributed
P_1 = load on one wheel of single axle
P_2 = load on one wheel of tandem axle

The live load plus impact moment is

$$E = 0.36(5.75) + 2.58 = 4.65 \text{ ft}$$

Use $E = 2.25 + 4.65/2 = 4.58$ ft

$$M_{LL+I} = 1.3\left(\frac{23.0}{4.58}\right)(0.2)(5.75) = 7508 \text{ lb-ft}$$

The required capacity of the overload vehicle at $\beta_L = 1.0$ is

$$M_u = 1.3[393 + 7508] = 10.3 \text{ k-ft}$$

Check for capacity:

$$a = \frac{0.62(60)}{0.85(3.25)(12)} = 1.12 \text{ in.}$$

$$\phi M_n = \phi \left[A_s f_y \left(d - \frac{a}{2} \right) \right]$$

$$= 0.80[(0.62)(60)(5.19 - 0.56)] (\tfrac{1}{12})$$

$$= 11.5 \text{ k-ft} > 10.3 \text{ k-ft} \quad \text{OK}$$

Check for maximum reinforcement in accordance with AASHTO Article 1.5.32(a) (1).

Maximum Reinforcement

$$0.75\rho_b = \left[\frac{0.85\beta_1 f_c'}{f_y} \left(\frac{87}{87 + f_y} \right) \right] (0.75)$$

$$= \left[\frac{(0.85)(0.85)(3.25)}{60} \left(\frac{87}{87 + 60} \right) \right] (0.75)$$

$$= 0.0174$$

$$\rho_{\max} = \frac{0.62}{12(5.19)} = 0.0100 < 0.0174 \quad \text{OK}$$

The AASHTO specifications require that distribution reinforcement be placed over the positive moment reinforcement. For main reinforcement perpendicular to traffic the requirement is that $220/\sqrt{S}$ % of the main reinforcement area be placed in the longitudinal direction, which gives 67% up to $S = 10.8$ ft.

The required distribution reinforcement is therefore

$$A_{s(\text{required})} = 0.67(0.62) = 0.415 \text{ in.}^2$$

Use #5 @ 8.96 in.

The main tranverse reinforcement consists of straight top and bottom bars alternating with bent type bars. This provides the maximum reinforcement in the critical areas for negative and positive moment. Location of the bend point for this bent bar is shown in Figure 4.32, which assumes parabolic moment envelopes for positive and negative moment. It also assumes that the maximum positive and negative moments equal 0.8 X simple span moment. The reinforcement requirement should be about half the maximum when the moment is half the maximum. Therefore, the bend point for alternate bars will be located at the average location on the two-moment envelopes for half the maximum moment.

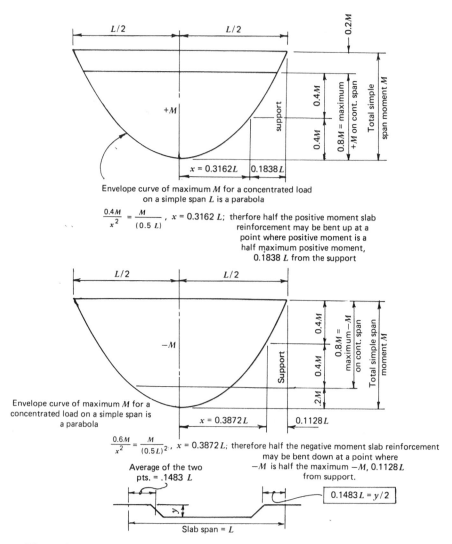

Figure 4.32   Determining bend points for transverse deck slab reinforcement.

Bend Point

$$A = 0.148(5.75) - \tfrac{1}{2}(0.32) = 0.691 \text{ ft}$$

$$\chi = 0.691 + 0.625 = 1.32 \text{ ft}$$

Use $\chi = 1$ ft 4 in. where χ = distance to ℄ girder.

The bend points are shown with the distribution reinforcement in Figure 4.33. Minimum reinforcement requirements are checked and are

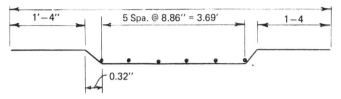

Figure 4.33   Bend point layout.

not significant. This is in accordance with AASHTO Article 1.5.7 in which the capacity must be greater than 1.2 × cracking moment.

Minimum Reinforcement

$$M_{cr} = 7.5\sqrt{3250} \; \frac{12(7.5)^2}{6} \left(\frac{1}{12}\right)$$

$$= 4008 \text{ lb-ft}; \; 1.2M_{cr} = 4810 \text{ k-ft}$$

$$A_{s(\text{required})} = \frac{4810(12)}{(5.19 - 0.56)(60)} = 0.21 \text{ in.}^2 \text{ (not significant)}$$

Example 4.4.   The deck slab in Example 4.3 is analyzed by the load factor design method. Because it is proper to apply the serviceability limits f_c = 1200 psi and f_s = 20,000 psi for a service load level to service load and load factor design and because these limits will usually control the design, it is obvious that there is no difference in the results between the two methods. To verify that this serviceability requirement controls the design the capacity requirements for the HS 20 vehicle can be computed and compared with the effective capacity already established. All other required computations are made under service load design criteria. For the HS 20 load

$$\#5 @ 6 \text{ in. } A_s = 0.62 \text{ in}^2, d = 7.5 - 2.0 - 0.31 = 5.19 \text{ in.}$$

$$M_u = 1.3(393) + 1.3(1.67)(4030) = 9260 \text{ lb-ft}$$

$$\phi M_n = 11.5 \text{ k-ft (Example 4.3)} > 9.3 \text{ k-ft}$$

Example 4.5.   The transversely reinforced deck shown in Figure 4.34 is designed by service load design procedures. This is the deck slab in Example 6.1 for the bridge detailed as a reinforced concrete box girder in Chapter 6.

The serviceability allowables f_c = 1200 psi, f_s = 20,000 psi, are discussed in Example 4.3.

This problem differs from Example 4.3 in that there is no specific

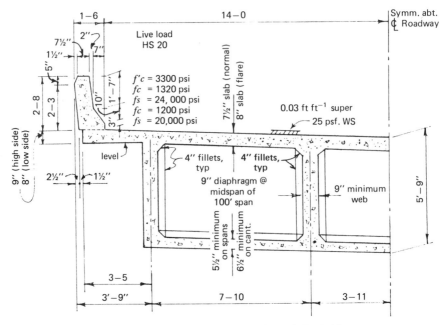

Figure 4.34   Roadway section: box girder bridge.

overload vehicle to be checked for capacity; a significant overhang also needs to be evaluated.

The span is taken as the clear distance between girder webs neglecting the fillets. The dead load, live load, and impact moments are calculated as done in Example 4.3 for the span between the girders and are computed as follows:

$$L = 7.83 - 0.75 = 7.08 \text{ ft}$$

Dead load

$$w = 0.625 \times 0.150 + 25 = 94 + 25 = 119 \text{ psf}$$

$$M_{DL} = \frac{119(7.08)^2}{10} = 597 \text{ lb-ft}$$

Live load + impact

$$M_{LL+I} = 1.3(0.8) \left(\frac{9.08}{32}\right)(16.0) = 4722 \text{ lb-ft}$$

The total design moment is

$$M_T = 597 + 4722 = 5319 \text{ lb-ft}$$

Preliminary analysis shows that #6 @ 7 in. with a $7\frac{1}{2}$-in. deck is reasonable. Check the stresses against serviceability allowables by using normal reinforced concrete stress formulas

$$f_s = \frac{M}{A_s jd} \quad \text{and} \quad f_c = \frac{f_s(k)}{n(1-k)}.$$

The charts in Appendix A can be used for k and j values based on the reinforcement ratio p:

$$\text{\#6 @ 7 in. } A_s = 0.75 \text{ in.}^2 \ 7\frac{1}{2}\text{-in. slab}$$

$$p = \frac{0.75}{12(5.12)} = 0.0122, \ d = 5.12 \text{ in.}$$

$$k = 0.385, j = 0.871$$

$$f_s = \frac{5319(12)}{0.75(0.871)(5.12)} = 19{,}000 \text{ psi} < 20{,}000 \text{ psi}$$

$$f_c = \frac{19000(0.385)}{10(0.615)} = 1195 \text{ psi} < 1200 \text{ psi} \quad \text{OK}$$

Next the overhang is checked for stress. Experience shows that an approximate-minimum 8-in. slab is required from the gutter line to the face of the girder support to handle the moment caused by rail live load plus dead load in accordance with AASHTO Article 1.3.2(H)(2). Because the deck is greater than 8 in. at the gutter line, it is necessary only to check the slab at the face of the girder where the minimum depth occurs. The slab overhang geometry is shown in Figure 4.35.

The parapet loading is $w = 366.3$ plf. The horizontal distance from the traffic face to the cg of the parapet is

$$\bar{x} = 0.91 \text{ ft (gutter line to cg parapet)}$$

The dead load shears and moments are

slab	$0.6937 \times 3.0417 \times 150 =$	$316.5 \times 1.52 =$	481
slab	$\frac{1}{2} \times 0.0563 \times 1.875 \times 150 =$	$7.9 \times 1.50 =$	12
slab	$1.1667 \times 0.0563 \times 150 =$	$9.9 \times 2.46 =$	24
parapet		$366.3 \times 2.78 =$	1018
WS	$1.875 \times 25 =$	$46.9 \times 0.94 =$	44
		747.5 lb	1579 lb-ft

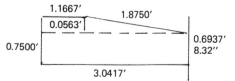

Figure 4.35   Slab overhand geometry.

The live load moment, according to AASHTO Article 1.3.2, is

live load + impact (wheel 1 ft from the curb)

$$E = 0.8(0.875) + 3.75 = 4.45 \text{ ft}$$

$$M_{LL+I} = 1.3 \left(\frac{16.0}{4.45}\right)(0.875) = 4090 \text{ lb-ft}$$

The second paragraph of AASHTO Article 1.3.2(b) does not apply because there is no sidewalk.

The rail live load moment, according to AASHTO Article 1.3.2(h)(2), is computed:

Rail Live Load

$$E = 0.8(2.78) + 5.0 = 7.22 \text{ ft}$$

$$M_{RLL} = \left(\frac{10.0}{7.22}\right)(3.0) = 4153 \text{ lb-ft}$$

Combinations of the two loadings are

$DL + LL + I$ (wheel 1 ft from the curb):

$$M_T = 1579 + 4090 = 5669 \text{ lb-ft}$$

$DL + RLL$:

$$M_T = 1579 + 4153 = 5732 \text{ lb-ft}$$

The stresses are checked in accordance with the established serviceability requirements:

$$d = 8.32 - 2.38 = 5.94 \text{ in.}$$

#6 @ 7 in. $A_s = 0.75$ reinforcement established in the span between girders

$$p = \frac{0.75}{12(5.94)} = 0.0105$$

From Appendix A, $k = 0.365$, $j = 0.878$; therefore

$$f_s = \frac{5732(12)}{0.75(0.878)(5.94)} = 17{,}886 \text{ psi} < 20{,}000 \text{ psi}$$

$$f_c = \frac{17{,}886(0.365)}{10(0.635)} = 1028 \text{ psi} < 1200 \text{ psi}$$

The required distribution reinforcement is

$$A_{s(\text{required})} = 0.67(0.75) = 0.50 \text{ in.}^2$$

Use #5 @ 7.40 in.

The girder-webflare varies in thickness; therefore the bend point is computed for the minimum and maximum web thicknesses and an average location is selected. Using Figure 4.32, we compute the bend point as

@ min web $t = 9$ in.

$$A = 0.148(7.08) - \tfrac{1}{2}(0.312) = 0.892$$

$$x = 0.892 + 0.375 = 1.267 \text{ ft}$$

@ max web $t = 13$ in. (interior panel)

$$A = 0.148(6.75) - \tfrac{1}{2}(0.312) = 0.843 \text{ ft}$$

$$x = 0.843 + 0.542 = 1.385 \text{ ft}$$

@ max web $t = 13$ (exterior panel – exterior girder)

$$A = 0.148(6.58) - \tfrac{1}{2}(0.312) = 0.818 \text{ ft}$$

$$x = 0.818 + 0.708 = 1.526 \text{ ft}$$

Use $x = 1$ ft 4 in. (1.333 ft).

The bent bar with distribution reinforcement is shown in Figure 4.36.

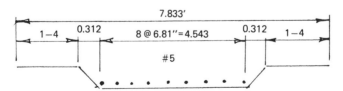

Figure 4.36   Bent bar distribution.

Example 4.6. The transverse slab in Example 4.5 is also designed by load factor design. Based on the serviceability computations given in Example 4.5, a $7\frac{1}{2}$ slab with #6 @ 7 in. between the girder is assumed.

The required capacity between the girders is

$$M_u = 1.3(597) + 1.3(1.67)(4722) = 11.0 \text{ k-ft}$$

The provided capacity between the girders is

$$a = \frac{0.75(60)}{0.85(3.3)(12)} = 1.34 \text{ in.,} \quad \frac{a}{2} = 0.67 \text{ in.}$$

$$\phi M_n = \phi \left[A_s f_y \left(d - \frac{a}{2} \right) \right]$$

$$= 0.80[(0.75)(60)(5.12 - 0.67)](\tfrac{1}{12})$$

$$= 13.4 \text{ k-ft} > 11.0 \text{ k-ft} \quad \text{OK}$$

The required capacity on the overhang is

$$M_u = 1.3(1579) + 1.3(1.67)(4153) = 11.1 \text{ k-ft}$$

The provided capacity on the overhang is

$$a = \frac{0.75(60)}{0.85(3.3)(12)} = 1.34 \text{ in.,} \quad \frac{a}{2} = 0.67 \text{ in.}$$

$$\phi M_n = \phi \left[A_s f_y \left(d - \frac{a}{2} \right) \right]$$

$$= 0.80[(0.75)(60)(5.94 - 0.67)](\tfrac{1}{12})$$

$$= 15.8 \text{ k-ft} > 11.1 \text{ k-ft} \quad \text{OK}$$

The bend points and distribution reinforcement requirements are the same as that for the service load in Example 4.5.

BIBLIOGRAPHY

1. *Continuous Concrete Bridges*, 2nd ed., Portland Cement Association.
2. *Concrete Members with Variable Moment of Inertia*, ST 103, Portland Cement Association.
3. *Beam Factors and Moment Coefficients for Members with Prismatic Haunches*, ST 81, Portland Cement Association.
4. *Beam Factors and Moment Coefficients for Members with Intermediate Expansion Hinges*, ST 75, Portland Cement Association.

5. *Frame Analysis Applied to Flat Slab Bridges*, ST 64-2, Portland Cement Association, 1944.

6. *Moments, Shears and Reactions for Continuous Highway Bridges*, American Institute of Steel Construction, 1966.

7. *Reinforced Concrete Design Handbook—Working Stress Method SP-3*, American Concrete Institute, 3rd ed., 1980.

8. P. Csagoly, M. Holowka, and R. Dorton, *The True Behavior of Thin Concrete Bridge Slabs*, Transportation Research Board Record 664 (Bridge Engineering—Vol. 1).

9. *AASHTO Standard Specifications for Highway Bridges, 1977*, with the 1978–1981 Interim Specifications, The American Association of State Highway and Transportation Officials.

10. *Ontario Highway Bridge Design Code, 1979*, Vols. 1 and 2, Ontario Ministry of Transportation and Communication.

11. R. Park and T. Paulay, *Reinforced Concrete Structures*, Wiley, New York, 1975.

12. Phil M. Ferguson, 4th ed., *Reinforced Concrete Fundamentals*, Wiley, New York, 1979.

13. *Notes on Load Factor Design for Reinforced Concrete Bridge Structures with Design Applications*, Portland Cement Association,

14. *ACI Building Code Requirements for Reinforced Concrete (ACI 318-77) and Commentary*, American Concrete Institute,

15. *Strength and Serviceability Criteria, Reinforced Concrete Bridge Members, Ultimate Design*, FHWA, October 1969.

FIVE

Design of Reinforced Concrete T-Beam Bridges

5.1 DESIGN PROCEDURES

T-beam construction consists of a vertical rectangular stem with a wide top flange. The wide top flange is usually the transversely reinforced deck slab and the riding surface for the traffic. Stem thicknesses vary from about 1 ft 2 in.–1 ft 10 in., set by the required horizontal spacing for the positive moment reinforcement.

Based on appropriate span lengths, T-beam construction should be considered as the next step above longitudinally reinforced slab construction. Therefore, it is reasonable to assume that T-beams should be used for spans of 45–95 ft. T-beams are also most effective in monolithic construction of the superstructure with the substructure. This framed monolithic construction has an advantage over continuous spans in that it uses more of the bending strength of the entire structure to support the loads.

A T-beam that is designed as an efficient section for positive moment will have some difficulty providing the required strength for negative moment. The reason is the loss of strength participation by the wide compression flange in the cracked tensile zone of the section. Some remedies for this problem are the following:

1. Thicken the stem in the negative moment areas.
2. Provide a haunch to get more depth in the negative moment areas.
3. Provide a bottom slab over a portion of the negative moment

areas. This type of construction, details of which are shown in Figure 5.1, was more commonly used some years ago.

4. Provide compression reinforcement combined with any one of the foregoing.

Older construction of reinforced concrete T-beams often had 3- or 4-in. fillets at the slab-stem intersection. Modern construction has eliminated their use and there is a lack of evidence of problems that may have resulted.

Although it is common to use 2 in. of cover elsewhere in bridge structures, $1\frac{1}{2}$ in. is common with T-beam stirrups, according to AASHTO Article 1.5.6(A). The spacing of reinforcement in the stems should correspond to AASHTO Article 1.5.5(A) in lieu of Article 1.5.5(C) because the confined area of T-beam stems would warrant a more open spacing.

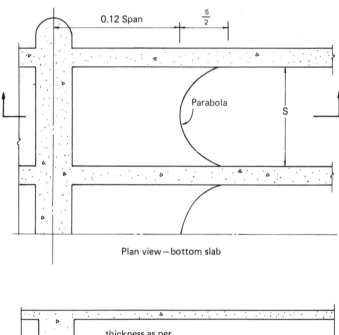

Plan view—bottom slab

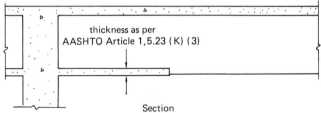

Section

Figure 5.1 Details for partial bottom slab.

For working stress design the normal k and j values must be taken from Appendix B, which is used expressly for T-beam design. The values of k and j in Appendix A are for rectangular sections and should be used for the negative moment section design because, in negative moment areas the slab area is in tension and the section for design becomes rectangular except when a partial slab exists. Concrete stress often is the most significant consideration in this negative moment design. For positive moment the section becomes a true T-beam and the concrete compressive stress is generally insignificant. The critical items for positive moment design are the area of reinforcement required and the ability to space that reinforcement effectively in the stem area.

5.2   DESIGN EXAMPLES

Example 5.1.   The T-beam superstructure for the bridge shown in Figures 5.2 and 5.3 is designed by service load design. The pier (Figure 5.4) for this bridge is designed in Example 10.2 and the deck slab, in Example 4.3. Both examples were also designed by service load procedures. The bridge in this example is designed by service load for an

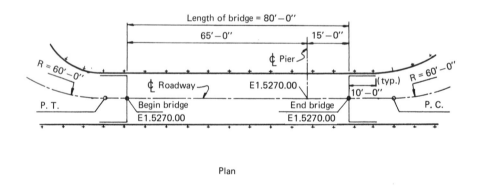

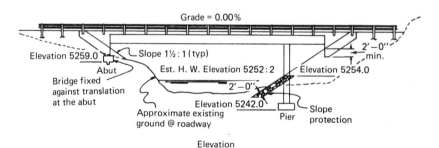

Figure 5.2   Plan and elevation: T-beam bridge.

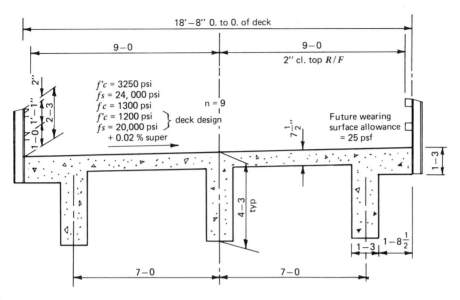

Figure 5.3 Roadway section: T-beam bridge.

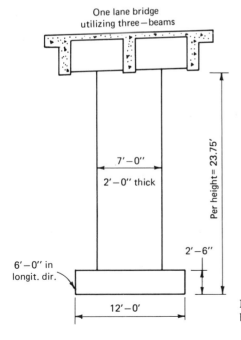

Figure 5.4 Pier elevation: T-beam bridge.

135

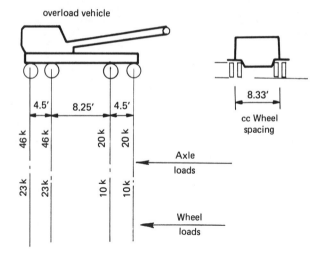

Designed by service load
Check overload capacity using overload
vehicle with $\beta_L = 1.0$

Figure 5.5   Design loadings.

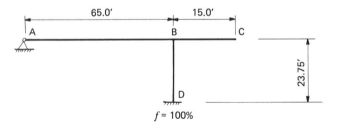

Figure 5.6   Structure layout.

HS 20 live load and checked for overload capacity with $\beta_L = 1.0$ and the overload vehicle in Figure 5.5.

The schematic layout of the structure for analysis purposes is shown in Figure 5.6. Note that this structure must be considered braced in the longitudinal direction.

After the design of the deck slab is completed (see Example 4.3), it is proper to compute the necessary frame constants for moment distribution. The gross concrete section is used for the superstructure and

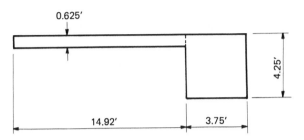

Figure 5.7   Gross superstructure section.

the transformed gross section is used for the substructure. The gross superstructure section can be modified as shown in Figure 5.7.

The moment of inertia for this section according to the equation $I_T = \Sigma(I_0 + A\bar{y}^2)$ is as follows:

I_T = gross moment of inertia

I_0 = moment of inertia for each element

A = gross area

$\bar{y}$ = distance from cg element to cg gross section

y_b = distance from bottom fiber to cg element

Element	A	y_b	A_{yb}	$\bar{y}$	$A\bar{y}^2$	I_0	I_T
Deck	9.32	3.94	36.70	1.15	12.33	0.30	12.63
Stem	15.94	2.12	33.87	0.67	7.16	23.99	31.15
	25.26		70.57				43.78

$\bar{y}_b$ = 2.79 ft. The relative superstructure stiffness (omitting E) is $3I/L$ for one end pinned and constant I. Therefore

$$K_{BA} = \frac{3(43.78)}{65.0} = 2.02$$

The cantilever span has no stiffness at point B.

The assumed pier section includes 13–#10 on each face in the longitudinal direction as shown in Figure 5.8:

The moment of inertia is computed for the concrete and the assumed reinforcement is then added:

$$I_{\text{conc}} = \frac{7.0(2.0)^3}{12} = 4.67 \text{ ft}^4$$

$$I_{R/F}^{\text{transf}} = 9(16.51)(2)(8.87)^2 \left(\tfrac{1}{12}\right) = 1.13 \text{ ft}^4$$

$$\sum I_{\text{pier}} = 1.13 + 4.67 = 5.80 \text{ ft}^4$$

The relative substructure stiffness (omitting E) is $4I/L$ for both ends fixed with a constant I:

$$K_{BD} = \frac{4(5.80)}{23.75} = 0.98$$

At B the distribution factors are computed with $D = K/\Sigma K$, where $\Sigma K = K_{BA} + K_{BD} = 0.98 + 2.02 = 3.00$; therefore

$$D_{BA} = \frac{2.02}{3.00} = 0.673$$

$$D_{BA} = \frac{0.98}{3.00} = 0.327$$

As previously noted, this frame is braced at the abutment longitudinally and sidesway computations should not be made. Thus the moment distribution is simple. The moment distribution for a unit fixed end moment at BA is shown in Table 5.1.

Table 5.2 gives the moment distribution for a unit fixed end moment at BC.

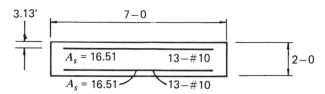

Figure 5.8 Pier section.

Table 5.1

				$C = 0.50$
		B		D
	BA 0.673	BC	BD 0.327	DB
FEM	−1000			
BAL	+673		+327	
CO				+164
Σ	−327		+327	+164

Table 5.2

				$C = 0.50$
		B		D
	BA 0.673	BC	BD 0.327	DB
FEM		+1000		
BAL	−673		−327	
CO				−164
Σ	−673	+1000	−327	−164

In girder design it is necessary to compute loads and moments separately for exterior and interior girders. Then when these moments are summed up it is determined which girder controls and at what particular point. All girders are detailed the same (for the maximum) because consistency also offers economy. This method is also consistent with AASHTO Article 1.3.1(b)(2)(a), which states that "In no case shall an exterior stringer have less carrying capacity than an interior stringer."

First, the dead loads and moments are computed for an exterior girder:

Exterior Girder

The uniform dead load of the superstructure, (assuming that the concrete weight is equal to 0.150 k-ft^{-3}) is

Slab	$0.625 \times 4.125 \times 0.150 = 0.39$
Overhang	$1.25 \times 1.71 \times 0.150 = 0.32$
Stem	$1.25 \times 3.62 \times 0.150 = 0.68$
WS	$5.83 \times .025 = 0.15$
Railing	$= \underline{0.03}$
	1.57 k-ft^{-1}

A diaphragm is placed at midspan as a concentrated load (assume 1.0 ft thick $\times$ 3.38 ft $\times$ 2.88 ft):

Conc @ Midspan

$$P = 1.0 \times 3.38 \times 2.88 \times 0.150 = 1.46 \text{ k}$$

The end beam and wing at the cantilever tip (point C) in Figure 5.9 are computed as P_1 = load from the end beam width tributary to the exterior girder = 7.88 k.

The end beam is 1 ft 3 in. thick and moments are summed around the tip at the cantilever to find the location of P_1, P_2, and P_3:

P_1	7.88 k $\times$ 0.625 ft =	-4.92
P_2	3.30 k $\times$ 5.00 ft =	$+16.50$
P_3	$\underline{5.25 \text{ k} \times 3.33 \text{ ft}} = $	$\underline{+17.48}$
	16.43 k	$+29.06$ k-ft

$$\bar{x} = \frac{29.06}{16.43} = 1.77 \text{ ft from cantilever tip}$$

The fixed end moment at BA with A pinned is computed by using

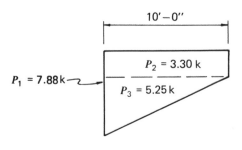

Figure 5.9   End beam details.

$M^F = wl^2/8$ for a uniform load and $M^F = 3Pl/16$ for a concentrated load at midspan:

$$M_{BA}^F = \frac{1.57(65.0)^2}{8} = 829 \left.\begin{array}{c} \\ \\ \end{array}\right\}$$

$$= \frac{3(1.46)(65.0)}{16} = 18 \qquad \text{sum} = 847 \text{ k-ft}$$

The cantilever fixed end moment at BC is computed by using $M^F = wl^2/2$ for a uniform load and $M^F = Pl$ for a concentrated load:

$$M_{BC}^F = 1.57(15.0)^2(\tfrac{1}{2}) = 177 \left.\begin{array}{c} \\ \\ \end{array}\right\}$$

$$= 16.43(15.0 + 1.77) = 275 \qquad \text{sum} = 453 \text{ k-ft}$$

Using these fixed end moments and the unit moment distribution already computed, we calculate the corner moments:

$$M_{BA} = 847(-0.327) + 453(-0.673) = -582 \text{ k-ft}$$

$$M_{BC} = -453 \text{ k-ft}$$

$$M_{BD} = 847(+0.327) + 453(-0.327) = +129 \text{ k-ft} \left.\begin{array}{c} \\ \end{array}\right\} \quad \begin{array}{l} \text{total moment} \\ \text{variation in} \end{array}$$

$$M_{DB} = 847(+0.164) + 453(-0.164) = +65 \text{ k-ft} \qquad \text{column} = +194$$

The moment at DB is modified to the top of the footing by assuming a straight-line variation in the moment between D and B:

$$M_{DB\,\text{col}} = +65 - \left(\frac{2.50}{23.75}\right)(194) = +45 \text{ k-ft}$$

The maximum positive moment occurs at $0.4AB$, which is approximately correct for all end spans of continuous units in which a nearly balanced span arrangement is used.

At $0.4AB$ the simple span moments for uniform and concentrated loads are combined with the negative moment effect that results from the corner moment at BA. This gives the dead load moment at $0.4AB$:

$$M_{0.4AB} = 0.120(1.57)(65.0)^2 = +796 \left.\begin{array}{c} \\ \\ \\ \end{array}\right\}$$

$$= 1.46(65.0)(0.20) = +19 \qquad \text{sum} = +582 \text{ k-ft}$$

$$= -0.4(582) = -233$$

Interior Girder

The loads and moments for the interior girder are computed like those for the exterior girder.

The uniform load is

$$\begin{array}{llr}
\text{Slab} & 0.625 \times 7.00 \times 0.150 = & 0.66 \\
\text{Stem} & 1.25 \times 3.62 \times 0.150 = & 0.68 \\
\text{WS} & 7.0 \times .025 = & \underline{0.18} \\
& & 1.52 \text{ k-ft}^{-1}
\end{array}$$

The diaphragm concentrated load at midspan is

$$P = 1.0 \times 3.38 \times 5.75 \times 0.150 = 2.91 \text{ k}$$

The concentrated load at the cantilever tip is due to the dead load of the end beam which is 9 ft 0 in. deep and 1 ft 3 in. thick. The entire dead load of the wings has been taken by the exterior girders. The dead load of the end beam is a rectangular block from which the previously computed uniform dead load over the thickness has been removed:

$$\left. \begin{array}{lr}
P = 9.0 \times 1.25 \times 7.00 \times 0.150 = 11.81 \\
-1.52 \times 1.25 \qquad\qquad = -1.90
\end{array} \right\} \text{algebraic sum} = 9.91 \text{ k}$$

The fixed end moments are

$$\left. \begin{array}{lr}
M_{BA}^F = 1.52(65.0)^2(\tfrac{1}{8}) = 803 \\
2.91(65.0)(\tfrac{3}{16}) = 35
\end{array} \right\} = 838 \text{ k-ft}$$

$$\left. \begin{array}{lr}
M_{BC}^F = 1.52(15.0)^2(\tfrac{1}{2}) \quad = 171 \\
9.91(15.0 - 0.625) = 142
\end{array} \right\} = 313 \text{ k-ft}$$

The distributed moments are

$$M_{BA} = 838(-0.327) + 313(-0.673) = -485 \text{ k-ft}$$

$$M_{BC} = -313 \text{ k-ft}$$

$$\left. \begin{array}{l}
M_{BD} = 838(+0.327) + 313(-0.327) = +172 \text{ k-ft} \\
M_{DB} = 838(+0.164) + 313(-0.164) = +86 \text{ k-ft}
\end{array} \right\} \begin{array}{l} \text{total moment} \\ \text{variation in the} \\ \text{column is } +258 \text{ k-ft} \end{array}$$

The modified moment for the pier shaft at the top of the footing is

$$M_{DB\,col} = +86 - \left(\frac{2.50}{23.75}\right)(+258) = +59 \text{ k-ft}$$

The positive moment at $0.4AB$ from the simple span moment and the modified corner moment is

$$M_{0.4AB} = 0.120(1.52)(65.0)^2 = +771 \left.\begin{matrix} \\ \\ \\ \end{matrix}\right.$$
$$2.91(65.1)(0.20) \quad = \ +38 \quad \right\} \quad \text{sum} = +615 \text{ k-ft}$$
$$-0.4(485) \qquad = -194$$

Next the critical live load moments are computed. Experience and visualization of the shape of the influence lines are used to locate the loadings for the critical conditions in which the HS 20 loadings are investigated.

The first case (Figure 5.10) should produce maximum positive moment at $0.4AB$. The fixed end moment is computed by using the table on Appendix O. The distributed moments are computed with the preceding unit moment distribution. The column loads and moments are also computed for use in Examples 3.1 and 10.2.

Case I. The fixed end moment is

$$M_{BA}^F = 16(65)\left(\frac{0.0871}{4} + 0.1680 + 0.1909\right) = 396 \text{ k-ft}$$

The distributed moments are computed by using the unit moment distribution:

$$M_{BA} = 396(-0.327) = -129 \text{ k-ft} \left.\begin{matrix} \\ \\ \\ \end{matrix}\right.$$
$$M_{BD} = 396(+0.327) = +129 \text{ k-ft} \quad \right\} \quad \text{the total variation in column}$$
$$M_{DB} = 396(+0.164) = \ +65 \text{ k-ft} \qquad \text{moment is } 194 \text{ ft}^{-1}$$

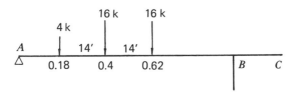

Figure 5.10 Maximum positive moment live loading: beam.

The modified pier shaft moment at the top of the footing is

$$M_{DB \, col} = +65 - \left(\frac{2.50}{23.75}\right)(194) = +45 \text{ k-ft}$$

The end shear at AB and the reaction at B are:

$$V_{AB} = 16\left(\frac{0.82}{4} + 0.60 + 0.38\right) - \frac{129}{65} = 19.0 - 2.0 = 17.0 \text{ k}$$

$$R_B = 16\left(\frac{0.18}{4} + 0.40 + 0.62\right) + \frac{129}{65} = 17.0 + 2.0 = 19.0 \text{ k}$$

The maximum positive moment at $0.4AB$ is computed by taking a free body from A to $0.4AB$ and summing the moments about $0.4AB$

$$\left.\begin{array}{l} M_{0.4AB} = 17.0(26.0) = +441 \\ \phantom{M_{0.4AB} =} -4.0(14.0) = -56 \end{array}\right\} = +385 \text{ k-ft}$$

Case IA loading (Figure 5.11) is placed so as to produce the maximum reaction combined with positive (rotational signs) column moments. The moments and loads are computed as in Case I. These are computed for use in Examples 3.1 and 10.5.

Case IA.

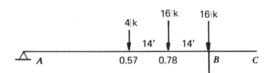

Figure 5.11   Maximum reaction live loading.

$$M_{BC}^F = 16(65)\left(\frac{0.1924}{4} + 0.1527\right) = 209 \text{ k-ft}$$

$$M_{BA} = 209(-0.327) = -68 \text{ k-ft}$$

$$M_{BD} = 209(+0.327) = +68 \text{ k-ft}$$

$$M_{DB} = 209(+0.164) = +34 \text{ k-ft}$$

$$R_B = 16.0\left(1.0 + 0.78 + \frac{0.57}{4}\right) + \frac{68}{65} = 31.8 \text{ k}$$

Case IB loading (Figure 5.12) is placed so as to produce the maximum positive (rotational signs) column moments, all loads and moments are computed as in Case I. These are computed for use in Examples 3.1 and 10.2. This same loading is a possibility for the maximum negative moment @BC.

Case IB.

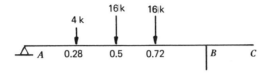

Figure 5.12 Maximum positive moment live loading: column.

$$M_{BC}^F = 16(65) \left(\frac{0.1290}{4} + 0.1875 + 0.1734 \right) = 409 \text{ k-ft}$$

$$M_{BA} = 409(-0.327) = -134 \text{ k-ft}$$

$$M_{BD} = 409(+0.327) = -134 \text{ k-ft}$$

$$M_{DB} = 409(+0.164) = +68 \text{ k-ft}$$

$$M_{DB \, col} = +68 - \left(\frac{2.50}{23.75} \right)(202) = +46 \text{ k-ft}$$

$$R_B = 16.0 \left(\frac{0.28}{4} + 0.5 + 0.72 \right) + \frac{137}{65} = 23.1 \text{ k}$$

Case II. Case II loading (Figure 5.13) is selected to produce maximum negative (rotational signs) column moments and is computed for use in Examples 3.1 and 10.2. This loading also produces the maximum negative bending moment at BC and is a possibility for maximum negative bending at BA. The fixed end moment at BC is a cantilever moment. All other moments and loads are computed as in Case I.

$$M_{BC}^F = 16(15 + 1) = -256 \text{ k-ft}$$

$$M_{BA} = 256(-0.673) = -172 \text{ k-ft}$$

$$M_{BC} = -256 \text{ k-ft}$$

$$M_{BD} = 256(-0.327) = -84 \text{ k-ft}$$

$$M_{DB} = 256(-0.164) = -42 \text{ k-ft}$$

$$M_{DB \, col} = -42 - \left(\frac{2.50}{23.75} \right)(-126) = -29 \text{ k-ft}$$

$$R_B = 32.0 + \frac{172}{65} = 34.6 \text{ k}$$

Figure 5.13   Maximum negative moment live loading: column.

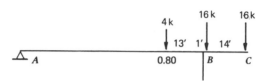

Figure 5.14   Maximum reaction live loading.

Case IIA. Case IIA loading (Figure 5.14) is placed to produce the maximum reaction combined with negative (rotational signs) column moments and is also computed for use in Examples 3.1 and 10.2.

$$M_{BA}^F = 4(65)(0.1440) = 37 \text{ k-ft}$$

$$M_{BC}^F = 256 \text{ k-ft}$$

$$M_{BA} = 37(-0.327) - 172 = -184 \text{ k-ft}$$

$$M_{BD} = 37(+0.327) - 84 = -72 \text{ k-ft}$$

$$M_{DB} = 37(+0.164) - 42 = -36 \text{ k-ft}$$

$$M_{DB\,col} = -36 - \left(\frac{2.50}{23.75}\right)(-108) = -25 \text{ k-ft}$$

$$R_B = 32.0 + 4(0.8) + \frac{184}{65} = 38.0 \text{ k}$$

Case III. Case III loading (Figure 5.15) is placed to produce the maximum negative moment at BA by using a truck loading:

$$M_{BA}^F = 16(65) \left(\frac{0.1918}{4} + 0.1568\right) = 213 \text{ k-ft}$$

$$M_{BC}^F = 16(15.0) = 240 \text{ k-ft}$$

$$M_{BA} = 213(0.327) + 240(0.673) = -231 \text{ k-ft}$$

$$M_{BC} = 213(0) + 240(1.00) = -240 \text{ k-ft}$$

Case IV. Case IV loading (Figure 5.16), a lane loading variation of HS 20 loading, is another possibility for producing the maximum nega-

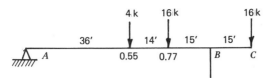

Figure 5.15 Maximum negative moment live loading.

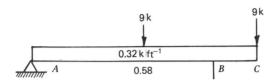

Figure 5.16 Maximum negative moment lane loading: column.

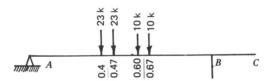

Figure 5.17 Maximum positive moment overloading: beam.

tive moment at BA. For a uniform load the fixed end moments at B are

$$M_{BA}^F = wl^2(0.125)$$

$$M_{BC}^F = wl^2(\tfrac{1}{2})$$

$$M_{BA}^F = 0.32(65.0)^2(0.125) + 9(65.0)(0.1925) = 282 \text{ k-ft}$$

$$M_{BC}^F = 0.32(15.0)^2(\tfrac{1}{2}) + 9(15.0) = 171 \text{ k-ft}$$

$$M_{BA} = 282(0.327) + 171(0.673) = -207 \text{ k-ft}$$

$$M_{BC} = -171 \text{ k-ft}$$

Next the moments and loads produced by the overload vehicle are computed. These loadings are similar to the various loadings computed for the HS 20 vehicle.

Case V. Case V loading (Figure 5.17) is designed primarily to produce maximum positive moment at $0.4AB$. The pier loads and moments are also computed for use in Examples 3.1 and 10.2.

$$M_{BA}^F = 23.0(65.0)(0.1680 + 0.1831) + 10(65.0)$$

$$\cdot (0.1920 + 0.1846) = 770 \text{ k-ft}$$

$$M_{BA} = 770(-0.327) = -252 \text{ k-ft}$$

$$M_{BD} = 770(+0.327) = +252 \text{ k-ft}$$

$$M_{DB} = 770(+0.164) = +126 \text{ k-ft}$$

$$M_{DB \, col} = +126 - \left(\frac{2.50}{23.75}\right)(378) = +86 \text{ k-ft}$$

$$R_B = 23.0(0.40 + 0.47) + 10.0(0.60 + 0.67) + \frac{252}{65.0}$$

$$= 20.0 + 12.7 + 3.9 = 36.6 \text{ k}$$

$$V_{AB} = 66.0 - 36.6 = 29.4 \text{ k}$$

$$M_{0.4AB} = 29.4(26.0) = +764 \text{ k-ft}$$

Case VA. Case VA loading (Figure 5.18) is placed to produce the maximum reaction combined with positive (rotational signs) column moments. This loading is computed for possible use in Examples 3.1 and 10.2 in which the design of the substructure is discussed.

$$M_{BA}^F = (23)(65)(0 + 0.0628) + 10(65)(0.1440 + 0.1705) = 298 \text{ k-ft}$$

$$M_{BA} = 298(-0.327) = -97 \text{ k-ft}$$

$$M_{BD} = 298(+0.327) = +97 \text{ k-ft}$$

$$M_{DB} = 298(+0.164) = +49 \text{ k-ft}$$

$$R_B = 23.0(1.0 + 0.93) + 10(0.80 + 0.73) + \frac{97}{65}$$

$$= 44.4 + 15.3 + 1.5 = 61.2 \text{ k}$$

Case VB. Case VB loading (Figure 5.19) is placed to produce the

Figure 5.18   Maximum reaction overloading.

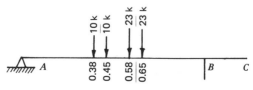

Figure 5.19   Maximum positive moment overloading: column.

maximum positive (rotational signs) column moments and is computed for use also in Examples 3.1 and 10.2. The same loading is a possibility for the maximum negative moment at BC.

$$M_{BA}^F = (23)(65)(0.1925 + 0.1877) + (10)(65)$$
$$\cdot (0.1626 + 0.1795) = 790 \text{ k-ft}$$

$$M_{BA} = 790(-0.327) = -258 \text{ k-ft}$$

$$M_{BD} = 790(+0.327) = +258 \text{ k-ft}$$

$$M_{DB} = 790(+0.164) = +129 \text{ k-ft}$$

$$M_{DB\,col} = +129 - \left(\frac{2.50}{23.75}\right)(387) = +88 \text{ k-ft}$$

$$R_B = 23(0.58 + 0.65) + 10(0.38 + 0.45) + \frac{258}{65}$$

$$= 28.3 + 8.3 + 4.0$$

$$= 40.6 \text{ k}$$

Case VI. Case VI loading (Figure 5.20) is selected to produce maximum negative (rotational signs) column moments and is computed for possible use in Examples 3.1 and 10.2. It also produces the maximum negative bending moment at BC and is a possibility for the maximum negative bending moment at BA.

$$M_{BC}^F = 23.0(15.0 + 10.5) = 586 \text{ k-ft}$$

$$M_{BA} = 586(-0.673) = -395 \text{ k-ft}$$

$$M_{BC} = -586 \text{ k-ft}$$

$$M_{BD} = 586(-0.327) = -192 \text{ k-ft}$$

$$M_{DB} = 586(-0.164) = -96 \text{ k-ft}$$

$$M_{DB\,col} = -96 + \left(\frac{2.50}{23.75}\right)(288) = -66 \text{ k-ft}$$

$$R_B = 46.0 + \frac{395}{65} = 52.1 \text{ k}$$

Figure 5.20 Maximum negative moment overloading: column.

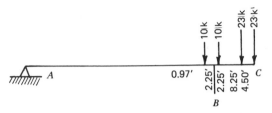

Figure 5.21   Maximum reaction overloading.

Case VIA. Case VIA loading (Figure 5.21) is placed to produce the maximum reaction combined with negative (rotational signs) column moment and is computed for use in Example 3.1 and 10.2.

$$M_{BA}^F = 10(65)(0.0287) = 19 \text{ k-ft}$$

$$M_{BC}^F = 23(15.0 + 10.5) + 10(2.25) = 609 \text{ k-ft}$$

$$M_{BA} = 19(-0.327) + 609(-0.673) = -416 \text{ k-ft}$$

$$M_{BC} = -609 \text{ k-ft}$$

$$M_{BD} = 19(+0.327) + 609(-0.327) = -193 \text{ k-ft}$$

$$M_{DB} = 19(+0.164) + 609(-0.164) = -96 \text{ k-ft}$$

$$M_{DB \text{ col}} = -96 - \left(\frac{2.50}{23.75}\right)(-289) = -66 \text{ k-ft}$$

$$R_B = 56.0 + 10(0.97) + \frac{416}{65} = 72.1 \text{ k}$$

The impact factor is computed for each span according to AASHTO Article 1.2.12. The average impact is computed for loads in both spans:

$$I = \frac{50}{65 + 125} = 0.263 \text{ in the span} \qquad I = 0.300 \text{ on the cantilever}$$

$$I_{\text{avg.}} = 0.282$$

The wheel distribution to each girder is computed in accordance with AASHTO Article 1.3.1(b) and coefficients are computed that include the effect of impact and wheel distribution. The averages are computed for loads in both spans. First they are computed for the HS 20 loading (Figure 5.22).

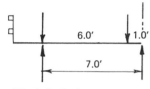

Wheel distribution = DF **Figure 5.22**   Exterior girder distribution factor.

HS 20. Exterior Girder

$$DF = \frac{7.0 + 1.0}{7.0} = 1.143$$

Coefficient = 1.263(1.143) = 1.443 in the span

1.300(1.143) = 1.486 on the cantilever   1.464 av

HS 20. Interior Girder

Because $S = 7.0 > 6.0$ AASHTO Table 1.3.1(B) indicates a simple beam distribution similar to the exterior girder. Therefore $D = 1.143$.

Coefficient = 1.263(1.143) = 1.443 on the span

1.300(1.143) = 1.486 on the cantilever   1.464 av

The same values are computed for the overload vehicle (Figure 5.23).

Overload. Exterior Girder

$$DF = 1.00$$

Coefficient = 1.263 on the span

1.300 on the cantilever   1.282 av

Figure 5.23   Interior girder distribution factor.

Overload. Interior Girder

$$DF = \frac{5.67}{7.00} = 0.810$$

Coefficient = 1.263(0.810) = 1.023 in the span

= 1.300(0.810) = 1.053 on the cantilever   1.038 av

The moments for the critical superstructure sections are then tabulated and summed. The live load moments for one line of wheels are computed for live load plus impact per girder.

The HS 20 service load moments are computed for the exterior and interior girders.

Service Load (HS 20) (Exterior Girder)

	DL	One Line Wheels-LL	$LL + I$	M_T (k-ft)
$M_{0.4AB}$	+582	+385 × 1.443 = +556		+1138
M_{BA}	−582	−231 × 1.464 = −338		−920
M_{BC}	−453	−256 × 1.486 = −380		−833

Service Load (HS 20) (Interior Girder)

	DL	One Line Wheels-LL	$LL + I$	M_T (k-ft)
$M_{0.4AB}$	+615	+385 × 1.443 = +556		+1171
M_{BA}	−485	−231 × 1.464 = −338		−823
M_{BC}	−313	−256 × 1.486 = −380		−693

The overload moments are tabulated, factored and summed.

Overload (Exterior Girder)

	DL	1.3DL	One Line Wheels-LL	$LL + I$	1.3 × $LL + I$	M_u (k-ft)
$M_{0.4AB}$	+582	+757	+764 × 1.263 = +965		+1254	+2011
M_{BA}	−582	−757	−395 × 1.300 = −514		−668	−1425
M_{BC}	−453	−589	−586 × 1.300 = −762		−990	−1580

Overload (Interior Girder)

	DL	1.3DL	One Line Wheels-*LL*	*LL* + *I*	1.3 × *LL* + *I*	M_u (k-ft)
$M_{0.4AB}$	+615	+800	+764 × 1.023 = +782	+1017		+1817
M_{BA}	−485	−630	−395 × 1.053 = −416	−541		−1171
M_{BC}	−313	−407	−586 × 1.053 = −617	−802		−1209

The minimum reinforcing requirements of AASHTO Article 1.5.7 are checked: Assume a rectangular section, 1 ft 3 in. × 4 ft 3 in.:

$$M_{cr} = 7.5\sqrt{f_c'}\,(S)$$

$$M_{cr} = 7.5\sqrt{3250}\,\left[\frac{15(51)^2}{6}\right]\left(\frac{1}{12}\right) = 232\ \text{k-ft}$$

$$1.2M_{cr} = 278\ \text{k-ft} \qquad d = 42.0\ \text{in.}{\pm} \qquad f_y = 60\ \text{ksi}$$

$$A_{s(\text{required})} = \frac{278(12)}{60(42.0{\pm})} = 1.32\ \text{in.}^2$$

therefore not significant.

Next design for the positive moment section @ 0.4*AB*. First approximate the required steel for the HS 20 service load.

$$\text{vehicle: } M_{0.4AB} = +1171\ \text{k-ft}$$

Assume $j = 0.90$ and $d = 45$ in.

$$\text{Approx } A_{s(\text{required})} = \frac{1171(12)}{24(0.90)(45)} = 14.5\ \text{in.}^2$$

Try 8-#11 and 1-#14; $A_s = 14.73$ in.^{2}

Σ moment about soffit:

$$3(1.56)(2.70)\ = 12.64$$

$$3(1.56)(6.20)\ = 29.02$$

$$2(1.56)(10.45) = 32.60$$

$$(2.25)(10.45) = \frac{23.51}{97.77 \text{ in.}^3}$$

$$\bar{y} = \frac{97.77}{14.73} = 6.64 \text{ in.}$$

$$d = 51.0 - 6.64 = 44.36 \text{ in.}$$

By using Appendix B

$$\frac{t}{d} = \frac{7.5}{44.36} = 0.169 \qquad pn = \frac{14.73(9)}{70(44.36)} = 0.0427 \qquad k = 0.268$$

$$j = 0.928$$

The computed stresses compared with the allowables are

$$f_s = \frac{1171(12)}{14.73(0.928)(44.36)} = 23.2 \text{ ksi} < 24.0 \text{ ksi}$$

$$f_c = \frac{23.2(0.268)}{9(0.732)} = 942 \text{ psi} < 1300 \text{ psi}$$

Next the overload condition is checked for capacity in which the required $M_u = +2011$ k-ft. First check the section to determine whether the analysis should be made for a rectangular or flanged section.

The area of reinforcement required to develop the compressive strength of the overhanging flanges is

$$A_{sf} = \frac{0.85 f_c'(b - b_w)}{f_y} \qquad h_f = \frac{0.85(3.25)(70.0 - 15)(7.5)}{60} = 18.99 \text{ in.}^2$$

$$A_{sf} = 18.99 \text{ in.}^2 > A_s \text{ provided} = 14.73 \text{ in.}^2$$

Therefore the section is rectangular for the purposes of analysis. The capacity for a rectangular section is then checked.

$$a = \frac{A_s f_y}{0.85 f_c' b} = \frac{14.73(60)}{0.85(3.25)(70.0)} = 4.57 \text{ in.}$$

$$\phi M_n = \phi \left[A_s f_y \left(d - \frac{a}{2} \right) \right]$$

$$= 0.90[(14.73)(60)(44.36 - 2.29)](\tfrac{1}{12})$$

$$2789 \text{ k-ft} > 2011 \text{ k-ft} = M_u \quad \text{OK}$$

Next the negative moment section at B is designed. For the HS 20 service load level the moment at BA is the most critical (M^μ_{BA} = -920 k-ft). For investigation of the capacity with the overload vehicle the moment at BC is the most critical (M^μ_{BC} = -1580 k-ft). The bridge is designed for service load and checked for overload capacity. The concrete compressive stress is normally critical for this section in the negative moment area. Therefore the compressive reinforcement is computed using $2n$. This analysis is as given in the ACI publication SP-3 "Reinforced Concrete Design Handbook." The design moment is M_{BA} = -920 k-ft.

The effective depth is

$$d = 51.0 - 2.0 - 0.625 = 47.74 \text{ in.}$$

The approximate required tensile reinforcement is

$$A_{s(\text{required})} = \frac{920(12)}{24(0.87)(47.74)} = 11.1 \text{ in.}^2$$

Try tensile reinforcement as follows: 9–#10, A_s = 11.43 in.^{2}
The approximate required depth based on $K = 194$ for balanced design is

$$d_{(\text{required})} = \sqrt{\frac{920(12)}{0.194(15)}} = 61.6 \text{ in.} > 47.74$$

Therefore needs compressive reinforcement.
Try 3–#11, A'_s = 4.68 in.^{2}
The cover to the cg of the compressive reinforcement (d') is

$$d' = 1.5 + 0.5 + 0.70 = 2.70 \text{ in.}$$

Then

$$\frac{d'}{d} = \frac{2.70}{47.74} = 0.057$$

The parameter m is

$$\frac{nA_s}{bd} + \frac{(2n-1)(A'_s)}{bd}$$

$$m = \frac{9(11.43)}{15(47.74)} + \frac{17(4.68)}{15(47.74)} = 0.144 + 0.111 = 0.258$$

The parameter q is

$$\frac{nA_s}{bd} + \frac{(2n - 1)(A'_s)(d')}{bd^2}$$

$$q = 0.144 + (0.111)(0.057) = 0.144 + 0.006 = 0.150$$

$$k = \sqrt{m^2 + 2q} - m$$

$$k = \sqrt{(0.258)^2 + 2(0.150)} - 0.258 = 0.347$$

$$\frac{1}{k} \times \frac{(2n - 1) A'_s}{bd} = \frac{0.111}{0.347} = 0.320$$

$$\frac{1}{k} \times \frac{d'}{d} = \frac{0.057}{0.347} = 0.164$$

$$z = \frac{0.167 + \{[(2n - 1)(A'_s)]/kbd\}(d'/kd)(1 - d'/kd)}{0.500 + \{[(2n - 1)(A'_s)]/kdb\}(1 - d'/kd)}$$

$$z = \frac{0.167 + (0.320)(0.164)(0.836)}{0.500 + (0.320)(0.836)} = 0.275$$

$$j = 1 - zk = 1 - (0.275)(0.347) = 0.905$$

The stresses based on the k and j values are computed and compared with the allowables.

$$f_s = \frac{M}{A_s jd}$$

$$f_s = \frac{920(12)}{11.43(0.905)(47.75)} = 22.4 \text{ ksi} < 24.0 \text{ ksi}$$

$$f_c = \frac{f_s(k)}{n(i - k)}$$

$$f_c = \frac{22.4(0.347)}{9(0.653)} = 1320 \text{ psi} > 1300 \text{ psi } 1.5\% \quad \text{OK}$$

$$f'_s = 2(f_s) \left(\frac{k - d'/d}{1 - k} \right)$$

$$f'_s = 2(22.4) \left(\frac{0.347 - 0.057}{0.653} \right) = 19.9 \text{ ksi} < 24.0 \text{ ksi}$$

The capacity check on this rectangular section for the overload vehicle is

$$a = \frac{A_s' f_y}{0.85 \, f_c' \, b} = \frac{11.43(60)}{0.85(3.25)(15)} = 16.55 \text{ in.}$$

$$\phi M_n = \phi \left[A_s f_y \left(d - \frac{a}{2} \right) \right]$$

$$= 0.90[(11.43)(60)(47.74 - 8.28)](\tfrac{1}{12}) = 2030 \text{ k-ft} > 1580 \text{ k-ft}$$

$$= M_u \text{ at } BC$$

According to ref. 11, side-face reinforcement should be placed in beams according to the following rules, as a revision of AASHTO Article 1.5.8(b):

$$s \leqslant d/10 \quad \text{or} \quad s \leqslant 12 \text{ in.}$$

36 in. $< d <$ 100 in. $\qquad \rho_{sk} \geqslant 0.00024(d - 30)$

$\qquad\qquad d >$ 100 in. $\qquad \rho_{sk} \geqslant 0.011 + 0.000058d$

36 in. $> d$ $\qquad \rho_{sk}$ has no requirement (provide temperature reinforcement @ 12 in. max spacing)

ρ_{sk} = skin reinforcement ratio defined as $\dfrac{\text{total } A_{sk}}{2(2c + d_b)(d/2)}$

A_{sk} is equally distributed over the beam faces at a distance $d/2$ adjacent to the main reinforcement at a spacing not to exceed the requirements for s.

A_{sk} = area of skin reinforcement
c = cover to skin reinforcement
d_b = bar diameter of skin reinforcement
d = effective depth
s = vertical spacing of skin reinforcement

Therefore for positive moment

$$s_{\text{max}} = \frac{d}{10} = \frac{44.36}{10} = 4.44 \text{ in.}$$

$$\rho_{sk} = 0.00024(44.36 - 30) = 0.0034 \text{ required}$$

$$\rho_{sk} = \frac{A_{sk}}{2(2c + d_b)} \left(\frac{d}{2} \right)$$

$$A_{sk} = \rho_{sk}\,2(2c + d_b)\ \left(\frac{d}{2}\right)$$

$$= (0.0034)(2)(4.0 + 0.50)(22.18) = 0.68 \text{ in.}^2$$

Use 8-#4, A_s = 1.60 in.^{2} near midspan, as shown in Figure 5.24.
For negative moment

negative moment d = 47.74 in.

$$s_{max} = \frac{d}{10} = 4.77 \text{ in.}$$

$$A_{sk} = (0.0043)(2)(4.0 + 0.50)(23.87) = 0.92 \text{ in.}^2$$

2-#8 should be carried full length along the top of the beam to control
any tensile stresses that might develop before the slab is placed. A
minimum size bar is #4, spaced at the maximum spacing with the #8
which gives 6-#4 and 2-#8, A_s = 2.78 in.^{2} > 0.92 in.^{2} OK. Details
of the steel layout are given in Figure 5.25.
  After completion of the initial superstructure design and before
beginning the pier design (Example 10.2) it is appropriate to design the
pier cap, which is part of the superstructure. The loads are carried into
the pier cap through the girders and then into the pier shaft through the
pier cap. The interior girder loads pass directly through the pier cap
without causing bending, shear, or torsion; the exterior girder loads
cause bending, shear, and torsion. Therefore the cap must be designed
for the loads and effects of the exterior girder. It is important to note
that the frame moments that enter the pier shaft from the exterior

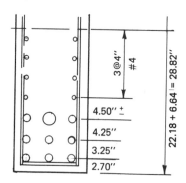

Figure 5.24   Steel details: midspan.

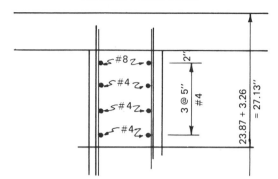

Figure 5.25   Steel details: pier.

girder must be transmitted by torsion through the cap. This significant effect is easily overlooked.

The loads and moments are then computed for an exterior girder. First the dead load reaction and pier moment are computed:

Dead Load (Exterior Girder)

$$R_B \, 1.57(32.5 + 15.0) + 1.46(0.5) + 16.43 + \frac{582}{65} = 100.7 \text{ k}$$

M_{BD} = +129 k-ft from a previous dead load calculation.

The dead load reaction and moment for the interior girder are computed for use in the pier design (Example 10.2).

Dead Load (Interior Girder)

$$R_B = 1.52(32.5 + 15.0) + 2.91(0.5) + 9.91 + \frac{485}{65}$$

$$= 72.2 + 1.46 + 9.91 + 7.46 = 91.0 \text{ k}$$

$$M_{BD} = +172 \text{ k-ft}$$

The live load plus impact loads and moments are tabulated for the various loadings already computed; that is, HS 20 and the overload vehicle. These loads and moments are first tabulated for one line of wheels and then for live load plus impact for one exterior girder by multiplying by the distribution plus impact coefficients.

Live Load + Impact (Exterior Girder)

HS 20

	One Line Wheels		$LL + I$	
Case	R_B (kips)	M_{BD} (k-ft)	R_B (kips)	M_{BD} (k-ft)
II	34.6	−84	51.4	−125
IA	31.8	+68	45.9	+98
IB	23.1	+137	33.3	+198
IIA	38.0	−72	56.5	−107

Overload

	One Line Wheels		$LL + I$	
Case	R_B (kips)	M_{BD} (k-ft)	R_B (kips)	M_{BD} (k-ft)
VI	52.1	−192	67.7	−250
VA	61.2	+97	77.3	+123
VB	40.6	+258	51.3	+326
VIA	72.1	−193	93.7	−251

These loads and moments are then combined with the dead load for design loads and moments. The overload values are factored loads and moments. Again these values are for one exterior girder.

Service Load (Exterior Girder)

HS 20

	R_B (kips)			M_{BD} (k-ft)		
Case	DL	$LL + I$	Σ	DL	$LL + I$	Σ
II	100.7	51.4	152.1	+129	−125	+4
IA	100.7	45.9	146.6	+129	+98	+227
IB	100.7	33.3	134.0	+129	+198	+327
IIA	100.7	56.5	157.2	+129	−107	+22

Overload

Case	R_B (kips)				M_{BD} (k-ft)			
	DL	LL + I	Σ	Ultimate Value $\times 1.3$	DL	LL + I	Σ	$\times 1.3 = M_u$
VI	100.7	67.7	168.4	218.9	+129	-250	-121	-157
VA	100.7	77.3	178.0	231.4	+129	+123	+252	+328
VB	100.7	51.3	152.0	197.6	+129	+326	+455	+592
VIA	100.7	93.7	194.4	252.7	+129	-251	-122	-159

The layout of the cap and the cap dead load shear and moment are shown in Figure 5.26. The cap cross section is 3 ft 3 in. wide × 4 ft 0 in. deep (3 ft $4\frac{1}{2}$ in. below the slab).

The maximum cap bending moment at a service load level is

$$M_T = 157.2(3.50) + 7 = 557 \text{ k-ft}$$

Assuming that $j = 0.87$ and $d = 45$ in.±, we obtain the approximate required reinforcement:

$$A_{s(required)} = \frac{557(12)}{24(0.87)(45)} = 7.12 \text{ in.}^2$$

Try 6-#10, $A_s = 7.62$ in.^{2} Therefore $d = 48 - 2 - 0.62 - 1.27 - 0.64 = 43.47$ in.

Using a rectangular section and Appendix A for k and j values, we compute the stresses (using $f_s = M/A_s jd$ and $f_c = f_s k/[n(1 - k)]$). These stresses are then compared with the allowables:

$$p = \frac{7.62}{(39)(43.47)} = 0.0045 \quad \text{from Appendix A}$$

$$k = 0.247 \qquad j = 0.917$$

$$f_s = \frac{557(12)}{7.62(0.917)(43.47)} = 22.0 \text{ ksi} < 24.0 \text{ ksi}$$

$$f_c = \frac{22.0(0.247)}{9(0.753)} = 802 \text{ psi} < 1300 \text{ psi} = 0.4f_c' \quad \text{OK}$$

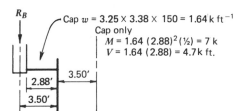

R_B

Cap w = 3.25 × 3.38 × 150 = 1.64 k ft^{-1}
Cap only
M = 1.64 (2.88)^2 (½) = 7 k
V = 1.64 (2.88) = 4.7 k ft.

3.50′

2.88′

3.50′

Figure 5.26 Pier cap loading.

For overload the capacity is

$$a = \frac{A_s f_y}{0.85 f'_c b} = \frac{7.62(60)}{0.85(3.25)(39)} = 4.24 \text{ in.}$$

$$\phi M_n = \phi \left[A_s f_y \left(d - \frac{a}{2} \right) \right]$$

$$= 0.90 \ [(7.62)(60)(43.47 - 2.12)] (\tfrac{1}{12}) = 1418 \text{ k-ft}$$

The required ultimate moment capacity is

$$M_u = (252.7 \times 3.50) + (7 \times 1.3) = 894 \text{ k-ft} < 1418 \text{ k-ft} \quad \text{OK}$$

The required skin reinforcement is computed as for the T-beam:

$$\rho_{sk} = 0.00024(43.47 - 30) = 0.0032$$

$$A_{sk} = (0.032)(2)(4.0 + 0.50)(21.74) = 0.63 \text{ in.}^2$$

$$s = \frac{d}{10} = \frac{43.47}{10} = 4.35 \text{ in.}$$

Use #4 @ $4\frac{1}{2}$ in.± in $d/2$.

Next the cap must be designed for shear and torsion. The stress due to direct shear is

$$v_L = \frac{V}{bd} = \frac{V}{39(43.47)} = \frac{V}{1695}$$

The stress due to torsion is

$$v_t = \frac{3M_T}{x^2 y} = \frac{3(M_T)(12)}{(39)^2 (48)} = \frac{M_t}{2028}$$

where M_t = torsional moment,
 x = smaller dimension of a rectangular shape,
 y = longer dimension of a rectangular shape.

The direct shear and the applied torsional moments are tabulated. The unit shears for each are computed and summed:

Service Load Stresses

Case	V (kips)			M_{BD} (k-ft)	v_L (psi)	v_t (psi)	Σv (psi)
II	152.1 + 4.7	=	156.8	+4	93	2	95
IA	146.6 + 4.7	=	151.3	+227	89	112	201
IB	134.0 + 4.7	=	138.7	+327	82	161	243
IIA	157.2 + 4.7	=	161.9	+22	96	11	107

The allowable shear to be taken by stirrups is $v_L - v_c = 4\sqrt{3250} = 228$ psi allowable.

$$v_c = 0.95\sqrt{f'_c} = 0.95\sqrt{3250} = 54 \text{ psi}$$

max total shear stress = 228 + 54 = 282 psi > 243 psi   OK.
    The required reinforcement for shear is

$$\text{shear } \frac{A_v}{s} = \frac{(v_L - v_c)(b)}{f_s} = \frac{(v_L - 54)(39)}{24{,}000} = \frac{v_L - 54}{615}$$

The required reinforcement for torsion is

$$\text{torsion } \frac{A_v}{s} = \frac{M_T}{0.8 f_s x_1 y_1} = \frac{M_T(12)}{0.8(24)(34.38)(43.38)} = \frac{M}{2386}$$

x_1 = smaller hoop dimension
y_1 = larger hoop dimension

The unit shears and moments are tabulated and expanded to obtain the required area of reinforcing per unit space.

Case	v_L (psi)	M_{BD} (k-ft)	Shear (A_v/s)	Torsion (A_v/s)
II	93	+4	0.0634	0.0017
IA	89	+227	0.0569	0.0951
IB	82	+327	0.0455	0.1370
IIA	96	+22	0.0683	0.0092

Assuming that a double hoop arrangement will be used that will overlap itself, similar to that shown in Figure 5.27, we assume four vertical stirrups legs for shear and two for torsion.

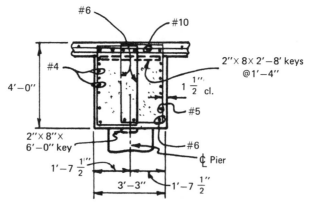

Figure 5.27   Pier cap steel details.

The reinforcement requirements are summed up on a "per leg" basis. The greatest value is selected and a required spacing is computed using #6 stirrup legs.

Case	Per Leg Shear (A_v/s)	Per Leg Torsion (A_v/s)	Per Leg (Ext) $(\Sigma A_v/s)$	$A_v = 0.44$ (s)
II	0.0159	0.0008	0.0167	
IA	0.0142	0.0476	0.0618	
IB	0.0114	0.0685	0.0799	5.5 in.
IIA	0.0171	0.0046	0.0217	

The minimum shear reinforcement requirements are

$$A_s = \frac{50 b_w s}{f_y} = \frac{50(15)(5.5)}{60,000} = 0.07 \text{ in.}^2$$

The maximum spacing is $d/2 = 43.47/2 = 21.7$ in. The overload vehicle is checked for capacity according to ACI 318-77(9).

First check the significance of torsion in computing the concrete shear strength. If the value

$$\phi \left(0.5\sqrt{f_c'} \ \Sigma x^2 y \right) = 0.85[(0.5)\sqrt{3250}(39)^2(48)](\tfrac{1}{12}) = 147 \text{ k-ft}$$

is less than M_{BD}, then torsion controls. If it can be seen that torsion does control, the equation that must be used from ACI 318-77 is eq. 11-5, which is

$$V_c = \frac{2\sqrt{f_c'}\ b_w d}{\sqrt{1 + [2.5 C_t (Tu/Vu)]^2}}$$

$$C_t = \frac{b_w d}{x^2 y} = \frac{39(43.47)}{(39)^2(48)} = 0.0232$$

$$V_c = \frac{2\sqrt{f'_c}\, b_w d}{\sqrt{1 + [2.5C_t(Tu/Vu)]^2}} = \frac{2\sqrt{3250}\,(39)(43.47)}{\sqrt{1 + [(2.5)(0.0232)(Tu/Vu)]^2}}$$

$$= \frac{193{,}297}{\sqrt{1 + [(0.580)(Tu/Vu)]^2}}$$

The required shear strength of the stirrups is

$$V_u \leqslant \phi(V_c + V_s) \quad \text{or} \quad V_u = R_B + (1.3 \times 4.7)$$

$$V_s = \frac{V_u}{\phi} - V_c \qquad \phi = 0.85$$

The required area of reinforcement for this shear strength is

$$V_s = \frac{A_v f_y d}{s}$$

$$\text{(per leg)} \quad \frac{A_v}{s} = \frac{V_s}{f_y d} = \frac{V_s}{60(43.47)(4)} = \frac{V_s}{10{,}433}$$

Based on these equations, reinforcement requirements for each live load case are computed and tabulated:

Case	T_u (k-ft)	V_u (kips)	T_u (k-in.)	V_c (kips)	$\dfrac{V_u}{\phi} - V_c$ (kips)	$\dfrac{A_v}{s}$ (per leg)
VI	157	225.0	1884	173.9	90.8	0.0087
VA	328	237.5	3936	139.4	140.0	0.0134
VB	592	203.7	7104	85.7	153.9	0.0148
VIA	159	258.8	1908	177.7	126.8	0.0122

The torsional strength required for the hoops is

$$T_s = \frac{T_u}{\phi} - V_c$$

By using ACI eq. 11-23, the area of reinforcement required per unit spacing is

$$T_s = \frac{A_t \alpha_t x_1 y_1 f_y}{s}$$

$x_1 = 34.38$ (shortest hoop dimension)
$y_1 = 43.38$ (longest hoop dimension)

where $\alpha_t = 0.66 + 0.33 \left(\dfrac{y_1}{x_1}\right),$

$$\alpha_t = 0.66 + 0.33 \left(\dfrac{43.38}{34.38}\right) = 1.08,$$

$$\frac{A_t}{s} = \frac{T_s}{\alpha_t x_1 y_1 f_y} = \frac{T_s}{1.08(34.38)(43.38)(60)} = \frac{T_s}{96643} \quad \text{per leg.}$$

The concrete torsional strength is

$$T_c = \frac{0.8\sqrt{f_c'}\,\sum x^2 y}{\sqrt{1 + (0.4 V_u / C_t T_u)^2}} = \frac{0.8\sqrt{3250}(39)^2(48)}{\sqrt{1 + (0.4 V_u / 0.0232 T_u)^2}}$$

$$= \frac{3330}{\sqrt{1 + [17.24(V_u / T_u)]^2}}$$

On the basis of these equations, the required area of reinforcement for each live load is calculated and tabulated:

Case	V_u (k)	T_u (k-in.)	T_c (k-in.)	$T_s = \dfrac{T_u}{\phi} - T_c$ (k-in.)	$\dfrac{A_t}{s}$ (per leg)
VI	225.0	1884	1455	761	0.0079
VA	237.5	3936	2308	2323	0.0240
VB	203.7	7104	2985	5373	0.0556
VIA	258.8	1908	1309	936	0.0097

The requirements for shear and torsion are summed and tabulated. The largest requirement, assuming #6 reinforcing bars, is expanded to the following spacing:

Case	$\dfrac{A_v}{s}$	$\dfrac{A_t}{s}$	$\sum\dfrac{A}{s}$	$\dfrac{A_v}{s} = 0.44$ in.^{2}
VI	0.0087	0.0079	0.0166	
VA	0.0134	0.0240	0.0374	
VB	0.0148	0.0556	0.0704	6.25 < 5.5 in. required with
VIA	0.0122	0.0097	0.0219	HS 20 service load

These results indicate that service load controls the design. The longitudinal reinforcement required for torsion is

$$A_l = 2A_t \left(\frac{x_1 + y_1}{s}\right) = 2(0.44) \left(\frac{34.38 + 43.38}{5.5}\right) = 12.44 \text{ in.}^2$$

The longitudinal reinforcement supplied is

$$\left.\begin{array}{l} 6\text{-}\#10, A_s = 7.62 \\ 4\text{-}\#6, A_s = 1.76 \\ 8\text{-}\#5, A_s = 2.48 \\ 4\text{-}\#4, A_s = 0.80 \end{array}\right\} \sum A_t = 12.66 \text{ in.}^2 > 12.44 \text{ in.}^2 \quad \text{OK}$$

After the substructure is designed the designer will have reached a point at which he is assured that the sections selected for the frame analysis are satisfactory. Therefore it is reasonable to complete the superstructure design.

The dead load moments are tabulated at the tenth points along the span and at the midpoint of the cantilever, as given in Table 5.1. The

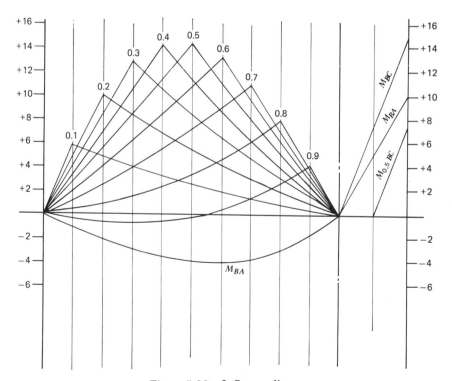

Figure 5.28 Influence lines.

Table 5.1 Exterior Girder

Point	+ M_{unif} (k-ft)	+ M_{conc} (k-ft)	Negative Moment (k-ft)	M_T (k-ft)
A			0	0
0.1	+298	+5	−58	+245
0.2	+531	+9	−116	+424
0.3	+696	+14	−175	+535
0.4	+796	+19	−233	+582
0.5	+829	+24	−291	+562
0.6	+796	+19	−349	+466
0.7	+696	+14	−407	+303
0.8	+531	+9	−466	+74
0.9	+298	+5	−524	−221
BA			−582	−582
BC			−453	−453
0.5BC			−196	−196
CB			−29	−29

Uniform wl^2 = 6633.
Concentrated $Pl/4$ = 23.7

$M_{0.5BC} = 1.57(7.5)^2(\tfrac{1}{2})$ = −44 k-ft   sum = −196
$16.43(7.5 + 1.77) = -152$ k-ft   k-ft

$M_{CB} = 16.43(1.77)$ = −29 k-ft

Interior Girder

Point	+ M_{unif} (k-ft)	+ M_{conc} (k-ft)	Negative Moment (k-ft)	M_T (k-ft)
AB				0
0.1	+289	+9	−48	+250
0.2	+514	+19	−97	+436
0.3	+674	+28	−146	+556
0.4	+771	+38	−194	+615
0.5	+803	+47	−242	+608
0.6	+771	+38	−291	+518
0.7	+674	+28	−340	+362
0.8	+514	+19	−388	+145
0.9	+298	+9	−436	−128
BA			−485	−485
BC			−313	−313
0.5BC			−111	−111
CB			0	0

Uniform wl^2 = 1.52(65.0)^2 = 6422
Concentrated $pl/4$ = 2.91(66.0)/4 = 47.3

$M_{0.5BC} = 1.52(7.5)^2(\tfrac{1}{2})$ = −43
$9.91(7.5 - 0.625) = -68$ sum = −111 k-ft

simple span positive moments are superimposed on the negative moments from the moment distribution to give the final dead load moment for exterior and interior girders. Note that the simple span moments are a function of wl^2 for the uniform dead load and of $Pl/4$ for the concentrated dead load.

The influence lines are computed as in Example 4.1 and are shown in Figure 5.28.

The maximum live load plus impact moments are obtained from the influence lines (Fig. 5.28) as done in Example 4.1 and are combined with the dead load moments to give the maximum for positive or negative moments for exterior or interior girders (see Table 5.2).

The initial design indicated that 8-#11 and 1-#14 for positive moment reinforcement at the maximum section should be used. Cutoffs are 1-#14, 3-#11, and 2-#11; 3-#11 extend into the negative moment area for the compressive reinforcement assumed in the initial design of the negative moment section.

Table 5.2   Maximum Moments. Service Load—HS 20[a]

Point	Exterior Girder			Interior Girder		
	M_{DL}	M_{LL+I}	M_T	M_{DL}	M_{LL+I}	M_T
Maximum Positive Moment						
A						
0.1	+245	+242	+487	+250	+242	+492
0.2	+424	+410	+834	+436	+410	+846
0.3	+535	+505	+1040	+556	+505	+1061
0.4	+582	+557	+1139	+615	+557	+1172
0.5	+562	+550	+1112	+608	+550	+1158
0.6	+466	+521	+987	+518	+521	+1039
0.7	+303	+430	+733	+362	+430	+792
0.8	+74	+294	+368	+145	+294	+439
0.9	-221	+120	-101	-128	+120	-8
B					Controls _____↑	
Maximum Negative Moment						
0.5	+562	-128		+608	-128	
0.6	+466	-153		+518	-153	
0.7	+303	-180	+123	+362	-180	+182
0.8	+74	-205	-131	+145	-205	-60
0.9	-221	-230	-451	-128	-230	-358
BA	-582	-343	-925	-485	-343	-828
BC	-453	-380	-833	-313	-380	-693
0.5BC	-196	-178	-374	-111	-178	-289
CB	-29	0	-29	-29	0	0
		↑_____ Controls				

[a] All moments are in kip-ft.

Computation of the positive resisting moments based on the allowable reinforcing steel stress follows:

Resisting Moments

Positive Moment

Use $j = 0.928$ for the maximum @ all sections (conservative).

Section 1.   8-#11 and 1-#14, $A_s = 14.73$ in.^{2}

$$cg = 6.64 \text{ in.}$$
$$d = 44.36 \text{ in.}$$
$$M_R = 14.73(0.928)(44.36)(24)(\tfrac{1}{12}) = 1213 \text{ k-ft}$$

Section 2.   8-#11, $A_s = 12.48$ in.^{2}

$$cg = 6(4.45) \quad = 26.70$$
$$2(10.45) = \underline{20.90}$$
$$47.60 \div 8 = 5.95 \text{ in.}$$
$$d = 51 - 5.95 = 45.05 \text{ in.}$$
$$M_R = 12.48(0.928)(45.05)(24)(\tfrac{1}{12}) = 1044 \text{ k-ft}$$

Section 3.   5-#11, $A_s = 7.80$ in.^{2}

$$cg = 3(2.70) = \quad 8.10$$
$$2(6.20) = \underline{12.40}$$
$$20.50 \div 5 = 4.10 \text{ in.}$$
$$d = 51 - 4.10 = 46.90 \text{ in.}$$
$$M_R = 7.8(0.928)(46.9)(24)(\tfrac{1}{12}) = 679 \text{ k-ft}$$

2−#11
3−#11   3.50″
             2.70″

Section 4.   3-#11, A_s = 4.68 in.^{2}

$$d = 51.0 - 2.70 = 48.30 \text{ in.}$$
$$M_R = 4.68(0.928)(48.30)(24)(\tfrac{1}{12}) = 420 \text{ k-ft}$$

3−#11   2.70″

On the basis of the initial design of 9-#10 for negative moment tensile steel and the allowable reinforcing steel stress, we compute the resisting moments for 1-#10 and then for a number (N) of #10 bars. The most conservative value of j is used for computing all areas of reinforcement.

Negative Moment

$$j = 0.95$$
$$d = 47.74 \text{ in.}$$
$$M_R = N(1.27)(0.905)(47.74)(24)(\tfrac{1}{12}) = 109.7N$$

N	M_C
9	988
8	878
6	658
4	439
2	219
1.24	136 $\longleftarrow$ 2-#8 = 1.24-#10

The required bar extension beyond the moment requirement, according to AASHTO Article 1.5.13, is computed next.

Bar Extensions

$$d = 3.8 \text{ ft}, \ 20\phi = 2.4 \text{ ft} \ (\#11), \ \frac{s}{20} = 3.2 \text{ ft}$$

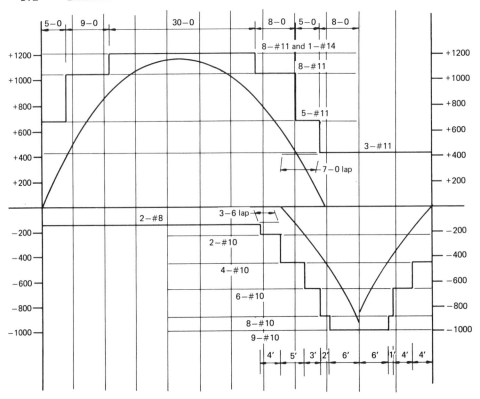

Figure 5.29   Maximum allowable moments.

Also AASHTO Article 1.5.13 also requires that one-third of the reinforcement extend into the simple support and one-quarter through the continuous support.

The maximum moments are plotted against the resisting moments (Figure 5.29) and cutoff locations are selected. Development length will not affect the results; therefore it is ignored. It can be seen that the interior girder controls for positive moment, the exterior girder, for negative moment. These are the moments plotted. All girders are made the same, however.

Similarly, the overload vehicle moments are taken from the influence lines and combined and factored with the dead load moments to give M_u. This is done for positive and negative moments and interior and exterior girders, as shown in Table 5.3.

The resisting capacity moments are computed for the same sections as for service load.

Table 5.3   Maximum Moments. Overload Vehicle[a]

Point	Exterior Girder			Interior Girder		
	M_{DL}	M_{LL+I}	$M_u = 1.3 \times M_T$	M_{DL}	M_{LL+I}	$M_u = 1.3 \times M_T$
			Maximum Positive Moment			
A						
0.1	+245	+419	+863	+250	+340	+767
0.2	+424	+714	+1479	+436	+578	+1318
0.3	+535	+889	+1851	+556	+720	+1659
0.4	+582	+976	+2025	+615	+791	+1828
0.5	+562	+993	+2022	+608	+804	+1836
0.6	+466	+897	+1772	+518	+726	+1617
0.7	+303	+760	+1382	+362	+616	+1272
0.8	+74	+532	+788	+145	+431	+749
0.9	-221	+225	+5	-128	+182	+70
B			↑___ Controls			
			Maximum Negative Moment			
0.5	+562	-256		+608	-207	
0.6	+466	-308	+205	+518	-250	
0.7	+303	-359	-73	+362	-291	+92
0.8	+74	-411	-438	+145	-333	-244
0.9	-221	-462	-888	-128	-374	-653
BA	-582	-543	-1462	-485	-440	-1203
BC	-453	-792	-1618	-313	-641	-1110
0.5BC	-196	-315	-664	-111	-255	-476
CB	-29	0	-38	0	0	0
			↑___ Controls			

[a]Moments are in kip-ft.

OVERLOAD: CAPACITY ANALYSIS

Resisting Moments

Positive Moment

$$a = \frac{A_s f_y}{0.85 f'_c b} = \frac{A_s(60)}{0.85(3.25)(70)} = 0.3103 A_s$$

Section 1.   8-#11 and 1-#14, $A_s = 14.73$ in.^{2}

$d = 44.36$ in.

$a = 0.3103(14.73) = 4.57$

$$\phi M_n = 0.90 \left[(60)(14.73) \left(44.36 - \frac{4.57}{2} \right) \right] \left(\frac{1}{12} \right) = 2789 \text{ k-ft}$$

Section 2.   8-#11, A_s = 12.48 in.^{2}

d = 45.05 in.

a = 0.3103(12.48) = 3.87 in.

$$\phi M_n = 0.90 \left[(60)(12.48) \left(45.05 - \frac{3.87}{2} \right) \right] \left(\frac{1}{12} \right) = 2421 \text{ k-ft}$$

Section 3.   5-#11, A_s = 7.80 in.^{2}

d = 46.90 in.

a = 0.3103(7.80) = 2.42 in.

$$\phi M_n = 0.90 \left[(60)(7.80) \left(46.90 - \frac{2.42}{2} \right) \right] \left(\frac{1}{12} \right) = 1604 \text{ k-ft}$$

Section 4.   3-#11, A_s = 4.68 in.^{2}

d = 48.30 in.

a = 0.3103(4.68) = 1.45 in.

$$\phi M_n = 0.90 \left[(60)(4.68) \left(48.30 - \frac{1.45}{2} \right) \right] \left(\frac{1}{12} \right) = 1002 \text{ k-ft}$$

Negative Moment

N = 9-#10

d = 47.74 in.

$$a = \frac{A_s (60)}{0.85(3.25)(15)} = 1.4480 A_s$$

A_s = 11.43 in.^{2}

a = 1.4480(11.43) = 16.55 in.

$$\phi M_n = 0.90 \left[(60)(11.43) \left(47.74 - \frac{16.55}{2} \right) \right] \left(\frac{1}{12} \right) = 2030 \text{ k-ft}$$

$N = 8 - \#10$

$A_s = 10.16 \text{ in.}^2$

$a = 1.4480(10.16) = 14.71 \text{ in.}$

$\phi M_n = 0.90 \left[(60)(10.16) \left(47.74 - \dfrac{14.71}{2} \right) \right] \left(\dfrac{1}{12} \right) = 1846 \text{ k-ft}$

$N = 6 - \#10$

$A_s = 7.62 \text{ in.}^2$

$a = 1.4480(7.62) = 11.03 \text{ in.}$

$\phi M_n = 0.90 \left[(60)(7.62) \left(47.74 - \dfrac{11.03}{2} \right) \right] \left(\dfrac{1}{12} \right) = 1448 \text{ k-ft}$

$N = 4 - \#10$

$A_s = 5.08 \text{ in.}^2$

$a = 1.4480(5.08) = 7.36 \text{ in.}$

$\phi M_n = 0.90 \left[(60)(5.08) \left(47.74 - \dfrac{7.36}{2} \right) \right] \left(\dfrac{1}{12} \right) = 1007 \text{ k-ft}$

$N = 2 - \#10$

$A_s = 2.54 \text{ in.}^2$

$a = 1.4480(2.54) = 3.68 \text{ in.}$

$\phi M_n = 0.90 \left[(60)(2.54) \left(47.74 - \dfrac{3.68}{2} \right) \right] \left(\dfrac{1}{12} \right) = 525 \text{ k-ft}$

$N = 1.24 - \#10 \ (2 - \#8)$

$A_s = 1.58 \text{ in.}^2$

$a = 1.4480(1.58) = 2.29 \text{ in.}$

$\phi M_n = \left[(60)(1.58) \left(47.74 - \dfrac{2.29}{2} \right) \right] \left(\dfrac{1}{12} \right) = 331 \text{ k-ft}$

The maximum values of M_u are plotted against the resisting capacities with the same bar extension and cutoffs used for service-load design

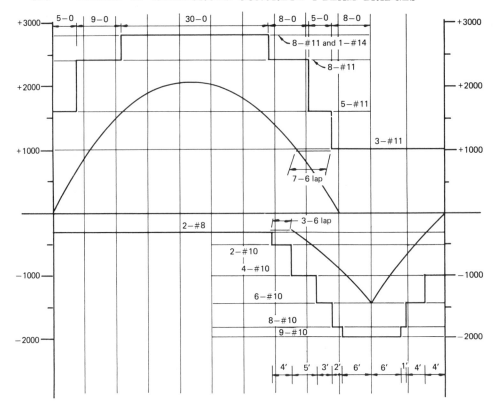

Figure 5.30   Maximum ultimate moments.

(Figure 5.30). It can be seen that the overload vehicle does not control the design. In this case the exterior girder controls both positive and negative moment; therefore these are the moments plotted. All girders are made the same.

The dead load shears are tabulated at the tenth points, at midlength of the cantilever, and at 1 ft BC (1 ft 0 in. from B toward C). The tabulation is done incrementally by listing the uniform dead load simple beam shears, the concentrated simple beam shears, and the shear due to unequal end moments. These increments are then summed to obtain the total dead load shears, which are computed for the exterior and interior girders, shown in Table 5.4.

Next, the live load shears are computed. For moderate or short span lengths the truck loading produces the maximum shear, which can be seen by examining the tables of HS 20 shears in Appendix A of the AASHTO specifications, where the truck loading controls up to a 120-ft span. We can easily establish the correct vehicle placement by

Table 5.4   Dead Load Shears

Point	Uniform (1.57 k-ft^{-1})	Concentrated	Moment V	V_{DL} (kips)
		Exterior Girder		
AB	+51.0	+0.7	−9.0	+42.7
0.1	+40.8	+0.7	−9.0	+32.5
0.2	+30.6	+0.7	−9.0	+22.3
0.3	+20.4	+0.7	−9.0	+12.1
0.4	+10.2	+0.7	−9.0	+1.9
0.5	0	±0.7	−9.0	−8.3, −9.7
0.6	−10.2	−0.7	−9.0	−19.9
0.7	−20.4	−0.7	−9.0	−30.1
0.8	−30.6	−0.7	−9.0	−40.3
0.9	−40.8	−0.7	−9.0	−50.5
BA	−51.0	−0.7	−9.0	−60.7
BC	+23.6	+16.4		+40.0
$1'BC$	+22.0	+16.4		+38.4
$0.5BC$	+11.8	+16.4		+28.2
CB	0	+16.4		+16.4
		Interior Girder		
AB	+49.4	+1.5	−7.5	+43.4
0.1	+39.5	+1.5	−7.5	+33.5
0.2	+29.6	+1.5	−7.5	+23.6
0.3	+19.8	+1.5	−7.5	+13.8
0.4	+9.9	+1.5	−7.5	+3.9
0.5	0	±1.5	−7.5	−6.0, −9.0
0.6	−9.9	−1.5	−7.5	−18.9
0.7	−19.8	−1.5	−7.5	−28.8
0.8	−29.6	−1.5	−7.5	−38.6
0.9	−39.5	−1.5	−7.5	−48.5
BA	−49.4	−1.5	−7.5	−58.4
BC	+22.8	9.9		+32.7
$1'BC$	+21.3	9.9		+31.2
$0.5BC$	+11.4	9.9		+21.3
CB	0	9.9		+9.9

visualizing the shape of the influence line for shear at any point. A typical line for shear at point x is shown in Figure 5.31.

For the curve shown it is obvious that the heaviest wheel should be placed at point x with the other wheels as close behind as possible toward support B. As we move closer to midspan, the truck direction should be reversed. The exact point for reversing truck direction depends on the magnitude and sign of the shear due to the different end

Figure 5.31   Influence line for shear.

moments. The space on the tables that follow indicates the location of the truck reversal.

With these principles in mind, the truck is placed on the influence lines and ordinates for the end moment are tabulated. The end moment and resulting shear are then computed (Table 5.5).

The simple beam shear and the shear due to unequal end moments are then superimposed to obtain the total live load shear per line of wheels, as given in Table 5.6. This is expanded to include the live load plus impact per exterior or interior girder by multiplying the shear by the distribution plus impact coefficients (1.443 and 1.486).

The maximum total shear, given in Table 5.7, is tabulated with the interior girder controlling up to $0.4AB$, where from $0.5AB$ to the cantilever tip the exterior girder controls. The unit shear is computed by using $v = V_T/b_w d$ with $b_w = 15$ in.

Required spacing $s = A_v f_v/[(v - v_c)(b_w)]$

$$v_c = 0.95\sqrt{f'_c} = 54 \text{ psi}$$

#4 stirrups, $A_v = 0.40$ in.^{2} (2 legs)

$$s = \frac{0.40(24,000)}{(v - v_c)(15)} = \frac{640}{v - v_c}$$

Table 5.5   Shear Design. Service Load—HS 20

Point	M_{BA} Ordinates			M_{BA} (kip-ft)	Moment V (kips)
	16 k	16 k	4 k		
AB		-2.2	-3.7	-50	-0.8
0.1	-1.05	-3.0	-4.1	-81	-1.2
0.2	-2.04	-3.6	-4.1	-107	-1.6
0.3	-2.90	-4.0	-3.6	-125	-1.9
0.4	-3.57	-4.1	-2.8	-134	-2.1
0.5	-3.99	-2.8	-0.8	-112	-1.7
0.6	-4.08	-3.3	-1.8	-125	-1.9
0.7	-3.79	-3.9	-2.7	-134	-2.1
0.8	-3.06	-4.1	-3.3	-128	-2.0
0.9	-1.82	-3.9	-3.9	-107	-1.6
BA		-3.2	-4.1	-68	-1.0

Table 5.6

Point	SBV	Moment V	One-Line Wheels V	Exterior and Interior V_{LL+I}
AB	+30.8	−0.8	+30.0	+43.3
0.1	+27.2	−1.2	+26.0	+37.5
0.2	+23.6	−1.6	+22.0	+31.7
0.3	+20.0	−1.9	+18.1	+26.1
0.4	+16.4	−2.1	+14.3	+20.6
0.5	−12.8	−1.7	−14.5	−20.9
0.6	−16.4	−1.9	−18.3	−26.4
0.7	−20.0	−2.1	−22.1	−31.9
0.8	−23.6	−2.0	−25.6	−36.9
0.9	−27.2	−1.6	−28.8	−41.6
BA	−30.8	−1.0	−31.8	−45.9
BC	+32.0 k		+32.0	+47.6
1'BC	+32.0, +16.0		+32.0, +16.0	+47.6, +23.8
0.5BC	+16.0		+16.0	+23.8
CB	+16.0		+16.0	+23.8

Table 5.7

Point	V_{DL}	V_{LL+I}	V_T (kips)	d (in.)	v (psi)
		Exterior Girder			
AB	+43.4	+43.3	+86.7	46.90	123
0.1	+33.5	+37.5	+71.0	45.05	105
0.2	+23.6	+31.7	+55.3	45.05	82
0.3	+13.8	+26.1	+39.9	44.36	60
0.4	+3.9	+20.6	+24.5	44.36	37
		Interior Girder			
0.5	−8.3, −9.7	−20.9	−30.6	44.36	46
0.6	−19.9	−26.4	−46.3	44.36	70
0.7	−30.1	−31.9	−62.0	45.05	92
0.8	−40.3	−36.9	−77.2	46.90	110
0.9	−50.5	−41.6	−92.1	47.74	127
BA	−60.7	−45.9	−106.6	47.74	149
BC	+40.0	+47.6	+87.6	47.74	122
1'BC	+38.4	+47.6, +23.8	+86.0, +62.2	47.74	120, 87
0.5BC	+28.2	+23.8	+52.0	47.74	73
CB	+16.4	+23.8	+40.2	47.74	56

On the basis of these formulas the spacing requirements are computed as

Point	$v - v_c$	#4 $s_{required}$
AB	69	9.3
0.1	51	12.5
0.2	28	22.9
0.3	6	
0.4	—	
0.5	—	
0.6	16	40.0
0.7	38	16.8
0.8	56	11.4
0.9	73	8.8
BA	95	6.7
BC	68	9.4
$1'BC$	66, 33	9.7, 19.4
$0.5BC$	19	33.7
CB	2	

The minimum reinforcement requirements are

$$s = \frac{A_v f_y}{50 b_w} = \frac{0.40(60,000)}{50(15)} = 32 \text{ in.}$$

$$\text{max spacing} = \frac{d}{2} = 1 \text{ ft.-10 in.}$$

The actual spacing is then plotted against the allowables, as shown in Figure 5.32. Note that this spacing is revised as the result of the overload shears on the cantilever.

The overload vehicle shears, are computed by using the procedure given for HS 20 shear, and are listed in Table 5.8.

These shears are tabulated, summed, factored, and converted to unit shears. The exterior girder controls for the overload vehicle. For unit shear

$$v_u = \frac{V_u}{\phi b_w d} \qquad \phi = 0.85 \qquad b_w = 15 \text{ in.}$$

Required spacing

$$s = \frac{A_v f_y}{(v_u - v_c) b_w} = \frac{0.40(60,000)}{(v_u - v_c)(15)} = \frac{1600}{v_u - v_c}$$

$$v_c = 2\sqrt{f_c'} = 2\sqrt{3250} = 114 \text{ psi}$$

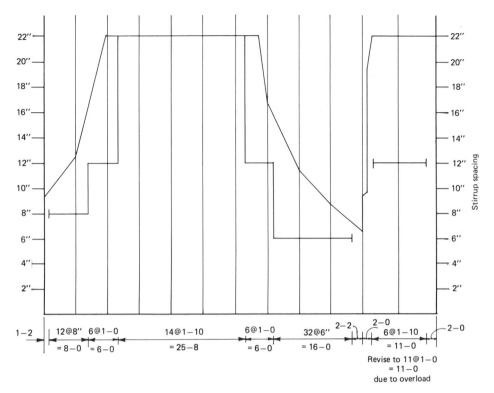

Figure 5.32 Stirrup spacing requirements: service load.

Table 5.8 Overload–Capacity

Point	V_{DL}	V_{LL+I}	V_T (kips)	$V_u = 1.3 \times V_T$ (kips)	d	$v_u = \dfrac{V_u}{\phi b_w d}$ (psi)
AB	+42.7	+74.3	+117.0	152.1	46.90	254
0.1	+32.5	+64.7	+97.2	126.4	45.05	220
0.2	+22.3	+55.3	+77.6	100.9	45.05	176
0.3	+12.1	+46.2	+58.3	75.8	44.36	134
0.4	+1.9	+37.4	+39.3	51.1	44.36	90
0.5	-9.7	-38.3	-48.0	62.4	44.36	110
0.6	-19.9	-47.1	-67.0	87.1	44.36	154
0.7	-30.1	-55.6	-85.7	111.4	45.05	194
0.8	-40.3	-63.4	-103.7	134.8	46.90	225
0.9	-50.5	-70.4	-121.2	157.6	48.30	256
BA	-60.7	-77.4	-138.1	179.5	47.74	295
BC	+40.0	+72.8	+112.8	146.6	47.74	241
$0.5BC$	+28.2	+59.8	+88.0	114.4	47.74	188
$10.5BC$	+23.5	+59.8, +29.9	+83.3, +53.4	108.3, 69.4	47.74	178, 114
CB	+16.4	+29.9	+46.3	60.2	47.74	99

181

On the basis of these equations the required spacing for the overload vehicle is computed in Table 5.9.

The actual spacing is then plotted against these requirements. Revised spacing on the cantilever for the overload vehicle is reflected on the spacing diagram for the HS 20 service load design. The overload plot is based on the capacity shown in Figure 5.33.

Example 5.2.   The T-beam superstructure designed by service load in Example 5.1 is designed here by load factor design in accordance with AASHTO Articles 1.5.30 to 1.5.40. The pier for this bridge is designed by load factor design in Example 10.1 and the deck slab, by load factor design in Example 4.4.   The service load level design vehicle for this structure is HS 20 and the overload design live load is 1.67 HS 20 loadings (β_L = 1.67) or 1.0 overload vehicle loadings (β_L = 1.0). The overload vehicle is shown in Figure 5.5.   The dimensions and layout of the bridge are shown in Figures 5.2 to 5.4.

The schematic layout, the frame constants, and the moment computation are the same as in Example 5.1, as are the wheel distribution and impact computations.   The service load level moments are computed for exterior and interior girders at the critical locations in Table 5.10.

The HS 20 moments are factored, summed with the factored dead load moments, and tabulated at the critical locations for exterior and interior girders (Table 5.11).   These ultimate moments are compared

Table 5.9

Point	$v_u - v_c$	#4 s
AB	140	11.4 in.
0.1	106	15.1
0.2	62	25.8
0.3	20	
0.4		
0.5		
0.6	40	40.0
0.7	80	20.0
0.8	111	14.4
0.9	142	11.3
BA	181	8.8 in.
BC	127	12.6 in.
0.5BC	74	21.6 in.
10.5BC	64, 0	25.0 in.
BC	0	

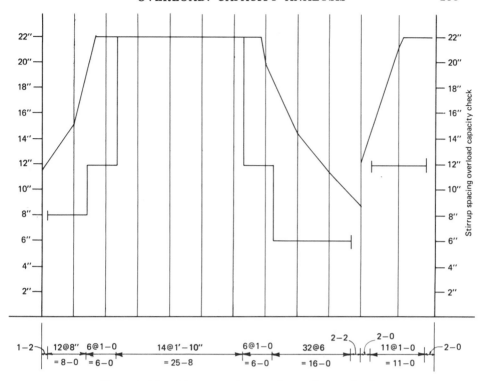

Figure 5.33 Stirrup spacing requirements: overload.

with the ultimate values computed by using the overload vehicle to determine the controlling loading. The overload M_u values are factored and tabulated at the critical locations in Table 5.12.

The minimum reinforcing requirements of AASHTO Article 1.5.7 were determined to be insignificant in Example 5.1.

Table 5.10

	DL	One Line Wheels-LL	Distr. & Imp. Coeff.		LL + I	M_T (k-ft)
Service Load Level (HS 20) (Exterior Girder)						
$M_{0.4AB}$	+582	+385	X 1.443	=	+556	+1138
M_{BA}	-582	-231	X 1.464	=	-338	-920
M_{BC}	-453	-256	X 1.486	=	-380	-833
Service Load Level (HS 20) (Interior Girder)						
$M_{0.4AB}$	+615	+385	X 1.443	=	+556	+1171
M_{BA}	-485	-231	X 1.464	=	-338	-823
M_{BC}	-313	-256	X 1.486	=	-380	-693

Table 5.11

	1.3DL	2.17 (LL + I)	M_u (k-ft)
Load Factor (HS 20) (Exterior Girder)			
$M_{0.4AB}$	+757	+1207	+1964
M_{BA}	-757	-733	-1490
M_{BC}	-589	-825	-1414
Load Factor (HS 20) (Interior Girder)			
$M_{0.4AB}$	+800	+1207	+2007
M_{BA}	-630	-733	-1363
M_{BC}	-407	-825	-1232

Table 5.12

	DL	1.3DL	One-Line Wheels-LL	Distr. & Imp. Coef.		LL + I	1.3 × LL + I	M_u (k-ft)
Overload (Exterior Girder)								
$M_{0.4AB}$	+582	+757	+764 ×	1.263	=	+965	+1254	+2011
M_{BA}	-582	-757	-395 ×	1.300	=	-514	-668	-1425
M_{BC}	-453	-589	-586 ×	1.300	=	-762	-990	-1580
Overload (Interior Girder)								
$M_{0.4AB}$	+615	+800	+764 ×	1.023	=	+782	+1017	+1817
M_{BA}	-485	-630	-395 ×	1.053	=	-416	-541	-1171
M_{BC}	-313	-407	-586 ×	1.053	=	-617	-802	-1209

The capacity analysis is started at 0.4AB by approximating the required tensile reinforcement. Assume $d - a/2 = 42.0$ in.±.

$$A_{s(required)} = \frac{M}{df_y \phi}$$

Positive Moment Section @ 0.4 AB ($M_u = +2011^{k-ft}$)

Capacity Analysis

$$\text{approx } A_{s(required)} = \frac{2011(12)}{(42.0\pm)(60)(0.90)} = 10.64 \text{ in.}^2$$

Try 9–#10, $A_s = 11.43$ in.^{2}

Next it is necessary to check the need to analyze the section as a T-section. The area of reinforcement required to develop the compressive strength of the overhanging flanges is

$$A_{sf} = \frac{0.85 f'_c (b - b_w) h_f}{f_y} = \frac{0.85(3.25)(70.0 - 15)(7.5)}{60}$$

$$= 18.99 \text{ in.}^2 > A_{s(\text{provided})} = 11.43 \text{ in.}^2$$

Therefore we have a rectangular section for analysis.

The capacity for a rectangular section is then checked. The reinforcing for the section is as shown: $d = 51.0 - 3.25 - 2.64 = 45.11$ in.

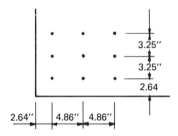

The capacity is

$$a = \frac{A_s f_y}{0.85 f'_c b} = \frac{11.43(60)}{0.85(3.25)(70.0)} = 3.54 \text{ in.}$$

$$\phi M_n = \phi \left[A_s f_y \left(d - \frac{a}{2} \right) \right]$$

$$= 0.90[(10.64)(60)(45.11 - 1.77)](\tfrac{1}{12})$$

$$= 2229 \text{ k-ft} > +2011 \text{ k-ft} = M_u \ @ \ 0.4AB \quad \text{OK}$$

The analysis of the section @ $0.4AB$ is continued by checking serviceability. The maximum and minimum service load level moments at $0.4AB$ are

$$M_{\max} = +1171 \text{ k-ft}$$

$$M_{\min} = +582 + 0.4(1.486 \times 172)$$

$$= +480 \text{ k-ft}$$

k and j are obtained by using Appendix B:

$$\frac{t}{d} = \frac{7.50}{45.11} = 0.166$$

$$pn = \frac{11.43(9)}{70(45.11)} = 0.0326 \qquad k = 0.234 \quad j = 0.932$$

The maximum and minimum steel stresses are

$$f_s = \frac{1171(12)}{11.43(0.932)(45.11)} = 29.2 \text{ ksi max}$$

$$f_s = \frac{480(12)}{11.43(0.932)(45.11)} = 12.0 \text{ ksi min}$$

The stress range is

$$f_f = 29.2 - 12.0 = 17.2 \text{ ksi}$$

The allowable stress range is

$$f_f = 21 - 0.33 f_{min} + 8 \left(\frac{r}{h}\right)$$

$$= 21 = 0.33(12.0) + 8(0.3) = 19.4 \text{ ksi} > 17.2 \text{ ksi allowable} \quad \text{OK}$$

The crack control criteria in AASHTO Article 1.5.39 is checked.

A = effective tension area of concrete surrounding the flexural tension reinforcement and having the same centroid as that reinment, divided by the number of bars or wires (in.^{2}). When the flexural reinforcement consists of several bar or wire sizes the number of bars or wires shall be computed as the total area of reinforcement divided by the area of the largest bar or wire used.

d_c = thickness of concrete cover measured from extreme tension fiber to center of bar or wire located closest thereto (in.).

Use $z = 130$ k-in.$^{-1}$

$$A = 15.0(11.78) = 176.70 \div 9 = 19.6 \text{ in.}^2$$

$$d_c = 2.64 \text{ in.}$$

$$f_s = \frac{z}{\sqrt[3]{d_c A}} \quad \text{(maximum allowable service load stress)}$$

$$f_s = \frac{130}{\sqrt[3]{2.64(19.6)}} = 34.9 \text{ ksi} > 29.2 \text{ ksi} \quad \text{OK}$$

The maximum reinforcement requirements in AASHTO Article

1.5.32(A)(1), which limits the reinforcement ratio to 75% of the balanced reinforcement ratio, are checked. Rectangular section

$$b = 70.0 \text{ in.}$$

$$0.75\rho_b = 0.0174$$

$$\rho = \frac{11.43}{70.0(45.11)} = 0.0036 < 0.0174 \quad \text{OK}$$

Next we design the negative moment section at B. The maximum ultimate moment is at BC, with the overload vehicle applied to the exterior girder (@ BC, $M_u = -1580$ k-ft).

An approximate required reinforcing area is computed by using $d - a/2 = 45.0$ in.$\pm$ and

$$\text{approx } A_{s(\text{required})} = \frac{M_u}{f_y \, d\phi} = \frac{1580(12)}{60(45.0\pm)(0.90)} = 7.80 \text{ in.}^2$$

Try 8-#9, $A_s = 8.00$ in.^{2}

With #9

$$d = 51.0 - 2.0 - 0.625 - 0.564 = 47.81 \text{ in.}$$

The actual capacity equals

$$a = \frac{A_s f_y}{0.85 f_c' b} = \frac{8.00(60)}{0.85(3.25)(15.0)} = 11.58 \text{ in.}$$

$$\phi M_n = \phi \left[A_s f_y \left(d - \frac{a}{2} \right) \right]$$

$$= 0.90[(8.00)(60)(47.81 - 5.79)](\tfrac{1}{12})$$

$$= 1513 \text{ k-ft} < 1580 \text{ k-ft} \quad \quad 4.4\% \text{ close enough} \quad \text{say OK}$$

For serviceability the maximum and minimum service load level moments are

$$M_{\text{max}} = -920 \text{ k-ft}$$

$$M_{\text{min}} = -582 \text{ k-ft}$$

The k and j values from Appendix A are

$$\rho = \frac{8.00}{15(47.81)} = 0.0112 \quad \quad k = 0.359 \quad j = 0.880$$

The maximum and minimum steel stress based on $f_s = M/A_s j_d$ is

$$f_s = \frac{920(12)}{8.00(0.880)(47.81)} = 32.8 \text{ ksi max}$$

$$f_s = \frac{582(12)}{8.00(0.880)(47.81)} = 20.7 \text{ ksi min}$$

The stress range is

$$f_f = 32.8 - 20.7 = 12.1 \text{ ksi}$$

The allowable stress range is

$$f_f = 21 - 0.33 f_{\min} + 8 \left(\frac{r}{h}\right)$$

$$= 21 - 0.33(12.1) + 8(0.3)$$

$$= 19.4 \text{ ksi} > 12.1 \text{ ksi} \quad \text{OK}$$

Next we check the crack control criteria in AASHTO Article 1.5.39. Assume that the reinforcement is spaced at 6 in.:

$$A = 6.38 \times 6.0 = 38.3 \text{ in.}^2$$

$$d_c = 3.19 \text{ in.}$$

The allowable stress for crack control equals

$$f_s = \frac{130}{\sqrt[3]{d_c A}}$$

$$f_s = \frac{130}{\sqrt[3]{3.19(38.3)}} = 26.2 \text{ ksi} < 32.8 \text{ ksi}$$

The maximum stress exceeds the allowable stress by an unreasonable amount; therefore it is necessary to increase the reinforcing. On the basis of an allowable of 26.2 ksi we calculate an approximate area of reinforcing:

$$\text{approx } A_{s(\text{required})} = \frac{920(12)}{26.2(0.880)(47.81)} = 10.02 \text{ in.}^2$$

Try 8-#10.

$$A_s = 10.16 \text{ in.}^2$$

$$A = 6.52(6.0) = 39.12 \text{ in.}^2$$

$$d_c = 3.26 \text{ in.}$$

Then

$$f_s = \frac{130}{\sqrt[3]{3.26(39.12)}} = 25.8 \text{ ksi}$$

(which is the new allowable for crack control).

A new maximum stress must be computed. The revised effective depth is

$$d = 51.0 - 3.26 = 47.74 \text{ in.}$$

The reinforcement ratio and k and j from Appendix A are

$$p = \frac{10.16}{(15)(47.74)} = 0.0142 \qquad k = 0.394 \quad j = 0.869$$

The revised steel stress is

$$f_s = \frac{920(12)}{10.16(0.869)(47.74)} = 26.2 \text{ ksi} > 25.8 \text{ ksi allowable}$$

$$1.5\% \text{ overstress} \quad \text{say OK}$$

Use 8-#10. The maximum reinforcement is checked:

$$\text{maximum } \rho = 0.75\rho_b = 0.0174$$

$$\text{actual } \rho = \frac{10.16}{15.0(47.74)} = 0.0142 < 0.0174 \quad \text{OK}$$

The skin reinforcing is computed in Example 5.1.

Before design of the pier shaft and footing we design the pier cap, which is an integral part of the superstructure and forms the monolithic connection between the superstructure and substructure. Again the interior girder loads and moments pass directly into the pier shaft and do not affect the cap. The exterior girder loads and moments cause

Table 5.13   Service Load Level (Exterior Girder) (HS 20)

	R_B (kips)			M_{BD} (kip-ft)		
Case	DL	LL + I	$\sum$	DL	LL + I	$\sum$
II	100.7	51.4	152.1	+129	−125	+4
IA	100.7	45.9	146.6	+129	+98	+227
IB	100.7	33.3	134.0	+129	+198	+327
IIA	100.7	56.5	157.2	+129	−107	+22

bending, shear, and torsion on the cap. Therefore the effects of the exterior girder on the pier cap must be considered.

The dead load and live load reactions and moments computed in Example 5.1 are summed and tabulated in Table 5.13, at a service load level. The column moment M_{BD} is the torsional moment that must be transmitted from the exterior girder through the cap to the pier shaft. These moments and loads are factored and summed in Table 5.14 for the various cases of live load. The first tabulation is for the HS 20 loading on the exterior girder. The next tabulation is for the overload vehicle loading on the exterior girder. (Table 5.15)

Table 5.14   Ultimate Loads—HS 20 (Exterior Girder)

	R_B (kips)			M_{BD} (kip-ft)		
Case	1.3 × DL	2.17 × LL + I	$\sum$	1.3 × DL	2.17 × LL + I	$\sum$
II	130.9	111.5	242.4	+168	−271	−103
IA	130.9	99.6	230.5	+168	+213	+381
IB	130.9	72.3	203.2	+168	+430	+598
IIA	130.9	122.6	253.5	+168	−232	−64

Table 5.15   Overload (Exterior Girder)

	R_B (kips)				M_{BD} (kip-ft)			
Case	DL	LL + I	$\sum$	×1.3	DL	LL + I	$\sum$	×1.3
VI	100.7	67.7	168.4	218.9	+129	−250	−127	−157
VA	100.7	77.3	178.0	213.4	+129	+123	+252	+328
VB	100.7	51.3	152.0	197.6	+129	+326	+455	+592
VIA	100.7	93.7	194.4	252.7	+129	−251	−122	−159

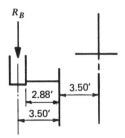

Figure 5.34   Pier cap dead load.

The layout of the cap and the cap dead load shear and moment are shown in Figure 5.34. The cap cross section is 3 ft 3 in. wide and 4 ft 0 in. deep. The depth below the slab is 3 ft $4\frac{1}{2}$ in. Cap $w = 3.25 \times 3.38 \times 150 = 1.64$ k-ft^{-1}. Cap only, $M = 1.64(2.88)^2 (\frac{1}{2}) = 7$ k-ft; $V = 1.64(2.88) = 4.7$ k.

It is reasonable to assume that the serviceability criteria for crack control will be critical because the cap reinforcing must be placed below several layers of deck reinforcing which gives a large value of d_c. A large value of d_c will give a low allowable f_s. Therefore we proportion the cap for crack control and check it for capacity and fatigue.

Assume #10 reinforcing. Assume 6 bars:

$$d = 48.0 - 2.0 - 0.62 - 1.27 - 0.64 = 43.47 \text{ in.}$$

$$d_c = 4.53 \text{ in.}$$

$$A = 2(4.53)(39.0) \div 6 = 58.9 \text{ in.}^2$$

$$f_s = \frac{z}{\sqrt[3]{d_c A}} = \frac{130}{[(4.53)(58.9)]^{1/3}}$$

$$= 20.2 \text{ ksi} < 24.0 \text{ ksi allowed for service load design}$$

Therefore use allowable $f_s = 24.0$ ksi.

The total cap bending moment at a service load level is

$$M_T = 157.2(3.50) = 550 \text{ k-ft} + 7 = 557 \text{ k-ft}$$

Using $d = 45$ in.$\pm$ and f_s allowable $= 24.0$ ksi, we compute an approximate required reinforcing area:

$$A_{s(\text{required})} = \frac{557(12)}{24(0.87)(45)} = 7.12 \text{ in.}^2$$

Try 6-#10, $A_s = 7.62$ in.^{2} The reinforcement ratio and the k and j values from Appendix A (rectangular section) are

$$d = 48.0 - 2.0 - 0.62 - 1.27 - 0.64 = 43.47 \text{ in.}$$

$$p = \frac{7.62}{(39)(43.47)} = 0.0045$$

$$k = 0.242 \quad j = 0.917$$

The service load level stress follows and is compared with our established allowable for crack control.

$$f_s = \frac{557(12)}{7.62(0.917)(43.47)} = 22.0 \text{ ksi} < 24.0 \text{ ksi} \quad \text{OK}$$

Next the capacity is checked:

$$a = \frac{A_s f_y}{0.85 f_c' b} = \frac{7.62(12)}{0.85(3.25)(39)} = 4.24 \text{ in.}$$

$$\phi M_n = \phi \left[A_s f_y \left(d - \frac{a}{2} \right) \right]$$

$$= 0.90[(7.62)(60)(43.47 - 2.12)]\left(\tfrac{1}{12}\right) = 1418 \text{ k-ft}$$

The required ultimate moment M_u is computed and compared with the capacity:

$$M_u = (253.5 \times 3.50) + (7 \times 1.3) = 896 \text{ k-ft} < 1418 \text{ k-ft} \quad \text{OK}$$

The maximum reinforcing requirement which limits the reinforcement ratio to 75% of the balanced ratio is checked:

$$0.75\rho_b = 0.0174$$

$$\rho = \frac{7.62}{(39)(43.47)} = 0.0045 < 0.0174 \quad \text{OK}$$

Next we check fatigue. The minimum moment is the dead load moment:

$$M_{DL} = (100.7 \times 3.50) + 7 = 359 \text{ k-ft}$$

The maximum stress was computed above:

$$f_s = 22.0 \text{ ksi}$$

The minimum stress is

$$f_s = \frac{359(12)}{7.62(0.917)(43.47)} = 14.2 \text{ ksi min}$$

The stress range is

$$f_f = 22.0 - 14.2 = 7.8 \text{ ksi}$$

The allowable stress range is

$$f_f = 21 - 0.33 f_{min} + 8 \left(\frac{r}{h}\right)$$

$$= 21 - 0.33(14.2) + 8(0.3)$$

$$= 18.7 \text{ ksi} > 7.8 \text{ ksi} \quad \text{OK}$$

The skin reinforcement requirement was computed in Example 5.1.

Next we design the pier cap hoop stirrups. This is done by using ACI 318-77 as recommended in AASHTO.

First we check the significance of torsion in computing the concrete shear strength. If the value $\phi(0.5\sqrt{f_c'}\ \Sigma x^2 y)$ is exceeded by the torsional moment, the equation accounting for torsion must be used to compute the concrete shear strength.

$$\phi \left(0.5\sqrt{f_c'}\ \Sigma x^2 y\right) = 0.85[(0.5)\sqrt{3250}(39.0)^2(48.0)](\tfrac{1}{12})$$

$$= 147 \text{ k-ft} < M_{BD} \text{ in all cases except II and IIA}$$

Therefore torsion controls. The equation that must then be used from ACI 318-77 is No. 11-5, which is

$$V_c = \frac{2\sqrt{f_c'}\ b_w d}{\sqrt{1 + [2.5 C_t (T_u/V_u)]^2}}$$

$$C_t = \frac{b_w d}{x^2 y} = \frac{39(43.47)}{(39)^2(48)} = 0.0232$$

$$V_c = \frac{2\sqrt{f_c'}\ b_w d}{\sqrt{1 + [2.5 C_t (T_u/V_u)]^2}} = \frac{2\sqrt{3250}(39)(43.47)}{\sqrt{1 + [(2.5)(0.0232)(T_u/V_u)]^2}}$$

$$= \frac{193,297}{\sqrt{1 + [(0.0580)(T_u/V_u)]^2}}$$

The ultimate shear is

$$V_u = R_B + (1.3 \times 4.7)$$

The concrete shear strength for direct shear only is

$$V_c = 2\sqrt{f_c'}\, b_w\, d = 2\sqrt{3250}\,(39)(43.47) = 193.3 \text{ k}$$

These are converted to the required stirrup shear strength for direct shear:

$$V_u \leqslant \phi(V_c + V_s)$$

$$V_s = \frac{V_u}{\phi} - V_c \qquad \phi = 0.85$$

The required stirrup shear strength V_s is then used to develop a required area of reinforcement:

$$V_s = \frac{A_v f_y d}{s}$$

$$\frac{A_v}{s} = \frac{V_s}{f_y d} = \frac{V_s}{60(43.47)} \overset{4 \text{ legs}}{\left(\frac{1}{4}\right)} = \frac{V_s}{10,433} \text{ per leg}$$

All of the above is then applied to the various cases of live load to develop a required area of reinforcing per unit space for direct shear (Table 5.16).

Table 5.16

Case	T_u (k-ft)	V_u (k)	T_u (k-in.)	V_c (k)	$V_s = V_u - V_c$ (k)	$\dfrac{A_v}{s}$ (per leg)
VI	157	225.0	1884	173.9	90.8	0.0087
VA	328	237.5	3936	139.4	140.0	0.0134
VB	592	203.7	7104	85.7	153.9	0.0148
VIA	159	258.8	1908	177.7	126.8	0.0122
II		248.5		193.3	99.1	0.0095
IA	381	236.6	4572	128.7	149.7	0.0143
IB	598	209.3	7176	86.8	159.4	0.0153
IIA		259.6		193.3	112.1	0.0107

For torsion the torsional strength required for the hoops is

$$T_s = \frac{T_u}{\phi} - T_c$$

Using ACI equation 11-23, we find that the area of reinforcing required per unit spacing for torsion is

$$T_s = \frac{A_t \alpha_t x_1 y_1 f_y}{s}$$

$x_1 = 34.38$ (shortest hoop dimension)
$y_1 = 43.38$ (longest hoop dimension)

where

$$\alpha_t = 0.66 + 0.33 \left(\frac{y_1}{x_1}\right),$$

$$\alpha_t = 0.66 + 0.33 \left(\frac{43.38}{34.38}\right) = 1.08,$$

$$\frac{A_t}{s} = \frac{T_s}{\alpha_t x_1 y_1 f_y} = \frac{T_s}{1.08(34.38)(43.38)(60)} = \frac{T_s}{96,643}$$

This again is per leg. The concrete torsional strength is

$$T_c = \frac{0.8\sqrt{f'_c}\ \Sigma x^2 y}{\sqrt{1 + (0.4 V_u/C_t T_u)^2}} = \frac{0.8\sqrt{3250}(39)^2(48)}{\sqrt{1 + (0.4 V_u/0.0232 T_u)^2}}$$

$$= \frac{3330}{\sqrt{1 + [17.24(V_u/T_u)]^2}}$$

On the basis of these equations, the required area of reinforcing for each case of live load is calculated and tabulated in Table 5.17.

The requirements for shear and torsion are summed and tabulated in Table 5.18. The largest requirement is expanded to a spacing that assumes #6 reinforcing bars.

The requirements for longitudinal reinforcement for torsion are covered in Example 5.1.

The dead load moments and the live load plus impact moments were computed in Example 5.1. The factored live load plus impact moments are compared between the HS 20 and the overload vehicle (Table 5.19).

The controlling values from the preceding tables are combined with

Table 5.17

Case	V_u (k)	T_u (k-in.)	T_c (k-in.)	$T_s = \dfrac{T_u}{\phi} - T_c$ (k-in.)	$\dfrac{A_t}{s}$
VI	225.0	1884	1455	261	0.0079
VA	237.5	3936	2308	2323	0.0240
VB	203.7	7104	2985	5373	0.0556
VIA	258.8	1908	1309	936	0.0097
II	248.5				
IA	236.6	4572	2485	2894	0.0299
IB	209.3	7176	2975	5467	0.0566
IIA	259.6				

Table 5.18

Case	$\dfrac{A_v}{s}$	$\dfrac{A_t}{s}$	$\sum \dfrac{A}{s}$	$A_v = 0.44$ in.^{2} s
VI	0.0097	0.0079	0.0166	
VA	0.0134	0.0240	0.0374	
VB	0.0148	0.0556	0.0704	
VIA	0.0122	0.0097	0.0219	
II	0.0095		0.0095	
IA	0.0143	0.0299	0.0442	
IB	0.0153	0.0566	0.0719	6.1 in.
IIA	0.0107		0.0107	

Use #6 ⊞ @ 6 in.

the dead load and factored by $\gamma = 1.3$ to convert the overload moments to ultimate moments (Table 5.20).

The maximum section for positive moment has 9-#10 for tensile reinforcement. This was determined in our initial section design. We cut off 1-#10, 3-#10, and then 2-#10; 3-#10 extend through the support. The resisting capacities for these various cutoff sections are the following:

Resisting Moments

Positive Moment

$$a = \frac{A_s f_y}{0.85 f_c' b} = \frac{A_s(60)}{0.85(3.75)(70)} = 0.3103 A_s$$

Table 5.19

Point	Live Load + Impact (HS 20) $\beta_L(LL+I)$ $\beta_L = 1.67$		Overload Vehicle $\beta_L = 1.0$	
	Exterior $\beta_L M_{LL+I}$	Interior $\beta_L M_{LL+I}$	Exterior $\beta_L M_{LL+I}$	Interior $\beta_L M_{LL+I}$
	Positive Moment			
AB				
0.1	+404	+404	+419	+340
0.2	+685	+685	+714	+578
0.3	+843	+843	+889	+720
0.4	+930	+930	+976	+791
0.5	+918	+918	+993	+804
0.6	+870	+870	+897	+726
0.7	+718	+718	+760	+616
0.8	+491	+491	+532	+431
0.9	+200	+200	+225	+182
BA				
	Negative Moment			
AB				
0.5	−214	−214	−256	−207
0.6	−256	−256	−308	−250
0.7	−301	−301	−359	−291
0.8	−342	−342	−411	−333
0.9	−384	−384	−462	−374
BA	−573	−573	−543	−440
BC	−635	−635	−792	−641
0.5BC	−297	−297	−315	−255
CB				

[a] All moments are in kip-ft.

Table 5.20

Point	Exterior Girder			Interior Girder		
	M_{DL}	$\beta_L M_{LL+I}$	$M_u = 1.3 \times M_T$	M_{DL}	$\beta_L M_{LL+I}$	$M_u = 1.3 \times M_T$
	Maximum Positive Moment					
AB						
0.1	+245	+419	+863	+250	+404	+850
0.2	+424	+714	+1479	+436	+685	+1457
0.3	+535	+889	+1851	+556	+843	+1819
0.4	+582	+976	+2025	+615	+930	+2008
0.5	+562	+993	+2022	+608	+918	+1984

Table 5.20   (Continued)

	Exterior Girder			Interior Girder		
Point	M_{DL}	$\beta_L M_{LL+I}$	$M_u = 1.3 \times M_T$	M_{DL}	$\beta_L M_{LL+I}$	$M_u = 1.3 \times M_T$
			Maximum Positive Moment			
0.6	+466	+897	+1772	+518	+870	+1804
0.7	+303	+760	+1382	+362	+718	+1404
0.8	+74	+532	+788	+145	+491	+826
0.9	−221	+225	+5	−128	+200	+94
BA						
			Maximum Negative Moment			
A						
0.5	+562	−256		+608	−214	
0.6	+466	−308	+205	+518	−256	
0.7	+303	−359	−73	+362	−301	+79
0.8	+74	−411	−438	+145	−342	−256
0.9	−221	−462	−888	−128	−384	−666
BA	−582	−543	−1462	−485	−573	−1375
BC	−453	−792	−1618	−313	−635	−1232
0.5BC	−196	−315	−664	−111	−297	−530
CB	−29	0	−38	0		0

[a] All moments are in kip-ft.

Spacing of Rows
of Reinforcement

Section 1

$\cdots$

9-#10, $A_s = 11.43$ in.^{2}

3.25 in.

cg = 5.89 in.

$d = 45.11$ in.

$\cdots$

$a = 3.54$ in.

3.25 in.

$\phi M_n = (0.90)[(11.43)(60.0)(45.11 - 1.77)](\tfrac{1}{12})$

$\cdots$

$= 2229$ k-ft

2.64 in.

Section 2

$\cdot$ $\cdot$

8-#10, $A_s = 10.16$ in.^{2}

3.25 in.

$\text{cg} = \dfrac{6(4.265) + 2(9.14)}{8} = 5.75$ in.

$\cdots$

$d = 51.0 - 5.75 = 45.25$ in.

3.25 in.

$\cdots$

2.64 in.

$a = 3.16$ in. $\dfrac{a}{2} = 1.58$ in.

$\phi M_n = (0.90)[(10.16)(60.0)(45.25 - 1.58)](\tfrac{1}{12})$

$= 1997$ k-ft

Section 3

$5\text{-}\#10, A_s = 6.35$ in.^{2}

3.25 in.

$\text{cg} = \dfrac{3(2.64) + 2(5.89)}{5} = 3.94$ in.

2.64 in.

$d = 51.0 - 3.94 = 47.06$ in.

$a = 1.98$ in. $\dfrac{a}{2} = 0.99$ in.

$\phi M_n = (0.90)[(6.35)(60.0)(47.06 - 0.99)](\tfrac{1}{12})$

$= 1316$ k-ft

Section 4

$3\text{-}\#10, A_s = 3.81$ in.^{2}

2.64 in.

$\text{cg} = 2.64$ in.

$d = 51.0 - 2.64 = 48.36$ in.

$a = 1.18$ in. $\dfrac{a}{2} = 0.59$ in.

$\phi M_n = (0.90)[(3.81)(60.0)(48.36 - 0.59)](\tfrac{1}{12})$

$= 819$ k-ft

Negative moment required 8-#9 as tensile reinforcement at the maximum section; 2-#8 extend full length in the girder stems beyond the last cutoff.

The various cutoffs and resisting moments are computed for negative moment:

Negative Moment: $d = 47.81$ in.

	a	$\dfrac{a}{2}$	$\left(d - \dfrac{a}{2}\right)\left(\dfrac{1}{12}\right)$	A_s	ϕM_n
8-#9	11.58	5.79	3.50	8.00	1513
6-#9	8.69	4.35	3.62	6.00	1173

Negative Moment: $d = 47.81$ in.

a	$\dfrac{a}{2}$	$\left(d - \dfrac{a}{2}\right)\left(\dfrac{1}{12}\right)$	A_s	ϕM_n
4–#9	5.79	2.90	3.74	4.00	808
2–#9	2.90	1.45	3.86	2.00	417
2–#8	2.29	1.14	3.89	1.58	332

Computation of the bar extensions in accordance with AASHTO Article 1.5.13 follows:

bar extensions: $d = 3.8$ ft $20\phi = 3.1$ ft $\dfrac{s}{20} = 3.2$ ft
 (#10)

AASHTO Article 1.5.13 also requires that one-third of the reinforcing extend into the simple support and one quarter of the reinforcing through the continuous support.

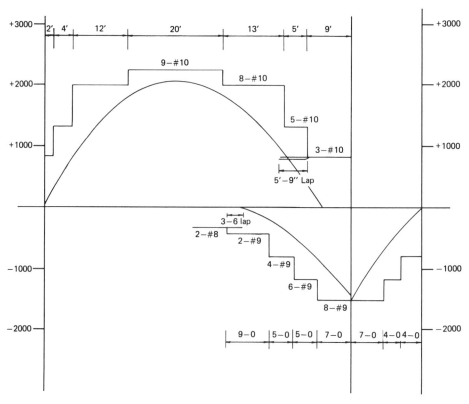

Figure 5.35   Ultimate moments.

The maximum ultimate moments are plotted against the resisting capacity moments and cutoff locations are selected (Figure 5.35). Development length will not affect the results; therefore it is ignored. The ultimate moments selected are the maximums for interior or exterior girders at any given point. All girders are made the same.

The exterior girder controls for shear. The shears for dead load plus live load plus impact were computed in Example 5.1. The factored live load plus impact shears are compared between the HS 20 and overload vehicles in Table 5.21.

Because 1 ft 0 in. away from the support the shear is not a significant item nor is the shear in the outer half of the cantilever, we use overload vehicle shears similar to those in Example 5.1. The unit shears and the required spacing are then the same as those in the overload shear design in Example 5.1, as are the minimum shear reinforcement and spacing. A revised spacing is plotted against the required spacing (Figure 5.36).

Table 5.21

Point	Overload Exterior Girder $\beta_L V_{LL+I}$	HS 20 Exterior Girder $\beta_L V_{LL+I}$
AB	+74.3	+72.3
0.1	+64.7	+62.6
0.2	+55.3	+52.9
0.3	+46.2	+43.6
0.4	+37.4	+34.4
0.5	−38.3	−34.9
0.6	−47.1	−44.1
0.7	−55.6	−53.3
0.8	−63.4	−61.6
0.9	−70.4	−69.5
BA	−77.4	−76.7
BC	+72.8	+79.5
$1'BC$	+59.8	+39.7
$0.5BC$	+29.9	+39.7
CB	+29.9	+39.7

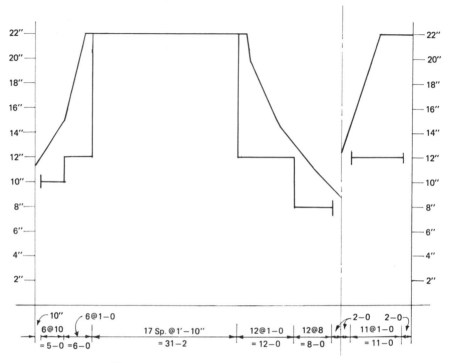

Figure 5.36   Stirrup spacing requirements.

BIBLIOGRAPHY

1. *Continuous Concrete Bridges*, 2nd ed., Portland Cement Association.
2. *Concrete Members with Variable Moment of Inertia*, ST 108, Portland Cement Association.
3. *Beam Factors and Moment Coefficients for Members with Prismatic Haunches*, ST 81, Portland Cement Association.
4. *Reinforced Concrete Design Handbook—Working Stress Method*, SP-3, 3rd ed., American Concrete Institute, 1980.
5. *AASHTO Standard Specifications for Highway Bridges, 1977*, with the 1978–1981 Interim Specifications, The American Association of State Highway and Transportation Officials.
6. R. Park and T. Paulay, *Reinforced Concrete Structures*, Wiley, New York, 1975.
7. Phil M. Ferguson, *Reinforced Concrete Fundamentals*, 4th ed., Wiley, New York, 1979.
8. *Notes on Load Factor Design for Reinforced Concrete Bridge Structures with Design Applications*, Portland Cement Association.
9. *ACI Building Code Requirements for Reinforced Concrete (ACI 318-77), and Commentary*, American Concrete Institute.
10. *Strength and Serviceability Criteria, Reinforced Concrete Bridge Members, Ultimate Design*, FHWA, October 1969.
11. G. C. Frantz and J. E. Breen, *Design Proposal for Side Face Crack Control Reinforcement for Large Reinforced Concrete Beams*, ACI Publication, *Concrete International*, 2 (10) (October 1980).

SIX

Design of Reinforced Concrete Box Girder Bridges

Reinforced concrete box girder construction contains a top slab that is usually wider than the box shape, which consists of vertical webs, and a bottom slab usually as broad as the out-to-out width of the girder webs. Stem widths of a nominal width of 9 in. are commonly used. Bottom slabs can be as thin as $5\frac{1}{2}$ in. The top slab, webs, and bottom slab are built to act as a unit, which means that full shear transfer must be provided between all parts of the section. Cast-in-place construction built on falsework is a typical construction procedure.

The reinforced concrete box girder becomes practical at approximately the maximum reasonable span length of a T-beam bridge. It is particularly appropriate for spans ranging between 95 and 140 ft. Beyond this span range it is probably more reasonable to go to a post-tensioned box girder because of the increase in massiveness required by material strengths readily available for this construction. This increase in massiveness results in higher dead loads that render the structure relatively inefficient at large span lengths. Post-tensioned box girders, discussed in Chapter 7, normally use higher strength materials and are more efficient at longer spans.

Reinforced concrete box girders have high torsional strength which makes their construction particularly suited to structures with significant curvature. Interchange ramp structures often utilize sharp curvature, which is one reason a number of reinforced concrete box girders are seen at interchanges on freeways throughout the United States. The high torsional strength also permits the bridge to be designed as a unit without considering individual girders.

This construction lends itself to aesthetic treatment. Sloping exterior webs are often used and haunching the girder soffit to maximum depths at interior piers is commonplace, typically on longer spans. Rounding the lower box corners has also been the choice on a number of occasions to minimize the massive appearance. Monolithic construction of substructure and superstructure is not only a common means of obtaining efficiency with concrete box girders but is also used for better appearance. Pier caps can be placed completely within the box and pier shafts can connect directly with the superstructure.

For the commonly used spans of 95–140 ft depth-to-span ratios of 0.055 are considered normal. For haunched structures the depth-to-span ratios might be about 0.05 at the minimum depth and 0.08 at the maximum.

6.1. DESIGN CODE AND PROCEDURES

Reinforced concrete box girders differ from T-beams in that they have a bottom flange; therefore for section analysis they should be designed as T-beams for positive and negative moments. This means that compression stresses are easily controlled in the areas of negative moment without the extensive compression reinforcement required in T-beam construction.

Thickening of the bottom slab is often used in negative moment regions to control compression stresses that are significant. Thickening of the girder webs is also used in areas adjacent to supports to control shear requirements. The specifications indicate that this change in web thickness should be tapered over a distance equal to a minimum distance of 12 times the difference in web thickness.

The specifications limit the minimum bottom flange thickness to $\frac{1}{16}$ of the clear span between girder webs or to a minimum of $5\frac{1}{2}$ in. The greater of these two values is needed. The bottom slab reinforcement required as a minimum is 0.4% of the flange area in the longitudinal direction and 0.5% of the flange area in the transverse direction. The transverse bottom slab reinforcement is to be distributed over both faces of the slab and hooked into the exterior girder webs.

The need for intermediate diaphragms in this type of construction has been questioned by many knowledgeable structural engineers. However, the specifications indicate diaphragms at a spacing of 60 ft maximum unless data indicates that diaphragms are not necessary. Diaphragms should be provided at main supports to provide load transfer of transverse wind loads to the substructure. On curved box girders the need for diaphragms and spacing requirements should be given special consideration.

For curved structures, torsion should be considered and exterior girder shears should be increased to account for torsion.

Main negative moment reinforcement should be placed against the underside of the top transverse reinforcement in the top slab. Positive moment reinforcement should be placed between the two layers of transverse reinforcement required by AASHTO. Reinforcement should be spaced equally across the flanges.

6.2   DESIGN EXAMPLES

Example 6.1.   The reinforced concrete box girder bridge shown in section in Figure 6.1 and in elevation in Figure 6.2 is designed by service load design. The transversely reinforced deck slab was designed for this same structure by the service load procedures given in Example 4.5.

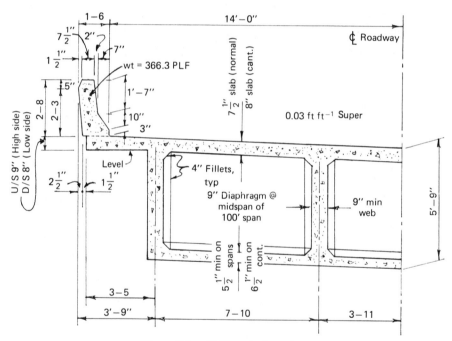

Figure 6.1   Half typical section.

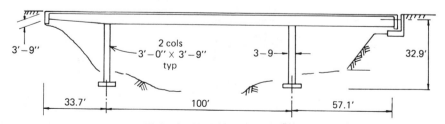

Figure 6.2   Elevation: box girder bridge.

In the design of this box girder it is assumed that only the flanges will resist bending and only the webs will resist the shear. For service load the following stress levels are used:

$$f_c' = 3300 \text{ psi}$$

$$f_c = 1320 \text{ psi}$$

$$f_s = 24,000 \text{ psi}$$

$$\left. \begin{array}{l} f_c = 1200 \text{ psi} \\ f_s = 20,000 \text{ psi} \end{array} \right\} \text{ for deck design only}$$

Here n is assumed to be 10 based on test results. The design live load is the AASHTO HS 20-44. The bridge is flared at the cantilever end, where the flare occurs in the roadway width and box girder web spacing (Figure 6.3). The flange thicknesses and steel requirements are determined according to the requirements of AASHTO Article 1.5.23(k);

Bottom Flange Thickness [Article 1.5.23(k)(3)(b)]. For normal girder spacing $S = 7.83-0.75 = 7.08$ ft.

$$\text{Required } t = \tfrac{1}{16}(7.08) = 0.442 \text{ ft, use } 5\tfrac{1}{2} \text{ in.}$$

For flared girder spacing on cantilever $S = 9.29-0.75 = 8.54$ ft.

$$\text{Required } t = \tfrac{1}{16}(8.54) = 0.534 \text{ ft, use } 6\tfrac{1}{2}.$$

Bottom Flange Reinforcement Parallel to Girders [Article 1.5.23(k)(4) (a)]. For normal girder spacing

$$A_{s(\text{required})} = 0.004(0.458)(24.25)(144) = 6.40 \text{ in.}^2$$

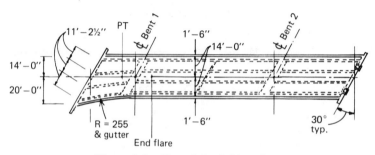

Figure 6.3   Box girder bridge plan.

For flared girder spacing

$$A_{s(\text{required})} = 0.004(0.542)(26.71)(144) = 8.34 \text{ in.}^2$$

Use a maximum of an 18 in. spacing.

Bottom Flange Reinforcement Normal to Girders [Article 1.5.23(k)(4) (b)]. For normal girder spacing (100-ft span, 56 ft 9 in. span).

$$A_{s(\text{required})} = 0.005(5.5)(12.0)$$
$$= 0.33 \text{ in.}^2 \quad \#5 @ 9 \text{ in. alternating top and bottom}$$
$$A_s = 0.41 \text{ in.}^2$$

For flared girder spacing

$$A_{s(\text{required})} = 0.005(6.5)(12.0)$$
$$= 0.39 \text{ in.}^2 \quad \#5 @ 9 \text{ in. alternating top and bottom}$$
$$A_s = 0.41 \text{ in.}^2$$

With the deck thickness determined, the section properties are computed for use in the frame analysis (Figure 6.4.) For the section shown the moment of inertia is computed by using $I_T = \Sigma (Ay^2 + I_0)$.

	A	y_b	Ay_b	$\bar{y}$	$A\bar{y}^2$	I_0	I_T
Tob slab	18.96	5.44	103.14	2.16	88.46	0.62	89.08
Webs	14.00	2.79	39.03	0.49	3.36	25.41	28.77
Bottom slab	11.11	0.23	2.56	3.05	103.35	0.19	103.54
	44.07 ft^2		144.73 ft^3				221.39 ft^4

$$\bar{y}_b = \frac{144.73}{44.07} = 3.28 \text{ ft}$$

Figure 6.5 is a schematic elevation of the monolithic structure.

The assumed column section for this structure is shown in Figure 6.6. The column stiffness, which uses the transformed reinforced section criteria for the two columns per bent, is computed using Figure 6.6 where the assumed reinforcement is computed as 1% of the gross column area for each face in the longitudinal direction

$$A_c = 3.0 \times 3.75 = 11.25 \text{ ft}^2$$

1% reinforcement of each face, $A_s = 0.1125 \text{ ft}^2$

$$nA_s = 1.125 \text{ ft}^2$$

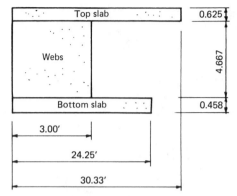

Figure 6.4 Section details.

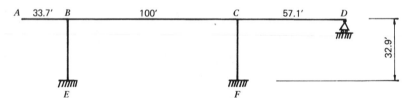

Figure 6.5 Schematic of structure.

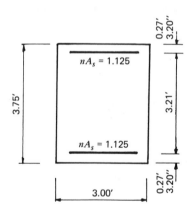

Figure 6.6 Column section.

The transformed moment of inertia for columns BE and CF is

$$I_C = \frac{3.00(3.75)^3}{12} = 13.18 \text{ ft}^4$$

$$I_S = [\,1.125(1.605)^2\,]\,2 = 5.80 \text{ ft}^4$$

$$I_T = 2(5.80 + 13.18) = 37.96 \text{ ft}^4$$

The relative stiffness coefficients are computed, assuming E is constant, as

$$K_{BC} = K_{CB} = 4\frac{I}{L} = \frac{4(221.39)}{100} = 8.86$$

$$K_{CD} = 3\frac{I}{L} = \frac{3(221.39)}{57.1} = 11.63$$

$$K_{BE} = K_{CF} = 4\frac{I}{L} = \frac{4(37.96)}{32.9} = 4.62$$

The distribution factors are computed next:

$$D_{BC} = \frac{K}{\sum K} = \frac{8.86}{13.48} = 0.657$$

$$D_{BE} = \frac{K}{\sum K} = \frac{4.62}{13.48} = 0.343$$

$$D_{CB} = \frac{K}{\sum K} = \frac{8.86}{25.11} = 0.353$$

$$D_{CF} = \frac{K}{\sum K} = \frac{4.62}{25.11} = 0.184$$

$$D_{CD} = \frac{K}{\sum K} = \frac{11.63}{25.11} = 0.463$$

To compute the final structure moments the effects of sidesway must be incorporated, as discussed in detail in Chapter 4. Because the substructure units are identical and fixed at the base, the fixed end moments can be assumed to be $M_{BE}^F = M_{EB}^F = M_{FC}^F = M_{CF}^F = +1000$ k-ft for some structural deflection Δ. The moment distribution for these fixed end moments is shown in Table 6.1.

The column shears that result from this sway are

$$\text{Col} \quad BE \quad V = \frac{700 + 848}{32.9} = 47.1 \text{ k}$$

$$\text{Col} \quad CF \quad V = \frac{932 + 868}{32.9} = 54.7 \text{ k}$$

Table 6.1 Sidesway

	B			E	F		C	
	BA	BC	BE	EB	FC	CF	CB	CD
		0.657	0.343			0.184	0.353	0.463
FEM			+1000	+1000	+1000	+1000		
BAL		-657	-343			-184	-353	-463
CO		-176		-172	-92		-328	
BAL		+116	+60			+60	+116	+152
CO		+58		+30	+30		+58	
BAL		-38	-20			-11	-20	-27
CO		-10		-10	-6		-19	
BAL		+7	+3			+3	+7	+9
M_Δ		-700	+700	+848	+932	+868	-539	-329

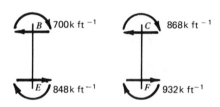

These column shears add up to give the unbalanced force of 101.8 k, which is to be applied to the structure as shown in Figure 6.7. For a fixed end moment of -1000 at *BA* the moment distribution is tabulated in Table 6.2. Note that the moments are corrected for sidesway (*SS* factor) on the basis of the unbalanced column shear.

The *SS* factor is computed as follows:

$$\text{Col} \quad BE \quad V = \frac{363 + 182}{32.9} = 16.6$$

$$\text{Col} \quad CF \quad V = \frac{-30 - 63}{32.9} = \frac{-2.8}{13.8 \text{ k (unbalanced shear)}}$$

$$SS \text{ factor} = \left(\frac{13.8}{101.8}\right) \times M_\Delta$$

A similar moment distribution is performed for a fixed end moment at *BC* equal to +1000 (Table 6.3).

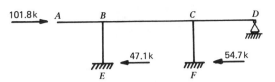

Figure 6.7   Sidesway forces.

Table 6.2

	B			E	F		C	
BA	BC	BE		EB	FC	CF	CB	CD
	0.657	0.343				0.184	0.353	0.463
FEM	−1000							
BAL		+657	+343					
CO				+172			+328	
BAL						−60	−116	−152
CO		−58			−30			
BAL		+38	+20					
CO				+10			+19	
BAL						−3	−7	−9
Σ	−1000	+637	+363	+182	−30	−63	+224	−161
SS		+94	−94	−114	−126	−117	+73	+44
M	−1000	+731	+269	+68	−156	−180	+297	−117

For the moment distribution the unbalanced column shear is the same as that for a fixed moment at BA, except that it is in the opposite direction. Therefore the sidesway corrections are the same but of opposite sign.

For a fixed end moment at CB of −1000 the moment distribution is performed in Table 6.4.

The sidesway correction is again based on column shear unbalance:

Col BE $V = \dfrac{-63 - 30}{32.9} = -2.8$

6.1 k (net shear unbalance)

Col CF $V = \dfrac{+98 + 195}{32.9} = +8.9$

Table 6.3

	B		E	F	C		
BA	BC	BE	EB	FC	CF	CB	CD
	0.657	0.343			0.184	0.353	0.463
FEM	+1000						
BAL	-657	-343					
CO			-172			-328	
BAL					+60	+116	+152
CO	+58			+30			
BAL	-38	-20					
CO			-10			-19	
BAL					+3	+7	+9
Σ	+363	-363	-182	+30	+63	-224	+161
SS	-94	+94	+114	+126	+117	-73	-44
M	+269	-269	-68	+156	+180	-297	+117

Table 6.4

	B		E	E	C		
BA	BC	BE	EB	FC	CF	CB	CD
	0.657	0.343			0.184	0.353	0.463
FEM						-1000	
BAL					+184	+353	+463
CO	+176			+92			
BAL	-116	-60					
COL			-30			-58	
BAL					+11	+20	+27
COL	+10			+6			
BAL	-7	-3					
Σ	+63	-63	-30	+98	+195	-685	+490
SS	+42	-42	-51	-56	-52	+32	+20
M	+105	-105	-81	+42	+143	-653	+510

Table 6.5

	B			E	F		C	
	BA	BC	BE	EB	FC	CF	CB	CF
		0.657	0.343			0.184	0.353	0.463
FEM								+1000
BAL						-184	-353	-463
CO		-176			-92			
BAL		+116	+60					
CO				+30			+58	
BAL						-11	-20	-27
CO		-10			-6			
BAL		+7	+3					
Σ		-63	+63	+30	-98	-195	-315	+510
SS		-42	+42	+51	+56	+52	-32	-20
M		-105	+105	+81	-42	-143	-347	+490

which is now proportioned to the preceding moment distribution for sidesway to obtain the moment corrections or

$$SS \text{ factor} = \frac{6.1}{101.8} \, M_\Delta$$

A similar moment distribution is performed for a fixed end moment at CD equal to +1000, as given in Table 6.5. For this moment distribution the unbalanced column shear is the same as that for a fixed end moment at CB, except that it is in the opposite direction. Therefore the sidesway corrections are the same but of opposite sign.

By dividing through by 1000 the resultant moment distribution for a fixed end moment at each joint is obtained (Table 6.6).

Table 6.6   Moment Distribution Summary

	M_{BA}	M_{BC}	M_{BE}	M_{EB}	M_{FC}	M_{CF}	M_{CB}	M_{CD}
M_{BA}^F	-1.000	+0.731	+0.269	+0.068	-0.156	-0.180	+0.297	-0.117
M_{BC}^F		+0.269	-0.269	-0.068	+0.156	+0.180	-0.297	+0.117
M_{CB}^F		+0.105	-0.105	-0.081	+0.042	+0.143	-0.653	+0.510
M_{CD}^F		-0.105	+0.105	+0.081	-0.042	-0.143	-0.347	+0.490

The next dead load on the structure is evaluated; the cantilever includes an additional 1 in. thickness on the bottom slab and on the webs and an additional $\frac{1}{2}$-in. thickness is used on the top slab.

The uniform dead load at B and A on cantilever AB are the following (assuming that the concrete weight equals 150 pcf):

Cantilever AB

@ B

Bottom slab	$0.625 \times 24.25 \times 150$	$= 2273$
Stems	$4.46 \times 3.33 \times 150$	$= 2228$
Top slab	$0.667 \times 24.25 \times 150$	$= 2425$
Top slab	$0.72 \times 3.04 \times 150$	$= 328$
Top slab	$0.70 \times 3.04 \times 150$	$= 319$
Parapet	2×366.3	$= 733$
Fillets	$12[\frac{1}{2} \times 0.33 \times 0.33 \times 150] =$	98
WS	28.0×25	$= 700$
Forms*	3×58	$= \underline{174}$
		9278 plf

@ A

Bottom slab	$0.625 \times 30.25 \times 150$	$= 2836$
Stems	$2.46 \times 3.33 \times 150$	$= 1230$
Top slab	$0.667 \times 30.25 \times 150$	$= 3025$
Top slab		$= 328$
Top slab		$= 319$
Parapets		$= 733$
Fillets		$= 98$
WS	34.0×25	$= 850$
Forms		$= \underline{174}$
		9593 plf

---

*Assuming 58 plf per 8 ft ± cell.

Also on the cantilever AB the concentrated load of A is the following:

End Beam + Wings

$$5.75 \times 1.0 \times 50.72 \times 150 = 43.75$$
$$\frac{-9.59}{34.2 \text{ k}}$$

The cantilever fixed end moment at BA is computed as follows:

$$\overline{V_{BA}}$$

$$9.278 \times 33.7 = 312.7 \quad \times \quad 16.85 \quad = 5268$$
$$\tfrac{1}{2} \times 0.315 \times 33.7 = \quad 5.3 \ \times (\tfrac{2}{3} \times 33.7) = \quad 119$$
$$\frac{34.2 \ \times \qquad 33.2 \qquad = 1135}{352.2 \text{ k} \qquad\qquad 6522 \text{ k-ft} = M_{BA}^{F}}$$

34.2 k

0.315 k ft^{-1}

9.278 k ft^{-1}

33.7′

A   B

The uniform dead load and concentrated dead load in spans BC and CD are computed as follows:

Span BC and CD Dead Load

Uniform

Bottom slab	$0.542 \times 24.25 \times 150$	$= 1972$
Stems	$4.58 \times \ \ 3.33 \times 150$	$= 2288$
Top slab	$0.625 \times 24.25 \times 150$	$= 2273$
Top slab	$0.72 \times \ \ 3.04 \times 150$	$= \ \ 328$
Top slab	$0.70 \times \ \ 3.04 \times 150$	$= \ \ 319$
Parapets	2×366.3	$= \ \ 733$
Fillets	$12[\tfrac{1}{2} \times 0.33 \times \ \ 0.33 \times 150] =$	$\ \ 98$
WS	28.0×25	$= \ \ 700$
Forms		$= \ \ 174$
		8885 plf

6805 plf

Concentrated Load @ 0.5BC

Interior Diaphragm

$$0.75 \times 4.58 \times 20.92 \times 150 = 10.8 \text{ k}$$

The fixed end moments due to dead load in spans BC and CD are computed:

$$M^F_{BC} = M^F_{CB} = 8.885(100.0)^2 \times \tfrac{1}{12} = 7404 \text{ k-ft} \left.\right\}$$
$$10.8(100.0) \times 0.1250 = 135 \text{ k-ft} \left.\right\} \Sigma = 7539 \text{ k-ft}$$

$$M^F_{CD} = 8.885(57.1)^2 \times \tfrac{1}{8} = 3621 \text{ k-ft}$$

By using the preceding tabulated moment distribution summary and multiplying these coefficients by the dead load fixed end moments we can compute all the distributed corner moments in Table 6.7. With these moments some of the critical sections are examined to make sure that the assumed basic dimensions for frame analysis are correct. Therefore preliminary dead load moments at $0.5BC$ and $0.6CD$ are computed;

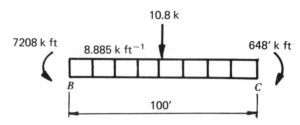

$$M_{0.5BC} = 8.885(100.0)^2 (0.125) = +11,106 \left.\right\}$$
$$10.8(100.0)(0.25) = +270 \left.\right\} = +4532 \text{ k-ft}$$
$$-\frac{7208 + 6481}{2} = -6844 \left.\right\}$$

Table 6.7

	M_{BA}	M_{BC}	M_{BE}	M_{EB}	M_{FC}	M_{CF}	M_{CB}	M_{CD}
$M^F_{BD} = 6522$	-6522	$+4768$	$+1754$	$+443$	-1017	-1174	$+1937$	-763
$M^F_{BC} = 7539$		$+2028$	-2028	-513	$+1176$	$+1357$	-2239	$+882$
$M^F_{CB} = 7539$		$+792$	-792	-611	$+317$	$+1078$	-4923	$+3845$
$M^F_{CD} = 3621$		-380	$+380$	$+293$	-152	-518	-1256	$+1774$
$\Sigma =$	-6522	$+7208$	-686	-388	$+324$	$+743$	-6481	$+5738$

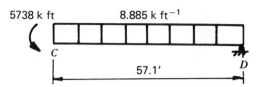

$$M_{0.6CD} = 8.885(57.1)^2(0.120) = +3476 \left.\begin{matrix} \\ \\ \end{matrix}\right\} = +1181 \text{ k-ft}$$
$$-0.4(5738) = -2295$$

The maximum dead load shears are computed next:

$$V_{BA} = 352.2 \text{ k}$$
$$V_{BC} = 8.885(50.0) = +444.2$$
$$0.5(10.8) = +5.4$$
$$\frac{7208 - 6481}{100.0} = +7.3$$
$$\overline{456.9 \text{ k}}$$

$$V_{CB} = 8.885(50.0) = -444.2$$
$$0.5(10.8) = -5.5$$
$$\frac{7208 - 6481}{100.0} = +7.3$$
$$\overline{-442.3 \text{ k}}$$

$$V_{CD} = 8.885\left(\frac{57.1}{2}\right) = +253.7$$
$$+\frac{5738}{57.1} = +100.5$$
$$\overline{+354.2 \text{ k}}$$

$$V_{DC} = 8.885\left(\frac{57.1}{2}\right) = -253.7$$
$$+\frac{5738}{57.1} = +100.5$$
$$\overline{-153.2 \text{ k}}$$

The impact coefficients for the different spans and the average for adjacent spans are computed according to AASHTO Article 1.2.12.

Impact

Cantilever AB $I = 0.300$

Span BC $I = \dfrac{50}{100.0 + 125} = 0.222$

Span CD $I = \dfrac{50}{57.1 + 125} = 0.274$

av $AB + BC$ $I = 0.261$

av $BC + CD$ $I = 0.248$

Because box girders are to be designed, the distribution factor as a unit for the exterior and interior girders given in AASHTO Article 1.3.1(b)(1) and (2) is combined. This is done by using the out-to-out slab width divided by seven to obtain the lines of wheels per box girder bridge; It should be noted that special attention must be given to long cantilever slabs in relation to live load distribution:

$$DF = \frac{30.33}{7} = 4.333 \text{ lines of wheels}$$

The distribution is now combined with impact to obtain $(LL + I)$ coefficients per total box girder:

Cantilever AB Coefficient $= 1.300(4.333) = 5.633$ av $= 5.464$

Span BC Coefficient $= 1.222(4.333) = 5.295$

Span CD Coefficient $= 1.274(4.333) = 5.521$ av $= 5.408$

These values are for a line of wheels.

The live load plus impact moments and shears are computed at the critical locations. First the maximum cantilever moment is calculated at BA by using the loading shown in Figure 6.8. For this loading the moments are

$$M_{BA} = 16 \left(33.7 + 19.7 + \frac{5.7}{4} \right) = -877 \text{ k-ft per line wheels}$$

$$M_{BA} = (-877)(5.633) = -4941 \text{ k-ft per box}$$

Figure 6.8   Maximum live load moment at BA.

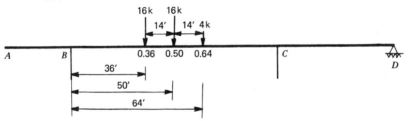

Figure 6.9   Maximum live load moment at 0.5 *BC*.

Next the maximum moment in *BC* is computed by using the same truck loading position given for computation of the moment at *BA*. Further study shows that the truck loading controls over the lane loading. By using the unit moment distribution summary the moment at *BC* is computed as Max M_{BC} (truck):

$$M_{BC} = (-4941)(0.731) \doteq -3612 \text{ k-ft per box}$$

Next the maximum moment at 0.5*BC* with the loading shown in Figure 6.9 is computed. For this loading the fixed end moments are (Appendix O)

$$M_{BC}^F = 16(100.0)\left(0.1475 + 0.1250 + \frac{0.0829}{4}\right) = 469 \text{ k-ft}$$

$$M_{CB}^F = 16(100.0)\left(0.0829 + 0.1250 + \frac{0.1475}{4}\right) = 392 \text{ k-ft}$$

Using the moment distribution summary, we obtain the corner moments

$$M_{BC} = 469(-0.269) + 392(-0.105) = -167 \text{ k-ft}$$
$$M_{CB} = 469(-0.297) + 392(-0.653) = -395 \text{ k-ft}$$

The simple beam shear at *BC* for the truck load is

$$SBV_{BC} = 16\left(0.64 + 0.50 + \frac{0.36}{4}\right) = 19.68 \text{ k}$$

The maximum moment at $0.5BC$ is computed:

$$
\begin{aligned}
M_{0.5BC} = \quad & 19.68(50.0) = +984 \\
& -16.0(14.0) = -224 \\
& -\left(\frac{167 + 395}{2}\right) = -281 \\
\hline
& \qquad\qquad +479 \text{ k-ft per line wheels} \\
= \quad & (+479)(5.295) = +2536 \text{ k-ft per box}
\end{aligned}
$$

The lane loading controls the negative moment at CB and CD as shown in Figure 6.10:

The fixed end moments for this loading are (see Appendixes O and P)

$$
M_{BC}^F = 0.32(100.0)^2\left(\tfrac{1}{12}\right) = 267 \left.\begin{array}{l}\\ \\ \end{array}\right\} 348 \text{ k-ft}
$$
$$
\qquad\qquad 9(100.0)(0.0895) = \;\;81
$$

$$
M_{CB}^F = 0.32(100.0)^2\left(\tfrac{1}{12}\right) = 267 \left.\begin{array}{l}\\ \\ \end{array}\right\} 398 \text{ k-ft}
$$
$$
\qquad\qquad 9(100.0)(0.1461) = 131
$$

$$
M_{CD}^F = \quad 0.32(57.1)^2\left(\tfrac{1}{8}\right) = 130 \left.\begin{array}{l}\\ \\ \end{array}\right\} 228 \text{ k-ft}
$$
$$
\qquad\qquad 9(57.1)(0.1909) = \;\;98
$$

Using the moment distribution summary and the live load plus impact coefficients we determine the moments at CB and CD:

$$
M_{CB} = 348(-0.297) + 398(-0.653) + 228(-0.347)
$$
$$
= -442 \text{ k-ft per line wheels}
$$
$$
M_{CD} = 348(-0.117) + 398(-0.510) + 228(-0.490)
$$
$$
= -355 \text{ k-ft per line wheels}
$$
$$
M_{CB} = 5.408(-442) = -2390 \text{ per box} \xleftarrow{\text{controls}}
$$
$$
M_{CD} = 5.408(-355) = -1922 \text{ per box} \xleftarrow{\text{controls}}
$$

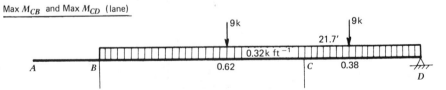

Figure 6.10   Maximum live load moment at CB and CD.

Max $M_{0.6CD}$

Figure 6.11   Maximum live load moment at 0.6 *CD*.

The maximum moment at 0.6*CD* is determined by using the loading shown in Figure 6.11. For which the fixed end moment at *CD* is (see Appendix O)

$$M_{CD}^F = 16(57.1) \left(0.1877 + 0.1680 + \frac{0.0733}{4}\right) = 342 \text{ k-ft}$$

Using the moment distribution summary, we find the corner moment at *CD*:

$$M_{CD} = 0.342(-0.490) = -168 \text{ k-ft}$$

The simple beam shear at *DC* is

$$SBV_{DC} = 16\left(0.35 + 0.6 + \frac{0.85}{4}\right) = 18.6 \text{ k}$$

The moment 0.6*CD*, then, is

$$M_{0.6CD} = 18.6(22.84) = +425$$
$$-4.0(14.0) = -56$$
$$-0.4(-168) = \underline{-67}$$
$$+302 \text{ k-ft per line wheels}$$

$$= 5.521(+302) = +1667 \text{ k-ft per box}$$

The loading for the maximum shear at *BA* is shown in Figure 6.12. The shear at *BA* for this loading is

$$V_{BA} = 5.633(36.0) = 202.8 \text{ k per box}$$

Max V_{BA}

Figure 6.12   Maximum live load shear at *BA*.

Figure 6.13   Maximum live load shear at BC.

The loading for the maximum shear at BC is shown in Figure 6.13. The fixed end moments due to this loading are (see Appendix O)

$$M_{BC}^F = 16(100.0)\left(0.1035 + \frac{0.1452}{4}\right) = 224 \text{ k-ft}$$

$$M_{CB}^F = 16(100.0)\left(0.0169 + \frac{0.0564}{4}\right) = 50 \text{ k-ft}$$

The corner moments for this loading are computed by using the moment distribution summary:

$$M_{BC} = 224(-0.269) + 50(-0.105) = -66 \text{ k-ft}$$
$$M_{CB} = 224(-0.297) + 50(-0.653) = -99 \text{ k-ft}$$

The shear at BC is computed as

$$V_{BC} = 16\left(1.0 + 0.86 + \frac{0.72}{4}\right) + \frac{66 - 99}{100.0} = 32.6 - 0.3$$

$$= 32.3 \text{ k per line wheel}$$

$$= 5.295(32.3) = 171.0 \text{ k per box}$$

The maximum live load plus impact shear at CB is computed in a manner similar to the shear at BC (Figure 6.14):

$$M_{BC}^F = 50 \text{ k-ft}$$

$$M_{CB}^F = 224 \text{ k-ft}$$

$$M_{BC} = 50(-0.269) + 224(-0.105) = -37 \text{ k-ft}$$

$$M_{CB} = 50(-0.297) + 224(-0.653) = -161 \text{ k-ft}$$

$$V_{CB} = 32.6 + \frac{161 - 37}{100.0} = 32.6 + 1.2 = 33.8 \text{ k per line wheels}$$

$$V_{CB} = 5.295(33.8) = 179.0 \text{ k per box}$$

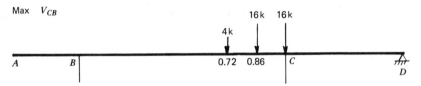

Figure 6.14 Maximum live load shear at CB.

The maximum live load plus impact shear at CD and DC is computed in a similar manner (Figure 6.15a and b):

$$M_{CD}^F = 16(57.1)\left(0.1641 + \frac{0.1887}{4}\right) = 193 \text{ k-ft}$$

$$M_{CD} = 193(-0.490) = -95 \text{ k-ft}$$

$$V_{CD} = 16\left(1.0 + 0.75 + \frac{0.51}{4}\right) + \frac{95}{57.1} = 30.0 + 1.7$$

$$= 31.7 \text{ k per line wheels}$$

$$= 5.521(31.7) = 175.0 \text{ k per box}$$

$$M_{CD}^F = 16(57.1)\left(\frac{0.1862}{4} + 0.1172\right) = 150 \text{ k-ft}$$

$$M_{CD} = 150(-0.490) = -73 \text{ k-ft}$$

$$V_{DC} = 30.0 - \frac{73}{57.1} = 30.0 - 1.3 = 28.7 \text{ k per line wheel}$$

$$= 5.521(28.7) = 158.6 \text{ k per box}$$

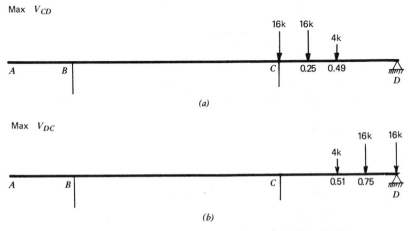

Figure 6.15 Maximum live load shear at CD and DC.

Table 6.8   Summary of Moments and Shears

	DL	LL + I	Σ
M_{BA}	-6522	-4941 ✓	-11,463 k-ft
M_{BC}	-7208	-3612 ✓	-10,820 k-ft
$M_{0.5BC}$	+4532	+2536 ✓	+7068 k-ft
M_{CB}	-6481	-2390 ✓	-8871 k-ft
M_{CD}	-5738	-1922 ✓	-7660 k-ft
$M_{0.6CD}$	+1181	+1667 ✓	+2848 k-ft
V_{BA}	352.2	202.8 ✓	555.0 k
V_{BC}	456.9	171.0	627.9 k
V_{CB}	442.3	179.0	621.3 k
V_{CD}	354.2	175.0	529.2 k
V_{DC}	153.2	158.6	311.8 k

A summary of the critical moments and shears is tabulated and summed for dead load plus live load and impact in Table 6.8.

With these moments the various sections can be designed. At section *BA* the shear and moments are $V = 555.0$ k and $M = 11,463$ k-ft. The approximate required area of reinforcement for negative moment is computed by assuming $j = 0.90$ and $d = \pm 65$ in.

$$A_s = \frac{11,463(12)}{24(0.90)(65)} = 98.0 \text{ in.}^{2\pm}$$

Try 44-#14, $A_s = 99.00$ in.^{2}

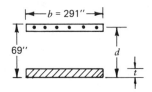

The width of the entire box (out-to-out of web) is $b = 291$ in. and $d = 69.00 - 2.0 - 0.75 - 0.85 = 65.40$ in.; $t = 7.00$ in. Using these assumed section dimensions and neglecting the webs (stems) we compute the stresses per Appendix B:

$$f_s = \frac{M}{A_s jd} \quad \text{and} \quad f_c = \frac{f_s(k)}{n(1 - k)}$$

$$pn = \frac{99.00(10)}{291(65.40)} = 0.0520 \quad \text{and} \quad \frac{t}{d} = \frac{7.0}{65.40} = 0.107$$

Therefore from Appendix B $k = 0.366$ $j = 0.949$.

$$f_s = \frac{11{,}463(12)}{99.00(0.949)(65.40)} = 22{,}387 \text{ psi} < 24{,}000 \text{ psi allowable}$$

$$f_c = \frac{22{,}434(0.366)}{10(0.634)} = 1292 \text{ psi} < 1320 \text{ psi allowable}$$

The design of the webs, which resist the entire shear, will consider the potential of developing a severe local stress concentration at the skew-supported end B because of the secondary shears due to torsion. To account for these influences the vertical shears are increased by the secant of the skew angle.

Therefore $V = 555.0(1.1547) = 640.9$ k. Assume 4-12 in. webs; $b = 4 \times 12 = 48$ in. Then

$$v = \frac{V}{bd} = \frac{640.9}{48(65.40)} = 204 \text{ psi}$$

Use $v_c = 0.95\sqrt{f'_c} = 55$ psi:

$$v - v_c = 204 - 55 = 149 \text{ psi} > 2\sqrt{f'_c} = 114 \text{ psi}$$
$$< 4\sqrt{f'_c} = 230 \text{ psi}$$

Maximum spacing = 12 in. (AASHTO Article 1.5.10 and 1.5.29).

Use #5 stirrups. 8-#5 stirrups legs; therefore $A_v = 8 \times 0.31 = 2.48$ in.^{2}, and the spacing is

$$s = \frac{A_v f_v}{(v - v_c)\, b} = \frac{2.48(24000)}{149(48)} = 8.3 \text{ in.}$$

The moment at BC is less than the moment at BA but the top deck is also thinner; therefore this section is analyzed by using the reinforcement at section BA and the information in Appendix B.

Section @ BC. $M = 10{,}820$ k-ft and $V = 627.9$ k.

$$44\text{-}\#14, \qquad A_s = 99.00 \text{ in.}^2$$
$$b = 291 \text{ in.} \qquad d = 65.40 \text{ in.} \qquad t = 6.50 \text{ in.}$$

$$pn = 0.0520 \quad \text{and} \quad \frac{t}{d} = \frac{6.50}{65.40} = 0.099$$

Therefore from Appendix B $k = 0.376$ $j = 0.953$. The stresses are

$$f_s = \frac{10,820(12)}{99.00(0.953)(65.40)} = 21,043 \text{ psi} < 24,000 \text{ psi}$$

$$f_c = \frac{21,043(0.376)}{10(0.624)} = 1268 \text{ psi} < 1320 \text{ psi}$$

Because the shear in this section is greater than the shear in BA, the webs are increased in thickness; 4–13 in. thick webs at BC:

$$b = 4 \times 13 = 52.0 \text{ in.}$$

Again the shear is magnified for the obtuse corner:

$$v = \frac{V}{b_w d} = \frac{627.9(1.1547)}{52.0(65.40)} = 213 \text{ psi}$$

$$v - v_c = 213 - 55 = 158 \text{ psi} > 2\sqrt{f_c'} = 114 \text{ psi}$$

$$< 4\sqrt{f_c'} = 230 \text{ psi}$$

Maximum spacing = 12 in. (AASHTO Article 1.5.10 and 1.5.29). Again use 8–#5 stirrup legs: $A_v = 2.48$ in.^{2} The required spacing is

$$s = \frac{A_v f_v}{(v - v_c) b} = \frac{2.48(24,000)}{158(52)} = 7.2 \text{ in.}$$

Use 7-in. spacing. Next the section at CB is designed. The shear and moment at service load are

Section @ CB.

$$M = 8871 \text{ k-ft} \qquad V = 621.3 \text{ k}$$

Assume $d = 65$ in.± and $j = 0.90$. The approximate area of reinforcment required is

$$A_s = \frac{8871(12)}{24(0.90)(65)} = 75.8 \text{ in.}^2$$

Try 48–#11, $A_s = 74.88$ in.^{2}

$b = 291$ in. $d = 69.00 - 2.00 - 0.75 - 0.70 = 65.55$ in. $t = 5.5$ in.

By using Appendix B k and j for T-beams are determined:

$$pn = \frac{74.88(10)}{291.0(65.55)} = 0.0392 \quad \text{and} \quad \frac{t}{d} = 5.5/65.55 = 0.0839$$

therefore $k = 0.344$ and $j = 0.958$. The actual stresses are computed and compared with the allowables:

$$f_s = \frac{8871(12)}{74.88(0.958)(65.55)} = 22,639 \text{ psi} < 24,000 \text{ psi}$$

$$f_c = \frac{22,639(0.344)}{10(0.656)} = 1187 \text{ psi} < 1320 \text{ psi}$$

The shear requirements at CB are computed in a manner similar to that performed at BC and BA. Again try 4 13-in. webs, $b = 52.0$ in. Compute the unit shear magnified for the obtuse corner by multiplying the shear by the secant of the angle:

$$v = \frac{V}{b_w d} = \frac{621.3(1.1547)}{52(65.55)} = 210 \text{ psi}$$

$$v - v_c = 210 - 55 = 155 \text{ psi} > 2\sqrt{f_c'} = 114 \text{ psi}$$
$$< 4\sqrt{f_c'} = 230 \text{ psi}$$

Maximum spacing is 12 in.
 #5 Stirrups. Use 8 legs, $A_v = 2.48$ in.^{2} The required spacing is

$$s = \frac{2.48(24,000)}{52(155)} = 7.4 \text{ in.}$$

Use 7-in. spacing.
 The design moments and shears at CD and DC are less than those already computed; therefore the two critical positive moment sections are designed next. First is the section at $0.5BC$ for which the moment is

$$M = 7068 \text{ k-ft}$$

Assume that $j = 0.90$ and $d = 66\pm$ in.; the approximate area of steel required is

$$A_{s(required)} = \frac{7068(12)}{24(0.90)(66)} = 59.5 \text{ in.}^2$$

Try 39-#11, A_s = 59.28 in.^{2}. Then

$$b = 364 \text{ in.} \quad d = 69.00 - 1.5 - 0.62 - 0.70 = 66.18 \text{ in.,} \quad t = 7.5 \text{ in.}$$

Again, using Appendix B, k and j are computed as

$$pn = \frac{59.28(10)}{364(66.18)} = 0.0246 \quad \text{and} \quad \frac{t}{d} = \frac{7.50}{66.18} = 0.113$$

Therefore $k = 0.226$ and $j = 0.950$.

The actual stresses are computed and compared with the allowables:

$$f_s = \frac{7068(12)}{59.28(0.950)(66.18)} = 22{,}757 \text{ psi} < 24{,}000 \text{ psi}$$

$$f_c = \frac{22{,}757(0.226)}{10(0.774)} = 664 \text{ psi} < 1320 \text{ psi}$$

Using the same procedure, we design the section at $0.6CD$:

$$M = +2848 \text{ k-ft}$$

$$\text{Approx } A_{s(\text{required})} = \frac{2848(12)}{24(0.90)(66)} = 24.0 \text{ in.}^2$$

Try 29-#8, A_s = 22.91 in.^{2}

$$b = 364 \text{ in.} \quad d = 69.00 - 1.5 - 0.62 - 0.40 = 66.48 \text{ in.}^2 \quad t = 7.5 \text{ in.}$$

$$pn = \frac{22.91(10)}{364(66.48)} = 0.0094$$

$$\frac{t}{d} = \frac{7.5}{66.48} = 0.113$$

Use $j = 0.95$ (conservative value):

$$f_s = \frac{28.48(12)}{22.91(0.95)(66.48)} = 23{,}620 \text{ psi}$$

Now that the superstructure sections have been designed it is proper to proceed with the design of the bent columns and footings in a manner similar to the approach used in Chapter 10. After completion of the substructure design the initial assumptions relating to the section properties and frame characteristics must be verified and then the superstructure design can be completed.

Next the bent #1 cap (support B), with $19'$-$6''$ cc of two columns to supporting the cap, is designed. For cap designs of this kind it is reasonable to neglect the monolithic column-to-cap connection because the stiffness of the cap is much greater than that of the column; thus the column acts as a simple support. The bent #2 cap (support C) is made the same as the bent #1 cap, which is the most critical.

The superstructure dead load can be spread equally over four girder webs; therefore the reaction per web = $(352.2 + 456.9)/4 = 202.3$ k, where $V_{BA} = 352.2$ k and $V_{BC} = 456.9$ k and the cap DB (assumed) equals 2.92 k-ft^{-1}.

The dead load is placed as shown in Figure 6.16. Note the different cantilever dimensions due to the girder flare. The dead load moments and shears at the critical locations can now be computed:

$$M_B = 202.3(4.33) \qquad = -876 \atop 2.92(4.75)(2.38) \quad = -29 \Bigg\} \quad -905 \text{ k-ft}$$

$$M_C = 202.3(3.83) \qquad = -775 \atop 2.92(4.25)(2.12) \quad = -26 \Bigg\} \quad -801 \text{ k-ft}$$

$$M_{0.5BC} = 202.3(5.22) \qquad = \qquad +1056 \text{ k-ft}$$
$$2.92(19.50)^2 \left(\tfrac{1}{8}\right) \quad = \qquad +138 \text{ k-ft} \quad \Bigg\} = +341 \text{ k-ft}$$
$$-\left(\frac{905 + 801}{2}\right) \qquad = \qquad -853 \text{ k-ft}$$

$$V_{BA} = 202.3 + 2.92(4.75) = \qquad 216.2 \text{ k}$$

$$V_{BC} = 202.3 + 9.75(2.92) = \qquad 230.8 \text{ k}$$

The maximum live load reaction at bent #1 occurs in the loading shown in Figure 6.17. With this loading corner moments can be computed by using the moment distribution summary:

$$M_{BA}^F = -877 \text{ k-ft}$$

$$M_{BA} = 877(-1.0) = -877 \text{ k-ft}$$

$$M_{BC} = 877(-0.731) = -641 \text{ k-ft}$$

$$M_{CB} = 877(+0.297) = +260 \text{ k-ft}$$

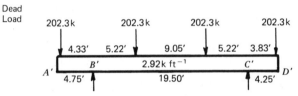

Figure 6.16 Dead load: bent cap.

Figure 6.17   Live load: bent cap.

The reaction at B' then equals

$$R_B = 36.0 + \frac{641 + 260}{100.0} = 36.0 + 9.0 = 45.0 \text{ k per line of wheels}$$

$$R_B = 45.0 \times 5.633 = 253.5 \text{ k per box}$$

$$R = 253.5 \div 4 = 63.4 \text{ k per girder web}$$

Assuming that the live load reaction for some of the adjacent webs is zero and only the 63.4 k reaction is concentrated over bent, the following two cases of live load plus impact (Figure 6.18) are assumed. The computed moments and shears are

$$M_B = -905 - 275 = -1180 \text{ k-ft}$$

$$V_{BA} = 216.2 + 63.4 = 279.6 \text{ k}$$

$$V_{BC} = 230.8 + 63.4 = 294.2 \text{ k}$$

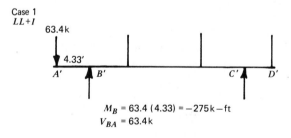

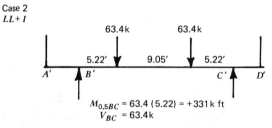

Figure 6.18   Live load reactions: bent cap.

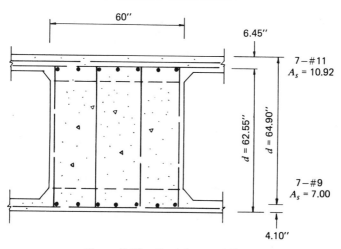

Figure 6.19   Bent 1 cap section.

The cap cross section is shown in Figure 6.19. The approximate required reinforcement is computed by assuming $j = 0.90$ and $d = 62$ in. ±:

$$\text{approx } A_{s(\text{required})} = \frac{1180(12)}{24(0.90)(62)} = 10.6 \text{ in.}^2$$

Try 7-#11, $A_s = 10.92$ in.^{2}; k and j are selected from Appendix A for rectangular sections, where $p = 10.92/[60(62.55)] = 0.0029$; therefore $k = 0.212$ and $j = 0.929$. The actual stresses are computed for negative moment at the support and compared with the allowables:

$$f_s = \frac{1180(12)}{10.92(0.929)(62.55)} = 22{,}315 \text{ psi} < 24{,}000 \text{ psi}$$

$$f_c = \frac{22{,}315(0.212)}{10(0.788)} = 600 \text{ psi} < 1320 \text{ psi}$$

The maximum unit shear at this section is

$$v = \frac{294.2}{69(62.55)} = 78 \text{ psi}$$

The moment that develops in each column is due to the differential bending moment at the support which induces a torque through the cap from one web on each side of the column. The amount of torsion that the cap must take is the moment for both columns divided by four.

For dead load the torsional moment is

$$M_{\text{tors}}^{DL} = \frac{-686}{4} = -172 \text{ k-ft}$$

For the cantilever $LL + I$ loading the torsional moment is

$$M_{\text{tors}}^{LL+I} = \frac{5.633(+236)}{4} = +332 \text{ k-ft}$$

The total torsional moment therefore is

$$M_{\text{tors}}^{DL+LL+I} = -172 + 332 = +160 \text{ k-ft}$$

The torsional unit shear is computed as

$$v_T = \frac{3M_{\text{tors}}}{x^2 y}$$

x = short side dimension

y = long side dimension

Then

$$v_T = \frac{3(160)(12)}{(60)^2(69)} = 23 \text{ psi}$$

For direct shear

$$\frac{A_v}{s} = \frac{(v - v_c)(b)}{f_v} = \frac{(78 - 55)(60)}{24,000} = 0.0575 \text{ in.}^2 \text{ in.}^{-1}$$

This could be spread over four stirrup legs.

$$\frac{A_v}{s} = \frac{0.0575}{4} = 0.0144 \text{ in.}^2 \text{ in.}^{-1} \text{ per leg}$$

For torsion from ref. 14

$$\frac{A_v}{s} = \frac{M_{\text{tors}}}{0.8(f_s) x_1 y_1}$$

x_1 = short side hoop length

y_1 = long side hoop length

$$\frac{A_v}{s} = \frac{160(12)}{0.8(24.0)(56)(60.3)} = 0.0296 \text{ in.}^2 \text{ in.}^{-1}$$

The requirement is spread over the two exterior stirrup legs:

$$\frac{A_v}{s} = \frac{0.0296}{2} = 0.0148 \text{ in.}^2 \text{ in.}^{-1} \text{ per leg}$$

Summing the stirrup requirement,

$$\sum \frac{A_v}{s} = 0.0144 + 0.0148 = 0.0292 \text{ in.}^2 \text{ in.}^{-1} \text{ per leg}$$

With #5 $A_v = 0.31$ in.^{2} in.$^{-1}$ the required spacing $s = 0.31/0.0292 = 10.6$ in. Use #5 ⌐⌐ @ 10 in.

Some longitudinal reinforcement should be spread equally around the torsional hoops. The required area is

$$A_{SL} = A_y \frac{(x_1 + y_1)}{s} = 0.62 \left(\frac{56 + 60.3}{10}\right) = 7.21 \text{ in.}^2$$

For positive moment the design moment is

$$M_{0.5BC} = +341 + 331 = +672 \text{ k-ft}$$

The approximate area of reinforcement required is

$$\text{approx } A_{s(\text{required})} = \frac{672(12)}{24(0.90)(65\pm)} = 5.74 \text{ in.}^2$$

Try 7-#9, $A_s = 7.00$ in.^{2} Using Appendix A for rectangular sections,

$$p = \frac{7.00}{60(64.90)} = 0.0017$$

Therefore $k = 0.164$ and $j = 0.945$.

The stresses are then computed and compared with the allowables;

$$f_s = \frac{672(12)}{7.00(0.945)(64.90)} = 18,783 \text{ psi} < 24,000 \text{ psi}$$

$$f_c = \text{small}$$

With the cap design completed, the detailed design of the longitudinal box can be started.

At BA the previously computed dead load shear and moment are

$$V_{BA} = -352.2 \text{ k} \qquad M_{BA} = -6522 \text{ k-ft}$$

Using the dead load diagram shown in the Fig. 6.20, we compute the cantilever dead load moment 14 ft from A and 28 ft from A;

@ $28'AB$

$$9.278 \times 28.0 = \quad 259.8 \times 14.00 = -3637$$
$$\tfrac{1}{2} \times 0.262 \times 28.0 = \quad 3.7 \times 18.67 = \quad -68$$
$$0.053 \times 28.0 = \quad 1.5 \times 14.00 = \quad -21$$
$$\underline{\qquad \qquad \quad 34.2 \times 27.50 = \quad -940}$$
$$-299.2 \text{ k} \qquad \qquad -4666 \text{ k-ft}$$

@ $14'AB$

$$9.278 \times 14.0 = \quad 129.9 \times \quad 7.00 = -909$$
$$\tfrac{1}{2} \times 0.131 \times 14.0 = \quad 0.9 \times \quad 9.33 = \quad -8$$
$$0.184 \times 14.0 = \quad 2.6 \times \quad 7.00 = \quad -18$$
$$\underline{\qquad \qquad \quad 34.2 \times 12.50 = \quad -462}$$
$$-167.6 \text{ k} \qquad \qquad -1397 \text{ k-ft}$$

and the shear at AB is

$$V_{AB} = 34.2 \text{ k}$$

Moment and shear at BC, as previously calculated are

$$M_{BC} = 7209 \text{ k-ft} \qquad V_{BC} = 456.9 \text{ k}$$

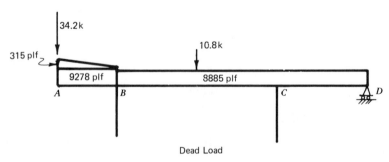

Figure 6.20   Dead load: beam.

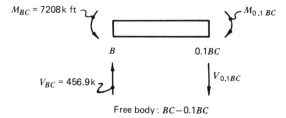

Free body: $BC - 0.1BC$

Figure 6.21 Free body: Section $B \sim 0.1\ BC$.

To calculate the shear and moment at each tenth point across the span BC a free body of the span from BC to the tenth points is used. An example of a free body from BC to $0.1BC$ is diagrammed in Figure 6.21.

Using the free body shown for BC to $0.1BC$, we sum the vertical forces to get $V_{0.1BC}$, just as the moments about $0.1BC$ are summed to get $M_{0.1BC}$:

@ 0.1BC

	V (k)	M (k-ft)
		-7208
	$+456.9 \times 10.0\ \text{ft} =$	$+4569$
$8.885 \times 10.0 =$	$-88.8 \times \quad 5.0\ \text{ft} =$	-444
	$+368.1\ \text{k}$	$-3083\ \text{k-ft}$

The same procedure is used to obtain V and M at the remaining points in span BC.

@ 0.2BC

		M
		-3083
	$+368.1 \times 10.0\ \text{ft} =$	$+3681$
$8.885 \times 10.0 =$	$-88.9 \times \quad 5.0\ \text{ft} =$	-444
	$+279.2\ \text{k}$	$+154\ \text{k-ft}$

@ 0.3BC

		M
		$+154$
	$+279.2 \times 10.0\ \text{ft} =$	$+2792$
$8.885 \times 10.0 =$	$-88.8 \times \quad 5.0\ \text{ft} =$	-444
	$+190.4\ \text{k}$	$+2502\ \text{k-ft}$

@ *0.4BC*

$$
\begin{array}{rr}
& +2502 \\
+190.4 \times 10.0 \text{ ft} = & +1904 \\
8.885 \times 10.0 = \underline{-88.9 \times 5.0 \text{ ft} =} & \underline{-444} \\
+101.5 \text{ k} & +3962 \text{ k-ft}
\end{array}
$$

@ *0.5BC*

$$
\begin{array}{rr}
& +3962 \\
+101.5 \times 10.0 \text{ ft} = & +1015 \\
8.885 \times 10.0 = \underline{-88.8 \times 5.0 \text{ ft} =} & \underline{-444} \\
+12.7 \text{ k} & +4533 \text{ k-ft}
\end{array}
$$

@ *0.6BC*

$$
\begin{array}{rr}
& +4533 \\
+12.7 \times 10.0 \text{ ft} = & +127 \\
8.885 \times 10.0 = -88.9 \times 5.0 \text{ ft} = & -444 \\
\underline{-10.8 \times 10.0 \text{ ft} =} & \underline{-108} \\
-87.0 \text{ k} & +4108 \text{ k-ft}
\end{array}
$$

@ *0.7BC*

$$
\begin{array}{rr}
& +4108 \\
-87.0 \times 10.0 \text{ ft} = & -870 \\
8.885 \times 10.0 = \underline{-88.8 \times 5.0 \text{ ft} =} & \underline{-444} \\
-175.8 \text{ k} & +2794 \text{ k-ft}
\end{array}
$$

@ *0.8BC*

$$
\begin{array}{rr}
& +2794 \\
-175.8 \times 10.0 \text{ ft} = & -1758 \\
8.885 \times 10.0 = \underline{-88.9 \times 5.0 \text{ ft} =} & \underline{-444} \\
-264.7 \text{ k} & +592 \text{ k-ft}
\end{array}
$$

@ *0.9BC*

$$
\begin{array}{rr}
& +592 \\
-264.7 \times 10.0 \text{ ft} = & -2647 \\
8.885 \times 10.0 = \underline{-88.8 \times 5.0 \text{ ft} =} & \underline{-444} \\
-353.5 \text{ k} & -2499 \text{ k-ft}
\end{array}
$$

@ *CB*

$$
\begin{array}{r}
-2499 \\
-353.5 \times 10.0 \text{ ft} = -3535 \\
8.885 \times 10.0 = \underline{-88.9} \times \quad 5.0 \text{ ft} = \underline{-444} \\
-442.4 \text{ k} \qquad -6478 \text{ k-ft}
\end{array}
$$

For span *CD* the same procedure is used to sum verticals and moments for each tenth of the span. The previously computed moment and shear at *CD* are

$$M_{CD} = -5738 \qquad V_{CD} = 354.2 \text{ k}$$

The moments and shears at each tenth point are the following:

@ *0.1CD*

$$
\begin{array}{r}
-5738 \\
+354.2 \times 5.71 \text{ ft} = +2022 \\
8.885 \times 5.71 = \underline{-50.7} \times 2.86 \text{ ft} = \underline{-145} \\
+303.5 \text{ k} \qquad -3861 \text{ k-ft}
\end{array}
$$

@ *0.2CD*

$$
\begin{array}{r}
-3861 \\
+303.5 \times 5.71 \text{ ft} = +1733 \\
\underline{-50.7} \times 2.85 \text{ ft} = \underline{-145} \\
+252.8 \text{ k} \qquad -2273 \text{ k-ft}
\end{array}
$$

@ *0.3CD*

$$
\begin{array}{r}
-2273 \\
+252.8 \times 5.71 \text{ ft} = +1443 \\
\underline{-50.7} \times 2.86 \text{ ft} = \underline{-145} \\
+202.1 \text{ k} \qquad -975 \text{ k-ft}
\end{array}
$$

@ *0.4CD*

$$
\begin{array}{r}
-975 \\
+202.1 \times 5.71 \text{ ft} = +1154 \\
\underline{-50.7} \times 2.85 \text{ ft} = \underline{-145} \\
+151.4 \text{ k} \qquad +34 \text{ k-ft}
\end{array}
$$

@ 0.5CD

$$
\begin{array}{ll}
 & +34 \\
+151.4 \times 5.71 \text{ ft} = & +864 \\
\underline{-50.7 \times 2.86 \text{ ft} = } & \underline{-145} \\
+100.7 \text{ k} & +753 \text{ k-ft}
\end{array}
$$

@ 0.6CD

$$
\begin{array}{ll}
 & +753 \\
+100.7 \times 5.71 \text{ ft} = & +575 \\
\underline{-50.7 \times 2.85 \text{ ft} = } & \underline{-145} \\
+50.0 \text{ k} & +1183 \text{ k-ft}
\end{array}
$$

@ 0.7CD

$$
\begin{array}{ll}
 & +1183 \\
+50.0 \times 5.71 \text{ ft} = & +286 \\
\underline{-50.7 \times 2.86 \text{ ft} = } & \underline{-145} \\
-0.7 \text{ k} & +1324 \text{ k-ft}
\end{array}
$$

@ 0.8 CD

$$
\begin{array}{ll}
 & +1324 \\
-0.7 \times 5.71 \text{ ft} = & -4 \\
\underline{-50.7 \times 2.85 \text{ ft} = } & \underline{-145} \\
-51.4 \text{ k} & +1175 \text{ k-ft}
\end{array}
$$

@ 0.9CD

$$
\begin{array}{ll}
 & +1175 \\
-51.4 \times 5.71 \text{ ft} = & -293 \\
\underline{-50.7 \times 2.86 \text{ ft} = } & \underline{-145} \\
-102.1 \text{ k} & +737 \text{ k-ft}
\end{array}
$$

@ DC

$$
\begin{array}{ll}
 & +737 \\
-102.1 \times 5.71 \text{ ft} = & -583 \\
\underline{-50.7 \times 2.85 \text{ ft} = } & \underline{-145} \\
-152.8 \text{ k} & +9 \text{ k-ft say 0}
\end{array}
$$

With the dead load actions evaluated, the influence line ordinates must be determined for LL effects. The fixed end moments are calculated for the spans by using the table in Appendix O, which is for spans with constant moments of inertia. This tabulation, including the computed corner moments which are calculated by using the fixed end moments and the moment distribution, appears in Table 6.9: for example with 1 k at $0.2BC$, $M_{BC} = 12.80(0.269) + 3.20(0.105) = -3.78$ k-ft or $M_{CB} = 12.80(0.297) + 3.20(0.653) = -5.89$ k-ft. The signs shown are bending signs contrasted to the rotational signs given in the moment distribution summary. The areas under the moment curves are tabulated as the sum of the ordinates times one tenth of the span length.

Table 6.9 lists the influence line ordinates for the corner moments.

Table 6.9   Influence Lines

Load @	M_{BA}^F	M_{BC}^F	M_{CB}^F	M_{CD}^F	M_{BA}	M_{BC}	M_{CB}	M_{CD}
A	33.70				-33.70	-24.63	+10.01	+3.94
$14'AB$	19.70				-19.70	-14.40	+5.85	+2.31
$28'AB$	5.70				-5.70	-4.17	+1.69	+0.67
B								
0.1		8.10	0.90			-2.27	-2.99	-1.41
0.2		12.80	3.20			-3.78	-5.89	-3.13
0.3		14.70	6.30			-4.62	-8.48	-4.93
0.4		14.40	9.60			-4.88	-10.55	-6.58
0.5		12.50	12.50			-4.68	-11.88	-7.84
0.6		9.60	14.40			-4.09	-12.25	-8.47
0.7		6.30	14.70			-3.24	-11.47	-8.23
0.8		3.20	12.80			-2.20	-9.31	-6.90
0.9		0.90	8.10			-1.09	-5.56	-4.24
C								
0.1				4.88		+0.51	-1.69	-2.39
0.2				8.22		+0.86	-2.85	-4.03
0.3				10.19		+1.07	-3.54	-4.99
0.4				10.96		+1.15	-3.80	-5.37
0.5				10.71		+1.12	-3.72	-5.25
0.6				9.59		+1.01	-3.33	-4.70
0.7				7.79		+0.82	-2.70	-3.82
0.8				5.48		+0.58	-1.90	-2.69
0.9				2.83		+0.30	-0.98	-1.39
D								
Area span AB					-568	-415	+169	+66
Area span BC					—	-308	-784	-517
Area span CD					—	+42	-140	-198
Total area					—	-681	-755	-649

To tabulate the influence lines for moments at points in the spans, where the loads are applied the simple span moments $(+M)$ are combined with corner moment effects $(-M)$ at the points of loading and moment diagram influence line ordinates are established as in Table 6.10. This procedure has been explained in more detail in Chapters 4 and 5.

The influence line ordinates for spans BC and CD are computed similarly and are listed in Tables 6.10 and 6.11. The influence lines are then plotted from Tables 6.10 and 6.11.

The maximum HS 20 live load plus impact moments are taken from the influence lines by placing the truck graphically at the maximum location and selecting the appropriate ordinates. Similarly, the maximum lane loading moments are determined by using ordinates for the concentrated loads and areas for the uniform loads.

The maximum ordinates that correspond to each wheel position for the HS 20 truck are tabulated for positive moment, multiplied by each concentrated wheel load, and summed to get the moment for one line of wheels (Table 6.12). When this is greater than the same moment obtained from lane loading, the moment is multiplied by the previously computed distribution plus impact coefficient to get the moment to be applied to the total box girder $(M_{LL + I})$.

The maximum positive moments for lane load are computed similarly and then compared with the truck moments. The lane loading is set equivalent to one line of wheels by setting $w = 0.640$ k-ft^{-1} equal to $w/2$ and then multiplying the resulting moment by the distribution plus impact coefficient to obtain the moment per entire box girder (Table 6.13).

Note that the truck loading controls all the critical locations for positive moment, which is typical for this span range. The same procedure is used for the negative HS 20 truck moments and the negative lane moments, as shown in Tables 6.14 and 6.15, respectively.

Note here that critical negative moments adjacent to the cantilever span are produced by truck loadings; critical negative moments over a continuous support are produced by lane loadings. Also note that for negative moments two concentrated loads in different spans are applied according to the AASHTO loading criteria for maximum moments.

The maximum positive and negative dead load plus live load plus impact moments can be obtained as listed in Table 6.16. These moments are applicable to the entire box girder acting as a unit.

Next, the resisting moments for different levels of reinforcement are computed. The design of the critical sections indicated that a maximum of 44–#14 at B and 48–#11 at C were needed. With these maxima, the bars are cut in increments of four or eight to provide a cutoff pattern consistent with the four girder webs. The j values computed for the maximum reinforcement give the minimum j value, hence conserva-

Table 6.10 Influence Lines: Span BC

Load @		BC	0.1	0.2	0.3	0.4	0.5	0.6	0.7	0.8	0.9	CB
0.1	+M		+9.00									
	−M	−2.27	−2.34									
	Σ		+6.66	+5.59	+4.52	+3.44	+2.37	+1.30	+0.23	−0.85	−1.92	−2.99
0.2	+M			+16.00								
	−M	−3.78		−4.20								
	Σ		+4.01	+11.80	+9.59	+7.38	+5.17	+2.95	+0.74	−1.47	−3.68	−5.89
0.3	+M				+21.00							
	−M	−4.62			−5.78							
	Σ		+1.99	+8.61	+15.22	+11.83	+8.45	+5.06	+1.68	−1.71	−5.09	−8.48
0.4	+M					+24.00						
	−M	−4.88				−7.15						
	Σ		+0.55	+5.99	+11.42	+16.85	+12.28	+7.72	+3.15	−1.42	−5.98	−10.55
0.5	+M						+25.00					
	−M	−4.68					−8.28					
	Σ		−0.40	+3.88	+8.16	+12.44	+16.72	+11.00	+5.28	−0.44	−6.16	−11.88
0.6	+M							+24.00				
	−M	−4.09						−8.99				
	Σ		−0.91	+2.28	+5.46	+8.64	+11.83	+15.01	+8.20	+1.38	−5.44	−12.25
0.7	+M								+21.00			
	−M	−3.24							−9.00			
	Σ		−1.06	+1.11	+3.29	+5.47	+7.65	+9.82	+12.00	+4.18	−3.65	−11.47
	+M									+16.00		

Table 6.10 (Continued)

Load @		BC	0.1	0.2	0.3	0.4	0.5	0.6	0.7	0.8	0.9	CB
0.8	−M	−2.20								−7.89		−9.31
	Σ		−0.91	+0.38	+1.67	+2.95	+4.24	+5.53	+6.82	+8.11	−0.60	
0.9	+M										+9.00	
	−M	−1.09									−5.11	−5.56
	Σ		−0.54	+0.02	+0.57	+1.12	+1.68	+2.23	+2.78	+3.34	+3.89	
Positive area			+132	+397	+599	+701	+704	+606	+409	+170	+39	
Negative area			−38							−59	−325	

Table 6.11 Influence Lines: Span CD

Load @		CD	0.1	0.2	0.3	0.4	0.5	0.6	0.7	0.8	0.9	DC
0.1	+M		+5.14									
	−M	−2.39	−2.15									0
	Σ		+2.99	+2.66	+2.33	+1.99	+1.66	+1.33	+1.00	+0.66	+0.33	
0.2	+M			+9.14								
	−M	−4.03		−3.22								
	Σ		+0.94	+5.92	+5.18	+4.44	+3.70	+2.96	+2.22	+1.48	+0.74	
	+M				+11.99							

		1	2	3	4	5	6	7	8	9	10
0.3	$-M$	−4.99			−3.49						
	Σ		−0.49	+4.00	+8.50	+7.29	+6.07	+4.86	+3.64	+2.43	+1.21
	$+M$										
0.4	$-M$	−5.37				−3.22					
	Σ		−1.41	+2.56	+6.52	+10.48	+8.73	+6.99	+5.24	+3.49	+1.75
	$+M$					+13.70					
0.5	$-M$	−5.25					−2.62				
	Σ		−1.87	+1.51	+4.90	+8.28	+11.66	+9.33	+7.00	+4.66	+2.33
	$+M$						+14.28				
0.6	$-M$	−4.70						−1.88			
	Σ		−1.95	+0.81	+3.56	+6.31	+9.07	+11.82	+8.86	+5.91	+2.96
	$+M$							+13.70			
0.7	$-M$	−3.82							−1.15		
	Σ		−1.73	+0.36	+2.46	+4.56	+6.65	+8.75	+10.84	+7.23	+3.61
	$+M$								+11.99		
0.8	$-M$	−2.69								−0.54	
	Σ		−1.28	+0.13	+1.54	+2.95	+4.37	+5.78	+7.19	+8.60	+4.30
	$+M$									+9.14	
0.9	$-M$	−1.39									−0.14
	Σ		−0.68	+0.03	+0.74	+1.45	+2.16	+2.87	+3.58	+4.29	+5.00
	$+M$										+5.14
Positive area			+22	+103	+204	+273	+309	+312	+283	+221	+127
Negative area			−54								

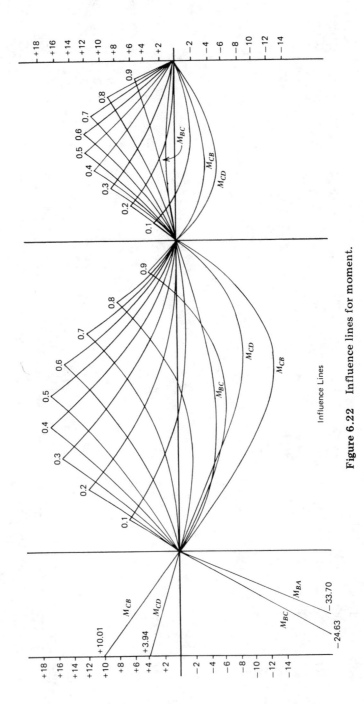

Figure 6.22 Influence lines for moment.

Table 6.12 Live Load and Impact: Positive Moment--Truck

Point	16 k	16 k	4 k	One Line Wheels M_{LL}	Total M_{LL+I} (kip-ft)
B					
0.1	+6.66	+3.2	+0.9	+161	+852
0.2	+11.80	+7.5	+4.3	+326	+1726
0.3	+15.22	+10.1	+7.5	+435	+2303
0.4	+16.85	+10.9	+10.0	+484	+2563
0.5	+16.72	+10.7	+10.0	+479	+2536
0.6	+15.01	+9.7	+8.1	+428	+2266
0.7	+12.00	+6.9	+5.1	+323	+1710
0.8	+8.11	+3.0	+1.9	+185	+980
0.9	+3.89	—	—	+62	
C					
0.1	+2.99	—	—	+48	
0.2	+5.92	+2.1	+0.4	+130	+718
0.3	+8.50	+4.3	+1.6	+211	+1165
0.4	+10.48	+5.6	+3.2	+270	+1491
0.5	+11.66	+5.8	+4.9	+299	+1651
0.6	+11.82	+6.0	+4.5	+303	+1673
0.7	+10.84	+6.1	+2.4	+281	+1551
0.8	+8.60	+5.5	+2.6	+236	+1303
0.9	+5.00	+3.3	+1.8	+140	+772
D					

Table 6.13 Live Load + Impact: Positive Moment--Lane

Point	0.32 k ft^{-1}	0.32 k ft^{-1}	32 k ft^{-1}	9 k	One Line Wheels M_{LL}	M_{LL+I} (kip-ft)
B						
0.1	+132		+24	+6.66	+110	
0.2	+397		+6	+11.80	+235	
0.3	+599			+15.22	+329	
0.4	+701			+16.85	+376	
0.5	+704			+16.72	+376	
0.6	+606			+15.01	+329	
0.7	+409			+12.00	+239	
0.8	+170	+52		+8.11	+144	
0.9	+39	+110		+3.89	+83	+439
C						
0.1	+59	+22		+2.99	+53	+293
0.2	+53	+103		+5.92	+103	

Table 6.13   (*Continued*)

Point	0.32 k ft^{-1}	0.32 k ft^{-1}	32 k ft^{-1}	9 k	One Line Wheels M_{LL}	M_{LL+I} (kip-ft)
0.3	+46	+204		+8.50	+157	
0.4	+40	+273		+10.48	+194	
0.5	+33	+309		+11.66	+214	
0.6	+26	+312		+11.82	+215	
0.7	+20	+283		+10.84	+195	
0.8	+13	+221		+8.60	+152	
0.9	+7	+127		+5.00	+88	
D						

Table 6.14   Live Load + Impact: Negative Moment—Truck

Point	16 k	16 k	4 k	One Line Wheels M_{LL}	M_{LL+I} (kip-ft)
AB					
14'AB	−14.00			−224	−1262
28'AB	−28.00	−14.00		−672	−3785
BA	−33.70	−19.70	−5.70	−877	−4940
BC	−24.63	−14.40	−4.16	−641	−3611
0.1				−551	−3104
0.2				−460	−2591
0.3				−370	−2084
0.4				−280	−1577
0.5				−190	−1070
0.6				−100	−563
0.7	−2.7	−2.7	−0.8	−90	−497
0.8	−3.1	−3.1	−1.4	−105	−580
0.9	−6.0	−6.0	−4.7	+661	
CB	−12.1	−12.1	−10.4	−429	
CD	−8.2	−8.2	−6.8	−290	
0.1				−261	−1382
0.2				−232	−1228
0.3				−203	−1075
0.4				−174	−921
0.5				−145	−768
0.6				−116	−614
0.7				−87	−461
0.8				−58	−307
0.9				−29	−154
DC					

Table 6.15   Live Load + Impact: Negative Moment—Lane

Point	0.32 k ft^{-1}	0.32 k ft^{-1}	0.32 k ft^{-1}	9 k	9 k	One Line Wheels M_{LL}	M_{LL+I} (kip-ft)
AB							
$14'AB$	−98			−14.0		−157	
$28'AB$	−392			−28.0		−377	
BA	−568			−33.70		−485	
BC	−415	−308		−24.63	−4.88	−497	
0.1	−357	−38		−21.17	−1.06	−326	
0.2	−298			−17.70		−255	
0.3	−240		−13	−14.24	−0.33	−212	
0.4	−181		−31	−10.77	−0.83	−172	
0.5	−123		−49	−7.31	−1.32	−133	
0.6	−65		−67	−3.85	−1.82	−93	
0.7	−6		−85	−0.38	−2.32	−53	
0.8		−59	−104	−1.71	−2.81	−93	
0.9		−325	−122	−6.16	−3.30	−228	−1233
CB		−784	−140	−12.25	−3.80	−440	−2380
CD		−517	−198	−8.47	−5.37	−353	−1909
0.1		−465	−54	−7.62	−1.95	−252	
0.2		−414		−6.78		−194	
0.3		−362		−5.93		−170	
0.4		−310		−5.08		−146	
0.5		−259		−4.24		−121	
0.6		−207		−3.39		−97	
0.7		−155		−2.54		−73	
0.8		−103		−1.69		−48	
0.9		−52		−0.85		−24	
DC							

Table 6.16   Maximum Moments[a]

Point	M_{DL}	$+M_{LL+I}$	$+M_T$	M_{DL}	$-M_{LL+I}$	$-M_T$
AB						
$14'AB$	−1397			−1397	−1262	−2659
$28'AB$	−4666			−4666	−3785	−8451
BA	−6522			−6522	−4940	−11462
BC	−7208			−7208	−3611	−10819
0.1	−3083	+852	−2231	−3083	−3104	−6187
0.2	+154	+1726	+1880	+154	−2591	−2437
0.3	+2502	+2303	+4805	+2502	−2084	+418
0.4	+3962	+2563	+6525	+3962	−1577	
0.5	+4533	+2536	+7069	+4533	−1070	
0.6	+4108	+2266	+6374	+4108	−563	
0.7	+2794	+1710	+4504	+2794	−497	

Table 6.16   (*Continued*)

Point	M_{DL}	$+M_{LL+I}$	$+M_T$	M_{DL}	$-M_{LL+I}$	$-M_T$
0.8	+592	+980	+1572	+592	−580	+8
0.9	−2499	+439	−2060	−2499	−1233	−3732
CB	−6478			−6478	−2380	−8858
CD	−5738			−5738	−1909	−7647
0.1	−3861			−3861	−1382	−5243
0.2	−2273	+718	−1555	−2273	−1228	−3501
0.3	−975	+1165	+190	−975	−1075	−2050
0.4	+34	+1491	+1525	+34	−921	−887
0.5	+753	+1651	+2404	+753	−768	−15
0.6	+1183	+1673	+2856	+1183	−614	+569
0.7	+1324	+1551	+2875	+1324	−461	
0.8	+1175	+1303	+2478	+1175	−307	
0.9	+737	+772	+1509	+737	−154	
DC						

^{a}Moments are in kip-ft.

tive resisting moments.   Because the cantilever has a 2′-0″ taper in depth, the resisting moments are computed at points A and B and a straight line is drawn between them.   Reinforcing bar extension is controlled by the $\frac{1}{20}$ span requirement.

The negative resisting moments are computed and tabulated as follows:

Resisting Moments

@ *A (44-#14)*

$$M_R = 24(N)(2.25)(0.949)(41.40)(\tfrac{1}{12}) = 176.80N$$

where N = number of bars.

N	M_R
44	7779
40	7072
32	5658
24	4243
16	2829
8	1414
8-#11	981

Resisting Moments

@ B (44-#14)

$$M_R = 24(N)(2.25)(0.949)(65.40)(\tfrac{1}{12}) = 279.29N$$

N	M_R
44	12,289
40	11,172
32	8,937
24	6,703
16	4,469
8	2,234
8-#11	1,549

$$\text{Bar extension } \frac{100}{20} = 5.0$$

Resisting Moments

@ C (48-#11)

$$M_R = (N)(1.56)(0.958)(65.55)(\tfrac{1}{12})(24) = 195.93N$$

N	M_R
48	9405
40	7837
32	6270
24	4702
16	3135
8	1567

$$\text{Bar extension } \frac{100}{20} = 5.0$$

The positive resisting moments are computed in the same manner.

Span BC: Resisting Moments

@ 0.5BC (38-#11)

$$M_R = 24(N)(1.56)(0.950)(66.18)(\tfrac{1}{12}) = 196.16N$$

N	M_R	N	M_R
38	7454	38	7454
32	6277	30	5884
26	5100	24	4708
20	3923	18	3531
14	2746	12	2354
8	1569	6	1177
8-#8	785		

Span CD: Resisting Moments

@ 0.6CD (29–#8)

$$M_R = 24(N)(0.79)(0.95)(66.48)(\tfrac{1}{12}) = 99.79N$$

N	M_R
29	2894
26	2594
20	1996
14	1397
8	798

The resisting moments are plotted against the maximum moments and cutoffs are established by using conservative values for bar extension beyond those required for moments (Figure 6.23).

Next, the maximum live load and impact shears across the entire bridge are combined with the dead load shears and used for plotting stirrup requirements. For span ranges included in this particular design it is reasonable to assume that the truck loading controls maximum shear and to ignore lane loading shears. The placement of the truck for maximum shear can be obtained by visualization of the standard shape of influence lines for shear (Figure 6.24).

To obtain the maximum live load and impact shears the corner moments must first be obtained. Then the shears due to corner moment effects can be computed; that is,

$$\text{Mom. } V = \frac{M_{CB} + M_{BC}}{100}$$

This is done first in span BC by using the influence line ordinates to compute moments (Table 6.17).

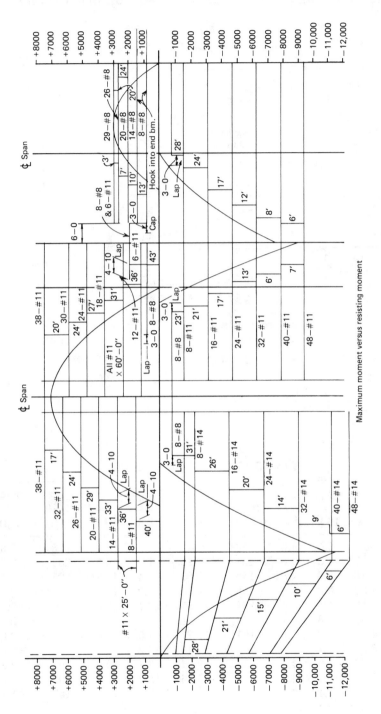

Figure 6.23 Resisting moments versus maximum moments.

251

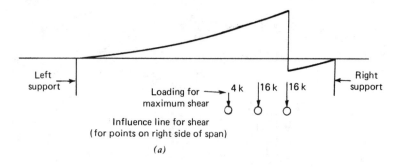

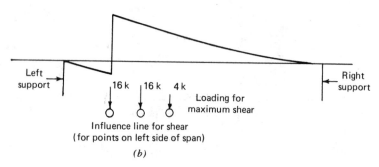

Figure 6.24   Influence line for shear.

Table 6.17   Loads in Span *BC*

Point	M_{BC} Ordinates 16 k	16 k	4 k	M_{BC} (kip-ft)	M_{CB} Ordinates 16 k	16 k	4 k	M_{CB} (kip-ft)	Moment V (kip)
BC	—	−3.0	−4.6	−66	—	−4.1	−8.0	−98	−0.3
0.1	−2.27	−4.3	−4.8	−124	−2.99	−7.0	−10.2	−201	−0.8
0.2	−3.78	−4.8	−4.8	−156	−5.89	−9.4	−11.7	−291	−1.4
0.3	−4.62	−4.8	−4.2	−168	−8.48	−11.2	−12.2	−364	−2.0
0.4	−4.88	−4.5	−3.4	−164	−10.55	−12.1	−12.0	−410	−2.5
0.5	−4.68	−4.8	−4.1	−168	−11.88	−9.8	−6.5	−373	−2.0
0.6	−4.09	−4.8	−4.8	−161	−12.25	−11.5	−9.0	−416	−2.6
0.7	−3.24	−4.4	−4.8	−141	−11.47	−12.2	−10.9	−422	−2.8
0.8	−2.20	−3.6	−4.6	−111	−9.31	−12.1	−12.0	−391	−2.8
0.9	−1.09	−2.6	−3.9	−75	−5.56	−10.2	−12.3	−301	−2.3
CB	—	−1.4	−3.0	−34	—	−7.0	−11.2	−157	−1.2

Table 6.18 Loads in Span CD

Point	M_{CD} Ordinates			M_{CD} (kip-ft)	Moment V (kip)
	16 k	16 k	4 k		
CD	—	-4.5	-5.3	-93	+1.6
0.1	-2.39	-5.2	-4.8	-141	+2.5
0.2	-4.03	-5.4	-4.0	-167	+2.9
0.3	-4.99	-5.0	-2.9	-171	+3.0
0.4	-5.37	-4.3	-1.6	-161	+2.8
0.5	-5.25	-3.4	-0.4	-140	+2.5
0.6	-4.70	-5.3	-2.7	-171	+3.0
0.7	-3.82	-5.4	-4.2	-164	+2.9
0.8	-2.69	-5.0	-5.1	-143	+2.5
0.9	-1.39	-4.2	-5.4	-111	+1.9
DC	—	-3.2	-5.2	-72	+1.3

Next, the loads in span CD are computed (Table 6.18), where

$$\text{Mom. } V = \frac{M_{CD} + M_{DC}}{57.1}.$$

The simple beam shears (SBV) for one line of wheels are combined with the moment shear effect M_V to obtain the total maximum shear for one line of wheels. An example of this procedure is shown in Figure 6.25. It is then expanded to the maximum shear on the entire box girder by multiplying the one line of wheel shears by the previously computed distribution and impact coefficient (Table 6.19). It should

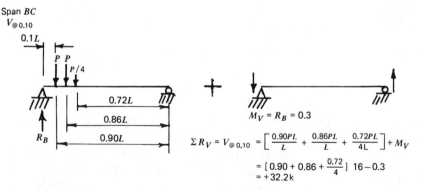

Figure 6.25 Maximum shears.

Table 6.19

Point	16 k	16 k	4 k	SBV	Moment V	One Line Wheels V_{LL}	Entire Box Girder V_{LL+I} (kips)
AB	1.0			−16.0		−16.0	−90.1
14'AB	1.0			−16.0		−16.0	−90.1
14'AB	1.0	1.0		−32.0		−32.0	−180.3
28'AB	1.0	1.0		−32.0		−32.0	−180.3
28'AB	1.0	1.0	1.0	−36.0		−36.0	−202.8
BA	1.0	1.0	1.0	−36.0		−36.0	−202.8
BC	1.0	0.86	0.72	+32.6	−0.3	+32.3	+171.0
0.1	0.9	0.76	0.62	+29.0	−0.8	+28.2	+149.3
0.2	0.8	0.66	0.52	+25.4	−1.4	+24.0	+127.1
0.3	0.7	0.56	0.42	+21.8	−2.0	+19.8	+104.8
0.4	0.6	0.46	0.32	+18.2	−2.5	+15.7	+83.1
0.5	0.5	0.36	0.22	−14.6	−2.0	−16.6	−87.9
0.6	0.6	0.46	0.32	−18.2	−2.6	−20.8	−110.1
0.7	0.7	0.56	0.42	−21.8	−2.8	−24.6	−130.3
0.8	0.8	0.66	0.52	−25.4	−2.8	−28.2	−149.3
0.9	0.9	0.76	0.62	−29.0	−2.3	−31.3	−165.7
CB	1.0	0.86	0.72	−32.6	−1.2	−33.8	−179.0
CD	1.0	0.755	0.510	+30.1	+1.6	+31.7	+175.0
0.1	0.9	0.655	0.410	+26.5	+2.5	+29.0	+160.1
0.2	0.8	0.555	0.310	+22.9	+2.9	+25.8	+142.4
0.3	0.7	0.455	0.210	+19.3	+3.0	+22.3	+123.1
0.4	0.6	0.355	0.110	+15.7	+2.8	+18.5	+102.1
0.5	0.5	0.255	0.010	+12.1	+2.5	+14.6	+80.6
0.6	0.6	0.355	0.110	−15.7	+3.0	−12.7	−70.1
0.7	0.7	0.455	0.210	−19.3	+2.9	−16.4	−90.5
0.8	0.8	0.555	0.310	−22.9	+2.5	−20.4	−112.6
0.9	0.9	0.655	0.410	−26.5	+1.9	−24.6	−135.8
DC	1.0	0.755	0.510	−30.1	+1.3	−28.8	−159.0

be noted that the order of wheel loads changes at the horizontal line locations.

The dead load and live load plus impact shears are combined and unit shears are computed and tabulated in Table 6.20. Note that these shears are magnified by the secant of the skew angle to approximate the increased shear at the obtuse corner.

Table 6.20

Point	V_{DL}	V_{LL+I}	V_{D+LL+I} (kips)	b	d	$v = \dfrac{V}{bd}(1.1547)$ (psi)
AB	−34.2	−90.1	−124.3	36.00	41.40	96
14′AB	−167.6	−90.1	−257.7	36.00	51.37	161
14′AB	−167.6	−180.3	−347.9	36.00	51.37	217
28′AB	−299.2	−180.3	−479.5	43.94	61.34	205
28′AB	−299.2	−202.8	−502.0	43.94	61.34	214
BA	−352.2	−202.8	−555.0	48.00	65.40	204
BC	+456.9	+171.0	+627.9	52.00	65.40	213
0.1	+368.1	+149.3	+517.4	45.60		200
0.2	+279.2	+127.1	+406.3	39.20		183
0.3	+190.4	+104.8	+295.2	36.00		145
0.4	+101.5	+83.1	+184.6	36.00	65.40	90
0.5	+12.7	−87.9	−65.2	36.00	65.55	32
0.6	−87.0	−110.1	−197.1	36.00		97
0.7	−175.8	−130.3	−306.1	36.00		149
0.8	−264.7	−149.3	−414.0	39.20		186
0.9	−353.5	−165.7	−519.2	45.60		200
CB	−442.4	−179.0	−621.4	52.00	65.55	211
CD	+354.2	+175.0	+529.2	44.00	65.55	212
0.1	+303.5	+160.1	+463.6	40.80		200
0.2	+252.8	+142.4	+395.2	37.60		185
0.3	+202.1	+123.1	+325.2	36.00		159
0.4	+151.4	+102.1	+253.5			124
0.5	+100.7	+80.6	+181.3			89
0.6	+50.0	−70.1	−20.1			10
0.7	−0.7	−90.5	−91.2			44
0.8	−51.4	−112.6	−164.0			80
0.9	−102.1	−135.8	−237.9			117
DC	−152.8	−159.0	−311.8	36.00	65.55	152

The resisting unit shears are computed for different stem widths and for different spacings of 8-#5 stirrups:

$$s = \frac{A_v f_v}{b(v - v_c)} = \frac{2.48(24,000)}{b(v - 55)}$$

Assume that v_c = 55 psi. Then

$$\text{when} \quad b = 36 \text{ in.} \quad v = \frac{1653}{s} + 55$$

$$\text{when} \quad b = 44 \text{ in.} \quad v = \frac{1353}{s} + 55$$

$$\text{when} \quad b = 48 \text{ in.} \quad v = \frac{1240}{s} + 55$$

$$\text{when} \quad b = 52 \text{ in.} \quad v = \frac{1145}{s} + 55$$

Spacings are compared with those previously calculated and the final resisting shears are tabulated:

With b = 36 in.		With b = 44 in.		With b = 48 in.		With b = 52 in.	
s	v	s	v	s	v	s	v
7"	291	8"	224	7"	232	7"	219
8"	262	1'-0"	168	1'-0"	153	1'-0"	150
9"	239	1'-6"	130	1'-6"	124	1'-6"	119
1'-0"	193						
1'-6"	147						

The resisting shears are plotted against the applied shears in Figure 6.26. Note that the increased web thickness varies from a maximum at the supports to 9-in. webs (b = 36 in.) at one-fourth the span from the support. On the basis of the data a proposed stirrup spacing is indicated in Figure 6.26.

The camber required to balance the dead load deflection can be computed by using the conjugate beam method described in preceding examples.

Example 6.2. The reinforced concrete box girder bridge designed by service load in Example 6.1 is designed by load factor design in accordance with AASHTO Articles 1.5.30–1.5.40. The deck slab for this bridge is designed by load factor design in Example 4.6. The service load level design loading for this bridge is HS 20-44 and the overload level design loading is 1.67 HS 20 loadings (β_L = 1.67). The roadway section and bridge layout are shown in Figures 6.1–6.3.

The schematic layout, the frame constants, and the moment computation are the same as those in Example 6.1. The wheel distribution and impact are also the same as in Example 6.1. The service load level

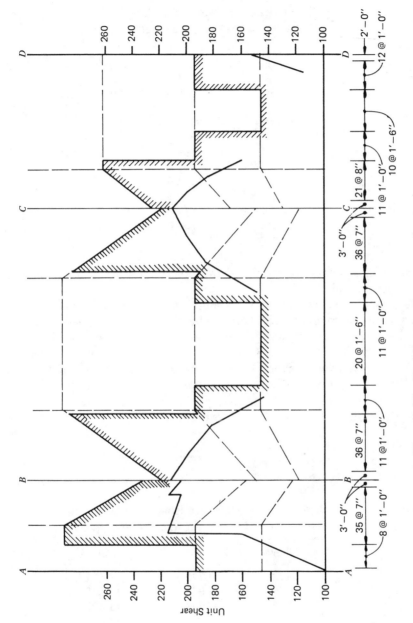

Figure 6.26 Resisting shears versus maximum shears.

257

Table 6.21   Service Load Level.
Summary of Moments and Shears

	DL	LL + I	Σ
M_{BA}	-6522	-4941	-11,463 k-ft
M_{BC}	-7208	-3612	-10,820 k-ft
$M_{0.5BC}$	+4532	+2536	+7068 k-ft
M_{CB}	-6481	-2390	-8871 k-ft
M_{CD}	-5738	-1922	-7660 k-ft
$M_{0.6CD}$	+1181	+1667	+2848 k-ft
V_{BA}	352.2	202.8	555.0 k
V_{BC}	456.9	171.0	627.9 k
V_{CB}	442.3	179.0	621.3 k
V_{CD}	354.2	175.0	529.2 k
V_{DC}	153.2	158.6	311.8 k

moments and shears are tabulated in Table 6.21 for the points considered most critical.

Next, the required ultimate moments and ultimate shears are computed at the same critical locations. The live load plus impact moments and shears are factored and summed with the factored dead load moments and then tabulated in Table 6.22 for the same critical locations.

Minimum reinforcing requirements of AASHTO Article 1.5.7 are not significant at the critical moment locations. The capacity analysis is examined first at 0.5BC by approximating a requirement for reinforcing at this point. Assume $(d - a/2) = 62$ in.±.

$$\text{Approx } A_{s(\text{required})} = \frac{M}{f_y(d - a/2)\,\phi}$$

Table 6.22   Ultimate Load Level.
Summary of Moments and Shears

	1.3 DL	2.17 (LL + I)	M_u or V_u
M_{BA}	-8479	-10,722	-19,201 k-ft
M_{BC}	-9370	-7,838	-17,208 k-ft
$M_{0.5BC}$	+5892	+5503	+11,395 k-ft
M_{CB}	-8425	-5186	-13,612 k-ft
M_{CD}	-7459	-4171	-11,630 k-ft
$M_{0.6CD}$	+1535	+3617	+5,153 k-ft
V_{BA}	457.9	440.1	898.0 k
V_{BC}	594.0	371.1	965.1 k
V_{CB}	575.0	388.4	963.4 k
V_{CD}	460.5	379.8	840.2 k
V_{DC}	199.2	344.2	543.4 k

Positive Moment Section at 0.5BC

$$(M_u = +11,395 \text{ k-ft})$$

Capacity Analysis

$$\text{Approx } A_{s(\text{required})} = \frac{11,395(12)}{(62.0\pm)(60)(0.90)} = 40.8 \text{ in.}^2$$

Try 44-#9, A_s 44.00 in.^{2}

The section is examined to determine whether the capacity should be based on a flanged section analysis or a rectangular section analysis. The area of reinforcement required to develop the compressive strength of the overhanging flanges is

$$A_{sf} = \frac{0.85f_c'(b - b_w)\,h_f}{f_y} = \frac{0.85(3.3)(364 - 36)(7.5)}{60}$$

$$= 115.0 \text{ in.}^2 > A_s(\text{provided}) = 44.0 \text{ in.}^2$$

Therefore the section is rectangular for the purposes of analysis. By using the rectangular section we compute the capacity:

$$d = 69.00 \text{ ft} - 1.50 \text{ in. (clear)} - 0.62 \text{ in. (#5)}$$

$$- 0.56 \text{ in. } (\tfrac{1}{2} \times \text{#9}) = 66.32 \text{ in.}$$

$$A_s = 44.00 \text{ in.}^2$$

$$a = \frac{A_s f_y}{0.85f_{cb}} = \frac{44.00(60)}{0.85(3.3)(364)} = 2.59 \text{ in.}$$

$$\frac{a}{2} = 1.29 \text{ in.}$$

$$\phi M_n = \phi \left[A_s f_y \left(d - \frac{a}{2} \right) \right]$$

$$= 0.90[(44.00)(60)(66.32 - 1.29)] \, (\tfrac{1}{12})$$

$$= +12,875 \text{ k-ft} > 11,395 \text{ k-ft required}$$

Therefore the section is satisfactory for capacity. The analysis of the section at 0.5BC is continued by checking serviceability. The maximum and minimum service load level moments at 0.5BC are

$$M_{\max} = +7068 \text{ k-ft}$$

$$M_{\min} = +4532 - 1070^* = +3462 \text{ k-ft}$$

*See Example 6.1.

k and j are obtained by using Appendix B:

$$\frac{t}{d} = \frac{7.50}{66.32} = 0.113$$

$$pn = \frac{44.00(10)}{364(66.32)} = 0.0182$$

Therefore $k = 0.20$ and $j = 0.95$.

The maximum and minimum steel stresses are

$$f_s = \frac{7068(12)}{44.00(0.95)(66.32)} = 30.6 \text{ ksi max}$$

$$f_s = \frac{3462(12)}{44.00(0.95)(66.32)} = 15.0 \text{ ksi min}$$

The stress range is

$$f_f = 30.6 - 15.0 = 15.6 \text{ ksi}$$

The allowable stress range is

$$f_f = 21 - 0.33f_{min} + 8\left(\frac{r}{h}\right) \quad \text{where } \left(\frac{r}{h}\right) = 0.30$$

$$= 21 - 0.33(15.0) + 8(0.3) = 18.4 \text{ ksi} > 15.6 \text{ ksi} \quad \text{OK}$$
$$\text{allowable}$$

The crack control criteria of AASHTO Article 1.5.39 is checked:

$$f_s = \frac{z}{\sqrt[3]{d_c A}}$$

Use $z = 130$ k in.$^{-1}$

Use $A = 291(5.36)/44 = 35.4$ in.^{2}

$$d_c = 2.68 \text{ in.}$$

where 44 is the number of bars,

$$\text{allowable } f_s = \frac{z}{\sqrt[3]{d_c A}} = \frac{130}{\sqrt[3]{(2.68)(35.4)}} = 28.5 \text{ ksi} < 30.6 \text{ ksi}$$
$$\text{allowable}$$

The reinforcing area must be increased to reduce the maximum service load level stress to 28.5 ksi or the allowable f_s must be increased to

30.6 ksi by using a closer spacing of reinforcement with the same area; therefore try the latter by using 56-#8, A_s = 44.24 in.^{2}

$$z = 130 \text{ k in.}^{-1}$$

$$d_c = 2.62 \text{ in.}$$

$$A = 291(5.24)/56 = 27.2 \text{ in.}^2$$

$$\text{allowable } f_s = \frac{z}{\sqrt[3]{d_c A}} = \frac{130}{\sqrt[3]{(2.62)(27.2)}}$$

$$= 31.3 \text{ ksi} > 30.6 \text{ ksi} \quad \text{OK}$$
$$\text{allowable}$$

The maximum reinforcing requirements of AASHTO Article 1.5.32 (A)(1), which limits the reinforcement ratio to 75% of the balanced reinforcement ratio, are now checked.

Assume a rectangular section with b = 364 in.

$$0.75\rho_B = \left[\frac{0.85\beta_1 f_c'}{f_y}\left(\frac{87000}{87000 + f_y}\right)\right] 0.75$$

$$= \left[\frac{0.85(0.85)(3.3)}{60}\left(\frac{87}{87 + 60}\right)\right] 0.75 = 0.0176$$

$$\rho = \frac{44.24}{364(66.32)} = 0.0018 < 0.0176 \quad \text{OK}$$

The capacity analysis of 0.6CD is performed as just completed for 0.5BC.

Positive Moment Selection @ 0.6CD

(M_u = +5153 k-ft). The approximate reinforcement required for capacity is

$$\text{Approx } A_{s(\text{required})} = \frac{M}{f_y d\phi}$$

$$= \frac{5153(12)}{60(62.0)(0.90)} = 18.5 \text{ in.}^2$$

Try 48-#6, A_s = 21.12 in.^{2}. The area of reinforcement required to develop the compressive strength of the overhanging flanges is

$$A_{sf} = \frac{0.85f_c'(b - b_w)h_f}{f_y} = \frac{0.85(3.3)(364 - 36)(7.5)}{60}$$

$$= 115.0 \text{ in.}^2 > A_{s(\text{provided})} = 21.12 \text{ in.}^2$$

Therefore a rectangular section is used for analysis purposes and now checked for capacity. The effective depth is

$$d = 69.00 - 1.50 \text{ in. (clear)} - 0.62 \text{ in. (\#5)}$$
$$- 0.38 \text{ in. } (\tfrac{1}{2} \times \#6) = 66.50 \text{ in.}$$

$$A_s = 21.12 \text{ in.}^2$$

$$a = \frac{A_s f_y}{0.85 f_c' b} = \frac{21.12(60)}{0.85(3.3)(3.64)} = 1.24 \text{ in.}$$

$$\frac{a}{2} = 0.62 \text{ in.}$$

$$\phi M_n = \phi \left[A_s f_y \left(d - \frac{a}{2} \right) \right]$$
$$= 0.90[(21.12)(60)(66.50 - 0.62)] \left(\tfrac{1}{12} \right)$$
$$= 6261 \text{ k-ft} > +5153 \text{ k-ft required}$$

Therefore the section is satisfactory for capacity. For serviceability computations the service load level maximum moment is

$$M_{\max} = +2848 \text{ k-ft}$$

and the minimum moment

$$M_{\min} = +1181 - 614^* = +567 \text{ k-ft}$$

Using Appendix B, we can conservatively say

$$k = 0.20 \quad \text{and} \quad j = 0.95$$

The maximum steel stress is

$$f_s = \frac{2848(12)}{21.12(0.95)(66.50)} = 25.6 \text{ ksi max}$$

The minimum steel stress is

$$f_s = \frac{567(12)}{21.12(0.95)(66.50)} = 5.1 \text{ ksi}$$

*See Example 6.1.

The stress range is

$$f_f = 25.6 - 5.1 = 20.5 \text{ ksi}$$

The allowable stress range is

$$f_f = 21 - 0.33 f_{\min} + 8 \left(\frac{r}{h}\right)$$

$$= 21 - 0.33(5.1) + (0.3) = 21.7 \text{ ksi} > 20.5 \text{ ksi} \quad \text{OK}$$

The crack control criteria are then checked:

$$z = 130 \text{ k in.}^{-1}$$

$$d_c = 1.50 + 0.62 + 0.38 = 2.50 \text{ in.}$$

$$A = (291 \times 5.00)/48 = 30.3 \text{ in.}^2$$

$$\text{allowable } f_s = \frac{z}{\sqrt[3]{d_c A}} = \frac{130}{\sqrt[3]{(2.50)(30.3)}}$$

$$= 30.7 \text{ ksi} > 25.6 \text{ ksi} \quad \text{OK} \\ \text{allowable}$$

There is no need to check the maximum reinforcement requirement which limits the reinforcing ratio to $0.75\rho_\beta$ because $0.6CD$ has less reinforcing than the section already checked at $0.5BC$.

Next the negative moment section at B is designed. The maximum ultimate moment at BA is

$$M_u^{BA} = -19,201 \text{ k-ft}$$

This is greater than the ultimate moment at BC; therefore it is used for design. An approximate required reinforcing area is computed by using $(d - a/2) = 62.0 \text{ in.}\pm$.

$$\text{approx } A_{s(\text{required})} = \frac{M_u}{f_y(d - a/2)\phi} = \frac{19,201(12)}{60(62.0)(0.90)} = 68.8 \text{ in.}^2$$

Try 56-#10, $A_s = 71.12 \text{ in.}^2$. With #10

$$d = 69.0 - 2.0(\text{clear}) - 0.75(\#6) - 0.64(\tfrac{1}{2} \times \#10)$$

$$= 65.61 \text{ in.}$$

The actual capacity is computed:

$$a = \frac{A_s f_y}{0.85 f'_c b} = \frac{71.12(60)}{0.85(3.3)(291)} = 5.22 \text{ in.}$$

Because a is less than the bottom flange thickness, the section is rectangular. Therefore

$$\frac{a}{2} = 2.61 \text{ in.}$$

$$\phi M_u = \phi \left[A_s f_y \left(d - \frac{a}{2} \right) \right]$$

$$= 0.90[(71.12)(60)(65.61 - 2.61)] \left(\tfrac{1}{12} \right)$$

$$= 20{,}163 \text{ k-ft} > 19{,}201 \text{ k-ft required}$$

For serviceability the maximum and minimum service load level moments are

$$M_{\max} = -11463 \text{ k-ft}$$

$$M_{\min} = -7208 + (5.521)(16.0) \left(1.1 + 1.7 + \frac{1.0}{4} \right)^*$$

$$= -7208 + 216$$

$$= -6992 \text{ k-ft}$$

The k and j terms from Appendix B are obtained by using

$$\frac{t}{d} = \frac{5.5}{65.61} = 0.84$$

$$pn = \frac{71.12(10)}{291(65.61)} = 0.037$$

Therefore $k = 0.32$ and $j = 0.95$.

The maximum and minimum steel stresses using $f_s = M/A_s jd$, are

$$f_s = \frac{11463(12)}{71.12(0.95)(65.61)} = 31.0 \text{ ksi}$$

$$f_s = \frac{6992(12)}{71.12(0.95)(65.61)} = 18.9 \text{ ksi}$$

*From influence lines, Example 6.1.

The stress range is

$$f_f = 31.0 - 18.9 = 12.1 \text{ ksi}$$

The allowable stress range is

$$f_f = 21 - 0.33(18.9) + 8(0.3) = 17.2 \text{ ksi} > 12.1 \text{ ksi} \quad \text{OK}$$
allowable

For crack control

$$d_c = 2.00 + 0.75 + 0.64 = 3.39 \text{ in.}$$
$$A = 6.78 \times 364/56 = 44.1 \text{ in.}^2$$

For crack control the maximum allowable steel stress is

$$f_s = \frac{z}{\sqrt[3]{d_c A}} = \frac{130}{\sqrt[3]{3.39(44.1)}} = 24.5 \text{ ksi}$$
$$= 24.5 \text{ ksi} < 31.0 \text{ ksi maximum}$$
allowable

A greater steel area is required for crack control. Try 60-#11, $A_s = 93.60$ in.^{2}

$$d_c = 2.00 + 0.75 + 0.70 = 3.45 \text{ in.}$$
$$A = 6.90 \times 364/60 = 41.9 \text{ in.}^2$$
$$\text{allowable } f_s = \frac{130}{\sqrt[3]{(3.45)(41.9)}} = 24.8 \text{ ksi}$$
$$\text{maximum } f_s = \frac{11,463(12)}{93.60(0.95)(65.56)} = 23.6 \text{ ksi} < 24.8 \text{ ksi} \quad \text{OK}$$

Use 60-#11 at B. Check for maximum reinforcement:

$$0.75\rho_B = 0.0176$$
$$\rho = \frac{93.60}{291(65.56)} = 0.0049 < 0.0176 \quad \text{OK}$$

Finally the negative moment section at C, in which the maximum ultimate moment at $CB = -13,612$ k-ft is designed. This is greater than the value of the maximum ultimate moment at CD; therefore it is our design moment.

Design of the section at B indicates that the serviceability criteria for crack control is the most critical for design. Therefore the design is governed by crack control and checked for capacity and fatigue. The maximum and minimum service load level moments are

maximum M_{CB} = -8871 k-ft

minimum M_{CB} = -6481 + 5.633(16)$\left(+10.01 + 5.85 + \dfrac{1.69}{4}\right)$*

$\qquad$ = -5013 k-ft

On the basis of the analysis in Section BC an approximate allowable of 25 ksi is assumed for determining the required reinforcement:

$$\text{approx } A_{s(\text{required})} = \frac{8871(12)}{25(0.95)(65.5\pm)} = 68.4 \text{ in.}^2$$

Try 56-#10, A_s = 71.12 in.^{2}. From the analysis of point BC

allowable f_s = 24.5 ksi

$$\text{actual } f_s = \frac{8871(12)}{71.12(0.95)(65.61)} = 24.0 \text{ ksi} < 24.5 \text{ ksi} \quad \text{OK}$$
$$\text{allowable}$$

Check fatigue:

$$\text{maximum } f_s = 24.0 \text{ ksi}$$
$$\text{minimum } f_s = \frac{5013}{8871} \times 24.0 = 13.6 \text{ ksi}$$

The stress range is

$$f_f = 24.0 - 13.6 = 10.4 \text{ ksi}$$

The allowable stress range is

$$f_f = 21 - 0.33(13.6) + 8(0.3) = 18.9 \text{ ksi}$$

Therefore the section is OK for fatigue. The capacity ϕM_n = 20,163 k-ft > 13,612 k-ft = M_u required; therefore the section is satisfactory for capacity. Maximum reinforcing requirements are less critical than at point BC already computed. Requirements for skin reinforcing can be computed like those in Example 5.1.

*From influence lines in Example 6.1.

Next, we check the shear requirements of BA. The ultimate shear V_u @ BA = 898.0 k. Assume 4-12 in. webs:

$$b_w = 48.0 \text{ in.}$$
$$d = 65.56 \text{ in.}$$
$$\phi = 0.85.$$

The ultimate unit shear is

$$v_u = \frac{V_u}{b_w d} = \frac{898.0}{48.0(65.56)} = 285 \text{ psi}$$

Magnify the shear for skewed bridge v_u = 1.1547*(285) = 329 psi. The required capacity unit shear is

$$v_n = \frac{v_u}{\phi} = \frac{329}{0.85} = 387 \text{ psi}$$

Assume that $v_c = 2\sqrt{3300} = 115$ psi.
Then

$$v_n - v_c = 387 - 115 = 272 \text{ psi} > 4\sqrt{f_c'} = 230 \text{ psi}$$
$$< 8\sqrt{f_c'} = 460 \text{ psi}$$

Therefore the maximum spacings given in AASHTO Article 1.5.10(c) should be reduced by one-half, which will probably not control. The spacing required is computed by using

$$s = \frac{A_v f_y}{(v_n - v_c) b_w}$$

Assume 8-#5 stirrup legs with A_v = 2.48 in.^{2} :

$$f_y = 60,000 \text{ psi}$$

Then

$$s = \frac{2.48(60,000)}{272(48.0)} = 11.4 \text{ in.}$$

*Secant of skew angle.

Use 8–#5 @ 10-in. spacing. The shear requirements @ BC are computed next. Again assume 4–12 in. webs:

$$b_w = 48.0 \text{ in.}$$

$$d = 65.56 \text{ in.}$$

$$\phi = 0.85$$

V_u @ $BC = 965.1$ k.   The ultimate unit shear is

$$v_u = \frac{V_u}{b_w d} = \frac{965.1}{(48.0(65.56)} = 307 \text{ psi}$$

Magnify for skewed bridge $v_u = 1.1547 \times 307 = 354$ psi. The required capacity unit shear is

$$v_n = \frac{v_u}{\phi} = \frac{354}{0.85} = 417 \text{ psi}$$

Assume $v_c = 2\sqrt{f_c'} = 115$ psi. Then

$$v_n - v_c = 417 - 115 = 302 \text{ psi} > 4\sqrt{f_c'} = 230 \text{ psi}$$
$$< 8\sqrt{f_c'} = 460 \text{ psi}$$

The maximum spacings given in AASHTO Article 1.5.10(c) should be reduced by one-half, which will probably not control. Therefore the section is within allowables. The required spacing is (assuming 8–#5 stirrup legs)

$$s = \frac{A_v f_y}{(v_n - v_c) b_w} = \frac{2.48(60,000)}{302(48.0)}$$

$$= 10.3 \text{ in.}$$

Use 8–#5 @ 10-in. spacing. The shear requirements at CB, CD, and DC are computed in the same manner.

Shear at CB. V_u @ $CB = 963.4$ k. This is close to the value of V_u at BC and the same section can be used; that is, 4–12-in. webs with 8–#5 @ 10-in. spacing.

Shear at CD. V_u @ $CD = 840.2$ k. Again assume that $b_w = 48.0$ in.

$$v_u = \frac{V_u}{b_w d} = \frac{840.2}{48.0(65.56)} = 267 \text{ psi}$$

In a similar manner magnify for the skew angle by multiplying by the secant of the skew.

$$v_u = 1.1547 \times 267 = 308 \text{ psi}$$

$$v_n = \frac{308}{0.85} = 363 \text{ psi}$$

$$v_n - v_c = 363 - 115 = 248 \text{ psi} > 4\sqrt{f_c'} = 230 \text{ psi}$$
$$< 8\sqrt{f_c'} = 460 \text{ psi}$$

The same minimum requirements control.

$$s = \frac{A_v f_y}{(v_n - v_c)\, b_w} = \frac{2.48(60,000)}{248(48.0)} = 12.5 \text{ in.}$$

Use 4–12-in. webs with 8-#5 @ 1 ft–0 in. spacing.

Shear at DC. V_u @ DC = 543.4 k.   Assume b_w = 36.0 in. (4–9 in. webs).

$$v_u = \frac{V_u}{b_w d} = \frac{543.4}{(36.0)(66.5)} = 227 \text{ psi}$$

Magnify for skew angle v_u = 1.1547 $\times$ 227 = 262 psi.

$$v_n = \frac{262}{0.85} = 308 \text{ psi}$$

$$v_n - v_c = 308 - 115 = 193 \text{ psi} < 4\sqrt{f_c'} = 230 \text{ psi}$$
$$< 8\sqrt{f_c'} = 460 \text{ psi}$$

$$s = \frac{A_v f_y}{(v_n - v_c)\, b_w} = \frac{2.48(60,000)}{193(36.0)} = 21.4 \text{ in.}$$

Use 4–9 in. webs with 8-#5 @ 1 ft–6 in. spacing.

At this point the design of the bent columns and footings, and the bent caps should be performed. For the bent caps, Example 6.1 is used.

At the support the cap service load level bending moments, shears, and torsional moments are

$$M_B = -1180 \text{ k-ft (negative maximum)}$$
$$V_{BA} = 279.6 \text{ k}$$
$$V_{BC} = 294.2 \text{ k}$$
$$M_{\text{tors}} = +160 \text{ k-ft}$$

These values are taken from Example 6.1 and are factored to give the ultimate values:

$$M_B = 1.3[(-905) + 1.67(-275)] = -1774 \text{ k-ft}$$

$$V_{BA} = 1.3[(216.2) + 1.67(63.4)] = 418.7 \text{ k}$$

$$V_{BC} = 1.3[(230.8) + 1.67(63.4)] = 437.7 \text{ k}$$

$$M_{\text{tors}} = 1.3[(-172) + 1.67(+332)] = +497 \text{ k-ft}$$

As shown in Figure 6.19, the cap is 5'-0" wide $\times$ 5'-9" deep, and because the main negative moment reinforcement is placed below several layers of reinforcing d_c used in the crack control criteria becomes very large. Therefore the allowable f_s is low. A minimum allowable f_s should be f_s = 24,000 psi because this is the service load allowable stress.

Experience will show that this minimum will govern for a crack control allowable. Therefore the section is proportioned, based on f_s = 24,000 psi. The section is then checked for capacity and fatigue.

Assume 7-#11 reinforcing in the top of the cap;

$$A_s = 10.92 \text{ in.}^2 \quad \text{and} \quad d = 62.55 \text{ in.}$$

k and j are selected from Appendix A for rectangular sections:

$$p = \frac{10.92}{(60)62.55} = 0.0029$$

and $k = 0.212$ and $j = 0.929$.

The service load level stress in the reinforcing is computed as follows and compared with the allowable:

$$f_s = \frac{1180(12)}{10.92(0.929)(62.55)} = 22,315 \text{ psi} < 24,000 \text{ psi}$$

therefore satisfactory for crack criteria.

Next check for capacity.

$$a = \frac{A_s f_y}{0.85 f'_c b} = \frac{10.92(60)}{0.85(3.3)(60)} = 3.89 \text{ in.}$$

$$\frac{a}{2} = 1.95 \text{ in.}$$

$$\phi M_u = \phi \left[A_s f_y \left(d - \frac{a}{2} \right) \right]$$

$$= 0.90[(10.92)(60)(62.55 - 1.95)] \left(\tfrac{1}{12}\right)$$

$$= 2978 \text{ k-ft} > 774 \text{ k-ft required}$$

The maximum reinforcing requirement, which limits the reinforcement ratio to 75% of the balanced ratio, is checked:

$$0.75\rho_\beta = 0.0176$$

$$\rho = \frac{10.92}{60(62.55)} = 0.0029 < 0.0176$$

Next check for fatigue, where the minimum moment is the dead load moment:

$$@ B \; M_{DL} = -905 \text{ k-ft}$$

The maximum reinforcing stress was computed as

$$\max f_s = 22{,}315 \text{ psi}$$

The minimum reinforcing stress is computed as

$$\min f_s = \frac{905(12)}{10.92(0.929)(62.55)} = 17.1 \text{ ksi}$$

The stress range is

$$f_f = 22.3 - 17.1 = 5.2 \text{ ksi}$$

The allowable stress range is

$$f_f = 21 - 0.33 f_{\min} + 8 \left(\frac{r}{h}\right)$$

$$= 21 - 0.33(17.1) + 8(0.3)$$

$$= 17.8 \text{ ksi} > 5.2 \text{ ksi}$$

Therefore OK for fatigue.
    At midspan the service load level moment is

$$M_{0.5BC} = +672 \text{ k}$$

The factored moment is

$$M_{0.5BC} = 1.3[(+341) + 1.67(+331)] = +1162 \text{ k-ft}$$

Again the minimum level stress for crack control will probably control the required area of reinforcing. The minimum stress is the service load allowable: $f_s = 24{,}000$ psi. For this stress the approximate required reinforcing is

$$\text{approx } A_{s\text{(required)}} = \frac{672(12)}{24(0.87)(65 \text{ in.}\pm)} = 5.94 \text{ in.}^2$$

Try 7-#9, $A_s = 7.00$ in.^{2} and $d = 64.90$ in.; k and j are obtained from Appendix A for rectangular sections:

$$\rho = \frac{7.00}{60(64.90)} = 0.0017 \qquad k = 0.164 \text{ and } j = 0.945$$

The maximum stress in the reinforcing is

$$\max f_s = \frac{672(12)}{7.00(0.945)(64.90)} = 18{,}783 \text{ psi} < 24{,}000 \text{ psi}$$

therefore OK for crack control.

The minimum moment at $0.5BC$ is

$$M_{\min} \text{ @ } 0.5BC = +341 - \tfrac{1}{2}(275) = +203 \text{ k-ft}$$

The minimum stress at $0.5BC$ is

$$\min f_s = \frac{203(12)}{7.00(0.945)(64.90)} = 5.7 \text{ ksi}$$

The stress range is

$$f_f = 18.8 - 5.7 = 13.1 \text{ ksi}$$

The allowable stress range is

$$f_f = 21 - 0.33 f_{\min} + 8 \left(\frac{r}{h}\right)$$

$$= 21 - 0.33(5.7) + 8(0.3)$$

$$= 21.5 \text{ ksi} > 13.1 \text{ ksi}$$

Therefore the section is satisfactory for fatigue.

The midspan capacity is then checked:

$$a = \frac{A_s f_y}{0.85 f'_c b} = \frac{7.00(60)}{0.85(3.3)(60)} = 2.50 \text{ in.}$$

$$\frac{a}{2} = 1.25 \text{ in.}$$

$$\phi M_u = \phi \left[A_s f_y \left(d - \frac{a}{2} \right) \right] \left(\frac{1}{12} \right)$$

$$= 0.90 \left[(7.00)(60)(64.90 - 1.25) \right] \left(\tfrac{1}{12} \right)$$

$$= 2005 \text{ k-ft} > 1162 \text{ k-ft}$$

Therefore the section is satisfactory for capacity.

Next we design the pier cap hoop stirrups according to ACI 318-77, as recommended in AASHTO.

First we check the significance of torsion in computing the concrete shear strength. If the value $\phi(0.5\sqrt{f_c'} \Sigma x^2 y)$ is exceeded by the torsional moment, the equation accounting for torsion must be used to compute the concrete shear strength:

$$\phi \left(0.5 \sqrt{f_c'} \ \Sigma x^2 y \right) = 0.85 \left[(0.5)\sqrt{3300} \ (60)^2 (69) \right] \left(\tfrac{1}{12} \right)$$

$$= 505 \text{ k-ft} > 497 \text{ k-ft}$$

This is close: therefore the effects of torsion are included. The calculation from ACI 318-77 that must then be used is eq. 11-5, which is

$$V_c = \frac{2\sqrt{f_c'} \ b_w d}{\sqrt{1 + [2.5C_t(T_u/V_u)]^2}}$$

$$C_t = \frac{b_w d}{x^2 y} = \frac{60(62.55)}{(60)^2(69)} = 0.0151$$

$$V_c = \frac{2\sqrt{f_c'} \ b_w d}{\sqrt{1 + [2.5C_t(T_u/V_u)]^2}}$$

$$= \frac{2\sqrt{3300} \ (60)(62.55)}{\sqrt{1 + \{(2.5)(0.0151)[497(12)/437.7]\}^2}}$$

$$= 383 \text{ k}$$

This is converted to the required stirrup shear strength for direct shear.

$$V_u \leqslant \phi(V_c + V_s)$$

$$V_s = \frac{V_u}{\phi} - V_c \qquad \phi = 0.85$$

$$V_s = \frac{438}{0.85} - 383 = 132 \text{ k}$$

The required stirrup shear strength V_s is then used to develop a required area of reinforcing.

$$V_s = \frac{A_v f_y d}{s}$$

$$\frac{A_v}{s} = \frac{V_s}{f_y d} = \frac{132}{60(62.55)} \left(\frac{1}{4}\right) = 0.0088 \text{ in.}^2 \text{ in.}^{-1} \text{ per leg}$$

The torsional strength required for the hoops is

$$T_s = \frac{T_u}{\phi} - T_c$$

Using the ACI equation (11-23), we obtain the area of reinforcing required per unit spacing for torsion:

$$T_s = \frac{A_t \alpha_t x_1 y_1 f_y}{s}$$

$$x_1 = 55.38 \text{ in. (shortest hoop dimension)}$$

$$y_1 = 60.34 \text{ in. (longest hoop dimension)}$$

where $\alpha_t = 0.66 + 0.33 \, (y_1/x_1)$; therefore

$$\alpha_t = 0.66 + 0.33 \left(\frac{60.34}{55.38}\right) = 1.02 \text{ per leg}$$

The concrete torsional strength is

$$T_c = \frac{0.8\sqrt{f_c'} \, \sum x^2 y}{\sqrt{1 + (0.4 V_u/C_t T_u)^2}}$$

$$= \frac{0.8\sqrt{3300} \, (60)^2 \, (69)}{\sqrt{1 + [(0.4)(438)/(0.0151)(497)(12)]^2}}$$

$$= 5219 \text{ k-in.} = 435 \text{ k-ft}$$

$$T_s = \frac{T_u}{\phi} - T_c = \frac{497}{0.85} - 435 = 150 \text{ k-ft}$$

The area of reinforcing required per unit spacing and per leg is

$$\sum \frac{A_t}{s} = \frac{T_s}{\alpha_t x_1 y_1 f_y} = \frac{150(12)}{1.02(55.38)(60.34)(60)} = 0.0088$$

The total A/s for shear and torsion is $A_v/s + A_t/s$, or

$$\sum \frac{A}{s} = 0.0088 + 0.0088 = 0.0176 \text{ in.}^2 \text{ in.}^{-1}$$

With #5 stirrups and $A_v = 0.31$ in.^{2} per leg

$$s(\text{required}) = \frac{0.31}{0.0176} = 17.6 \text{ in.}$$

Use a nominal spacing of 12 in.

$$\#5 \ \boxplus \ @ \ 12''$$

Table 6.23 Live Load + Impact[a,b]

Point	$+\beta_L(M_{LL+I})$	$-\beta_L(M_{LL+I})$
AB		
14'AB		-2107
28'AB		-6321
BA		-8250
BC		-6030
0.1	+1423	-5184
0.2	+2882	-4327
0.3	+3846	-3480
0.4	+4280	-2634
0.5	+4235	-1787
0.6	+3784	-940
0.7	+2856	-830
0.8	+1637	-969
0.9	+733	-2059
CB		-3975
CD		-3188
0.1		-2308
0.2	+1199	-2051
0.3	+1946	-1795
0.4	+2490	-1538
0.5	+2757	-1282
0.6	+2794	-1025
0.7	+2590	-770
0.8	+2176	-513
0.9	+1289	-257
DC		

[a] Moments are in kip-ft.
[b] $\beta_L = 1.67$.

Table 6.24   Ultimate Moments[a]

Point	M_{DL}	$+\beta_L M_{LL+I}$	$+M_u$ (1.3 × ΣM)	M_{DL}	$-\beta_L M_{LL+I}$	$-M_u$ (1.3 ΣM)
AB						
14′*AB*	−1397			−1397	−2107	−4556
28′*AB*	−4666			−4666	−6321	−14283
BA	−6522			−6522	−8250	−19203
BC	−7208			−7208	−6030	−17210
0.1	−3083	+1423	−2158	−3083	−5184	−10747
0.2	+154	+2882	+3947	+154	−4327	−5425
0.3	+2502	+3846	+8252	+2502	−3480	−1272
0.4	+3962	+4280	+10715	+3962	−2634	+1727
0.5	+4533	+4235	+11399	+4533	−1787	
0.6	+4108	+3784	+10260	+4108	−940	
0.7	+2794	+2856	+7345	+2794	−830	+4072
0.8	+592	+1637	+2897	+592	−969	−490
0.9	−2499	+733	−2296	−2499	−2059	−5926
CB	−6478			−6478	−3975	−13588
CD	−5738			−5738	−3188	−11604
0.1	−3861			−3861	−2308	−8020
0.2	−2273	+1199	−1396	−2273	−2051	−5621
0.3	−975	+1946	+1262	−975	−1795	−3601
0.4	+34	+2490	+3281	+34	−1538	−1955
0.5	+753	+2757	+4563	+753	−1282	−688
0.6	+1183	+2794	+5170	+1183	−1025	+205
0.7	+1324	+2590	+5088	+1324	−770	
0.8	+1175	+2176	+4356	+1175	−513	
0.9	+737	+1289	+2634	+737	−257	
DC						

[a]Moments are in kip-ft.

The requirements for longitudinal reinforcement for torsion are covered in Example 6.1.

After completion of the pier shaft and footing design it is appropriate to establish main reinforcing cutoffs and stirrup spacing, which are obtained by plotting the ultimate moments against the capacity moments. The main box girder sections are designed by computing the factored dead load moments and the live load plus impact moments which were calculated for service load in Example 6.1. These results are listed in Table 6.23.

The dead load and factored live load plus impact are combined and factored by using $\gamma = 1.3$ to obtain the ultimate moments in Table 6.24.

Using the maximum reinforcing requirements, we can compute the

Table 6.25   Resisting Moments

Positive Moment

Span *BC*.   56-#8

$$d = 69.00 - 1.50 - 0.62 - 0.50 = 66.38 \text{ in.}$$
$$b = 364 \text{ in.} \qquad f_y = 60 \text{ ksi} \qquad \phi = 0.90$$

	a	$a/2$	$(d - a/2)(\frac{1}{12})$	A_s	ϕM_n
56-#8	2.60	1.30	5.42	44.24	12,956
44-#8	2.04	1.02	5.45	34.76	10,223
32-#8	1.49	0.74	5.47	25.28	7,467
20-#8	0.93	0.46	5.49	15.80	4,686
38-#8	0.37	0.19	5.52	6.32	1,883

Span *CD*.   48-#6

$$d = 66.50 \text{ in.} \qquad b = 364 \text{ in.} \qquad f_y = 60 \text{ ksi} \qquad \phi = 0.90$$

	a	$a/2$	$(d - a/2)(\frac{1}{12})$	A_s	ϕM_n
48-#6	1.24	0.62	5.49	21.12	6261
36-#6	0.93	0.47	5.50	15.84	4706
24-#6	0.62	0.31	5.52	10.56	3146
12-#6	0.31	0.16	5.53	5.28	1577

Negative Moment

@ *A*.   60-#11

$$d = 41.56 \text{ in.} \qquad b = 291 \text{ in.} \qquad f_y = 60 \text{ ksi} \qquad \phi = 0.90$$

	a	$a/2$	$(d - a/2)(\frac{1}{12})$	A_s	ϕM_n
60-#11	6.88	3.44	3.18	93.60	16,056
44-#11	5.04	2.52	3.25	68.64	12,059
28-#11	3.21	1.60	3.33	43.68	7,855
12-#11	1.38	0.69	3.39	18.72	3,425

@ *B*.   60-#11

$$d = 65.56 \text{ in.} \qquad b = 291 \text{ in.} \qquad f_y = 60 \text{ ksi} \qquad \phi = 0.90$$

	a	$a/2$	$(d - a/2)(\frac{1}{12})$	A_s	ϕM_n
60-#11	6.88	3.44	5.18	93.60	26,165
44-#11	5.05	2.52	5.25	68.64	19,471
28-#11	3.21	1.61	5.33	43.68	12,571
12-#11	1.38	0.69	5.41	18.72	5,465

@ *C*.   56-#10

$$d = 65.61 \text{ in.} \qquad b = 291 \text{ in.} \qquad f_y = 60 \text{ ksi} \qquad \phi = 0.90$$

	a	$a/2$	$(d - a/2)(\frac{1}{12})$	A_s	ϕM_n
56-#10	5.23	2.61	5.25	71.12	20,161
40-#10	3.73	1.87	5.31	50.80	14,572
24-#10	2.24	1.12	5.37	30.48	8,845
8-#10	0.75	0.37	5.44	10.16	2,983

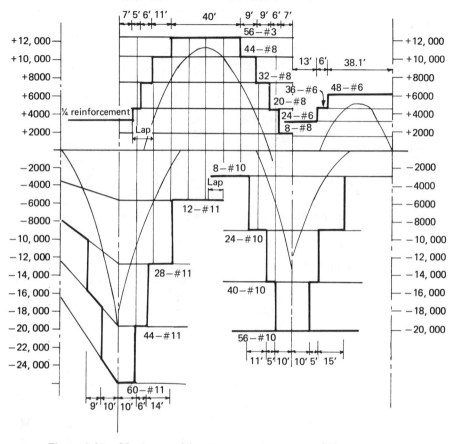

Figure 6.27   Maximum ultimate moments versus resisting moments.

resisting moment capacities for different levels of reinforcing in a tabular form for each location (Table 6.25).

A bar extension approximately equal to d of 5.5 ft is used. AASHTO Article 1.5.13 requires that one-third of the reinforcing should extend into the simple support and one-fourth should extend through the continuous support. The maximum ultimate moments are plotted against the resisting capacity moments in Figure 6.27 and cutoff locations are selected. Development length does not affect the results and is ignored.

Next it is necessary to complete the computation of the shear requirements for the entire length of the bridge. This is done by expanding the requirements computed at the critical locations. The dead load shears and live load plus impact shears were computed in Example 6.1 and the factored values of these shears are tabulated, summed, and con-

Table 6.26

Point	V_{LL}	$\beta_L V_{LL+I}$	V_{OL}	$V_u = \gamma V_{OL}$ (kips)	b	d	$v_n = \frac{V_u}{\phi b d}(1.1547)$ (psi)
AB	34.2	150.5	184.7	240.1	36.00	41.56	218
$14'AB$	167.6	150.5	318.1	413.5	40.99	51.53	266
$14'AB$	167.6	301.1	468.7	609.3	40.99	51.53	392
$28'AB$	299.2	301.1	600.3	780.4	45.97	61.50	375
$28'AB$	299.2	338.7	637.9	829.2	45.97	61.50	398
BA	352.2	338.7	690.9	898.1	48.00	65.56	388
BC	456.9	285.6	742.5	965.2	48.00	65.56	417
0.1	368.1	249.3	617.4	802.7	43.20	65.56	385
0.2	279.2	212.3	491.5	638.9	38.40	66.50	340
0.3	190.4	175.0	365.4	475.0	36.00	↑	270
0.4	101.5	138.8	240.3	312.4	36.00		117
0.5	12.7	146.8	159.5	207.3	36.00		118
0.6	87.0	183.9	270.9	352.1	36.00		200
0.7	175.8	217.6	393.4	511.4	36.00	↓	290
0.8	264.7	249.3	514.0	668.2	38.40	66.50	355
0.9	353.5	276.7	630.2	819.3	43.20	65.61	393
CB	442.4	298.9	741.3	963.7	48.00	65.61	416
CD	354.2	292.2	646.4	840.4	48.00	65.61	363
0.1	303.5	267.4	570.9	742.1	43.20	65.61	356
0.2	252.8	237.8	490.6	637.8	38.40	66.50	339
0.3	202.1	205.6	407.7	530.0	36.00	↑	301
0.4	151.4	170.5	321.9	418.5	↑		237
0.5	100.7	134.6	235.3	305.9			174
0.6	50.0	117.1	167.1	217.2			123
0.7	0.7	151.1	151.8	197.4			112
0.8	51.4	188.0	239.4	311.3		↓	177
0.9	102.1	226.8	328.9	427.6	↓		242
DC	152.8	265.5	418.3	543.8	36.00	66.50	308

verted to unit shears in Table 6.26, where

$$\beta_L = 1.67$$

$$\gamma = 1.3$$

$$\phi = 0.85$$

$$V_{OL} = \text{overload shear (kips)}$$

$$V_u = \text{ultimate shear (kips)}$$

Table 6.27

Point	v_n (psi)	$v_n - v_c$ (psi)	b_w (in.)	Required s (in.)
AB	218	103	36.00	40.1
$14'AB$	266	151	40.99	24.0
$14'AB$	392	277	40.99	13.1
$28'AB$	372	260	45.97	12.5
$28'AB$	398	283	45.97	11.4
BA	388	273	48.00	11.4
BC	417	302	48.00	10.3
0.1	385	270	43.20	12.8
0.2	340	225	38.40	17.2
0.3	270	155	36.00	26.7
0.4	177	62	36.00	66.4
0.5	118	3	36.00	—
0.6	200	85	36.00	48.7
0.7	290	175	36.00	23.6
0.8	355	240	38.40	16.1
0.9	393	278	43.20	12.4
CB	416	301	48.00	10.3
CD	363	248	48.00	12.5
0.1	356	241	43.20	14.3
0.2	339	224	38.40	17.3
0.3	301	186	36.00	22.3
0.4	237	122		33.7
0.5	174	59		70.6
0.6	123	8		—
0.7	112	-3		—
0.8	177	62		67.0
0.9	242	128		32.4
DC	308	193	36.00	21.4

Again note that the unit shears are magnified by the secant of the skew angle to approximate the increased shear at the obtuse corner.

Using

$$s = \frac{A_v f_y}{(v_n - v_c)(b_w)}$$

we can compute the required spacing listed in Table 6.27. Assume →
$v_c = 2\sqrt{3300} = 115$ psi. $A_v f_y = 2.48(60,000) = 148,800$ lb.

The required spacing is plotted against the capacity of an actual spacing in Figure 6.28. Note that a $1'$-$6''$ spacing is used as a maximum and that the girder web flares vary from the 12-in. width at B and C to a 9-in. minimum width at the quarter points adjacent to B and C. The re-

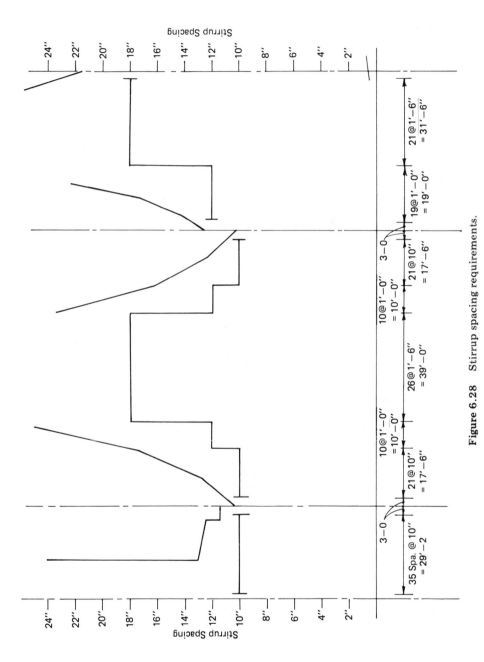

Figure 6.28 Stirrup spacing requirements.

sisting shear spacings are plotted for a 10-, 12-, and 18-in. spacing for 8-#5.

BIBLIOGRAPHY

1. *Continuous Concrete Bridges*, 2nd ed., Portland Cement Association.
2. *Concrete Members with Variable Moment of Inertia*, ST 103, Portland Cement Association.
3. *Beam Factors and Moment Coefficients for Members with Prismatic Haunches*, ST 81, Portland Cement Association.
4. *Reinforced Concrete Design Handbook—Working Stress Method*, SP-3, 3rd. ed., American Concrete Institute, 1980.
5. *AASHTO Standard Specifications for Highway Bridges, 1977*, with the 1978–1981 Interim Specifications, The American Association of State Highway and Transportation Officials.
6. R. Park and T. Paulay, *Reinforced Concrete Structures*, Wiley, New York, 1975.
7. Phil M. Ferguson, *Reinforced Concrete Fundamentals*, 4th ed., Wiley, New York, 1979.
8. *Notes on Load Factor Design for Reinforced Concrete Bridge Structures with Design Applications*, Portland Cement Association.
9. *ACI Building Code Requirements for Reinforced Concrete (ACI 318-77)* and *Commentary*, American Concrete Institute.
10. *Design Proposal for Side Face Crack Control Reinforcement for Large Reinforced Concrete Beams*, Frantz and Breen, ACI publication, *Concrete International*, Vol. 2, No. 10, October 1980.
11. *Strength and Serviceability Criteria, Reinforced Concrete Bridge Members, Ultimate Design*, FHWA, October 1969.
12. *Continuous Hollow Girder Concrete Bridges*, Portland Cement Association.
13. Phillip H. Lovering, *Hollow Box Girder Viaduct Vital for Complex Traffic Interchange*, Modern Developments in Reinforced Concrete, RC 28, Portland Cement Association.
14. *Design of Beams Subject to Torsion Related to the New Australian Code*, Henry J. Cowan, Journal of the American Concrete Institute, January 1960.
15. C. P. Heins, *Bending and Torsional Design in Structural Members*, Lexington Books, D. C. Heath and Company.

SEVEN

Design of Posttensioned Concrete Box Girder Bridges

Posttensioned concrete box girder construction involves procedures similar to those discussed in Chapter 6 for reinforced concrete box girders. Both normally use cast-in-place construction built on falsework. The main differences are in the prestressing and span ranges. Posttensioned box girder construction is most practical in spans where reinforced box girders become inefficient, say at 140 ft. Long spans at reasonable depth-to-span ratios can be obtained because high-strength materials are normally used with posttensioned construction.

Bottom slab thicknesses of $5\frac{1}{2}$ in. or more are acceptable. This thickness is often increased in negative moment areas to obtain a larger lower flange. This also will lower the concrete center of gravity, and thereby increase the eccentricity of the prestressing force. Girder webs should normally be $1'-0''$ or wider to accommodate the posttensioning ducts.

Several things have contributed in recent years to the more common application of this form of construction. It has become feasible because larger tendons (up to $55\frac{1}{2}$ in. ϕ strands) are readily available with the jacks required to jack such large tendons. Also, greater tendon lengths are practicable when low friction rigid conduits are used. Higher strength cast-in-place concrete is now normally used to produce high efficiency.

The torsional and aesthetic attributes mentioned in Chapter 6 for reinforced concrete box girders are applicable as well to posttensioned box girders. In addition, a smaller depth requirement for long spans gives this type of structure a slender and graceful appearance.

7.1 DESIGN CODE AND PROCEDURES

Design of posttensioned bridges differs primarily in two areas from the design of the pretensioned bridges described in Chapter 8. First the secondary effect of posttensioning with draped cables over continuous supports must be examined. This normally produces increased prestress in negative moment regions and decreased prestress in positive moment regions. Frictional losses from interaction of the strands and ducts must also be accounted for by use of the equation

$$T_x = \frac{T_0}{e^{(KL + \mu\alpha)}}$$

where T_x = stress at any point away from the jack,
   T_0 = stress at the jack,
   e = base of Naperian logarithms,
   K = friction wobble coefficient,
   L = length from jack to point x,
   μ = friction curvature coefficient,
   α = angle change of prestressing between jack and point x.

Posttensioned reinforcing as a rule is draped through the girder webs, but sometimes it is appropriate to place short tendons in the top and bottom slabs in the regions of high negative and positive moments. Full-length tendons are normally feasible for lengths of 600 ft. Jacking of tendons at both ends but not necessarily simultaneously is a means of using full-length tendons effectively over long distances.

AASHTO limits the allowable anchorage bearing stress to 3000 psi. In no case should this stress exceed $0.9f'_{ci}$. It is essential that it be distributed over a substantial area to control "bursting stresses." This is done with reinforcing grids or spirals behind the anchorage. At anchorages, and also at couplers, it is necessary to specify the amount of anchor set to take place. These anchorages and couplers must develop at least 95% of the minimum specified ultimate strength of the prestressing steel without exceeding this specified set.

AASHTO specifies that diaphragms be spaced at 80-ft intervals, maximum. Special consideration should be given to the placement of diaphragms in curved girders.

7.2 DESIGN EXAMPLE

Example 7.1. Cast-in-Place Posttensioned Design. The bridge to be designed is shown in Figure 7.1 in section near midspan and the pier.

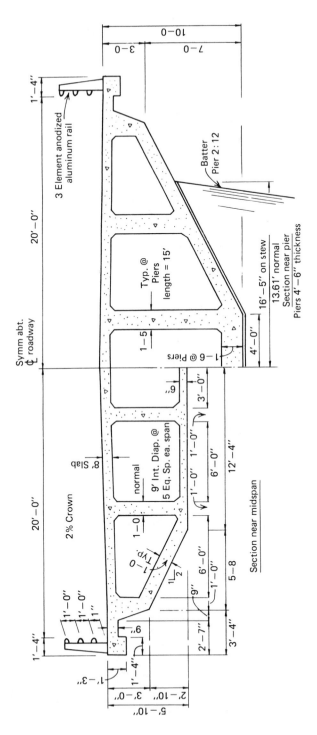

Figure 7.1 Typical section.

285

Figure 7.2 is an elevation and plan of the proposed design. Important data follow:

Live load   HS 20-44

Prestress (superstructure)

$f_c' = 4000$ psi

$f_c = 1600$ psi

$f_s = 24,000$ psi (Grade 60 A615)

Prestressing steel $\frac{1}{2}$ in. ϕ 270 k strands

Substructure

$f_c' = 3300$ psi

$f_c = 1320$ psi

$f_s = 24,000$ psi

For 4000 psi

$$E_c = 57000\sqrt{4000}$$

$$= 3,605,000 \text{ psi}$$

$$n = \frac{29}{3.605} = 8$$

For $f_c' = 3300$ psi   use $n = 9$

Galvanized rigid conduit

$k = 0.0002$

$\mu = 0.25$

Jack at both ends (not simultaneous)

34°   skew

$\sin 34° = 0.55919290$

$\cos 34° = 0.82903757$

$\sec 34° = 1.20621795$

$\tan 34° = 0.67450852$

The roadway slab is designed by the procedures outlined in Chapter 4. The deck slab thickness, as designed, is shown in the typical section in Figure 7-1. The depth configuration for the box girder is parabolic (Figure 7.3).

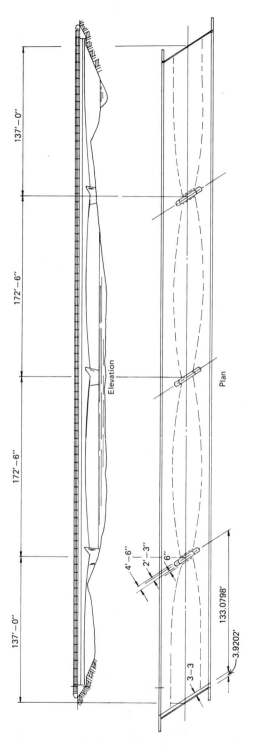

Figure 7.2 Elevation and plan views.

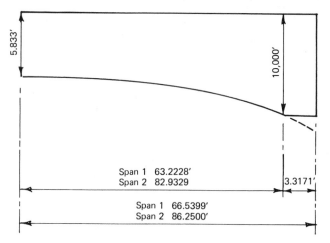

Figure 7.3   Girder configuration.

The equation for a parabola is $y = ax^2$, where $a = \text{rise}/(\text{length})^2$

	a
Span 1	0.00104242
Span 2	0.00060581

For spans 1 and 2 the haunch and girder depths are computed at various locations:

Span 1

Point	x	$y = ax^2$	Depth (ft)
0.5AB	0	0	5.8333
0.6	13.308	0.1846	6.0179
0.7	26.616	0.7385	6.5718
0.8	39.924	1.6615	7.4949
0.9	53.232	2.9538	8.7872
BA	—	—	10.0000

Span 2

Point	x	$y = ax^2$	Depth (ft)
BC	—	—	10.0000
0.1	69.00	2.8843	8.7176
0.2	51.75	1.6224	7.4557

Point	x	$y = ax^2$	Depth (ft)
0.3	34.50	0.7211	6.5544
0.4	17.25	0.1803	6.0136
0.5	0	0	5.8333

Because the haunch depths for both spans are similar at tenth points, the same values of depth are assumed. These values are tabulated:

Average Depth: All Spans (ft)

5.83 @ 0.5   7.48 @ 0.2 or 0.8
6.02 @ 0.4 or 0.6   8.75 @ 0.1 or 0.9
6.56 @ 0.3 or 0.7   10.00 @ support

The roadway sections are laid out in rectangles and triangles for computation of the moment of inertia at the various tenth points (Figure 7.4).

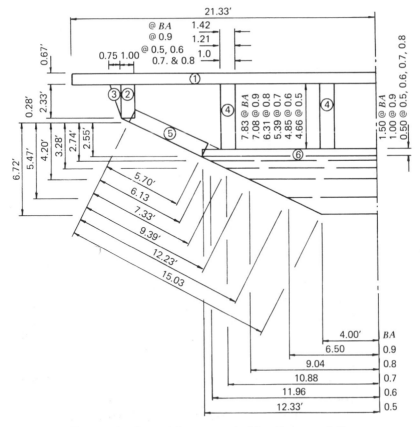

Figure 7.4 Layout for moment of inertia computations.

The moment of inertia is computed at various tenth points along the span, using the equation $I_T = \Sigma (I_0 + A_y^{-2})$, in addition to the section modulus at the top and bottom. The values $(12/S)$ which are noted, are ratios to be used as multipliers times moments in kip-feet to obtain stresses. These computations are tabulated in Table 7.1.

Table 7.1

Element	A	y_b	A_{yb}	$\bar{y}$	A_y^{-2}	I_0	I_T

@ 0.0AB to 0.5AB   D = 5.83 ft

Element	A	y_b	A_{yb}	$\bar{y}$	A_y^{-2}	I_0	I_T
1	28.44	5.50	156.42	2.22	140.16	1.05	141.21
2	4.67	4.00	18.68	0.72	2.42	2.12	4.54
3	1.75	4.39	7.68	1.11	2.16	0.53	2.69
4	18.64	2.83	52.75	0.45	3.77	33.73	37.50
5	11.40	1.28	14.60	2.00	45.60	6.17	51.77
6	12.33	0.25	3.08	3.03	113.20	0.26	113.46
(11,121)	77.23 ft²		253.21 ft³				351.17 ft⁴

y_b = 3.28 ft (39.36 in.) S_b = 185,006 ft³; $12/S_b$ = 0.0649 1/in.²-ft
y_t = 2.55 ft (30.60 in.) S_t = 237,969 ft³; $12/S_t$ = 0.0504 1/in.²-ft

@ 0.6AB   D = 6.02 ft

Element	A	y_b	A_{yb}	$\bar{y}$	A_y^{-2}	I_0	I_T
1	28.44	5.69	161.82	2.30	150.45	1.05	151.50
2	4.67	4.19	19.57	0.80	2.99	2.12	5.11
3	1.75	4.58	8.02	1.19	2.48	0.53	3.01
4	19.40	2.92	56.65	0.47	4.29	38.03	42.32
5	12.26	1.37	16.80	2.02	50.03	7.67	57.70
6	11.96	0.25	2.99	3.14	117.92	0.25	118.17
(11,301)	78.48 ft²		265.85 ft³				377.81 ft⁴

y_b = 3.39 ft (40.68 in.) S_b = 192,583 ft³; $12/S_b$ = 0.0623 1/in.²-ft
y_t = 2.63 ft (31.56 in.) S_t = 248,234 ft³; $12/S_t$ = 0.0483 1/in.²-ft

@ 0.7AB   D = 6.56 ft

Element	A	y_b	A_{yb}	$\bar{y}$	A_y^{-2}	I_0	I_T
1	28.44	6.23	177.18	2.52	180.61	1.05	181.66
2	4.67	4.73	22.09	1.02	4.86	2.12	6.98
3	1.75	5.12	8.96	1.41	3.48	0.53	4.01
4	21.56	3.20	68.99	0.51	5.61	52.20	57.81
5	14.66	1.64	24.04	2.07	62.86	13.15	75.97
6	10.88	0.25	2.72	3.46	130.25	0.23	130.48
(11,802)	81.96 ft²		303.98 ft³				456.91 ft⁴

y_b = 3.71 ft (44.52 in.) S_b = 212,814 ft³; $12/S_b$ = 0.0564 1/in.²-ft
y_t = 2.85 ft (34.20 in.) S_t = 277,032 ft³; $12/S_t$ = 0.0433 1/in.²-ft

Table 7.1 (*Continued*)

Element	A	y_b	A_{yb}	$\bar{y}$	A_y^{-2}	I_0	I_T

@ 0.8AB   D = 7.48

Element	A	y_b	A_{yb}	$\bar{y}$	A_y^{-2}	I_0	I_T
1	28.44	7.15	203.35	2.89	237.53	1.05	238.58
2	4.67	5.65	26.39	1.39	9.02	2.12	11.14
3	1.75	6.04	10.57	1.78	5.54	0.53	6.07
4	25.24	3.66	92.38	0.60	9.08	83.75	92.84
5	18.78	2.10	39.44	2.16	87.62	27.61	115.23
6	9.04	0.25	2.26	4.01	145.36	0.19	145.55
(12,660)	87.92 ft²		374.39 ft³				609.41 ft⁴

y_b = 4.26 ft (51.12 in.) S_b = 247,197 ft³; $12/S_b$ = 0.0485 1/in.²-ft
y_t = 3.22 ft (38.65 in.) S_t = 327,037 ft³; $12/S_t$ = 0.0367 1/in.²-ft

@ 0.9AB   D = 8.75

Element	A	y_b	A_{yb}	$\bar{y}$	A_y^{-2}	I_0	I_T
1	28.44	8.42	239.46	3.60	368.58	1.05	369.63
2	4.67	6.92	32.32	2.10	20.59	2.12	22.71
3	1.75	7.31	12.79	2.49	10.85	0.53	11.38
4	34.22	4.54	155.36	0.28	2.68	143.14	145.82
5	24.46	2.74	67.02	2.08	105.82	60.99	166.81
6	13.00	0.50	6.50	4.32	242.61	1.08	243.69
(15,342)	106.54 ft²		513.45 ft³				960.04 ft⁴

y_b = 4.82 ft (57.84 in.) S_b = 344,180 ft³; $12/S_b$ = 0.0349 1/in.²-ft
y_t = 3.93 ft (47.16 in.) S_t = 422,144 ft³; $12/S_t$ = 0.0284 1/in.²-ft

@ AB   D = 10.00

Element	A	y_b	A_{yb}	$\bar{y}$	A_y^{-2}	I_0	I_T
1	28.44	9.67	275.01	4.07	471.11	1.05	472.16
2	4.67	8.17	38.15	2.57	30.84	2.12	32.96
3	1.75	8.56	14.98	2.96	15.33	0.53	15.86
4	44.37	5.42	240.49	0.18	1.44	227.22	228.66
5	30.06	3.36	101.00	2.24	150.83	113.09	263.92
6	12.00	0.75	9.00	4.85	282.27	2.25	284.52
(17,466)	121.29 ft²		678.63 ft³				1298.08 ft⁴

y_b = 5.60 ft (67.20 in.) S_b = 400,550 ft³; $12/S_b$ = 0.0300 1/in.²-ft
y_t = 4.40 ft (52.80 in.) S_t = 509,791 ft³; $12/S_t$ = 0.0235 1/in.²-ft

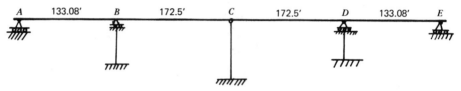

Figure 7.5   Structure schematic.

With these properties calculated, the girder stiffnesses can be determined; a schematic of the structure is shown in Figure 7.5.

Because the section is parabolic, the frame constants given in Appendix J are used; the r value is determined by applying the computed moments of inertia:

$$r = \sqrt[3]{\frac{I_{\max}}{I_{\min}}} - 1 \qquad r = \sqrt[3]{\frac{1298.08}{351.17}} - 1 = 0.548$$

From the curves

$$C_{AB} = 0.705 \quad \text{and} \quad k_{AB} = 4.48 \qquad C_{BA} = 0.452 \quad \text{and} \quad k_{BA} = 7.0$$

k_{BA} is modified for a hinge at AB:

$$\text{mod } k_{BA} = 7.0(1 - 0.705 \times 0.452) \qquad = 4.77$$

Again, from the curves,

$$C_{BC} = C_{CB} = 0.628 \qquad k_{BC} = k_{CB} = 8.05$$

The stiffness factors at B and C are

$$K_{BA} = \frac{4.77(351.17)}{133.08} = 12.59$$

$$K_{BC} = \frac{8.05(351.17)}{172.5} = 16.39$$

$$K_{CB} = 16.39$$

The distribution factors at B and C are

$$D_{BA} = \frac{12.59}{28.98} = 0.434$$

$$@ B$$

$$D_{BC} = \frac{16.39}{28.98} = 0.566$$

$$D_{CB} = D_{CD} = 0.500 \qquad @ C$$

A fixed end moment of −1000 is applied at *BA* and distributed throughout the structure by moment distribution:

	0.628			0.628	

	B		*C*		*D*	
	BA	*BC*	*CB*	*CD*	*DC*	*DE*
	0.434	0.566	0.500	0.500	0.566	0.434
FEM	−1000					
Bal	+434	+566				
CO			+355			
Bal			−178	−178		
CO		−112			−112	
Bal	+49	+63			+63	+49
CO			+40	+40		
Bal			−40	−40		
CO		−25			−25	
Bal	+11	+14			+14	+11
Distributed moment	−506	+506	+178	−178	−60	+60

A fixed end moment of +1000 is applied at *BC* and distributed throughout the structure by moment distribution:

	0.628			0.628	

	B		*C*		*D*	
	BA	*BC*	*CB*	*CD*	*DC*	*DE*
	0.434	0.566	0.500	0.500	0.566	0.434
FEM		+1000				
Bal	−434	−566				
CO			−355			
Bal			+178	+178		
CO		+112			+112	
Bal	−49	−63			−63	−49
CO			−40	−40		
Bal			+40	+40		
CO		+25			+25	
Bal	−11	−14			−14	−11
M	−494	+494	−178	+178	+60	−60

A fixed end moment of -1000 is applied at CB and distributed throughout the structure by moment distribution:

	B		C		D	
	0.628			0.628		
	BA	BC	CB	CD	DC	DE
	0.434	0.566	0.500	0.500	0.566	0.434
FEM			−1000			
Bal			+500	+500		
CO		+314			+314	
Bal	−136	−178			−178	−136
CO			−112	−112		
Bal			+112	+112		
CO		+70			+70	
Bal	−30	−40			−40	−30
CO			−25	−25		
Bal			+25	+25		
M	−166	+166	−500	+500	+166	−166

The application of a unit fixed end moment at DE, DC, and CD yields similar results with opposite rotational signs. After the distributed moments are divided by 1000 a summary of the unit moment distribution is tabulated to convert the signs to a bending convention.

Moment Distribution Summary

	M_B	M_C	M_D
M_{BA}^F	−0.506	+0.178	−0.060
M_{BC}^F	−0.494	−0.178	+0.060
M_{CB}^F	−0.166	−0.500	+0.166
M_{CD}^F	+0.166	−0.500	−0.166
M_{DC}^F	+0.060	−0.178	−0.494
M_{DE}^F	−0.060	+0.178	−0.506

The dead load concentrations for the diaphragms in Span AB are calculated by using the dimensions in Figure 7.1 and a unit weight of concrete equal to 0.150 k ft^3.

@ 0.2, @ 0.4   D = 5.83

$$P = \quad 3(4.67 \times 6.00 \times 0.75 \times 0.150) = \quad 9.46$$
$$2(1.5 \times 6.0 \times 0.75 \times 0.150) = \quad 2.03$$
$$2(\tfrac{1}{2} \times 3.33 \times 6.0 \times 0.75 \times 0.150) = \quad \underline{2.25}$$
$$13.7 \text{ k}$$

@ 0.6   D = 6.02

$$P = \quad 3(4.86 \times 6.00 \times 0.75 \times 0.150) = \quad 9.84$$
$$2(1.5 \times 6.0 \times 0.75 \times 0.150) = \quad 2.03$$
$$2(\tfrac{1}{2} \times 3.33 \times 6.0 \times 0.75 \times 0.150) = \quad \underline{2.25}$$
$$14.1 \text{ k}$$

@ 0.8   D = 7.48

$$P = \quad 3(6.32 \times 6.00 \times 0.75 \times 0.150) = 12.80$$
$$2(1.5 \times 6.0 \times 0.75 \times 0.150) = \quad 2.03$$
$$2(\tfrac{1}{2} \times 3.33 \times 6.0 \times 0.75 \times 0.150) = \quad \underline{2.25}$$
$$17.1 \text{ k}$$

The uniform dead load that approximates the effect of the diaphragm concentrations equals twice the concentrations distributed over the length of the span:

$$\text{Span } AB = w = \frac{2(13.7 + 14.1 + 17.1)}{133.08} = 0.44 \text{ k ft}^{-1}$$

$$\text{Span } BC = w = \frac{2(14.1 + 17.1)}{172.5} = 0.36 \text{ k ft}^{-1}$$

Conservatively use $w = 0.45$ k-ft^{-1}.

Note that in computing dead and live loads the box is considered a unit. Therefore computation of the dead load action for the entire box is based on the action of the entire section.

The uniform dead load is tabulated and summed by using the dimensions given in Figures 7.1 and 7.4

Uniform Load

Section @ 0.5	$A = 77.23 \times 0.150 = 11.58$
Extra thickness on overhang	$2(0.083 \times 2.58 \times 0.150) = \quad 0.03$
Edge beam	$2(0.50 \times 1.33 \times 0.150) = \quad 0.20$
Railing (45 plf each)	$2 \times 0.045 = \quad 0.09$
Crown	$\tfrac{1}{2} \times 0.06 \times 6.0 \times 0.150 = \quad 0.03$
Fillets	$18(\tfrac{1}{2} \times 0.25 \times 0.25 \times 0.150) = \quad 0.08$
Forms (6 psf)	$36 \times 0.006 = \quad 0.22$
Diaphragm	0.45
WS (25 psf)	$40 \times 0.25 = \quad \underline{1.00}$
	13.68 k ft^{-1}

The haunch load at the support is computed by using the difference between the maximum and minimum cross-section areas.

Haunch Load

Use $w = (121.29 - 77.23) \times 0.150 = 6.61$ k ft^{-1}

Using the coefficients taken from the charts in Appendix J, we compute the fixed end moments:

$$M_{AB}^F = \left.\begin{array}{l} (0.0705)(13.68)(133.08)^2 = 17,081 \\ (0.00195)(6.61)(133.08)^2 = 228 \end{array}\right\} \quad 17,309 \text{ k-ft}$$

$$M_{BA}^F = \left.\begin{array}{l} (0.1120)(13.68)(133.08)^2 = 27,135 \\ (0.0161)(6.61)(133.08)^2 = 1,885 \end{array}\right\} \quad 29,020 \text{ k-ft}$$

$$M_{BC} = \left.\begin{array}{l} (0.096)(13.68)(172.5)^2 = 39,078 \\ (0.0154 + 0.0030)(6.61)(172.5)^2 = 3,619 \end{array}\right\} \quad 42,697 \text{ k-ft}$$

The fixed end moment at BA is modified for AB being hinged; therefore,

$$\text{Mod } M_{BA}^F = M_{BA}^F + M_{AB}^F(C_{AB})$$

$$\text{Mod } M_{BA}^F = 29,020 + 17,309(0.705) = 41,222 \text{ k-ft}$$

The dead load moments at B and C are computed in Table 7.2 by using the moment distribution summary previously given.

The simple span actions for uniform and haunch loads are combined with the negative moments from corner restraint computed by the moment distribution in Table 7.2. The haunch simple span moments can be taken from the charts in Appendix J. These moments are tabulated and summed in Table 7.3.

Next the ordinates for influence lines are computed.

For span AB the fixed end moments are computed and modified for

Table 7.2   Moment Distribution[a]

	M_B	M_c
$M_{BA}^F = 41,222$	$-20,858$	$+7,338$
$M_{BC}^F = 42,697$	$-21,226$	$-7,648$
$M_{CB}^F = 42,697$	$-7,133$	$-21,484$
$M_{CD}^F = 42,697$	$+7,133$	$-21,848$
$M_{DC}^F = 42,697$	$+2,578$	$-7,648$
$M_{DE}^F = 41,222$	$-2,473$	$-7,338$
	$-41,979$	$-43,588$

[a] All moments are in k-ft.

Table 7.3 Moments: Dead Load[a]

Point	Unit +M	Haunch +M	Σ +M	$-M$	M_{DL}
A					
0.1	+10,902	+246	+11,148	-4,198	+6,950
0.2	+19,382	+492	+19,874	-8,396	+11,478
0.3	+25,439	+738	+26,177	-12,594	+13,583
0.4	+29,073	+972	+30,045	-16,792	+13,253
0.5	+30,285	+1,217	+31,502	-20,990	+10,512
0.6	+29,073	+1,452	+30,525	-25,187	+5,338
0.7	+25,439	+1,651	+27,090	-29,385	-2,295
0.8	+19,382	+1,639	+21,021	-33,583	-12,562
0.9	+10,902	+1,182	+12,084	-37,781	-25,697
B				-41,979	-41,979
0.1	+18,318	+2,439	+20,757	-42,140	-21,383
0.2	+32,565	+3,580	+36,145	-42,301	-6,156
0.3	+42,742	+3,993	+46,735	-42,462	+4,273
0.4	+48,848	+4,071	+52,919	-42,623	+10,296
0.5	+50,883	+4,091	+54,974	-42,784	+12,190
0.6	+48,848	+4,071	+52,919	-42,944	+9,975
0.7	+42,742	+3,993	+46,735	-43,105	+3,630
0.8	+32,565	+3,580	+36,145	-43,266	-7,121
0.9	+18,318	+2,439	+20,757	-43,427	-22,670
C				-43,588	-43,588

[a]Moments are in k-ft.

one end (AB) hinged. Appendix J is used for computations with the equation Mod. $M_{BA}^{F} = M_{BA}^{F} + M_{AB}^{F}(C_{AB})$. The corner moments are computed by using the moment distribution summary and are equal to the following:

Load @	M_{AB}^{F}	M_{BA}^{F}	Mod M_{BA}^{F} (kip-ft)	M_B' (kip-ft)	M_c (kip-ft)	M_D (kip-ft)
A						
0.1	10.51	1.86	9.27	-4.69	+1.65	-0.56
0.2	16.10	6.51	17.86	-9.04	+3.18	-1.07
0.3	17.83	12.78	25.35	-12.83	+4.51	-1.52
0.4	16.57	18.90	30.58	-15.47	+5.44	-1.83
0.5	13.44	24.09	33.57	-16.99	+5.98	-2.01
0.6	9.38	26.75	33.36	-16.88	+5.94	-2.00
0.7	5.46	25.68	29.53	-14.94	+5.26	-1.77
0.8	2.53	20.69	22.47	-11.37	+4.00	-1.35
0.9	0.60	11.95	12.37	-6.26	+2.20	-0.74
B						
Area				-1444	+508	-171

These areas are the areas under the influence curve for span AB.

The fixed end moments are computed for span *BC* by using Appendix J. The influence line ordinates and areas are tabulated for span *BC* by using the moment distribution summary equations:

Load @	M_{BC}^F (kip-ft)	M_{CB}^F (kip-ft)	M_B (kip-ft)	M_C (kip-ft)	M_D (kip-ft)
B					
0.1	15.20	1.24	−7.71	−3.33	+1.12
0.2	25.74	4.83	−13.52	−7.00	+2.35
0.3	30.88	11.21	−17.12	−11.10	+3.71
0.4	30.70	18.80	−18.29	−14.86	+4.96
0.5	26.05	26.05	−17.19	−17.66	+5.89
0.6	18.80	30.70	−14.38	−18.70	+6.22
0.7	11.21	30.88	−10.66	−17.44	+5.80
0.8	4.83	25.74	−6.66	−13.73	+4.56
0.9	1.24	15.20	−3.14	−7.82	+2.60
C					
Area			−1875	−1926	+642

The bridge is symmetrical; therefore the influence lines for spans *CD* and *DE* are symmetrical to spans *BC* and *DB*, respectively as listed in Table 7.4.

The influence lines for the corner moments are plotted in Figure 7.6.

The influence line ordinates for points in the loaded span can be computed by the following procedure.

1. The peak moment for each loading of 1 k is computed by subtracting the distributed moment (−*M*) from the simple span moment (+*M*) at the point of loading.

2. The moment diagram for each loading is tabulated across the span horizontally, using a straight line distribution, from the peak.

3. These tabulated moment diagrams are listed in the order of applied vertical loading.

4. The results tabulated are the moment diagrams (horizontal values) and influence lines (vertical values).

These results and the area under each influence line are listed in Table 7.5 for span *AB* and in Table 7.6 for span *BC*.

The influence lines for points and loads in span *AB* are plotted in Figure 7.7.

The influence lines for points and loads in span *BC* are plotted in Figure 7.8.

It is assumed that the live load distribution equals the out-to-out

Table 7.4^a

Load @	M_B	M_C	M_D
Span *CD*			
C			
0.1	+2.60	-7.82	-3.14
0.2	+4.56	-13.73	-6.66
0.3	+5.80	-17.44	-10.66
0.4	+6.22	-18.70	-14.38
0.5	+5.89	-17.66	-17.19
0.6	+4.96	-14.86	-18.29
0.7	+3.71	-11.10	-17.12
0.8	+2.35	-7.00	-13.57
0.9	+1.12	-3.33	-7.71
D			
Area	+642	-1926	-1875

Load @	M_B	M_C	M_D
Span *DE*			
D			
0.1	-0.74	+2.20	-6.26
0.2	-1.35	+4.00	-11.37
0.3	-1.77	+5.26	-14.94
0.4	-2.00	+5.94	-16.88
0.5	-2.01	+5.98	-16.99
0.6	-1.83	+5.44	-15.47
0.7	-1.52	+4.51	-12.83
0.8	-1.07	+3.18	-9.04
0.9	-0.56	+1.65	-4.69
E			
Area	-171	+508	-1444

^{a}Moments are in k-ft.

width of deck divided by 7, which represents the number of lines of wheels to be applied to the box. The analysis does not differentiate between the interior and exterior girders, but it is consistent with the AASHTO specifications. The *LL* distribution = 42.67/7 = 6.095 lines of wheels.

For span *AB* the impact factor is

$$I = \frac{50}{133.08 + 125} = 0.194$$

and the combined distribution and impact coefficient is

coefficient = 1.194(6.095) = 7.277 lines of wheels

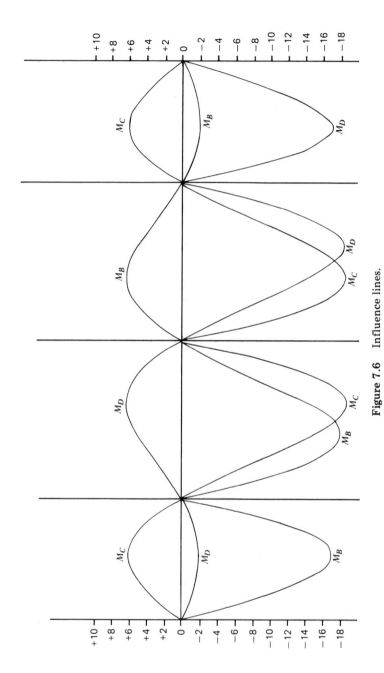

Figure 7.6 Influence lines.

Table 7.5 Influence Lines: Span AB. L = 133.08 ft

Load @		AB	0.1	0.2	0.3	0.4	0.5	0.6	0.7	0.8	0.9	BA
0.1	+M	0	+11.98									
	−M		−0.47									
	Σ		+11.51	+9.71	+7.91	+6.11	+4.31	+2.51	+0.71	−1.09	−2.89	−4.69
0.2	+M	0		+21.29								
	−M			−1.81								
	Σ		+9.74	+19.48	+15.92	+12.35	+8.78	+5.22	+1.66	−1.91	−5.48	−9.04
0.3	+M	0			+27.95							
	−M				−3.85							
	Σ		+8.03	+16.07	+24.10	+18.82	+13.55	+8.27	+3.00	−2.28	−7.55	−12.83
0.4	+M	0				+31.94						
	−M					−6.19						
	Σ		+6.43	+12.88	+19.31	+25.75	+18.88	+12.01	+5.14	−1.73	−8.60	−15.47
0.5	+M	0					+33.27					
	−M						−8.50					
	Σ		+4.95	+9.91	+14.86	+19.82	+24.77	+16.42	+8.07	−0.29	−8.64	−16.99
0.6	+M	0						+31.94				
	−M							−10.13				
	Σ		+3.64	+7.27	+10.90	+14.54	+18.18	+21.81	+12.14	+2.46	−7.21	−16.88
0.7	+M	0							+27.95			
	−M								−10.46			
	Σ		+2.50	+5.00	+7.50	+9.99	+12.49	+14.99	+17.49	+6.68	−4.13	−14.94
0.8	+M	0								+21.29		
	−M									−9.10		
	Σ		+1.52	+3.05	+4.57	+6.10	+7.62	+9.14	+10.67	+12.19	+0.41	−11.37
0.9	+M	0									+11.98	
	−M										−5.63	
	Σ		+0.71	+1.41	+2.12	+2.82	+3.53	+4.23	+4.94	+5.64	+6.35	−6.26
Positive area			+652	+1128	+1426	+1548	+1492	+1259	+849	+359	+90	
Negative area										−97	−592	

Table 7.6 Influence Lines: Span BC. L = 172.5 ft

Load @		B	0.1	0.2	0.3	0.4	0.5	0.6	0.7	0.8	0.9	C
0.1	$+M$		+15.52									
	$-M$	-7.71	-7.27									
	Σ		+8.25	+6.96	+5.68	+4.39	+3.10	+1.82	+0.53	-0.75	-2.04	-3.33
0.2	$+M$			+27.60								
	$-M$	-13.52		-12.22								
	Σ		+0.93	+15.38	+12.58	+9.78	+6.99	+4.19	+1.39	-1.40	-4.20	-7.00
0.3	$+M$				+36.22							
	$-M$	-17.12			-15.31							
	Σ		-4.44	+8.23	+20.91	+16.34	+11.76	+7.19	+2.62	-1.95	-6.53	-11.10
0.4	$+M$					+41.40						
	$-M$	-18.29				-16.92						
	Σ		-7.60	+3.10	+13.79	+24.48	+17.92	+11.37	+4.81	-1.75	-8.30	-14.86
0.5	$+M$						+43.12					
	$-M$	-17.19					-17.42					
	Σ		-8.61	-0.03	+8.54	+17.12	+25.70	+17.03	+8.36	-0.32	-8.99	-17.66
0.6	$+M$							+41.40				
	$-M$	-14.38						-16.97				
	Σ		-7.91	-1.44	+5.02	+11.49	+17.96	+24.43	+13.65	+2.86	-7.92	-18.70
0.7	$+M$								+36.22			
	$-M$	-10.66							-15.41			
	Σ		-6.16	-1.67	+2.83	+7.32	+11.82	+16.31	+20.81	+8.06	-4.69	-17.44
0.8	$+M$									+27.60		
	$-M$	-6.66								-12.32		
	Σ		-3.92	-1.18	+1.57	+4.31	+7.05	+9.80	+12.54	+15.28	+6.91	-13.73
0.9	$+M$										+15.52	
	$-M$	-3.14									-7.35	
	Σ		-1.88	-0.63	+0.63	+1.89	+3.14	+4.40	+5.66	+6.91	+8.17	-7.82
Positive area			+158	+581	+1234	+1675	+1819	+1665	+1214	+571	+155	
Negative area			-699	-85						-106	-736	

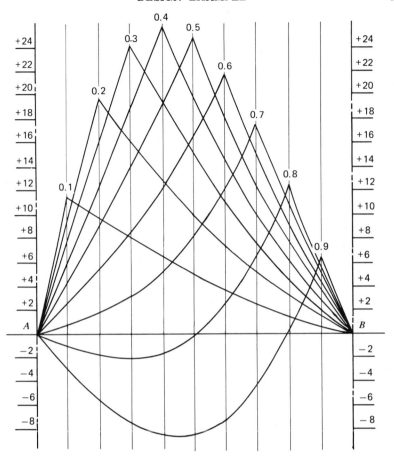

Figure 7.7 Influence lines for Span AB.

For span BC these values are

$$I = \frac{50}{172.5 + 125} = 0.168$$

coefficient = 1.168(6.095) = 7.119 lines of wheels

The same values as an average of spans AB and BC are

$$I = 0.181$$

coefficient = 7.198 lines of wheels

Using the HS 20 truck loading and the computed influence lines we

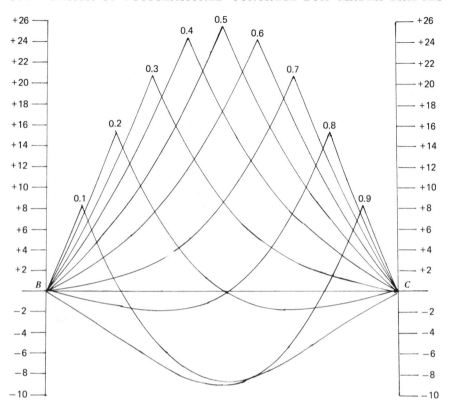

Figure 7.8   Influence lines for Span *BC*.

compute the maximum positive moments for one line of wheels and then expand to the total moment per box girder by multiplying by the distribution plus impact coefficients (Table 7.7).

Using the HS 20 lane loading and the tabulated influence lines, we compute the maximum positive moments in Table 7.8.

The HS 20 truck loadings will not control for negative moment, but the HS 20 lane loadings will result in the maximum negative moments in Table 7.9.

The moments are converted to stresses for the dead load and the maximum positive and maximum negative live load plus impact. These stresses are tabulated and summed to obtain the stresses for maximum bending in each direction. The stresses are computed by using the multiplier coefficients calculated with the section properties. The stresses are listed in Table 7.10 for the *bottom fiber* (in psi).

The stresses are computed (in psi) for the *top fiber*, as given in Table 7.11.

Table 7.7 Positive Moment: Truck[a]

Point	One Line of Wheels M_{LL}	M_{LL+I}[b]
AB		
0.1	+371	+2700
0.2	+620	+4512
0.3	+755	+5494
0.4	+802	
0.5	+768	
0.6	+665	
0.7	+514	
0.8	+318	
0.9	+200	
B	+222	
0.1	+174	
0.2	+432	+3075
0.3	+630	+4485
0.4	+756	
0.5	+795	
0.6	+756	
0.7	+626	
0.8	+409	
0.9	+170	
C	+214	

[a]Moments are in k-ft.
[b]Governing values.

Table 7.8 Positive Moment: Lane[a]

Point	One Line of Wheels M_{LL}	M_{LL+I}[b]
AB		
0.1	+333	
0.2	+577	
0.3	+735	
0.4	+810	+5894
0.5	+803	+5843
0.6	+722	+5254
0.7	+572	+4162
0.8	+389	+2831
0.9	+271	+1972

Table 7.8 (*Continued*)

Point	One Line of Wheels M_{LL}	M_{LL+I}[b]
B	+261	+1858
0.1	+248	+1766
0.2	+365	
0.3	+593	
0.4	+788	+5610
0.5	+867	+6172
0.6	+828	+5895
0.7	+674	+4798
0.8	+478	+3403
0.9	+365	+2598
C	+379	+2698

[a]Moments are in k-ft.
[b]Governing values.

Table 7.9 Negative Moment: Lane[a]

Point	One Line of Wheels M_{LL}	M_{LL+I}
AB	0	
0.1	−84	−605
0.2	−167	−1202
0.3	−251	−1807
0.4	−335	−2411
0.5	−418	−3009
0.6	−502	−3613
0.7	−586	−4218
0.8	−707	−5089
0.9	−1005	−7234
B	−1435	−10,329
0.1	−866	−6233
0.2	−503	−3621
0.3	−422	−3038
0.4	−439	−3160
0.5	−461	−3318
0.6	−482	−3469
0.7	−504	−3628
0.8	−627	−4513
0.9	−997	−7176
C	−1570	−11,301

[a]Moments are in k-ft.

Table 7.10   Bottom Fiber Stresses[a]

	Maximum Positive Moment			Maximum Negative Moment		
Point	f_{DL}	f_{LL+I}	$f_{DL+LL+I}$	f_{DL}	f_{LL+I}	$f_{DL+LL+I}$
A	0	0	0	0	0	0
0.1	-451	-175	-626	-451	+39	-412
0.2	-745	-293	-1038	-745	+78	-667
0.3	-882	-357	-1239	-882	+117	-765
0.4	-860	-383	-1243	-860	+156	-704
0.5	-682	-379	-1061	-682	+195	-487
0.6	-333	-327	-660	-333	+225	-108
0.7	+129	-235	-106	+129	+238	+367
0.8	+609	-137	+472	+609	+247	+856
0.9	+897	-69	+828	+897	+252	+1149
B	+1259	-56	+1203	+1259	+310	+1569
0.1	+746	-62	+808	+746	+218	+964
0.2	+299	-149	+150	+299	+176	+475
0.3	-241	-253	-494	-241	+171	-70
0.4	-641	-350	-991	-641	+197	-444
0.5	-791	-401	-1192	-791	+215	-576
0.6	-621	-367	-988	-621	+216	-405
0.7	-205	-271	-476	-205	+205	0
0.8	+345	-165	+180	+345	+219	+564
0.9	+791	-91	+700	+791	+250	+1041
C	+1308	-81	+1227	+1308	+339	+1647

[a] + = compression, - = tension.

Because of poor foundation conditions it is assumed that the abutments may settle 1 in. This settlement therefore influences the superstructure and must be included in the design. We examine this effect by using the conjugate beam analysis for span AB and assuming a fixed end moment of 1000 which results from some unknown deflection at the abutment. The moment diagram, the moments of inertia, and the M/EI diagram (assuming that $E = 1$) are shown in Figure 7.9. Note that the real beam is a cantilever in one direction, the conjugate beam a cantilever in the opposite direction. The moment at A on the conjugate beam is the deflection of A on the real beam. Summing the moments about A yields a conjugate moment (real deflection with $E = 1$) at B equal to 10,375.38.

Correction of the deflection for the proper E values gives

$$E_c = 57,000\sqrt{4000}\ (0.144) = 519,000 \text{ ksi}$$

$$\Delta_{AB} = \frac{10,375.38}{519,000} = 0.0200 \text{ ft}$$

Table 7.11   Top Fiber Stresses

Point	Maximum Positive Moment			Maximum Negative Moment		
	f_{DL}	f_{LL+I}	$f_{DL+LL+I}$	f_{DL}	f_{LL+I}	$f_{DL+LL+I}$
A	0	0	0	0	0	
0.1	+350	+136	+486	+350	−30	+320
0.2	+578	+227	+805	+578	−61	+517
0.3	+685	+277	+962	+685	−91	+594
0.4	+668	+297	+965	+668	−122	+546
0.5	+530	+294	+824	+530	−152	+378
0.6	+258	+254	+512	+258	−175	+83
0.7	−99	+180	+81	−99	−183	−282
0.8	−461	+104	−357	−461	−187	−648
0.9	−730	+56	−674	−730	−205	−935
B	−987	+44	−943	−987	−243	−1230
0.1	−607	+50	−557	−607	−177	−784
0.2	−226	+113	−113	−226	−133	−359
0.3	+185	+194	+379	+185	−132	+53
0.4	+497	+271	+768	+497	−153	+344
0.5	+614	+311	+925	+614	−167	+447
0.6	+482	+285	+767	+482	−168	+314
0.7	+157	+208	+365	+157	−157	0
0.8	−261	+125	−136	−261	−166	−427
0.9	−644	+74	−570	−644	−204	−848
C	−1024	+63	−961	−1024	−266	−1290

For Δ_{AB} = 1 in. We can proportion the values to obtain M_{BA}^F .

$$M_{BA}^F = \frac{0.0833}{0.0200} \times (1000) = 4167 \text{ k-ft}$$

Applying the deflection at one abutment and then at the other and summing produces the influence from deflection at both ends:

	M_{BA}^F = 4167	M_{DE}^F = 4167	Δ = 1 in. (both abutments)
M_B	−2109	−250	−2359
M_C	+742	+742	+1484
M_D	−250	−2109	−2359

The moments for a 1-in. deflection at both abutments are converted to stresses by using the computed multipliers taken from the section properties. The stresses are listed in Table 7.12.

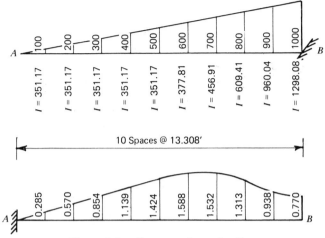

Figure 7.9 Conugate beam loading.

Table 7.12

Point	Δ = 1 in. [both abutments; M_Δ (kip ft)]	Section Multipliers Bottom Fiber	Section Multipliers Top Fiber	f_b (psi)	f_t (psi)
A	0	0.0649	0.0504	0	0
0.1	-236			+15	-12
0.2	-472			+31	-24
0.3	-708			+46	-36
0.4	-944			+61	-48
0.5	-1180	0.0649	0.0504	+77	-59
0.6	-1415	0.0623	0.0483	+88	-68
0.7	-1651	0.0564	0.0433	+93	-71
0.8	-1887	0.0485	0.0367	+92	-69
0.9	-2123	0.0349	0.0284	+74	-60
B	-2359	0.0300	0.0235	+71	-55
0.1	-1975	0.0349	0.0284	+69	-56
0.2	-1590	0.0485	0.0367	+77	-58
0.3	-1206	0.0564	0.0433	+68	-52
0.4	-821	0.0623	0.0483	+51	-40
0.5	-438	0.0649	0.0504	+28	-22
0.6	-53	0.0623	0.0483	+3	-3
0.7	+331	0.0564	0.0433	-19	+14
0.8	+715	0.0485	0.0367	-35	+26
0.9	+1100	0.0349	0.0284	-38	+31
C	+1484	0.0300	0.0235	-45	+35

309

Table 7.13

Point	$M_\Delta{}^a$ (kip-ft)	Multipliers Bottom Fiber	Top Fiber	f_b (psi)	f_t (psi)
B	−250	0.0300	0.0235	+8	−6
0.1	−151	0.0349	0.0284	+5	−4
0.2	−52	0.0485	0.0367	+3	−2
0.3	+48	0.0564	0.0433	−3	+2
0.4	+147	0.0623	0.0483	−9	+7
0.5	+246	0.0649	0.0504	−16	+12
0.6	+345	0.0623	0.0483	−21	+17
0.7	+444	0.0564	0.0433	−25	+19
0.8	+544	0.0485	0.0367	−26	+20
0.9	+643	0.0349	0.0284	−22	+18
C	+742	0.0300	0.0235	−22	+17

^{a}Abutment 2 Δ = 1 in.

The stresses in span BC are computed similarly for a 1-in. deflection at abutment 2 only, as listed in Table 7.13.

The bottom fiber stresses due to abutment deflection can be summed with the maximum dead load plus live load plus impact stresses. We use the most critical deflection cases to obtain the results in Table 7.14.

The resulting top fiber stresses are listed in Table 7.15.

With these data it is possible to design for the number of tendons. From experience the approximate number of tendons required in the exterior webs is assumed to be one; three tendons are in each of the interior webs. In addition, 4-in. diameter ducts are assumed. The lowest location for a duct in the exterior web is shown in Figure 7.10.

It should be noted that the section properties in Span AB are constant from the end support to $0.5AB$. In order not to develop end moments at the supports the ducts are positioned at the cg of the box. Therefore, with a total of two ducts in the exterior webs and 12 ducts in the interior webs, the location of the ducts in the interior webs at the cg of the box is

$$0 \text{ in.} = \frac{2(4.62) - 12(\bar{y})}{14} \qquad \bar{y} = 0.77 \text{ in.}$$

The tendon location at the jack is shown in Figure 7.11.

The center of gravity of the strands will vary in location from the center of gravity ducts. Shown in Figure 7.12 is the location of the strands at midspan; at the interior support they would be reversed.

Table 7.14 Summary of Stresses, Including a 1-in. Abutment Settlement[a]

Bottom Fiber	Positive Moment			Negative Moment		
Point	f^b_{D+L+I}	f^b_Δ	$f^b_{D+L+I+\Delta}$	f^b_{D+L+I}	f^b_Δ	$f^b_{D+L+I+\Delta}$
A	0			0	0	0
0.1	-626			-412	$+15$	-397
0.2	-1038			-667	$+31$	-636
0.3	-1239			-765	$+46$	-719
0.4	-1243			-704	$+61$	-643
0.5	-1061			-487	$+77$	-410
0.6	-660			-108	$+88$	-20
0.7	-106			$+367$	$+93$	$+460$
0.8	$+472$			$+856$	$+92$	$+948$
0.9	$+828$			$+1149$	$+74$	$+1223$
B	$+1203$			$+1569$	$+71$	$+1640$
0.1	$+808$			$+964$	$+69$	$+1033$
0.2	$+150$			$+475$	$+77$	$+552$
0.3	-494	-3	-497	-70	$+68$	-2
0.4	-991	-9	-1000	-444	$+15$	-393
0.5	-1192	-16	-1208	-576	$+28$	-548
0.6	-988	-21	-1009	-405	$+3$	-402
0.7	-476	-25	-501	0		0
0.8	$+180$	-35	$+145$	$+564$		$+564$
0.9	$+700$	-38	$+662$	$+1041$		$+1041$
C	$+1227$	-45	$+1182$	$+1647$		$+1647$

[a]Stresses in psi.

Table 7.15

Top Fiber	Positive Moment			Negative Moment		
Point	f^t_{D+L+I}	f^t_Δ	$f^t_{D+L+I+\Delta}$	f^t_{D+L+I}	f^t_Δ	$f^t_{D+L+I+\Delta}$
A	0			0	0	0
0.1	$+486$			$+320$	-12	$+308$
0.2	$+805$			$+517$	-24	$+493$
0.3	$+962$			$+594$	-36	$+558$
0.4	$+965$			$+546$	-48	$+498$
0.5	$+824$			$+378$	-59	$+319$
0.6	$+512$			$+83$	-68	$+17$
0.7	$+81$			-282	-71	-211
0.8	-357			-648	-69	-579
0.9	-674			-935	-60	-875
B	-943			-1230	-55	-1175
0.1	-557			-784	-56	-840
0.2	-113			-359	-58	-417
0.3	$+379$	$+2$	$+381$	$+53$	-52	$+1$

Table 7.15 *(Continued)*

Top Fiber	Positive Moment			Negative Moment		
Point	f^t_{D+L+I}	f^t_Δ	$f^t_{D+L+I+\Delta}$	f^t_{D+L+I}	f^t_Δ	$f^t_{D+L+I+\Delta}$
0.4	+768	+7	+775	+344	-40	+304
0.5	+925	+12	+937	+447	-22	+425
0.6	+767	+17	+784	+314	-3	+311
0.7	+365	+19	+384	0		0
0.8	-136	+26	-110	-427		-427
0.9	-570	+31	-539	-848		-848
C	-961	+35	-926	-1290		-1290

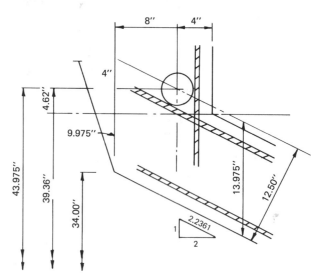

Figure 7.10 Exterior web tendon.

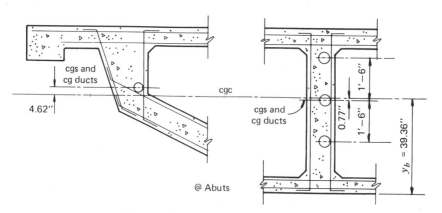

Figure 7.11 Tendons at jacking ends.

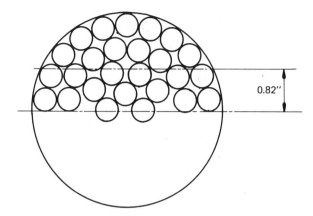

Figure 7.12 Strand location at midspan.

The lowest location of the tendons to resist maximum induced positive moment (Figure 7.13) is at the minimum depth section and is computed as

$$e = \frac{2(5.44) + 12(26.78)}{14}$$

$$= 23.73 \text{ in. } (1.98 \text{ ft})$$

Note that the prestressing force times e produces an induced negative moment.

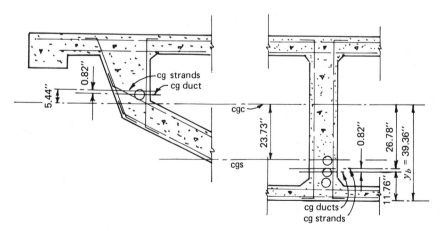

Figure 7.13 Lowest tendon location.

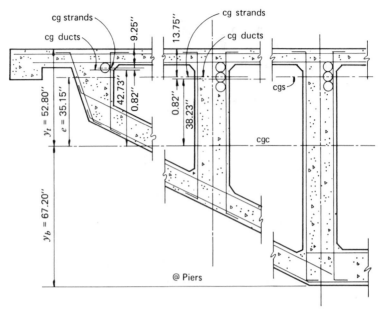

Figure 7.14   Highest tendon location.

Similarly, the highest tendon location to resist maximum induced negative moment (Figure 7.14) is computed as

$$e = \frac{2(42.73) + 12(38.23)}{14}$$

$$= 38.87 \text{ in. } (3.24 \text{ ft})$$

and is located at the maximum depth section of the interior piers.

The vertical cg of the steel is plotted for span AB in Figure 7.15 by locating it at the cg of the concrete at support A and at the lowest point at $0.4AB$, and at the highest point at pier B. An inflection point is established at $0.9AB$ by drawing a straight line from $0.4AB$ to B, and setting the cg of the steel for this point on that line. This ensures that the parabolas from $0.4AB$ to $0.9AB$ and from B back to $0.9AB$ will have the same slope because the slope of the parabola is twice the slope of the straight line. Similarly, the vertical cg of the steel profile is plotted for span BC, except that the lowest point is selected at $0.5BC$ rather than at 0.4 of the span, as used previously (Figure 7.16).

The slope of the cg of the steel at each tenth point is tabulated in Table 7.16. These values are used later to compute shears.

The total angle change α is computed in increments along the span to the point of symmetry, which is at the interior pier at point C. The angle change for each increment is equal to [2× vertical offset $(V)/$

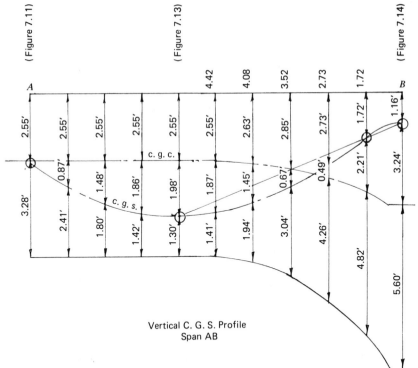

Figure 7.15   Vertical CGS Profile Span *AB*.

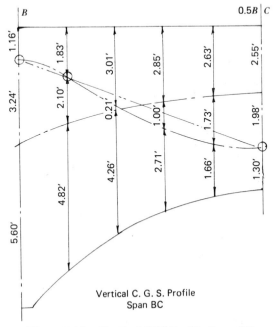

Figure 7.16   Vertical CGS Profile Span *BC*.

Table 7.16

Point	Ordinate	L	Slope = 2 × ordinate/L
A	1.98	53.23	0.0744
0.1	1.13	39.92	0.0566
0.2	0.52	26.62	0.0391
0.3	0.14	13.31	0.0210
0.4	0	0	0
0.5	0.11	13.31	0.0165
0.6	0.45	26.62	0.0338
0.7	1.01	39.92	0.0506
0.8	1.80	53.23	0.0676
0.9	2.81	66.54	0.0845
B	0	0	0
0.1	2.70	69.00	0.0783
0.2	1.52	51.75	0.0587
0.3	0.68	34.50	0.0394
0.4	0.17	17.25	0.0197
0.5			0

length (L)] in radians. These values are used to compute the effects of friction losses along the cable.

Angle Change (α)

A to 0.4AB

$$V = 1.98 \text{ ft}$$
$$L = 53.23 \text{ ft} \qquad \Delta\angle = \frac{2(1.98)}{53.23} = 0.0744$$

0.4AB to 0.9AB

$$V = 2.81 \text{ ft}$$
$$L = 66.54 \text{ ft} \qquad \Delta\angle = \frac{2(2.81)}{66.54} = 0.0845$$

0.9AB to B

$$V = 0.56 \text{ ft}$$
$$L = 13.31 \text{ ft} \qquad \Delta\angle = \frac{2(0.56)}{13.31} = 0.0841$$

B to 0.1BC

$$V = 0.67 \text{ ft}$$
$$L = 17.25 \text{ ft} \qquad \Delta\angle = \frac{2(0.67)}{17.25} = 0.0777$$

0.1BC to 0.5BC

$$V = 2.70 \text{ ft}$$
$$L = 69.00 \text{ ft}$$
$$\Delta \measuredangle = \frac{2(2.70)}{69.00} = 0.0783$$

0.5BC to 0.9BC $\Delta \measuredangle$ $= 0.0783$

0.9BC to C $\Delta \measuredangle$ $= 0.0777$
$$\alpha = 0.5550 \text{ rad}$$

According to AASHTO Article 1.6.7(A), the friction losses at C are

$$T \times C = \frac{T_0}{e^{(KL + \mu\alpha)}}$$

where $K = 0.0002$, $L = 305.58$ ft, $\mu = 0.25$, and $\alpha = 0.5550$; therefore

$$T_C = \frac{T_0}{e^{\{(0.0002)(305.58) + (0.25)(0.5550)\}}}$$

$$T_C = \frac{T_0}{e^{\{0.1999\}}} = \frac{T_0}{1.2213} = 0.8188 \, T_0$$

where T_0 represents the stress at the jack. AASHTO allows overstressing the tendons to $0.80f_s'$ for short periods. After the anchorage is seated the stress is only permitted to be $0.70f_s'$. With this in mind, it is reasonable to stress the cables initially to $0.75f_s'$ and to allow an anchor set of $\frac{5}{8}$ in. The combination of these values gives a maximum stress after seating of $0.70f_s'$, which can be seen when the stresses are plotted in the tendon stress and primary moment diagrams. The given values are

$$\Delta L = \tfrac{5}{8} \text{ in. anchor set}$$
$$f_{\text{jack}} = 0.75(270) = 202.5 \text{ ksi}$$
$$E_s = 27,000 \text{ ksi}$$

The initial steel stress at C is therefore,

$$f_s' = (0.8188)(202.5) = 165.8 \text{ ksi}$$

The frictional loss to C is the value $d = 202.5 - 165.8 = 36.7$ ksi or $0.1812f_{\text{jack}}$. The length over which the anchor set affects the stress is x, which equals

$$x = \sqrt{\frac{E_c(\Delta L)(L)}{12(d)}}$$

$$x = \sqrt{\frac{27,000(0.625)(305.58)}{12(36.7)}} = 108.21 \text{ ft}$$

The loss due to anchor set is Δf:

$$\Delta f = \frac{2(d)(x)}{L}$$

$$\Delta f = \frac{2(36.7)(108.21)}{305.58} = 26.0 \text{ ksi} = 0.1284 f_{\text{jack}}$$

Using the lump sum losses as given in AASHTO Article 1.6.7(B)(2), we obtain

$$\text{losses} = 32 \text{ ksi} = 0.1588 f_{\text{jack}}$$

The initial losses due to elastic shortening can be used when initial stress is computed:

$$\text{initial losses} = 13.5 \text{ ksi} = 0.0667 f_{\text{jack}}$$

These values are plotted in Figure 7.17 for half the bridge because the bridge is symmetrical. The frictional losses are assumed to vary in a straight line to the point of symmetry because the true variation from a straight line is insignificant. The stress variation is plotted for initial stress, for the initial stress minus the elastic shortening, and for the final stress after all losses have occurred. All results are plotted in terms of the stress at the jack f_j. The final stress values are tabulated at the tenth points and are listed as C_f values, which are coefficients of the stress at the jack. The eccentricity is tabulated at the same point and multiplied by C_f to give the C_m values, which are the coefficients of the primary moment due to posttensioning. The primary moment is that moment produced by the eccentric placement of the tendons.

Note that the maximum initial stress is $0.9358 f_j$ and because $f_j = 0.75 f_s'$ the maximum stress equals $0.75(0.9358) f_s' = 0.70 f_s'$ which is the AASHTO allowable for maximum initial stress in the prestressing steel.

The secondary moment results from the primary moment and slope discontinuity over the interior supports. AASHTO Article 1.6.12(B) requires the inclusion of calculation of the secondary moment to equalize the slope of adjacent spans across interior supports. To obtain these values the slope on adjacent spans at the interior supports, which are

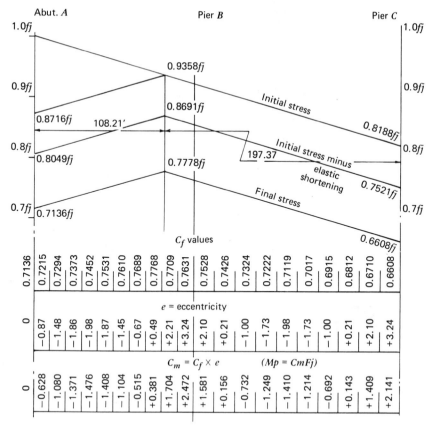

Figure 7.17   Tendon stresses and primary moments.

the result of the primary moment, must be computed. This is done for span AB in Figure 7.18, where the primary moment is plotted and tabulated with the moment of inertia.

The slopes are obtained by using conjugate beams. To develop a conjugate beam for slope analysis an M/I diagram is provided in Figure 7.18. Again, all of this is in terms of stress at the jack.

The actual slope at BA equals the shear on the conjugate beam at BA, which is computed by summing the moments about A on the conjugate beam (Table 7.17).

The constant value of E is divided into this end shear to obtain the slope; therefore

$$\theta_{BA} = \frac{68.27F_j}{E}$$

Next the fixed end moment required to bring θ_{BA} to zero is computed. For span AB the plot of this fixed end moment and the loading

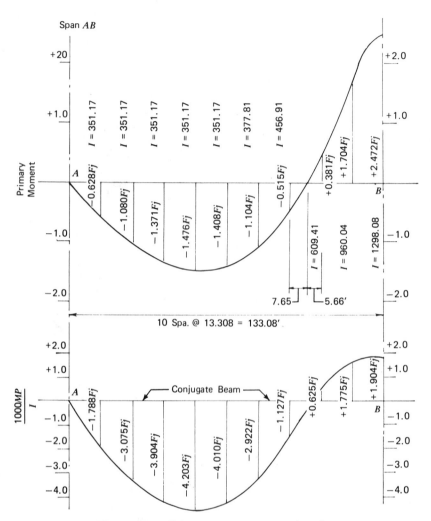

Figure 7.18 Primary moment versus inertia.

Table 7.17 Span _AB_

ΣM_A		V		M
$\frac{1}{2} \times (-1.788) \times 13.308 =$	$-11.90 \times$	$8.87 =$	-105.53
$(-1.788) \times 13.308 =$	$-23.79 \times$	$19.96 =$	-474.94
$\frac{1}{2} \times (-1.287) \times 13.308 =$	$-8.56 \times$	$22.18 =$	-189.94
$(-3.075) \times 13.308 =$	$-40.92 \times$	$33.27 =$	-1361.48
$\frac{1}{2} \times (-0.829) \times 13.308 =$	$-5.52 \times$	$35.49 =$	-195.77
$(-3.904) \times 13.308 =$	$-51.95 \times$	$46.58 =$	-2420.04
$\frac{1}{2} \times (-0.299) \times 13.308 =$	$-1.99 \times$	$48.80 =$	-97.09
$(-4.010) \times 13.308 =$	$-53.37 \times$	$59.89 =$	-3196.03

320

Table 7.17   (Continued)

ΣM_A	V	M
$\frac{1}{2} \times (-0.193) \times 13.308 =$	$-1.28 \times 57.67 =$	-74.06
$(-2.922) \times 13.308 =$	$-38.89 \times 73.19 =$	-2846.06
$\frac{1}{2} \times (-1.088) \times 13.308 =$	$-7.24 \times 70.98 =$	-513.86
$(-1.127) \times 13.308 =$	$-15.00 \times 86.50 =$	-1297.34
$\frac{1}{2} \times (-1.795) \times 13.308 =$	$-11.94 \times 84.28 =$	-1006.63
$\frac{1}{2} \times (-1.127) \times 7.65 =$	$-4.31 \times 95.71 =$	-412.51
$\frac{1}{2} \times (+0.625) \times 5.66 =$	$+1.77 \times 104.58 =$	$+185.11$
$(+0.625) \times 13.308 =$	$+8.32 \times 113.12 =$	$+940.88$
$\frac{1}{2} \times (+1.150) \times 13.308 =$	$+7.65 \times 115.34 =$	$+882.59$
$(+1.775) \times 13.308 =$	$+23.62 \times 126.43 =$	$+2986.49$
$\frac{1}{2} \times (+0.129) \times 13.308 =$	$+0.86 \times 128.64 =$	$+110.42$
	-233.44	9085.79

$$V_{BA} = \frac{9085.79}{133.08} = 68.27F_j$$

on the conjugate beam are shown in Figure 7.19. Again the actual slope at BA is computed by summing the moments about A on the conjugate beam to obtain the shear at BA, which is the actual slope on the real beam (Table 7.18).

The constant value at E is divided into the end shear to find the end slope at BA:

$$\theta_{BA} = \frac{77.94 M_{BA}^F}{E}$$

The same procedure is used for span BC. The plot and tabulation of the primary moment, the moment of inertia, and the conjugate beam loading are shown in Figure 7.20.

The end shears and the slopes that result from the primary moment are computed by summing the moments about B as shown in Table 7.19 for span BC.

A fixed end moment must be applied at BC and CB to bring the slopes to zero. For BC the plot and the tabulation of the fixed end moment, the moment of inertia, and the conjugate beam loading are shown in Figure 7.21.

The end shears on the conjugate beam and the slopes are computed by summing the moment on the conjugate beam about B as given in Table 7.20 for span BC.

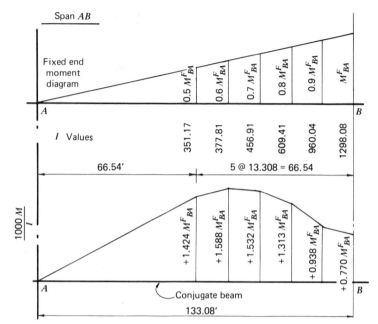

Figure 7.19 Fixed end moment and conjugate beam.

Table 7.18   Span AB

$$\Sigma M_A = 0$$

$\frac{1}{2} \times 1.424 \times 66.54 =$	$47.38 \times$	$44.36 =$	2101.62
$1.424 \times 13.308 =$	$18.95 \times$	$73.19 =$	1386.99
$\frac{1}{2} \times 0.164 \times 13.308 =$	$1.09 \times$	$75.41 =$	82.29
$1.532 \times 13.308 =$	$20.39 \times$	$86.50 =$	1763.55
$\frac{1}{2} \times 0.056 \times 13.308 =$	$0.37 \times$	$84.28 =$	31.40
$1.313 \times 13.308 =$	$17.47 \times$	$99.81 =$	1744.02
$\frac{1}{2} \times 0.219 \times 13.308 =$	$1.46 \times$	$97.59 =$	142.21
$0.938 \times 13.308 =$	$12.48 \times$	$113.12 =$	1412.07
$\frac{1}{2} \times 0.375 \times 13.308 =$	$2.50 \times$	$110.90 =$	276.72
$0.770 \times 13.308 =$	$10.25 \times$	$126.43 =$	1295.55
$\frac{1}{2} \times 0.168 \times 13.308 =$	$\underline{1.12} \times$	$121.21 =$	$\underline{135.50}$
	133.46		$10,371.92$

$$V_{AB} = \frac{10,371.92}{133.08} = 77.94 M_{BA}^F$$

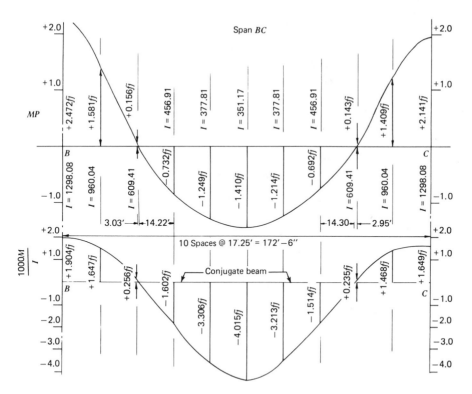

Figure 7.20 Primary moment and conjugate beam.

Table 7.19   Span *BC*

		$\Sigma M_B = 0$		
(+1.647) × 17.25 =	+28.41 ×	8.62 =	+244.90
$\frac{1}{2}$ × (+0.257) × 17.25 =	+2.22 ×	5.75 =	+12.75
(+0.256) × 17.25 =	+4.42 ×	25.88 =	+114.29
$\frac{1}{2}$ × (+1.391) × 17.25 =	+12.00 ×	23.00 =	+275.94
$\frac{1}{2}$ × (+0.256) × 3.03 =	+0.39 ×	35.51 =	+13.77
$\frac{1}{2}$ × (−1.602) × 14.22 =	+11.39 ×	47.01 =	−535.45
(−1.602) × 17.25 =	−27.63 ×	60.38 =	−1688.57
$\frac{1}{2}$ × (−1.704) × 17.25 =	−14.70 ×	63.25 =	−929.59
(−3.306) × 17.25 =	−57.03 ×	77.62 =	−4426.55
$\frac{1}{2}$ × (−0.709) × 17.25 =	−6.12 ×	80.50 =	−492.27
(−3.213) × 17.25 =	−55.42 ×	94.88 =	−5258.65
$\frac{1}{2}$ × (−0.802) × 17.25 =	−6.92 ×	92.00 =	−636.39
(−1.514) × 17.25 =	−26.12 ×	112.12 =	−2928.18

323

Table 7.19 *(Continued)*

$$\Sigma M_B = 0$$

$$\tfrac{1}{2} \times (-1.699) \times 17.25 = \quad -14.65 \times 109.25 = \quad -1600.94$$

$$\tfrac{1}{2} \times (-1.514) \times 14.30 = \quad -10.82 \times 125.52 = \quad -1378.77$$

$$\tfrac{1}{2} \times (+0.235) \times \; 2.95 = \quad +0.35 \times 137.02 = \quad +47.49$$

$$(+0.235) \times 17.25 = \quad +4.05 \times 146.62 = \quad +594.36$$

$$\tfrac{1}{2} \times (+1.233) \times 17.25 = \quad +10.63 \times 149.50 = \quad +1589.88$$

$$(+1.468) \times 17.25 = \quad +25.32 \times 163.88 = \quad +4149.93$$

$$\tfrac{1}{2} \times (+0.181) \times 17.25 = \quad \underline{+1.56} \times 166.75 = \quad \underline{+260.32}$$

$$-141.45 \qquad\qquad -12{,}551.73$$

$$V_{CB} = \frac{12{,}551.73}{172.5} = 72.76F_j \qquad\qquad \theta_{CB} = \frac{72.76F_j}{E}$$

$$V_{BC} = 141.45 - 72.76 = 68.69F_j \qquad\qquad \theta_{BC} = \frac{68.69F_j}{E}$$

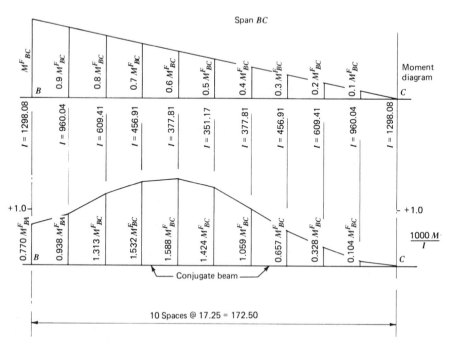

Figure 7.21 Fixed end moment and conjugate beam.

Table 7.20   Span BC

		$\Sigma M_B = 0$	
$0.770 \times 17.25 =$	$13.28 \times$	$8.62 =$	114.50
$\frac{1}{2} \times 0.168 \times 17.25 =$	$1.45 \times$	$11.50 =$	16.66
$0.938 \times 17.25 =$	$16.18 \times$	$25.88 =$	418.75
$\frac{1}{2} \times 0.375 \times 17.25 =$	$3.23 \times$	$28.75 =$	92.99
$1.313 \times 17.25 =$	$22.65 \times$	$43.12 =$	976.64
$\frac{1}{2} \times 0.219 \times 17.25 =$	$1.89 \times$	$46.00 =$	86.89
$1.532 \times 17.25 =$	$26.43 \times$	$60.38 =$	1595.66
$\frac{1}{2} \times 0.056 \times 17.25 =$	$0.48 \times$	$63.25 =$	30.55
$1.424 \times 17.25 =$	$24.56 \times$	$77.62 =$	1906.66
$\frac{1}{2} \times 0.164 \times 17.25 =$	$1.41 \times$	$74.75 =$	105.73
$1.059 \times 17.25 =$	$18.27 \times$	$94.88 =$	1733.24
$\frac{1}{2} \times 0.365 \times 17.25 =$	$3.15 \times$	$92.00 =$	289.63
$0.657 \times 17.25 =$	$11.33 \times$	$112.12 =$	1270.68
$\frac{1}{2} \times 0.402 \times 17.25 =$	$3.47 \times$	$109.25 =$	378.00
$0.328 \times 17.25 =$	$5.66 \times$	$129.38 =$	732.03
$\frac{1}{2} \times 0.329 \times 17.25 =$	$2.84 \times$	$126.50 =$	358.96
$0.104 \times 17.25 =$	$1.79 \times$	$146.62 =$	263.04
$\frac{1}{2} \times 0.224 \times 17.25 =$	$1.93 \times$	$143.75 =$	277.72
$\frac{1}{2} \times 0.104 \times 17.25 =$	$0.90 \times$	$161.00 =$	144.42
	160.90		$10{,}793.55$

$$V_{CB} = \frac{10{,}793.55}{172.5} = 62.57 M_{BC}^F$$

$$\theta_{CB} = \frac{62.57 M_{BC}^F}{E}$$

$$V_{BC} = 160.90 - 62.57 = 98.33 M_{BC}^F$$

$$\theta_{BC} = \frac{98.33 M_{BC}^F}{E}$$

A fixed end moment at CB would produce a mirror image of the fixed end moment at BC. Therefore for a fixed end moment at CB the slopes would be

$$\theta_{CB} = \frac{98.33 M_{CB}^F}{E} \qquad\qquad \theta_{BC} = \frac{62.57 M_{CB}^F}{E}$$

The slopes from a fixed end moment at BC and CB are summed:

$$\theta_{BC} = \frac{98.33 M^F_{BC} + 62.57 M^F_{CB}}{E}$$

$$\theta_{CB} = \frac{62.57 M^F_{BC} + 98.33 M^F_{CB}}{E}$$

The actual slopes from the primary moment are equated to the slopes from the fixed end moments to obtain fixed end moments in terms of force at the jack:

M^F_{BA}

$$\frac{68.27 F_j}{E} = \frac{77.94 M^F_{BA}}{E}$$

$$M^F_{BA} = 0.876 F_j$$

M^F_{BC} and M^F_{CB}

$$\frac{68.69 F_j}{E} = \frac{98.33 M^F_{BC} + 62.57 M^F_{CB}}{E}$$

$$\frac{72.76 F_j}{E} = \frac{62.57 M^F_{CB} + 98.33 M^F_{CB}}{E}$$

$$1.0978 F_j = 1.5715 M^F_{BC} + M^F_{CB}$$

$$\underline{0.7400 F_j = 0.6363 M^F_{BC} + M^F_{CB}}$$

$$0.3578 F_j = 0.9352 M^F_{BC}$$

$$M^F_{BC} = 0.3826 F_j$$

$$M^F_{CB} = 1.0978 F_j - 1.5715(0.3826) F_j = 0.4965 F_j$$

Using the moment distribution summary tabulated, we compute the secondary moments at the interior supports as a function of the coefficients F_j (Table 7.21).

The secondary and primary moments are computed and summed to give the total prestressed moment (Table 7.22).

The midspan section is analyzed for the maximum required post-tensioning force. The maximum applied midspan stresses are

@ 0.4AB

$$f_b = -1243 \text{ psi}$$
$$f_t = +965 \text{ psi}$$

@ 0.5BC

$$f_b = -1208 \text{ psi}$$
$$f_t = +937 \text{ psi}$$

The allowable stresses are

$$\text{allowable tension} = 3\sqrt{4000} = -190 \text{ psi}$$
$$\text{allowable compression} = 0.4(4000) = +1600 \text{ psi}$$

Table 7.21

	M_B	M_C	M_D
$M_{BA}^F = 0.876$	+0.443	−0.156	+0.053
$M_{BC}^F = 0.383$	+0.189	+0.068	−0.023
$M_{CB}^F = 0.496$	+0.082	+0.248	−0.082
$M_{CD}^F = 0.496$	−0.082	+0.248	+0.082
$M_{DC}^F = 0.383$	−0.023	+0.068	+0.189
$M_{DE}^F = 0.876$	+0.053	−0.156	+0.443
	+0.662	+0.320	+0.662
	+0.662F_j	+0.320F_j	+0.662F_j

Table 7.22

Point	M_p	M_s	M_{p+s}	Point	M_p	M_s	M_{p+s}
A				B	+2.472F_j	+0.662F_j	+3.134F_j
0.1	−0.628F_j	+0.066F_j	−0.562F_j	0.1	+1.581F_j	+0.628F_j	+2.209F_j
0.2	−1.080F_j	+0.132F_j	−0.948F_j	0.2	+0.156F_j	+0.594F_j	+0.750F_j
0.3	−1.371F_j	+0.199F_j	−1.172F_j	0.3	−0.732F_j	+0.559F_j	−0.173F_j
0.4	−1.476F_j	+0.265F_j	−1.211F_j	0.4	−1.249F_j	+0.525F_j	−0.724F_j
0.5	−1.408F_j	+0.331F_j	−1.077F_j	0.5	−1.410F_j	+0.491F_j	−0.919F_j
0.6	−1.104F_j	+0.397F_j	−0.707F_j	0.6	−1.214F_j	+0.457F_j	−0.757F_j
0.7	−0.515F_j	+0.463F_j	−0.052F_j	0.7	−0.692F_j	+0.423F_j	−0.269F_j
0.8	+0.381F_j	+0.530F_j	+0.911F_j	0.8	+0.143F_j	+0.388F_j	+0.531F_j
0.9	+1.704F_j	+0.596F_j	+2.300F_j	0.9	+1.409F_j	+0.354F_j	+1.763F_j
B	+2.472F_j	+0.662F_j	+3.134F_j	C	+2.141F_j	+0.320F_j	+2.461F_j

Using the applied and allowable stresses with the computed post-tensioning forces and moments, we calculate the required force at the jack:

$$f_b = \text{allowable tension stress} = \frac{P}{A} + \frac{M}{S} + \text{stresses from applied loads}$$

$$f_b = -190 = \frac{0.7119F_j}{11,121} + \frac{0.919F_j(12)}{185,006} - 1208$$

$$-190 = 0.00006401F_j + 0.00005961F_j - 1208$$

Therefore $F_j = 8235$ k.

$$f_t = \text{allowable compression stress}$$

$$= \frac{P}{A} + \frac{M}{S} + \text{stresses from applied loads}$$

$$f_t = +1600 = \frac{0.7119F_j}{11,121} - \frac{0.919F_j(12)}{237,969} + 937$$

$$+663 = 0.00006401F_j - 0.00004634F_j$$

F_j = does not control (a large force is required to reach allowable stress). The same conditions are checked at $0.3AB$:

@ 0.3AB

$$f_b = -1239 \text{ psi}$$

$$f_t = +962 \text{ psi}$$

$$f_b = -190 = \frac{0.7373F_j}{11,121} + \frac{1.172F_j(12)}{185,006} - 1239$$

$$+1049 = 0.00006630F_j + 0.00005956F_j$$

$F_j = 8335$ k

$$f_t = +1600 = \frac{0.7373F_j}{11,121} - \frac{1.172F_j(12)}{237,969} + 962$$

$$+638 = 0.00006630F_j - 0.00005910F_j$$

F_j = does not control (a large force is required to reach allowable stress). The required force should also be checked at the interior supports in the area of maximum negative moment. The applied stresses at these points are

@ B

$$f_b = +1640 \text{ psi}$$
$$f_t = -1175 \text{ psi}$$

@ C

$$f_b = +1647 \text{ psi}$$
$$f_t = -1290 \text{ psi}$$

Assume that point C controls because the prestress force due to larger frictional losses is smaller.

Using these applied and allowable stresses with the computed post-tensioning forces and moments, we again calculate the required force at the jack:

@ C

$$f_b = +1600 = \frac{0.6608F_j}{17,466} - \frac{2.461F_j(12)}{400,550} + 1647$$

$$-47 = 0.00003783F_j - 0.00007373F_j$$

$F_j = 1309 \text{ k min}$

$$f_t = -190 = \frac{0.6608F_j}{17,466} + \frac{2.461F_j(12)}{509,791} - 1290$$

$$+1100 = 0.00003783F_j + 0.00005793F_j$$

$F_j = 11,487 \text{ k controls.}$

The required number of strands is computed by using

$$f_j = 0.75f_s'$$
$$f_s' = 270 \text{ ksi}$$
$$A_s = 0.1531 \text{ in.}^2 \text{ per strand}$$

strand value (@ jack) = $0.75(270)(0.1531) = 31.0$ k per strand

$$\text{No. strands required} = \frac{11,487}{31.0} = 370$$

370 strands $\div$ 14 ducts = 26.4 strands per duct.

Try 26-$\frac{1}{2}$ in. ϕ, 270 k strands per duct.

$$\text{No. strands} = 14(26) = 364$$

$$F_j = 364(31.0) = 11,284 \text{ k}$$

On the basis of this force at the jack the final bottom fiber stress (psi) due to posttensioning is computed and tabulated at the tenth points (Table 7.23).

The top fiber stress is computed similarly (Table 7.24).

The total applied stresses are combined with the posttensioning stresses and tabulated for maximum positive and negative moments. These stresses are listed for the bottom fiber in psi (Table 7.25).

Table 7.23   Bottom Fiber[a]

Point	C_f	A	$\dfrac{C_f F_j}{A}$	C_m	S_b	$\dfrac{C_m F_j(12)}{S_b}$	f_b (psi)
A	0.7136	11,121	+724	0	185,006	0	+724
0.1	0.7215	↑	+732	−0.562	↑	+411	+1143
0.2	0.7294		+740	−0.948		+694	+1434
0.3	0.7373		+748	−1.172		+858	+1606
0.4	0.7452	↓	+756	−1.211	↓	+886	+1642
0.5	0.7531	11,121	+764	−1.077	185,006	+788	+1552
0.6	0.7610	11,301	+760	−0.707	192,583	+497	+1257
0.7	0.7689	11,802	+735	−0.052	212,814	+33	+768
0.8	0.7768	12,660	+692	+0.911	247,197	−499	+193
0.9	0.7709	15,342	+567	+2.300	344,180	−905	−338
B	0.7631	17,466	+493	+3.134	400,500	−1060	−567
0.1	0.7528	15,342	+554	+2.209	344,180	−869	−315
0.2	0.7426	12,660	+662	+0.750	247,197	−411	+251
0.3	0.7324	11,802	+700	−0.173	212,814	+110	+810
0.4	0.7222	11,301	+721	−0.724	192,583	+509	+1230
0.5	0.7119	11,121	+722	−0.919	185,006	+673	+1395
0.6	0.7017	11,301	+701	−0.757	192,583	+532	+1233
0.7	0.6915	11,802	+661	−0.269	212,814	+171	+832
0.8	0.6812	12,660	+607	+0.531	247,197	−291	+316
0.9	0.6710	15,342	+494	+1.763	344,180	−694	−200
C	0.6608	17,466	+427	+2.461	400,550	−832	−405

[a]F_j = 11,284 k.

Table 7.24 Top Fiber[a]

Point	C_f	A	$\dfrac{C_f F_j}{A}$	C_m	S_t	$\dfrac{C_m F_j(12)}{S_t}$	f_t
A			+724	0	237,969	0	+724
0.1			+732	-0.562	↑	-320	+412
0.2			+740	-0.948		-539	+201
0.3			+748	-1.172		-667	+81
0.4			+756	-1.211	↓	-689	+67
0.5			+764	-1.077	237,969	-613	+151
0.6			+760	-0.707	248,234	-386	+374
0.7			+735	-0.052	277,032	-25	+710
0.8			+692	+0.911	327,037	+377	+1069
0.9			+567	+2.300	422,144	+738	+1305
B			+493	+3.134	509,791	+832	+1325
0.1			+554	+2.209	422,144	+709	+1263
0.2			+662	+0.750	327,037	+310	+972
0.3			+700	-0.173	277,032	-85	+615
0.4			+721	-0.724	248,234	-395	+326
0.5			+722	-0.919	237,969	-523	+199
0.6			+701	-0.757	248,234	-413	+288
0.7			+661	-0.269	277,032	-132	+529
0.8			+607	+0.531	327,037	+220	+827
0.9			+494	+1.763	422,144	+566	+1060
C			+427	+2.461	509,791	+654	+1081

[a] $F_j = 11,284$ k.

Table 7.25 Final Stresses: Bottom Fiber[a]

Point	Maximum Positive Moment			Maximum Negative Moment		
	Prestress	$DL + LL + I + \Delta$	f_b	Prestress	$DL + LL + I + \Delta$	f_b
A	+724		+724	+724		+724
0.1	+1143	-626	+517	+1143	-397	+746
0.2	+1434	-1038	+396	+1434	-636	+798
0.3	+1606	-1239	+367	+1606	-719	+887
0.4	+1642	-1243	+399	+1642	-643	+999
0.5	+1552	-1061	+491	+1552	-410	+1142
0.6	+1257	-660	+797	+1257	-20	+1237
0.7	+768	-106	+662	+768	+460	+1228
0.8	+193	+472	+665	+193	+948	+1141
0.9	-338	+828	+490	-338	+1223	+890
B	-567	+1203	+636	-567	+1640	+1073
0.1	-315	+808	+493	-315	+1033	+718
0.2	+251	+150	+401	+251	+552	+803
0.3	+810	-497	+313	+810	-2	+808
0.4	+1230	-1000	+230	+1230	-393	+837
0.5	+1395	-1208	+187	+1395	-548	+847

331

Table 7.25 (Continued)

	Maximum Positive Moment			Maximum Negative Moment		
Point	Prestress	$DL + LL + I + \Delta$	f_b	Prestress	$DL + LL + I + \Delta$	f_b
0.6	+1233	−1009	+224	+1233	−402	+831
0.7	+832	−501	+331	+832	0	+832
0.8	+316	+145	+461	+316	+564	+880
0.9	−200	+662	+462	−200	+1041	+841
C	−405	+1182	+777	−405	+1647	+1242

[a] Allowable compression = 1600 psi > 1242 psi OK.

The stresses in the top fiber (psi) are computed as shown in Table 7.26.

Next it is necessary to check the initial stresses. The secondary moments are recomputed in the same manner as for the final stresses.

The initial moments due to the combined primary and secondary effects are tabulated in Table 7.27.

Table 7.26 Final Stresses: Top Fiber[a]

	Maximum Positive Moment			Maximum Negative Moment		
Point	Prestress	$DL + LL + I + \Delta$	f_t	Prestress	$DL + LL + I + \Delta$	f_t
A	+724	0	+724	+724	0	+724
0.1	+412	+486	+898	+412	+308	+720
0.2	+201	+805	+1006	+201	+493	+694
0.3	+81	+962	+1043	+81	+558	+639
0.4	+67	+965	+1032	+67	+498	+565
0.5	+151	+824	+975	+151	+319	+470
0.6	+374	+512	+886	+374	+17	+391
0.7	+710	+81	+791	+710	−211	+499
0.8	+1069	−357	+712	+1069	−579	+490
0.9	+1305	−674	+631	+1305	−875	+430
B	+1325	−943	+382	+1325	−1175	+150
0.1	+1263	−557	+706	+1263	−840	+423
0.2	+972	−113	+859	+972	−417	+555
0.3	+615	+381	+996	+615	+1	+616
0.4	+326	+775	+1101	+326	+304	+630
0.5	+199	+937	+1136	+199	+425	+624
0.6	+288	+784	+1072	+288	+311	+599
0.7	+529	+384	+913	+529	0	+529
0.8	+827	−110	+717	+827	−427	+400
0.9	+1060	−539	+521	+1060	−848	+212
C	+1081	−926	+155	+1081	−1290	−209

[a] Allowable tension = 190 psi < 209 psi; allowable compression = 1600 psi > 1136 psi. This is close enough to be considered satisfactory.

Table 7.27

Point	M_p	M_s	M_{p+s}	Point	M_p	M_s	M_{p+s}
A	0	0	0	B	$+2.768F_j$	$+0.748F_j$	$+3.516F_j$
0.1	$-0.707F_j$	$+0.075F_j$	$-0.632F_j$	0.1	$+1.773F_j$	$+0.709F_j$	$+2.482F_j$
0.2	$-1.215F_j$	$+0.150F_j$	$-1.065F_j$	0.2	$+0.175F_j$	$+0.670F_j$	$+0.845F_j$
0.3	$-1.541F_j$	$+0.224F_j$	$-1.317F_j$	0.3	$-0.824F_j$	$+0.631F_j$	$-0.193F_j$
0.4	$-1.656F_j$	$+0.299F_j$	$-1.357F_j$	0.4	$-1.407F_j$	$+0.592F_j$	$-0.815F_j$
0.5	$-1.579F_j$	$+0.374F_j$	$-1.205F_j$	0.5	$-1.590F_j$	$+0.553F_j$	$-1.037F_j$
0.6	$-1.236F_j$	$+0.449F_j$	$-0.787F_j$	0.6	$-1.372F_j$	$+0.514F_j$	$-0.858F_j$
0.7	$-0.576F_j$	$+0.524F_j$	$-0.052F_j$	0.7	$-0.783F_j$	$+0.475F_j$	$-0.308F_j$
0.8	$+0.425F_j$	$+0.598F_j$	$+1.023F_j$	0.8	$+0.162F_j$	$+0.436F_j$	$+0.598F_j$
0.9	$+1.906F_j$	$+0.673F_j$	$+2.579F_j$	0.9	$+1.601F_j$	$+0.397F_j$	$+1.998F_j$
B	$+2.768F_j$	$+0.748F_j$	$+3.516F_j$	C	$+2.437F_j$	$+0.358F_j$	$+2.795F_j$

C_i values (initial load coefficient for force at the jack) are tabulated by using the computed stress diagram (Figure 7.17). The initial bottom fiber stresses (psi) are tabulated in Table 7.28. Note that the C_i values are for the initial stress minus elastic shortening.

The same is performed for the top fiber (Table 7.29).

Table 7.28   Initial Prestress: Bottom Fiber[a]

Point	C_i	C_iF_j / A	A	C_m	S_b	$C_mF_j(12)$ / S_b	f_b
A	0.8049	11,121	+817	0	185,006	0	+817
0.1	0.8128	↑	+825	-0.632	↑	+462	+1287
0.2	0.8207		+833	-1.065		+780	+1613
0.3	0.8286		+841	-1.317		+964	+1805
0.4	0.8365	↓	+849	-1.357	↓	+993	+1842
0.5	0.8444	11,121	+857	-1.205	185,006	+882	+1739
0.6	0.8523	11,301	+851	-0.787	192,583	+553	+1404
0.7	0.8602	11,802	+822	-0.052	212,814	+33	+855
0.8	0.8681	12,660	+774	+1.023	247,197	-560	+214
0.9	0.8622	15,342	+634	+2.579	344,180	-1015	-381
B	0.8544	17,466	+552	+3.516	400,550	-1189	-637
0.1	0.8441	15,342	+621	+2.482	344,180	-977	-356
0.2	0.8339	12,660	+743	+0.845	247,197	-463	+280
0.3	0.8237	11,802	+788	-0.193	212,814	+123	+911
0.4	0.8135	11,301	+812	-0.815	192,583	+573	+1385
0.5	0.8032	11,121	+815	-1.037	185,006	+759	+1574
0.6	0.7930	11,301	+792	-0.858	192,583	+603	+1395
0.7	0.7828	11,802	+748	-0.308	212,814	+196	+944
0.8	0.7725	12,660	+688	+0.598	247,197	-328	+360
0.9	0.7623	15,342	+561	+1.998	344,180	-786	-225
C	0.7521	17,466	+486	+2.795	400,550	-945	-459

[a]$F_j = 11,284$ k.

Table 7.29   Initial Stress: Top Fiber[a]

	$C_i F_j$			$C_m F_j (12)$	
Point	A	C_m	S_t	S_t	f_t
A	+817	0	237,969	0	+817
0.1	+825	−0.632	↑	−360	+465
0.2	+833	−1.065		−606	+227
0.3	+841	−1.317		−749	+92
0.4	+849	−1.357	↓	−772	+77
0.5	+857	−1.205	237,969	−686	+171
0.6	+851	−0.787	248,234	−429	+422
0.7	+822	−0.052	277,032	−25	+797
0.8	+774	+1.023	327,037	+424	+1198
0.9	+634	+2.579	422,144	+827	+1461
B	+552	+3.516	509,791	+934	+1486
0.1	+621	+2.482	422,144	+796	+1417
0.2	+743	+0.845	327,037	+350	+1093
0.3	+788	+0.193	277,032	−94	+694
0.4	+812	−0.815	248,234	−445	+367
0.5	+815	−1.037	237,969	−590	+225
0.6	+792	−0.858	248,234	−468	+324
0.7	+748	−0.308	277,032	−150	+598
0.8	+688	+0.598	327,037	+248	+936
0.9	+561	+1.998	422,144	+641	+1202
C	+486	+2.795	509,791	+742	+1228

[a] F_j = 11,284 k.

The initial posttensioning stresses are combined with the dead load stresses and tabulated (Table 7.30). Examination of the data indicates that none of the stresses is close to being critical.

On the basis of the stresses just computed, we estimate the elastic shortening that takes place initially and on a long-term basis. This information is useful in determining joint requirements and for setting the bearings. To accommodate these two requirements the average of the top and bottom fiber stresses is used to evaluate the shortening:

The average bottom fiber stress is f_b^i = 764 psi

The average top fiber stress is f_t^i = 727 psi

The average of these two values gives the average girder stress:

$$f_{av}^i = 746 \text{ psi}$$

The modulus is

$$E_{ci} = 57,000 \sqrt{4000} = 3,605,000 \text{ psi}$$

Table 7.30   Initial Stresses

Point	Bottom Fiber			Top Fiber		
	Prestress	DL	f_b	Prestress	DL	f_t
A	+817	0	+817	+817	0	+817
0.1	+1287	-451	+836	+465	+350	+815
0.2	+1613	-745	+868	+227	+578	+805
0.3	+1805	-882	+923	+92	+685	+777
0.4	+1842	-860	+982	+77	+668	+745
0.5	+1739	-682	+1057	+171	+530	+701
0.6	+1404	-333	+1071	+422	+258	+680
0.7	+855	+129	+726	+797	-99	+698
0.8	+214	+609	+823	+1198	-461	+737
0.9	-381	+897	+516	+1461	-730	+731
B	-637	+1259	+622	+1486	-987	+499
0.1	-356	+746	+390	+1417	-607	+810
0.2	+280	+299	+579	+1093	-226	+867
0.3	+911	-241	+670	+694	+185	+879
0.4	+1385	-641	+744	+367	+497	+864
0.5	+1574	-791	+783	+225	+614	+839
0.6	+1395	-621	+774	+324	+482	+806
0.7	+944	-205	+739	+598	+157	+755
0.8	+360	+345	+705	+936	-261	+675
0.9	-225	+791	+566	+1202	-644	+558
C	-459	+1308	+849	+1228	-1024	+204

Half the bridge length will effect the shortening at each end:

length = 133.08 + 172.5 = 305.58 ft = 3667 in.

The average initial shortening is therefore

$$\Delta L = \left(\frac{746}{3,605,000} \right) 3667 = 0.759 \text{ in.}$$

Use creep factor = 2.5 and assume that half the creep takes place before backwall construction. The initial movement at the abutment = 0.759 + 1.5($\frac{1}{2}$)(0.759) = 1.328 in. (1$\frac{3}{8}$ in.)   The remaining creep then equals 0.75(0.759) = 0.57 in. ($\frac{5}{8}$ in)

The maximum shortening on a sliding plate joint with a 3$\frac{1}{2}$-in. nominal plate lap is computed by combining the remaining creep with the temperature drop for a moderate climate:

$$\text{temperature drop} = \overset{\alpha}{0.000006} \times \overset{L}{305.58} \times \overset{\Delta t}{40} = 0.073 \text{ ft } (\tfrac{7}{8} \text{ in.})$$

maximum shortening = $\frac{5}{8}$ + $\frac{7}{8}$ = 1$\frac{1}{2}$ in.

Therefore the 3$\frac{1}{2}$ in. lap of plates is satisfactory.

Table 7.31   Required Ultimate Moment:
Maximum Positive Moment[a]

Point	1.3DL	1.3Δ	2.17(L + I)	1.0M_s	Positive Moment $M_{u\,(required)}$
A	0		0	0	0
0.1	+9,035		+5,859	+747	+15,641
0.2	+14,921		+9,791	+1494	+26,206
0.3	+17,658		+11,922	+2241	+31,821
0.4	+17,229		+12,790	+2988	+33,007
0.5	+13,666		+12,679	+3735	+30,080
0.6	+6,939		+11,401	+4482	+22,822
0.7	−2,984		+9,032	+5229	+11,277
0.8	−16,331		+6,143	+5976	
0.9	−33,406		+4,279	+6723	
B	−54,573		+4,032	+7470	
0.1	−27,798		+3,832	+7084	
0.2	−8,003		+6,673	+6698	+5,368
0.3	+5,555	+62	+9,732	+6312	+21,661
0.4	+13,385	+191	+12,173	+5926	+31,675
0.5	+15,847	+320	+13,393	+5540	+35,100
0.6	+12,968	+449	+12,792	+5155	+31,364
0.7	+4,719	+577	+10,412	+4769	+20,477
0.8	−9,257	+930	+7,385	+4383	+3,441
0.9	−29,471	+1430	+5,638	+3997	
C	−56,664	+1929	+5,855	+3611	

[a]Moments are in k-ft.

The required ultimate moment, when considering maximum positive bending, is tabulated in Table 7.31 for dead load, abutment settlement, live load and impact, and secondary post-tensioning moments.

The required ultimate moment, when considering maximum negative bending, is tabulated in Table 7.32 for the same items of loading with the exception that the secondary post-tensioning moment is neglected because it is positive and it would be conservative to ignore this effect.

Next the ultimate moment capacities are compared with the required ultimate moments just computed. First the positive moment capacities are computed by using the ultimate stress diagram in Figure 7.22.

The d values have already been computed. The area of steel is $A_s = 364 \times (0.1531) = 55.728$ in.^{2} and $b_t = 512$ in.^{2} (the width of the top flange). The factor $p = A_s/b_d$ and the stress in the prestressing steel at failure $f'_{su} = f'_s[1 - 0.5p(f'_s/f'_c)]$ can also be computed.

It is important initially to check the value $p(f'_{su}/f'_c)$ to ensure that it is less than 0.30. This value is checked in Table 7.33.

Table 7.32   Required Ultimate Moment: Maximum Negative Moment[a]

Point	1.3DL	1.3Δ	2.17(L + I)	Negative Moment $M_{u\,(required)}$
A	0	0	0	0
0.1	+9,035	−307	−1,313	
0.2	+14,921	−614	−2,608	
0.3	+17,658	−920	−3,921	
0.4	+17,229	−1227	−5,232	
0.5	+13,666	−1534	−6,530	
0.6	+6,939	−1840	−7,840	−2,741
0.7	−2,984	−2146	−9,153	−14,283
0.8	−16,331	−2453	−11,043	−29,827
0.9	−33,406	−2760	−15,698	−51,864
B	−54,573	−3067	−22,413	−80,053
0.1	−27,798	−2568	−13,526	−43,892
0.2	−8,003	−2067	−7,858	−17,928
0.3	+5,555	−1568	−6,592	−2,605
0.4	+13,385	−1067	−6,857	
0.5	+15,847	−569	−7,200	
0.6	+12,968	−69	−7,528	
0.7	+4,719		−7,873	−3,154
0.8	−9,257		−9,793	−19,050
0.9	−29,471		−15,572	−45,043
C	−56,664		−24,523	−81,187

[a]Moments are in k-ft.

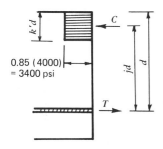

Figure 7.22 Positive moment ultimate stress diagram.

The ultimate capacity is calculated by using

$$T = f'_{su} \cdot A_s$$

$$T = C$$

$$A_c = \frac{C}{0.85f'_c}$$

$$k'd = \frac{A_c}{b}$$

$$jd = d - \frac{k'd}{2}$$

$$\phi M_n = T - \frac{jd}{12}\ \phi$$

These values are listed in Table 7.34.

Note that these values are in excess of the required; therefore the structure is adequate and checked for positive moment capacity.

Next the negative moment capacity is compared with the required ultimate moment by using the negative ultimate moment stress diagram in Figure 7.23.

Table 7.33[a]

			$b = 512$ in.			
Point	d	A_s	(p)	$0.5p(f_s'/f_c')$	f_{su}'	$p(f_{su}'/f_c')$
A	30.60	55.728	0.00356	0.120	237.6	0.212
0.1	41.04		0.00265	0.089	245.9	0.163
0.2	48.36		0.00225	0.076	249.5	0.140
0.3	52.92		0.00206	0.070	251.1	0.129
0.4	54.36		0.00200	0.068	251.6	0.126
0.5	53.04		0.00205	0.069	251.4	0.129
0.6	48.96		0.00222	0.075	249.8	0.139
0.7	42.24		0.00258	0.087	246.5	0.159
0.8						
0.9						
B						
0.1						
0.2	36.12		0.00301	0.102	242.5	0.183
0.3	46.20		0.00236	0.080	248.4	0.147
0.4	52.32		0.00208	0.070	251.1	0.131
0.5	54.36		0.00200	0.068	251.6	0.126
0.6	52.32		0.00208	0.070	251.1	0.131
0.7	46.20		0.00236	0.080	248.4	0.147
0.8	36.12		0.00301	0.102	242.5	0.183
0.9						<0.30
C		55.728				$\therefore$ OK

[a]Top flange area = $8(512) = 4096$ in.^{2} A_c slightly exceeds 4096 in a few places, but it is close enough to ignore any area except the flange; therefore a rectangular section can be assumed.

Table 7.34 Ultimate Moment: Positive Moment

Point	T	A_c	$k'd$	jd	$\left(\dfrac{jd}{12}\right)$	$\phi = 0.95$ ϕM_n
A	13,241	3894	7.61	26.80	2.23	28,051
0.1	13,704	4031	7.87	37.10	3.09	40,228
0.2	13,904	4089	7.99	44.36	3.70	48,873
0.3	13,993	4115	8.00	48.92	4.08	54,236
0.4	14,021	4123	8.00	50.36	4.20	55,944
0.5	14,010	4121	8.00	49.04	4.09	54,436
0.6	13,921	4094	8.00	44.96	3.75	49,594
0.7	13,737	4040	7.89	38.24	3.19	41,630
0.8						
0.9						
B						
0.1						
0.2	13,514	3975	7.76	32.24	2.69	34,490
0.3	13,843	4071	7.95	42.22	3.52	46,291
0.4	13,993	4116	8.00	48.32	4.03	53,572
0.5	14,021	4123	8.00	50.36	4.20	55,944
0.6	13,993	4116	8.00	48.32	4.03	53,572
0.7	13,843	4071	7.95	42.22	3.52	46,291
0.8	13,514	3975	7.76	32.24	2.69	34,490
0.9						
C						

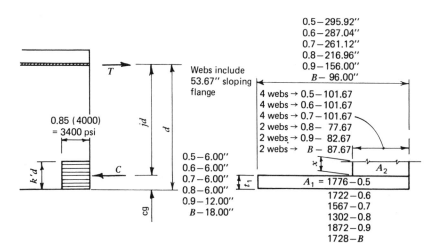

Figure 7.23 Negative ultimate moment stress diagram.

Table 7.35

Point	d	A_s	b	p	$0.5p(f'_s/f'_c)$	f'_{su}	$[p(f'_{su}/f'_c)]$
0.6	23.28	55.728	287	0.00834	0.281	194.1	0.405
0.7	36.48	↑	261	0.00585	0.197	216.8	0.317
0.8	57.00		217	0.00450	0.152	229.0	0.258
0.9	84.36		156	0.00424	0.143	231.4	0.245
B	106.08		96	0.00547	0.185	220.0	0.301
0.1	83.04		156	0.00430	0.145	230.8	0.248
0.2	53.64		217	0.00479	0.162	226.3	0.271
0.3	32.52		261	0.00657	0.222	210.1	0.345
0.4							
0.5							
0.6							
0.7	32.52		261	0.00657	0.222	210.1	0.345
0.8	53.64		217	0.00479	0.162	226.3	0.271
0.9	83.04	↓	156	0.00430	0.145	230.8	0.248
C	106.08	55.728	96	0.00547	0.185	220.0	0.301

Next compute the term $[p(f'_{su}/f'_c)]$ for negative bending in a manner similar to that performed for positive bending (Table 7.35).

Note that a number of points exceed the allowable 0.30, which indicates that some sections may be overreinforced. Further studies will be made to examine these effects by using the procedure given in ref. (9). This procedure requires the computation of the steel strain at ultimate ϵ_{su}, which is the sum of the following three items:

1. ϵ_{se}: the steel strain due to effective prestress.
2. ϵ_{ce}: the concrete strain at the level of the steel due to prestressing.
3. ϵ_{cu}: the concrete strain at the level of steel at ultimate load.

First we compute ϵ_{cu} assuming that the ultimate concrete compressive strain equals 0.0034, where $k'd = x + t_1$ (Table 7.36).

Next we compute ϵ_{ce} (Table 7.37), based on the concrete stress at the steel f_{cgs}. The strain equals f_{cgs}/E_c, where $E_c = 3,605,000$ psi. The stress f_{cgs} is obtained by interpolation.

The value ϵ_{se} is obtained by using the final value of f_s after all losses, where $E_s = 27,000$ ksi. After obtaining and tabulating ϵ_{se} we add $\epsilon_{se} + \epsilon_{ce} + \epsilon_{cu}$ to obtain the total steel strain at ultimate. The value (f'_{su}/E_s) is then computed as the steel strain at ultimate when failure occurs due to yielding of the steel. When f'_{su}/E is greater than ϵ_{su} the section is overreinforced (Table 7.38).

Again the negative moment capacity is computed by using the preceding ultimate moment stress diagram for negative moment and yield-

Table 7.36

Point	x	$k'd$	$c = k'd/0.85$	d	$\epsilon_{cu} = [(d - c)/c]$ (0.0034)
0.6	14.35	20.35	23.94	23.28	-0.0001
0.7	19.97	25.97	30.55	36.48	0.0007
0.8	31.57	37.57	44.20	57.00	0.0010
0.9	23.24	35.24	41.46	84.36	0.0035
B	21.42	39.42	46.38	106.08	0.0044
0.1	23.12	35.12	41.32	83.04	0.0034
0.2	30.99	36.99	43.52	53.64	0.0008
0.3	18.46	24.46	28.78	32.52	0.0004
0.4					
0.5					
0.6					
0.7	18.46	24.46	28.78	32.52	0.0004
0.8	30.99	36.99	43.52	53.64	0.0008
0.9	23.12	35.12	41.32	83.04	0.0034
C	21.42	39.42	46.38	106.08	0.0044

ing in the steel at failure. The values from Figure 7.23 are computed in Table 7.39.

The following values for cg are computed:

$$\text{cg} = \frac{A_1(t_1/2) + A_2\left(t_1 + \dfrac{x}{2}\right)}{A_c}$$

Table 7.37

Point	f_b	f_t	y_b	D	f_{cgs}	ϵ_{ce}
0.6	+1257	+374	1.94	6.02	+974	0.0003
0.7	+768	+710	3.04	6.56	+741	0.0002
0.8	+193	+1069	4.75	7.48	+749	0.0002
0.9	-338	+1305	7.03	8.75	+982	0.0003
B	-567	+1325	8.84	10.00	+1105	0.0003
0.1	-315	+1263	6.92	8.75	+933	0.0003
0.2	+251	+972	4.47	7.48	+682	0.0002
0.3	+810	+615	2.71	6.56	+729	0.0002
0.7	+832	+529	2.71	6.56	+707	0.0002
0.8	+316	+827	4.47	7.48	+621	0.0002
0.9	-200	+1060	6.92	8.75	+796	0.0002
C	-405	+1081	8.84	10.00	+909	0.0003

Table 7.38

Point	f_s final	$\epsilon_{se} = f_s$ final$/E$	ϵ_{ce}	ϵ_{cu}	$\Sigma = \epsilon_{su}$	f'_{su}/E
0.6	154.1	0.0057	0.0003	0.0001	0.0059	0.0072^a
0.7	155.7	0.0058	0.0002	0.0007	0.0067	0.0080^a
0.8	157.3	0.0058	0.0002	0.0010	0.0070	0.0085^a
0.9	156.1	0.0058	0.0003	0.0035	0.0096	0.0086
B	154.5	0.0057	0.0003	0.0044	0.0104	0.0081
0.1	152.4	0.0056	0.0003	0.0034	0.0093	0.0085
0.2	150.4	0.0056	0.0002	0.0008	0.0066	0.0084^a
0.3	148.3	0.0055	0.0002	0.0004	0.0061	0.0078^a
0.4						
0.5						
0.6						
0.7	140.0	0.0052	0.0002	0.0004	0.0058	0.0078^a
0.8	137.9	0.0051	0.0002	0.0008	0.0061	0.0084^a
0.9	135.9	0.0050	0.0002	0.0034	0.0086	0.0085
C	133.8	0.0050	0.0003	0.0044	0.0097	0.0081

[a]Overreinforced. Use modified AASHTO formulas.

@ 0.6AB

$$cg = \frac{1722(3.0) + 1459(6.0 + 7.18)}{3181} = 7.67 \text{ in.}$$

@ 0.7AB

$$cg = \frac{1567(3.0) + 1986(6.0 + 9.78)}{3553} = 10.14 \text{ in.}$$

@ 0.8AB

$$cg = \frac{1302(3.0) + 2452(6.0 + 15.78)}{3754} = 15.27 \text{ in.}$$

@ 0.9AB

$$cg = \frac{1872(6.0) + 1921(12.0 + 11.62)}{3793} = 14.92 \text{ in.}$$

@ B

$$cg = \frac{1728(9.0) + 1878(18.0 + 10.71)}{3606} = 19.26 \text{ in.}$$

Table 7.39   Ultimate Moment: Negative Moment

Point	T	A_c	A_1	A_2	x	cg^a
0.6	10,817	3181	1722	1459	14.35	7.67
0.7	12,081	3553	1567	1986	19.57	10.14
0.8	12,762	3754	1302	2452	31.57	15.27
0.9	12,895	3793	1872	1921	23.24	14.92
B	12,260	3606	1728	1878	21.42	19.26
0.1	12,862	3783	1872	1911	23.12	14.87
0.2	12,611	3709	1302	2407	30.99	15.01
0.3	11,708	3444	1567	1877	18.46	9.67
0.4						
0.5						
0.6						
0.7	11,708	3444	1567	1877	18.46	9.67
0.8	12,611	3709	1302	2407	30.99	15.01
0.9	12,862	3783	1872	1911	23.12	14.87
C	12,260	3606	1728	1878	21.42	19.26

^{a}See the following computations.

@ 0.1BC

$$cg = \frac{1872(6.0) + 1911(12.0 + 11.56)}{3783} = 14.87 \text{ in.}$$

@ 0.2BC

$$cg = \frac{1302(3.0) + 2407(6.0 + 15.50)}{3709} = 15.01 \text{ in.}$$

@ 0.3BC

$$cg = \frac{1567(3.0) + 1877(6.0 + 9.23)}{3444} = 9.67 \text{ in.}$$

The values of jd and M_n are computed and listed in Table 7.40. Note that the ultimate moment at 0.6AB, 0.7AB, 0.8AB, 0.2BC, 0.3BC, 0.7BC and 0.8BC are modified, which are the points that were over-reinforced. These modifications are made before the capacity reduction factor ϕ is applied (Table 7.40).

It is interesting to note the reduction in capacity due to an overreinforced section. These modified moment capacities are computed ac-

Table 7.40

Point	jd	$\left(\dfrac{jd}{12}\right)$	T	M_n	Mod M_n	$\phi = 0.95$ ϕM_n
0.6	15.61	1.30	10,817		10,970	10,422
0.7	26.34	2.20	12,081		20,325	19,309
0.8	41.73	3.48	12,762		33,789	32,100
0.9	69.44	5.79	12,895	74,662		70,929
B	86.82	7.24	12,260	88,762		84,324
0.1	68.17	5.68	12,862	73,056		69,403
0.2	38.63	3.22	12,611		30,589	29,060
0.3	22.85	1.90	11,708		16,939	16,092
0.4						
0.5						
0.6						
0.7	22.85	1.90	11,708		16,939	16,092
0.8	38.63	3.22	12,611		30,589	29,060
0.9	68.17	5.68	12,862	73,056		69,403
C	86.82	7.24	12,260	88,762		84,324

cording to AASHTO Article 1.6.10(A) and the equations supplied therein:

@ 0.6AB

$$\text{Mult} = [0.25(4000)(101.67)(23.28)^2$$
$$+ 0.85(4000)(185)(6)(20.28)]\ 0.95$$
$$= 0.95[55101 + 76537] = [131{,}638\ \text{k-in.}]\ 0.95$$
$$= [10{,}970\ \text{k-ft}]\ 0.95$$
$$= 10{,}422\ \text{k-ft} > 2741\ \text{k-ft required}$$

@ 0.7AB

$$\text{Mult} = [0.25(4000)(101.67)(36.48)^2$$
$$+ 0.85(4000)(159)(6)(33.48)]\ 0.95$$
$$= 0.95[135{,}301 + 108{,}596] = [248{,}897\ \text{k-in.}]\ 0.95$$
$$= 0.95[33{,}789\ \text{k-ft}]$$
$$= 32{,}100\ \text{k-ft} > 29{,}827\ \text{k-ft required}$$

@ 0.2BC & 0.8BC

$$\text{Mult} = [0.25(4000)(101.67)(53.64)^2$$
$$+ 0.85(4000)(139)(6)(50.64)]\ 0.95$$

$$= 0.95[223,476 + 143,595] = [367,071 \text{ k-in.}]\ 0.95$$

$$= 0.95[30,589 \text{ k-ft}]$$

$$= 29,060 \text{ k-ft} > 19,050 \text{ k-ft required}$$

@ 0.3BC & 0.7BC

$$\text{Mult} = [0.25(4000)(101.67)(32.52)^2$$

$$+ 0.85(4000)(159)(6)(29.52)]\ 0.95$$

$$= 0.95[107,521 + 95,751] = [203,272 \text{ k-in.}]\ 0.95$$

$$= 0.95[16,939 \text{ k-ft}]$$

$$= 16,092 \text{ k-ft} > 3154 \text{ k-ft required}$$

The dead load shears due to the secondary moments and the maximum shears due to abutment settlement are computed at the tenth points (Table 7.41).

The maximum live load plus impact shears with the HS 20 truck and lane loadings are tabulated in Table 7.42. These values are computed

Table 7.41

Point	V_{DL}	Prestress V_{ms}	Δ = 1-in. Abutments V_Δ
A	+633	+56	−18
0.1	+450		
0.2	+268		
0.3	+86		
0.4	−96		
0.5	−278		
0.6	−462		
0.7	−650		
0.8	−867		
0.9	−1081		
BA	−1336	+56	−18
BC	+1361	−22	+22
0.1	+1031		
0.2	+738		
0.3	+473		
0.4	+229		
0.5	0		
0.6	−247		
0.7	−491		
0.8	−756		
0.9	−1049		
CB	−1379	−22	+22

Table 7.42

Point	7.277 7.119 V_{LL+I}
AB	+237[a]
0.1	+202[a]
0.2	+169[a]
0.3	+135[a]
0.4	+108[a]
0.5	−146[a]
0.6	−176
0.7	−207
0.8	−241
0.9	−275
BA	−311
BC	+323
0.1	+278
0.2	+236
0.3	+195
0.4	+158
0.5	+125
0.6	−159
0.7	−195
0.8	−236
0.9	−279
CB	−324

[a]HS 20 truck moment controls the design.

by using the influence lines for shear and correction due to corner moments.

The prestress force component V_{p_f} which reduces the applied shear is computed at the tenth points by multiplying the final prestress force at each point by the slope of the cgs at that point (Table 7.43).

The prestress shear requirements are computed according to the 1977 AASHTO requirements, which are discussed in some detail in Chapter 8. An approximation of the depth and width resisting shear is used when this complicated section is analyzed. This procedure is checked and evaluated later when a more precise shear flow concept at one of the critical points is used. In particular, the shear flow concept is useful in evaluating the shear design requirements. The ultimate shears are computed along the girder (Table 7.44).

This computation requires factoring of the computed shears and their sum. The sum of the applied shears is then magnified by the secant of the skew angle to account for a possible increase in shear on the

Table 7.43

Point	P_f	Slope	V_{P_f}
AB	8052	0.0744	599
0.1	8141	0.0566	461
0.2	8230	0.0391	322
0.3	8320	0.0210	175
0.4	8409	0	0
0.5	8498	0.0165	140
0.6	8587	0.0338	290
0.7	8676	0.0506	439
0.8	8765	0.0676	593
0.9	8699	0.0845	735
BA	8611	0	0
BC	8611	0	0
0.1	8496	0.0783	665
0.2	8379	0.0587	492
0.3	8264	0.0394	326
0.4	8149	0.0197	161
0.5	8033	0	0
0.6	7918	0.0197	156
0.7	7802	0.0394	307
0.8	7687	0.0587	451
0.9	7572	0.0783	593
CB	7456	0	0

individual webs due to the skewed relationship of the bearings. The ultimate shear V_u is then obtained by subtracting the prestress force component. This shear (in kips) is spread equally over the four interior webs, assuming that the bottom and top shear flanges do not contribute to the shear resistance (Table 7.44). The influence of all elements (webs and flanges) in resisting the bending shear is described later.

The concrete shear strength is computed by using a shear strength of 180 psi:

$$V_c = 180 \; bjd$$

$$= 180(12)(jd) = 2.16jd \; @ \; AB \text{ to } 0.8AB, \; 0.2BC \text{ to } 0.8BC$$

$$= 180(14.5)(jd) = 2.61jd \; @ \; 0.9AB, \; 0.1BC \text{ and } 0.9BC$$

$$= 180(17)(jd) = 3.06jd \; @ \; BA, BC, CB$$

A reasonable assumption that will simplify shear computations uses $j = 0.9$ and $d = 0.8D$; then $jd = 0.72D$.

Assume that there are two legs of #5 stirrups in each girder web. Then $2A_v f_y' = 2(0.62)(60) = 74.4$ k.

Table 7.44

Point	$1.44 \times V_{DL}$	$2.41 \times V_{LL+I}$	$2.41 \times V_\Delta$	$1.0 \times V_{Ms}$	Applied ΣV	Sec. 34° = $1.206 \times \Sigma V$	V_{Pf}	V_u	Per Web $V_u/4$
AB	+912	+571		+56	1539	1856	599	1257	314
0.1	+648	+487		+56	1191	1436	461	975	244
0.2	+386	+407		+56	849	1023	322	701	175
0.3	+124	+325		+56	505	609	175	434	108
0.4	-138	+260		+56	178	215	0	215	54
0.5	-400	-352	-43		795	959	140	819	205
0.6	-665	-424	-43		1132	1365	290	1075	269
0.7	-936	-499	-43		1478	1782	439	1343	336
0.8	-1248	-581	-43		1872	2258	593	1665	416
0.9	-1557	-663	-43		2263	2729	735	1994	498
BA	-1924	-750	-43		2717	3277	0	3277	819
BC	+1960	+778	+53		2791	3366	0	3366	842
0.1	+1485	+670	+53		2208	2663	665	1998	500
0.2	+1063	+569	+53		1685	2032	492	1540	385
0.3	+681	+470	+53		1204	1452	326	1126	282
0.4	+330	+381	+53		764	921	161	760	190
0.5	0	+301	+53		354	427	0	427	107
0.6	-356	-383		-22	761	918	156	762	190
0.7	-707	-470		-22	1199	1446	307	1139	285
0.8	-1089	-569		-22	1680	2026	451	1575	394
0.9	-1511	-672		-22	2205	2659	593	2066	516
CB	-1986	-781		-22	2789	3364	0	3364	841

The spacing according to 1977 AASHTO Section (1.6) is $s = 2A_v f_y' jd / (V_u - V_c)$. The minimum spacing is $s = 372/12 = 31.0$ in. The tie requirements according to the same section are #5 @ 2 ft-10 in. A reasonable maximum spacing might be limited to #5 @ 1 ft-6 in. by exercising engineering judgment. Using these criteria, we compute the spacing at the tenth points. At each point also the value $(V_u - V_c)$ is compared with the allowable of $8\sqrt{f_c'}$ listed in Table 7.45.

The allowable shear stirrup spacing is plotted against an actual shear stirrup spacing for completion of the basic stirrup design (Figure 7.24).

As mentioned earlier, an exact analysis of shear forces due to bending and torsion is given in which two loading cases are examined:

Case A. This loading case considers the following:

$$\text{full } (DL) + \text{full } (LL + I)$$

Table 7.45

Point	Av Web Depth D (in.)	jd (in.)	V_c (kips)	$V_u - V_c$ (kips)	$\dfrac{jd}{V_u - V_c}$	s (in.)	$v_u = V_u/bd$ (psi)	$v_c = 180$ $v_u - v_c$ (psi)
AB	70.0	50.4	109	205	0.246	18.3	467	287
0.1	↑	↑	109	135	0.373	27.8	363	183
0.2			109	66	0.764	56.8	260	80
0.3								
0.4	↓	↓						
0.5	70.0	50.4	109	96	0.525	39.1	305	125
0.6	72.2	52.0	112	159	0.327	24.3	388	208
0.7	78.7	56.7	122	214	0.265	19.7	445	265
0.8	85.4	68.3	148	268	0.255	19.0	508	327
0.9	93.0	74.4	194	304	0.245	18.2	462	282
BA	100.5	80.4	246	573	0.140	10.4	599	419
BC	100.5	80.4	246	596	0.135	10.0	616	436
0.1	93.0	74.4	194	306	0.243	18.1	463	284
0.2	85.4	68.3	148	237	0.288	21.4	470	290
0.3	78.7	56.7	122	160	0.354	26.3	373	193
0.4	72.2	52.0	112	78	0.667	49.6	274	94
0.5	70.0	50.4						
0.6	72.2	52.0	112	78	0.667	49.6	274	94
0.7	78.7	56.7	122	163	0.348	25.9	377	197
0.8	85.4	68.3	148	246	0.278	20.7	481	301
0.9	93.0	74.4	194	322	0.231	17.2	478	298
CB	100.5	80.4	246	595	0.135	10.1	615	435

$v_u - v_c < 8\sqrt{4000} = 506$ psi   OK.

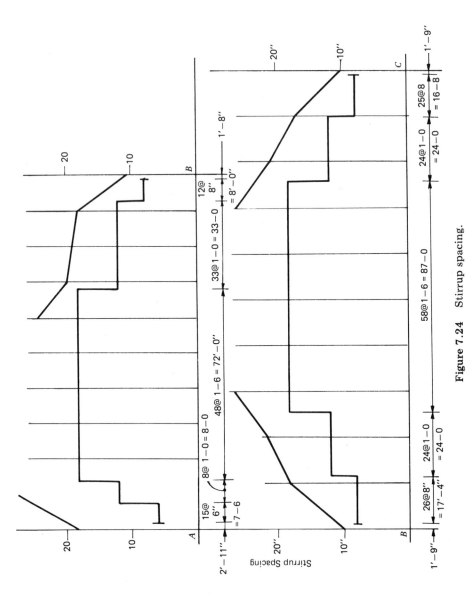

Figure 7.24 Stirrup spacing.

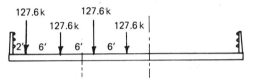

Figure 7.25   Bending and torsional loading.

where the truck is positioned-concentric with the section, and has no torsional effects. The resulting maximum shear is $V_u^{BC} = 3366$ k.

Case B.   This case involves both bending and torsional effects in which the live load is placed eccentrically on the bridge (Figure 7.25).

Using the $(LL + I)$ distribution factor $= 6.095$ and the new $(LL + I)$ as four lines of wheels, we obtain the maximum shear:

$$V_u^{BC} = 1960 + 53 + \left(\frac{4}{6.095}\right)(778)(1.206) = 3043 \text{ k}$$

The torsional moment is therefore

$$M_{LL + I} = 127.6(6 + 12 + 18) = 4595 \text{ k-ft} = 55,140,000 \text{ lb-in.}$$

The resistance of a closed box section to bending shears can be determined by a procedure in ref. (9). This technique first assumes the closed section is open, then the bending shears are computed using (VQ/I). The section is then closed to restore continuity at the cut by assuming the following:

$$\oint \frac{q_0}{GT} \, ds + q_r \oint \frac{ds}{GT} - \sum_k q_k \int_{ik} \frac{ds}{GT} = 0$$

where $i = $ cell $1, 2, \cdots, n$,
   $k = $ cell adjacent to cell i,
   $q_0 = $ shear flow of open profile $\sim \left(\dfrac{VQ}{I}\right)$,
   $q_r = $ redundant or closure shear flow.

The determinate shear flow (VQ/I) of the open section (Figure 7.26) is computed. First the moment of inertia is computed and listed in Table 7.46.

The (V/I_t) term therefore equals

$$\text{case A:} \quad \frac{V}{I_t} = 0.1458 \text{ lb in.}^{-4}$$

$$\text{case B:} \quad \frac{V}{I_t} = 0.1350 \text{ lb in.}^{-4}$$

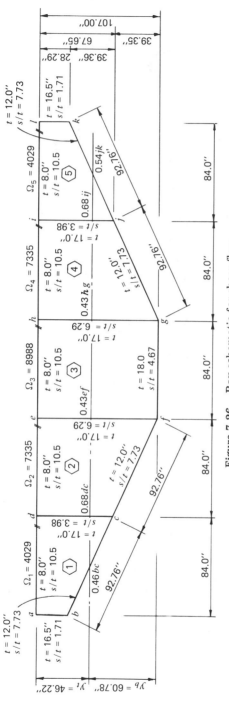

Figure 7.26 Box schematic for shear flow.

Table 7.46[a]

Element	Element Dimensions	A	y_b	$A\bar{y}_b$	$\bar{y}$	$A\bar{y}^2$	I_0	I_t
al	420.0×8.0	3,360	107.00	359,520.00	46.22	7,177,929	17,920	7,195,849
ab and kl	28.29×16.5	933.57	92.86	86,691.31	32.08	960,762	62,263	1,023,025
dc and ij	67.65×17.0	2,300.10	73.18	168,321.32	12.40	353,663	877,205	1,230,868
ef and gh	107.00×17.0	3,638.00	53.50	194,633.00	7.28	192,808	3,470,955	3,663,763
bf and gk	78.71×28.28	4,451.84	39.35	175,179.81	21.43	2,044,485	2,298,359	4,342,844
fg	84×18	1,512.00	0	0	60.78	5,585,643	40,824	5,626,467
		16,195.51 in.²		984,345.44 in.³				23,082,816 in.⁴

[a] $y_b = 60.78$ in; $y_t = 46.22$ in.

Table 7.47

Point	A	y	Q_z	Δq_0	q_0
$\parallel$					0
	672	46.22	31,060	4529	
a					4529
	467	32.08	14,981	2184	
b					6713
	512	8.96	4,587	669	
$0.46bc$					7382
	601	10.72	6,443	939	
c					6443
	368	10.72	3,945	575	
$0.32cd$					
	782	23.11	18,072	2635	
d					

The determinate shear flows q_0 around each cell are now computed (Case A), as shown in Figure 7.26.

First for cell (1) the determinate shears are computed and tabulated in Table 7.47.

This is followed by a similar computation for all other cells (Table 7.48).

These shear flows are plotted in Figure 7.27.

The correction or redundant shears q_r can be evaluated by applying the previously given equation around each cell;

Cell 1

$$\oint_{dL/dR} q_0 \frac{ds}{t} + q_1 \int_{dL/dR} \frac{ds}{t} - q_2 \int_{c/d} \frac{ds}{t} = 0$$

$\{[\tfrac{1}{2}(4529)](10.5) + [(4529) + \tfrac{1}{2}(2184)](1.71) + [(6443)$

$\quad + \tfrac{1}{2}(270) + \tfrac{2}{3}(793)](7.73) - [(4529) + \tfrac{1}{2}(2060)$

$\quad + \tfrac{2}{3}(1234)](3.98)\} + q_1[10.5 + 1.17$

$\quad + 7.73 + 3.98] - q_2[3.98] = 0$

Reduction of terms yields

$$q_1(23.92) - 3.98q_2 = -62,924 \tag{1}$$

Cell 2

$$\oint_{eL/eR} q_0 \frac{ds}{t} + q_2 \int_{eL/eR} \frac{ds}{t} - q_3 \int_{f/e} \frac{ds}{t} - q_1 \int_{c/d} \frac{ds}{t} = 0$$

$\{23,777 + 25,399 + 1[(6356) + \tfrac{1}{2}(6676)$

$\quad + \tfrac{2}{3}(799)](7.73) - [(2569) + \tfrac{1}{2}(1960) + \tfrac{2}{3}(3478)](6.29)\}$

$\quad + q_2[10.5 + 3.98 + 7.73 + 6.29] - 6.29q_3 - 3.98q_1 = 0$

Table 7.48

Point	A	y	Q_z	Δq_0	q_0
Cell 2					
‖					0
d	672	46.22	31,060	4529	
	782	23.11	18,072	2635	4,529
$0.68dc$					7,164
c	368	10.72	3,945	575	
	557	31.27	17,417	2539	6,589 + 6,443 = 13,032
$0.5cf$					10,493
f	557	50.94	28,374	4137	
	1037	30.39	31,514	4595	6,356
$0.57fe$					
e	782	23.11	18,072	2635	
Cell 3					
‖					0
e	672	46.22	31,060	4,529	
	782	23.11	18,072	2,635	4529
$0.43ef$					7164
f	1037	30.39	31,514	4,595	
	1512	60.78	91,899	13,399	2569 + 6356 = 8925
g					
	1037	30.39	31,514	4,595	4474
$0.57gh$					
h	782	23.11	18,072	2,635	
Cell 4					
‖					0
h	672	46.22	31,060	4529	
	782	23.11	18,072	2635	4529
$0.43hg$					7164
	1037	30.39	31,514	4595	
g	557	50.94	28,374	4137	2569 − 4474 = 1905
$0.5gl$					6042
j	557	31.27	17,417	2539	
	368	10.72	3,945	575	8581
$0.32ji$					
i	782	23.11	18,072	2635	
Cell 5					
‖					0
i	672	46.22	31,060	4529	
	782	23.11	18,072	2635	4529
$0.68ij$					7164
j	368	10.72	3,945	575	
	601	10.72	6,443	939	6589 − 8581 = 1992
$0.54jk$					2931
k	512	8.96	4,587	669	
	467	32.08	14,981	2184	2262
l					78 say 0

Reduction of terms yields

$$-q_1(3.98) + q_2(28.50) - q_3(6.29) = -91,320 \qquad (2)$$

Cell 3

$$\oint_{hL/hR} q_0 \frac{ds}{t} + q_3 \int_{hL/hR} \frac{ds}{t} - q_2 \int_{f/e} \frac{ds}{t} - q_4 \int_{g/h} \frac{ds}{t} = 0$$

$$\{23,777 + 36,908 + [\tfrac{1}{2}(8925)(0.67) - \tfrac{1}{2}(4474)(0.33)] \,(4.67)$$
$$- 36,908\} + q_3[10.5 + 6.29 + 4.67 + 6.29]$$
$$- q_2(6.29) - q_4(6.29) = 0$$

Reduction of terms yields

$$-q_2(6.29) - q_4(6.29) + q_3(27.75) = -34,292 \qquad (3)$$

Cell 4

$$\oint_{iL/iR} q_0 \frac{ds}{t} + q_4 \int_{iL/iR} \frac{ds}{t} - q_3 \int_{g/h} \frac{ds}{t} - q_5 \int_{j/i} \frac{ds}{t} = 0$$

$$\{23,777 + 36,908 - [(1905) + \tfrac{1}{2}(6676) + \tfrac{2}{3}(799)] \,(7.73) - 25,399\}$$
$$+ q_4[10.5 + 3.98 + 7.73 + 6.29] - q_3(6.29) - q_5(3.98) = 0$$

Reduction of terms yields

$$-q_3(6.29) + q_4(28.50) - q_5(3.98) = -9360 \qquad (4)$$

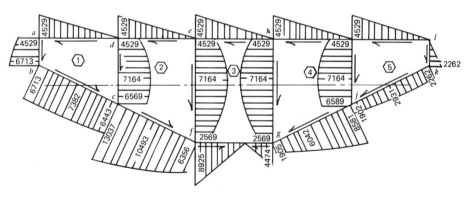

Case A open shears

Figure 7.27   Case A: Shear flow determinate.

Cell 5

$$\oint_{\iota L/\iota R} q_0 \frac{ds}{t} + q_5 \int_{\iota L/\iota R} \frac{ds}{t} - q_4 \int_{j/i} \frac{ds}{t} = 0$$

$$\{23{,}777 + 25{,}399[(1992) + \tfrac{1}{2}(270 + \tfrac{2}{3}(793)]\,(7.73)$$

$$- [(\tfrac{1}{2})(2262)]\,(1.71)\} + q_5[10.5 + 1.71 + 7.73 + 3.98]$$

$$- q_4(3.98) = 0$$

Reduction of terms yields

$$-q_4(3.98) + q_5(23.92) = -26714 \tag{5}$$

Solution of eqs. (1)–(5) simultaneously produces the redundant shear flows plotted in Figure 7.28:

$$q_1 = -3322 \text{ k in.}^{-1}$$
$$q_2 = -4166 \text{ k in.}^{-1}$$
$$q_3 = -2255 \text{ k in.}^{-1}$$
$$q_4 = -333 \text{ k in.}^{-1}$$
$$q_5 = -1172 \text{ k in.}^{-1}$$

The sum of determinate and redundant shear flows ($q_0 + q_r$) gives the final shear flow shown in Figure 7.29.

Use of the diagram for Case *A* and a modification of the (V/I) term using a ratio of

$$\left(\frac{V}{I}\right)_B \Big/ \left(\frac{V}{I}\right)_A = \frac{0.1350}{0.1458} = 0.904$$

times the results in Case *A* give the final bending shear flows for Case *B* in Figure 7.30.

Case A warping correction

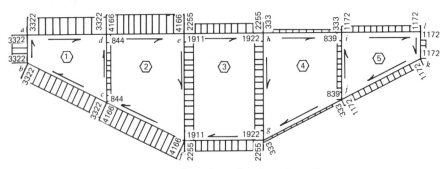

Figure 7.28   Case *A*: redundant shear flows.

Case A Final shears

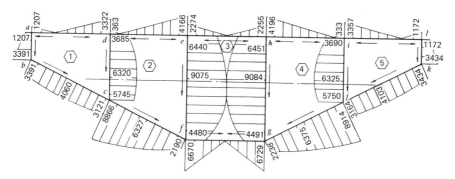

Figure 7.29 Case A: final bending shear flows.

Summ obt

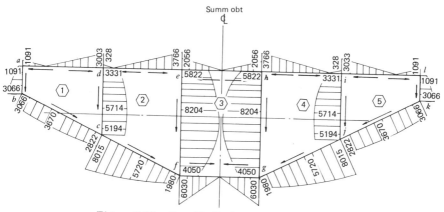

Figure 7.30 Case B: final bending shear flows.

The induced shear stress or flows due to pure torsional effects or moment (Case *B*) can also be determined (9) by using the following equation:

$$q_i \oint_i \frac{ds}{t} - \sum_k q_k \int_{ik} \frac{ds}{t} = \sum \left(\frac{M_z}{K_T} \right) A_{0_i}$$

where M_z = total applied torque applied to section,
   A_{0_i} = areas enclosed by cell *i*,
   $\oint_i$ = contour integral of shear stress along cell *i* walls,
   i = cell 1, 2, 3, $\cdots$, *m*,
   k = cells adjacent to cell *i*.

The torsional constant K_T of the entire section can be determined after the preceding equations are solved for q_i, where, in general,

$$q_i = \bar{C}_i \left(\frac{M_z}{K_T} \right)$$

Therefore

$$K_T = 2 \sum_{i=1}^{n} \bar{C}_i A_i$$

$$M_z = 2 \sum_{i=1}^{n} q_i A_i$$

Application of these equations gives

Cell 1

$$q_1(23.92) - q_2(3.98) = 2(4029) \frac{M_z}{K_T}$$

Cell 2

$$-q_1(3.98) + q_2(28.5) - q_3(6.29) = 2(7335) \frac{M_z}{K_T}$$

Cell 3

$$-q_2(6.29) + q_3(28.50) - q_4(3.98) = 2(8988) \frac{M_z}{K_T}$$

Cell 4

$$-q_3(6.29) + q_4(28.50) - q_5(3.98) = 2(7335) \frac{M_z}{K_T}$$

Cell 5

$$-q_4(3.98) + q_5(23.92) = 2(4029) \frac{M_z}{K_T}$$

Solution of these equations gives

$$q_1 = 6.53 \times 10^8 \frac{M_z}{K_T}$$

$$q_2 = 10.04 \times 10^8 \frac{M_z}{K_T}$$

$$q_3 = 13.89 \times 10^8 \frac{M_z}{K_T}$$

$$q_4 = 10.04 \times 10^8 \frac{M_z}{K_T}$$

$$q_5 = 6.53 \times 10^8 \frac{M_z}{K_T}$$

$$q_{12} = -2.98 \times 10^8 \frac{M_z}{K_T}$$

$$q_{23} = -3.27 \times 10^8 \frac{M_z}{K_T}$$

$$q_{34} = -2.98 \times 10^8 \frac{M_z}{K_T}$$

The torsional constant K_T for the entire cell is computed from

$$K_T = 2 \sum \bar{C}_i A_i$$

or

$$K_T = 2[6.53(2 \times 4029) + 10.04(7335 \times 2) + 13.89(8988)] \times 10^8$$
$$K_T = 649.4 \times 10^{11} \text{ in.}^4$$

Therefore the

$$\frac{M_z}{K_T} = \frac{55.14 \times 10^6}{64.9 \times 10^{12}} = 84.9 \times 10^{-8} \text{ lb in.}^{-3}$$

We multiply this constant by the q terms to obtain the torsional shear flows in Figure 7.31.

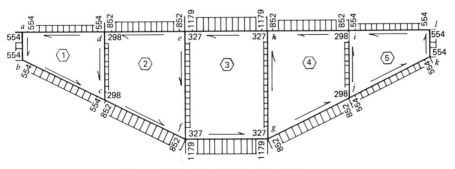

Figure 7.31 Case B: Torsional shears.

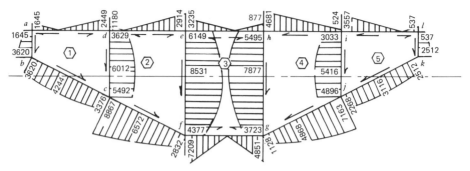

Figure 7.32 Case B: total shear.

The torsional shears (Figure 7.31) are combined with the bending shears (Figure 7.30) to give the total shears shown in Figure 7.32 for Case *B*.

A comparison of the three computations for shear requirements at *BC* follows:

approximate method using AASHTO $\qquad V_u = 616$ psi

shear flow using AASHTO live load distribution $\qquad V_u = \dfrac{9084}{17} = 534$ psi

shear flow using eccentric live load and torsion $\qquad V_u = \dfrac{8531}{17} = 502$ psi

This summary shows that the AASHTO live load distribution is a good approximation and that the assumptions made by using the AASHTO shear criteria are conservative. It is important to note that the shear requirements in the sloping soffit are significant. A consideration of the maximum shear flow (Case *A* governs)

$$q = 8914 \text{ lb in.}^{-1}$$

$$v_u = \frac{8914}{12} = 743 \text{ psi}$$

$$v_u - v_c = 743 - 180 = 563 \text{ psi} > 8\sqrt{4000} = 506 \text{ psi}$$

Revise concrete strength to $f'_c = 4500$ psi – $8\sqrt{4500} = 537$ psi. This would be within 5% and probably OK.

$$\frac{Vs}{d} = 563(12) = 6756 \text{ lb in.}^{-1}$$

$$s = \frac{2(0.88)(60)}{6.756}$$

$$= 15.6 \text{ in. (AASHTO)} \quad 7.8 \text{ in. (ACI)} \quad \#6 @ 8 \text{ in. OK to use}$$

Table 7.49

Point	Prestress	DL	Δ
AB	0	0	
0.1	+0.0912	−0.0870	+0.0042
0.2	+0.1677	−0.1581	+0.0096
0.3	+0.2191	−0.2024	+0.0167
0.4	+0.2391	−0.2146	+0.0245
0.5	+0.2268	−0.1956	+0.0312
0.6	+0.1862	−0.1522	+0.0340
0.7	+0.1286	−0.0974	+0.0312
0.8	+0.0691	−0.0465	+0.0226
0.9	+0.0235	−0.0119	+0.0116
B	0	0	
0.1	+0.0038	−0.0258	−0.0220
0.2	+0.0423	−0.0830	−0.0407
0.3	+0.0994	−0.1534	−0.0540
0.4	+0.1510	−0.2113	−0.0603
0.5	+0.1739	−0.2326	−0.0587
0.6	+0.1580	−0.2080	−0.0500
0.7	+0.1119	−0.1480	−0.0361
0.8	+0.0567	−0.0772	−0.0205
0.9	+0.0148	−0.0219	−0.0071
C			

The camber due to final prestress and the dead load deflection are computed by using the conjugate beam method described in the examples in Chapter 8. These deflections are listed in Table 7.49.

BIBLIOGRAPHY

1. *Continuous Concrete Bridges*, Portland Cement Association.
2. *AASHTO Standard Specifications for Highway Bridges*, 1977, with the 1978 and 1981 Interim Specifications, The American Association of State Highway and Transportation Officials.
3. *Post-Tensioned Box Girder Bridge Manual*, Post-Tensioning Institute, 1978.
4. *Post-Tensioned Box Girder Bridges—Design and Construction*, Western Concrete Reinforcing Steel Institute, 1969.
5. T. Y. Lin and Ned H. Burns, *Design of Prestressed Concrete Structures*, Wiley, New York, 1981.
6. T. Y. Lin, *Prestressed Concrete Structures*, Wiley, New York, 1955.
7. *Post-Tensioning Manual*, Post-Tensioning Institute, 1976.
8. James R. Libby, *Prestressed Concrete—Design and Construction*, Van Nostrand-Reinhold, New York, 1961.
9. C. P. Heins, *Bending and Torsional Design in Structural Members*, Heath, Lexington, MA, 1975.

EIGHT

Standard Precast/ Prestressed Concrete Bridges

The precasting industry has in recent years become a well-established entity throughout the United States. Efficient management and outstanding quality control procedures have awarded the industry a highly competitive position in the construction market. Precasting plants service the entire country in a cost-effective manner. Precast concrete superstructures generally eliminate the need for construction falsework which has always been economically advantageous. The major limitation to precast/prestress construction when it comes to hauling and erecting has been the length and weight limits. Segmental construction, described in Chapter 9, has helped to solve this problem.

For bridge structures two basic precast sections are produced: I-beam sections to be used with a cast-in-place deck and the multibeam sections that include an integral riding surface. These multibeams interact with one another by the use of common shear keys and transverse post-tensioning. The multibeams are often (although not always) used on shorter span structures and are commonly the choice of agencies with inadequate field staffs because they require little construction inspection.

Concrete strengths in the order of 5000–6500 psi are used, but plants with good quality-control procedures often produce very high strengths. Strengths at transfer of prestress usually range from 4000 to 5000 psi; $\frac{1}{2}$ in. diameter seven-wire, 270 k strand is commonly used today as the major prestressing reinforcement. In many areas strands are bundled near the midspan area of beams to obtain a greater eccentricity.

The normal design procedure for precast/prestressed concrete is to design for allowable working stresses and to check initial stresses and ultimate moment capacity. Most designs for this type of section utilize pretensioned prestressing in which the strands are stressed before the concrete is placed around the strands. When the concrete has cured to sufficient strength the strands are cut and the prestressing is transferred to the beams. Some post-tensioning is used on occasion with precast sections. The ducts are cast in the section and strands are installed after concrete placement and curing. After the concrete cures the strands are jacked, anchored, and grouted.

Camber needs to be estimated during design and accounted for when the riding surface is established. On some low-volume, low-speed, single-span structures camber is sometimes ignored. However, if camber is not accounted for in multispan structures, the "roller coaster" riding surface produced is a hazard to the traveling public.

8.1 DESIGN CODE

The AASHTO specifications cover the design of pretensioned and post-tensioned structures in Section 1.6, which requires that all critical sections be designed for ultimate strength and allowable stresses. These design specifications are generally intended for use with concrete strengths of 5000–6000 psi. It is reasonable, however, to use the same specifications for higher strength concrete if caution is exercised in design and quality control in the fabrication of prestressed beams.

Prestress losses for pretensioned precast beams are covered in AASHTO Article 1.6.7(B)(1) and (2). For normal design it is proper to use the lump-sum tosses provided in (2) when the design concrete strengths are within 500 psi, $\pm$, of the 5000 psi strength given. For strengths greater than $f'_c = 5500$ psi the losses should be computed by formulas for the individual loss sources given in (1). It is proper to subtract the elastic shortening loss from the initial prestress force when computing initial stresses. When using the lump-sum losses a reasonable approximation of the elastic shortening loss is 0.07×270 ksi $= 18.9$ ksi.

AASHTO Article 1.6.12(c) covers the design of precast pretensioned I-beam bridges erected as simple spans for noncomposite loads and made continuous for composite loadings. This section indicates the potential need for considering the shortening effects (i.e., creep and shinkage). When significant they can be included in the design by following the procedure given in ref. (2). Positive moment at the piers is most likely to occur in this type of design in which more than two spans exist. This is the result of very little dead load moment on the continuous spans and insufficient dead load negative moment to negate the effect of a live load placed in a span beyond those adjacent to the support. Paragraph (2) of this article permits the reinforcement in this section to be designed for an allowable working stress equal to $0.6f_y$. This article also permits an increase in the allowable compression stress on the bottom fiber adjacent to the pier. The allowable stress is increased to $0.6f'_c$ when the stress due to prestressing is included.

Paragraph (3) of AASHTO Article 1.6.12(c) states that the reinforced concrete section over the piers of bridges built with prestressed beams made continuous for live load should be designed by load factor design. Because the rest of the superstructure is designed for load factor and working stress design, a similar approach is appropriate for this reinforced concrete section.

The 1980 interim specifications revised the shear design for prestressed concrete. This revision includes a method of computing concrete shear strength developed for uniform loads on building floors. It is not likely that a designer would obtain consistent results with this procedure. Therefore it is reasonable that the 1977 specification for shear be used in accordance with recommendations by the PCI Committee on Bridges. No known problems exist from the use of this method; therefore it should be considered safe.

8.2 DESIGN EXAMPLES

Example 8.1 Simple Span Prestressed I-Beam Design. The bridge shown in Figures 8.1 and 8.2 is designed in this example. The normal

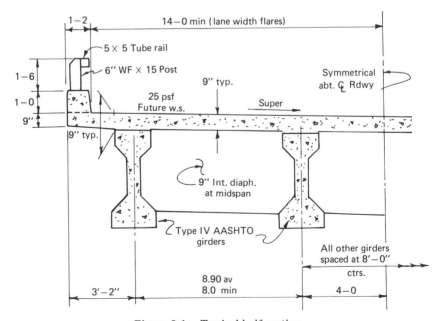

Figure 8.1 Typical half section.

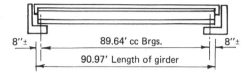

8''± | 89.64' cc Brgs. | 8''±

90.97' Length of girder

Figure 8.2 Longitudinal section.

design procedure would be to check the interior and exterior girders to determine the most critical. For purposes of this example it is assumed that the interior girders control and that the exterior girders will be made the same.

The slab is designed as indicated in Chapter 4. The girder is designed as a simple span with the girder section carrying the dead load of slab and girder and the composite girder and slab section carrying all superimposed dead load with live load plus impact. The live load is HS 20 with a check for overload based on extra heavy loadings.

The girder concrete strength is

$$f'_c = 5000 \text{ psi}$$

$$f'_c = 4500 \text{ psi at transfer of prestress}$$

$$\tfrac{1}{2} \text{ in. 270 k strands are used}$$

Strength of concrete other than prestressed is $f'_c = 3300$ psi and mild reinforcing is A615 Grade 60 with an allowable service load stress of $f_s = 24,000$ psi.

The section properties for a Type IV AASHTO (Appendix S) are

$$w = 822 \text{ plf}$$

$$A = 789 \text{ in.}^2$$

$$I = 260,730 \text{ in.}^4$$

$$y_b = 24.73 \text{ in.}$$

$$y_t = 29.27 \text{ in.}$$

$$S_b = 10,543 \text{ in.}^3$$

$$S_t = 8908 \text{ in.}^3$$

It is then appropriate to compute a multiplier to convert moments in k-ft to stresses in psi:

$$\frac{12,000}{S_b} = 1.138 \left(\frac{\text{lb-in.}}{\text{k-ft}} \times \frac{1}{\text{in.}^3} \right), \quad \frac{12,000}{S_t} = 1.347 \left(\frac{\text{lb-in.}}{\text{k-ft}} \times \frac{1}{\text{in.}^3} \right)$$

The composite section is shown in Figure 8.3.

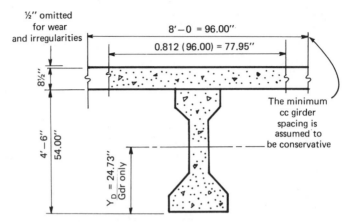

Figure 8.3   Composite section.

Note that the effective flange width is controlled by the center-to-center girder spacing [AASHTO Article 1.6.23(A)]. This effective width is then modified by the modular ratio n, which compares the modulus of elasticity for the slab concrete with the modulus of elasticity for the girder concrete.

Using $E_c = 57,000\sqrt{f_c'}$, we obtain

$$\text{slab } f_c' = 3300 \text{ psi}$$

$$E_c = 57,000\sqrt{3300}$$

$$= 3,274,000 \text{ psi}$$

$$\text{girder } f_c' = 5000 \text{ psi}$$

$$E_c = 57,000\sqrt{5000}$$

$$= 4,031,000 \text{ psi}$$

$$n = \frac{3274}{4031} = 0.812$$

The moment of inertia and other section properties are computed for this composite section by using $I_T = \Sigma(A\bar{y}^2 + I_0)$

Element	A	y_b	A_{yb}	$\bar{y}$	$A\bar{y}^2$	I_0	I_T
Slab	663	58.25	38,595	18.23	220,337	3,989	224,326
Girder	789	24.73	19,512	15.29	184,456	260,730	445,186
	1452 in.²		58,107 in.³				669,512 in.⁴

$$\bar{y}_b = 40.02 \text{ in.} \qquad S_b = 16,729 \text{ in.}^3$$

$$\bar{y}_t = 13.98 \text{ in.} \qquad S_t = 47,891 \text{ in.}^3$$

$$y_{\text{slab}} = 22.48 \text{ in.} \qquad S_{\text{slab}} = 29,783 \text{ in.}^3$$

Again a multiplier is computed to convert moments in k-ft to stresses in psi:

$$\frac{12,000}{S_b} = 0.717\left(\frac{\text{lb-in.}}{\text{k-ft}} \times \frac{1}{\text{in.}^3}\right)$$

$$\frac{12,000}{S_t} = 0.251\left(\frac{\text{lb-in.}}{\text{k-ft}} \times \frac{1}{\text{in.}^3}\right)$$

$$\frac{12,000}{S_{\text{slab}}} = 0.403 \times 0.812 = 0.327\left(\frac{\text{lb-in.}}{\text{k-ft}} \times \frac{1}{\text{in.}^3}\right)$$

Using the girder dead load $w = 822$ lb-ft^{-1}, we determine the dead load shears at the support and quarter point with the moment at intermediate points in the span:

Point	V_{DL} (k)	M_{DL} (k-ft)
A	36.8	
0.25	18.4	+619
0.4		+793
0.5		+826

The uniform slab dead load is computed by using a concrete unit weight of 0.150 k-ft^3:

$$\left.\begin{aligned} w = 0.75 \times 8.45 \text{ ft} \times 0.150 = 0.951 \\ 1.67 \times 0.12 \text{ ft} \times 0.150 = 0.030 \end{aligned}\right\} \text{ sum} = 0.981 \text{ k-ft}^{-1}$$

where the 0.030 k-ft^{-1} is an approximate allowance for the fillet over the girder.

The diaphragm dead load at midspan is

$$P = 0.75 \times 7.78 \times 3.83 \text{ ft} \times 0.150 = 3.35 \text{ k}$$

The diaphragm and slab dead load are applied simultaneously on the girder section; hence the shears and moments are combined. The tabulation of these shears and moments follows:

Point	V_S (k)	V_D (k)	V_{S+D} (k)	M_S (k-ft)	M_D (k-ft)	M_{S+D} (k-ft)
A	44.0	1.7	45.7			
0.25	22.0	1.7	23.7	+739	+38	+777
0.4		1.7		+946	+60	+1006
0.5		1.7		+985	+75	+1060

The superimposed dead load on the composite section consists of wearing surface only on an interior girder if the traffic barrier dead load is applied to the exterior girder only. The superimposed dead load for an interior girder is

$$w = 8.45' \times 25$$

$$= 211 \text{ lb-ft}^{-1} \text{ (assuming 25 psf typical for wearing surface)}$$

These superimposed dead load shears and moments are tabulated:

Point	V_{SDL} (k)	M_{SDL} (k-ft)
A	9.5	
0.25	4.7	+159
0.4		+203
0.5		+212

For a span of 89.64 ft the maximum HS 20 live load shear and moment are obtained from Appendix A in the AASHTO specifications and are equal to 64.5 k and 1337.9 k-ft, respectively. Using an average girder spacing, we compute the live load distribution per girder according to AASHTO Article 1.3.1(B):

$$DF = \frac{S}{5.5}$$

$$DF = \frac{8.45}{5.5} = 1.536$$

The impact according to AASHTO Article 1.2.12(C) is

$$I = \frac{50}{89.64 + 125} = 0.233$$

The combined live load distribution and impact coefficient is

$$\text{coefficient} = 1.233(1.536) = 1.894$$

The critical live load plus impact shears for the interior girder are

$$V_{AB} = 1.894(64.5)(\tfrac{1}{2}) = 61.1 \text{ k}$$

$$V_{0.25AB} = 1.894(16.0)\left(0.75 + 0.594 + \frac{0.438}{4}\right) = 44.0 \text{ k}$$

The critical live load plus impact moments for the interior girder are computed by assuming that the maximum moment occurs at midspan and has a parabolic distribution. These critical moments are

$$M_{0.5} = 1.894(1337.9)(\tfrac{1}{2}) = +1267 \text{ k-ft}$$

$$M_{0.4} = 0.96(1267) = +1216 \text{ k-ft}$$

$$M_{0.25} = 0.75(1267) = +950 \text{ k-ft}$$

The axle loads for the overload vehicle are given in Figure 8.4. Using the same wheel distribution per girder computed for the HS 20 vehicle and the same impact value, we calculate the critical overload shears:

$$V_{AB} = 1.894[30.0(1.0 + 0.944) + 22.5(0.799 + 0.743)] = 176.2 \text{ k}$$

$$V_{0.25} = 1.894[30.0(0.75 + 0.694) + 22.5(0.549 + 0.493)] = 126.5 \text{ k}$$

The maximum midspan moment is determined by computing the shear for the most critical loading (Figure 8.4):

$$V_{AB} = 30(0.556 + 0.50) + 22.5(0.355 + 0.299) = 46.40 \text{ k}$$

The maximum midspan moment is

$$M_{0.5} = [(46.40)(44.82) - 30.0(5.0)](1.894) = +3654 \text{ k-ft}$$

Again a parabolic distribution of the maximum live load plus impact moments is assumed in order to obtain maxima at intermediate points:

$$M_{0.4} = 0.96(3654) = +3508 \text{ k-ft}$$

$$M_{0.25} = 0.75(3654) = +2740 \text{ k-ft}$$

This completes the computation of the moments.

Using the section properties, we convert these moments to stresses at the top and bottom girder fibers by multiplying the $12{,}000/S$ values computed with the section properties by the moment, which gives stresses in psi. The tabulated values refer to Gdr-Girder dead load, S + diaphragm = slab + diaphragm, SDL = superimposed dead load,

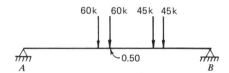

Figure 8.4 Critical loading.

HS 20 = live load plus impact, and the sum of effects from these loadings is Σf_b or Σf_t.

$$\text{Bottom fiber } \frac{12{,}000}{S_b} = 1.138 \qquad\qquad \frac{12{,}000}{S_b} = 0.717$$

Point	Gdr	S + Diaphragm	SDL	HS 20	Σf_b (psi)
0.25	-704	-884	-114	-681	-2383
0.4	-902	-1145	-146	-872	-3065
0.5	-940	-1206	-152	-908	-3206

$$\text{Top fiber } \frac{12{,}000}{S_t} = 1.347 \qquad\qquad \frac{12{,}000}{S_t} = 0.251$$

Point	Gdr	S + Diaphragm	SDL	HS 20	Σf_t (psi)
0.25	+834	+1047	+40	+238	+2159
0.4	+1068	+1355	+51	+305	+2779
0.5	+1113	+1428	+53	+318	+2912

The required prestress reinforcement can be determined. First the number of strands required is computed based on the allowable tension in the bottom fiber = $6\sqrt{f_c'} = 6\sqrt{5000} = 424$ psi. Assume that the strand value = $0.1531[(0.7)(270) - 45.0)] = 22.05$ k per strand based on the area of each strand = 0.1531 in.^{2}

$$f_s' = 270 \text{ ksi for strands}$$

$$0.7f_s' = \text{initial force to each strand}$$

$$45.0 \text{ ksi} = \text{assumed total losses}$$

Assume that $e = 20.0$ in. $\mp$. Then

$$\text{allowable tension} = \sum f_b(DL + LL + I) + \frac{P}{A} + \frac{Pe}{S_b}$$

$$-424 = -3206 + \frac{P}{789} + \frac{P(20.0)}{10543}, P = 879 \text{ k}$$

$$\text{No. strands} = \frac{879}{22.05} = 39.9 \text{ strands required}$$

Next the number of strands required, based on an allowable top girder compressive stress, $0.4(f_c) = 0.4(5000) = +2000$ psi:

$$\text{allowable compression} = \sum f_t(DL + LL + I) + \frac{P}{A} - \frac{Pe}{S_t}$$

$$+2000 = +2912 + \frac{P}{789} - \frac{P(20.0)}{8908}, P = 933 \text{ k}$$

$$\text{number of strands required} = \frac{933}{22.05} = 42.3 \text{ strands}$$

Finally, the number of strands required for ultimate moment is determined. For this it is necessary to compare the HS 20 and overload moments at midspan, which are based on factored moments:

$$1.67 \times \text{HS 20} = 2116 \text{ k-ft}$$

$$1.0 \times \text{overload} = 3654 \text{ k-ft}$$

Therefore the overload controls and required factored total ultimate moments are

$$M_u = 1.3(826 + 1060 + 212 + 3654) = 7478 \text{ k-ft}$$

Approximate $f'_{su} = 240 \pm$ ksi; the strand value at ultimate load equals $0.1531(240.0) = 36.7$ k. The approximate jd value at ultimate equals 4.75 ft; therefore the required tensile force in the strands equals 7478 k-ft/4.75 = 1574 k. The approximated number of strands required equals 1574/36.7 k/strand = 42.9 strands. Try $44\frac{1}{2}\phi$ 270 strands. These strands should be arranged in the bottom flange to obtain the maximum eccentricity at midspan. Our trial strand locations are shown in Figure 8.5 in the bottom flange.

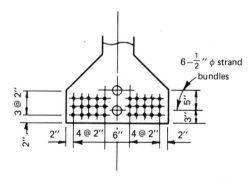

Figure 8.5   Trial strand locations: bottom flange.

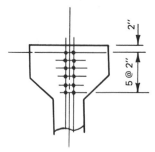

Figure 8.6   Trial strand locations: top flange/end.

Some strands should be deflected up at the end of the girder to eliminate the possibility of tension or excessive stresses at that point. The 12 bundled strands are deflected up and splayed out (Figure 8.6).

These deflected strands should be held down at the bottom of the girder until about the 0.4 point because the applied moments at the 0.4 point are about equal to those at the midspan 0.5 point. Deflected strands with only one hold down at midspan can result in an overstress at the 0.4 point.

Also when deflected strands are used, the uplift force at the deflecting point, which represents a component of the prestress force, should be checked. Most prestressing beds used to manufacture prestressed I-girders are capable of 25 k. If the uplift allowable is exceeded, a staggered hold down should be provided where the deflection point is separated into two deflecting points at about 5 ft-0 in. centers. A staggered deflection location is used in this example (Figure 8.7). Figure

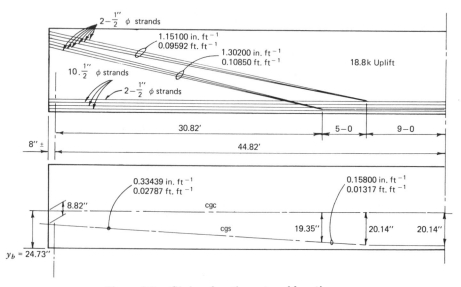

Figure 8.7   Girder elevation: strand locations.

8.7 also contains a girder elevation that illustrates the strand deflection pattern; the cgs profile is also shown in the lower girder elevation. The eccentricities at the significant points are given with the slope of the cgs.

The strands are stressed initially to $0.7f'_s$ or 28.91 k per strand. A 19,000-psi reduction in the initial force is allowed to accommodate elastic shortening when the initial stresses are checked. This gives a strand value of 26.04 k per strand for initial stress computation. An allowable loss of 45,000 psi is assumed as a lump-sum loss for final conditions [AASHTO Article 1.6.7(B)]. This gives a final strand value of 22.05 k per strand.

For $44\frac{1}{2}$ in. strands the prestress force values are

initial $\qquad P_i = 44(28.91) = 1272.0$ k

initial minus
elastic shortening $\qquad P_i - E_s = 44(26.04) = 1145.8$ k

final $\qquad P_f = 44(22.05) = 970.2$ k

By using these forces and the eccentricities obtained from Figure 8.7, the stresses due to prestress can be computed and tabulated at the significant points:

					Final		Int – ES	
					---	---	---	---
Point	$\dfrac{P_f}{A}$	e	$\dfrac{P_f e}{S_b}$	$\dfrac{P_f e}{S_t}$	f_b (psi)	f_t (psi)	f_b (psi)	f_t (psi)
A	+1230	9.04	+832	−985	+2062	+245	+2435	+289
0.25	+1230	16.54	+1522	−1801	+2752	−571	+3250	−674
0.4	+1230	20.14	+1853	−2193	+3083	−963	+3641	−1137
0.5	+1230	20.14	+1853	−2193	+3083	−963	+3641	−1137

The initial stresses are tabulated and summed with the girder dead load stress superimposed on the stresses from the prestress force:

	Bottom Fiber			Top Fiber		
Point	P/S	Girder	f_b (psi)	P/S	Girder	f_t (psi)
A	+2435		+2435	+289		+289
0.25	+3250	−704	+2546	−674	+834	+160
0.4	+3641	−902	+2739	−1137	+1068	−69
0.5	+3641	−940	+2701	−1137	+1113	−24

The allowable initial stresses according to AASHTO art. 1.6.6(B) are:

$$\text{allowable compression } 0.6(4500) = 2700 \text{ psi}$$

$$\text{allowable tension w/o reinforcement } 3\sqrt{4500} = 201 \text{ psi}$$

$$\text{allowable tension w/reinforcement } 7.5\sqrt{4500} = 503 \text{ psi}$$

It can be seen that the actual compression stress is at the allowable stress and no reinforcement is required for tension; therefore the girder is satisfactory for initial conditions. The girder stresses are then checked at various stages of loading for the final conditions:

	Bottom Fiber (psi)			Top Fiber (psi)		
	0.25	0.4	0.5	0.25	0.4	0.5
Prestress	+2752	+3083	+3083	−571	−963	−963
Girder *DL*	−704	−902	−940	+834	+1063	+1113
Σ	+2048	+2181	+2143	+263	+102	+150
Slab + Diaphragm *DL*	−884	−1145	−1206	+1047	+1355	+1428
Σ	+1164	+1036	+937	+1310	+1457	+1578
Superimposed *DL*	−114	−146	−152	+40	+51	+53
Σ	+1050	+890	+785	+1350	+1508	+1631
HS 20 (*LL* + *I*)	−681	−872	−908	+238	+305	+318
Σ	+369	+18	−123	+1583	+1813	+1949

The allowable stresses according to AASHTO Article 1.6.6(B) are

$$\text{allowable compression} = 0.4(5000) = 2000 \text{ psi}$$

$$\text{allowable tension} = 6\sqrt{5000} = 424 \text{ psi}$$

All stresses are within these allowables; therefore the design can be considered satisfactory for stress. Next it is required that ultimate capacity be checked and the overload condition that will control, determined; that is, 1.67 × HS 20 live loads or the overload live load:

$$1.67 \times \text{HS 20 moment} = 2116 \text{ k-ft}$$

$$1.0 \times \text{overload moment} = 3654 \text{ k-ft}$$

The overload moment will control the required factored ultimate moment; therefore

Girder *DL*   Slab + Diaphragm *DL*   Superimposed *DL*   *L* + *I*
$$= 1.3 (+ \quad 826 \quad + \quad 1060 \quad + \quad 212 \quad + 3654) = +7131 \text{ k-ft}$$

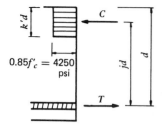

Figure 8.8 Ultimate stress diagram.

The ultimate capacity can be computed by using the sketched stress diagram (Figure 8.8) which simulates the ultimate condition. The reinforcement ratio pf'_{su}/f'_c must be below 0.30 to ensure failure in the reinforcement. It is then checked at midspan:

$e = 20.14$ in. $d = 57.91$ in. $A_s = 6.7364$ in.[2] $b = 77.95$ in.

Then

$$p = 0.001492$$

$$0.5 \frac{f'_s}{f'_c} = 0.0403$$

The average steel stress at failure is computed:

$$f'_{su} = f'_s\left(1 - 0.5 \frac{f'_s}{f'_c}\right)$$

$$f'_{su} = 270(1 - 0.0403) = 259.1 \text{ ksi}$$

$$p \frac{f'_{su}}{f'_c} = 0.077 < 0.30 \quad \text{OK}$$

$$T = 6.7364(259.1) = 1745 \text{ k}$$

$$k'd = \frac{1745}{4.25(77.95)}$$

$$= 5.27 \text{ in.} < \text{slab thickness; therefore a rectangular section}$$

$$d = 57.91 \text{ in.}$$

$$jd = 57.91 - \frac{5.27}{2} = 55.28 \text{ in.} = 4.606 \text{ ft}$$

$$M_n = 1745(4.606) = 8033 \text{ k-ft} > 7131 \text{ k-ft required}$$

The section is therefore satisfactory. Shear is now checked at the 0.25 point in accordance with 1977 AASHTO Article 1.6.13.

Again the HS 20 live load and impact shear is compared with the overload shear:

$$1.67 \times \text{HS 20 shear} = 73.5 \text{ k}$$

$$1.0 \times \text{overload shear} = 126.5 \text{ k}$$

and again the overload controls. $\phi = 0.90$. The required ultimate shear then equals

$$\frac{(1.3)}{(0.90)} [\overset{\text{Girder } DL}{18.4} + \overset{\text{Slab + diaphragm } DL}{23.7} + \overset{\text{Superimposed } DL}{4.7} \overset{LL + I}{+ 182.2}] = 249.6 \text{ k}$$

Prestress shear component = 27.0 k

$$V_c = 180 \, (bjd)$$

Assume $j = 0.90$, $d = 0.80 \times$ total depth.

$$V_c = 180(8)(0.90)(0.80)(62.5) = 64.8 \text{ k}$$

$$V_u - V_c = 249.6 - 27.0 - 64.8 = 157.8 \text{ k}$$

Assume 2-#4 stirrup legs, $A_v = 0.40$ in.^{2}, $f_y = 60$ ksi. The required spacing is

$$s = \frac{2A_v f_y jd}{V_u - V_c} = \frac{2(0.40)(60)(0.90)(0.80)(62.5)}{157.8} = 13.7 \text{ in.}$$

The maximum spacing according to AASHTO Article 1.6.13 is 240/8 = 30 in. and the tie requirements are #4 @ 1'-10". Therefore use #4 @ 12 in. The required end block reinforcement to restrain bursting stresses according to AASHTO Article 1.6.15 is

$$A_s = \frac{0.04(1272.0)}{20} = 2.54 \text{ in.}^2$$

Use six legs of #6. This reinforcing must be placed in a distance $d/4$ from the end of the beam:

$$\frac{d}{4} = 0.80(54)(\tfrac{1}{4}) = 10.8 \text{ in.}$$

Elastic shortening is checked along the girder bottom fiber because it is desirable to compensate for the shortening in the bearing area.

The average initial f_b = +2582 psi.
The length of girder = 1126 in.
Assume E_{ci} = 57000$\sqrt{4500}$ = 3,824,000 psi.
Creep = 2.25 × (elastic shortening).
Assume that 50% of creep takes place before erection.
The initial shortening ΔL = [2582(1126)]/3,824,000 = 0.7603 in.

The shortening at erection is (0.7603) + 0.50(2.25 - 1.0)(0.7603) = 1.236 in. Therefore cast girders $1\frac{1}{4}$ in. longer to account for elastic shortening and creep. Last, it is necessary to compute the anticipated girder camber and dead load deflections to set the concrete fillets correctly between the girder and slab.

First compute camber due to prestress by examining the moment diagram for the prestress for half the symmetric beam in Figure 8.9. Assume that EI = 1 and, using a conjugate beam analysis, compute the midspan deflection by summing the moments about the ₵ span of the conjugate beam:

$$
\begin{array}{rl}
958 \times 30.82 = 29,526 \times 29.82 = & -880,452 \\
\tfrac{1}{2} \times 1093 \times 30.81 = 16,843 \times 24.27 = & -408,783 \\
2051 \times 5.00 = 10,255 \times 11.50 = & -117,932 \\
\tfrac{1}{2} \times 84 \times 5.00 = 210 \times 10.67 = & 2,241 \\
2135 \times 9.00 = \overline{19,215} \times 4.50 = & -86,468 \\
& \overline{76,049} \times 44.82 = +3,408,516 \\
& \overline{1,912,640}
\end{array}
$$

Note that the end reaction is equal to the weight of half the span where the actual initial E = 550,700 ksf and the actual I = 12.5738 ft^4. Therefore the initial prestress camber is

$$ \Delta_p = \frac{1,912,640}{(12.5738)(550,700)} = 0.2762 \text{ ft (upward)} $$

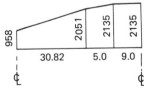

Support   Span   **Figure 8.9**   Prestress moment diagram.

Using $\Delta = 5wl^4/384EI$ for a uniform load deflection, we find that the girder dead load deflection is

$$\Delta_G = \frac{5(0.822)(89.64)^4}{384(550,700)(12.5738)} = 0.0998 \text{ ft}$$

With creep characteristics similar to those assumed for girder shortening the anticipated camber at erection is

$$\Delta_{p-g} = (0.2762 - 0.0998) + \tfrac{1}{2}(2.25 - 1.0)(0.2762 - 0.0998)$$
$$= 0.2867 \text{ ft } (3\tfrac{1}{2} \text{ in.})$$

The assumed deflection due to the diaphragm at midspan equals $PL^3/48EI$:

$$\Delta_D = \frac{3.35(89.64)^3}{48(580,400)(12.5738)} = 0.0069 \text{ ft}$$

The girder deflection due to slab DL is

$$\Delta_s = \frac{5(0.981)(89.64)^4}{384(580,400)(12.5738)} = 0.1130 \text{ ft}$$
$$\Delta_s + \Delta_D = 0.1199 \text{ ft } (1\tfrac{1}{2} \text{ in.})$$

If the girder fillet is zero at midspan, the required fillet at the bearings can be anticipated as $3\tfrac{1}{2} - 1\tfrac{1}{2}$ in. or 2 in. The camber diagram is shown in Figure 8.10.

Example 8.2. Continuous pretensioned I-beam design. The roadway section for the two-span I-beam design is shown in Figure 8.11. The I-beams are erected as simple spans for the girder and slab dead loads

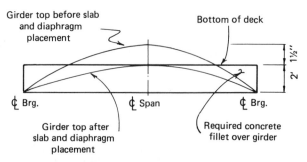

Figure 8.10   Camber diagram.

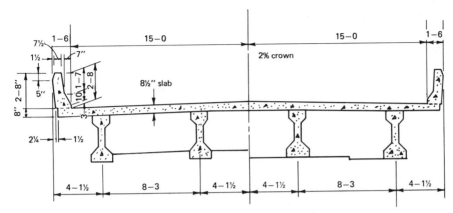

Figure 8.11 Typical roadway section.

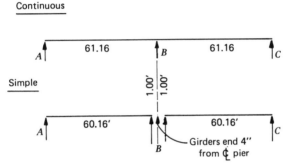

Figure 8.12 Span arrangements.

and are made continuous for any additional superimposed dead load
and live load plus impact loadings.

Individual elastomeric bearings are provided at the pier as separate
simple span supports; therefore the simple spans are slightly shorter than
the continuous spans. The span arrangements are shown in Figure 8.12.

Note.   Tenth points are located on the continuous spans.
Live loading, girder type, and material strengths are as follows:

HS 20 Live Load, AASHTO Type III Girders

$f'_c = 5000$ psi $\left.\right]$ Prestress
$f'_{ci} = 4500$ psi

270 k $\frac{1}{2}\phi$ strands
$f'_c = 3300$ psi
$f_c = 1320$ psi $\left.\right]$ Reinforced
$f_s = 24,000$ psi
$n = 9$

The transversely reinforced deck slab design is explained in Chapter 4. The section properties for the AASHTO Type III girder are taken from Appendix S:

Girder Only Type III

$$A = 560 \text{ in.}^2 \qquad W = 583 \text{ plf}$$
$$I = 125{,}000 \text{ in.}^4 \qquad S_b = 6190 \text{ in.}^3$$
$$y_b = 20.26 \text{ in.} \qquad S_t = 5069 \text{ in.}^3$$
$$y_t = 24.74 \text{ in.}$$

The section moduli are converted into moment multipliers to obtain stresses.

Bottom Fiber

$$\frac{12{,}000}{S_b} = 1.939 \left(\frac{\text{lb-in.}}{\text{k-ft}} \times \frac{1}{\text{in.}^3} \right)$$

Top Fiber

$$\frac{12{,}000}{S_t} = 2.367 \left(\frac{\text{lb-in.}}{\text{k-ft}} \times \frac{1}{\text{in.}^3} \right)$$

The center-to-center girder spacing used as the effective flange width for the composite section is modified for the different concrete modulus of elasticity between the girder and the slab.

Slab

$$f'_c = 3300 \text{ psi}$$
$$E_c = 3{,}274{,}000 \text{ psi}$$

Girder

$$f'_c = 5000 \text{ psi}$$
$$E_c = 4{,}031{,}000 \text{ psi}$$
$$n = \frac{3274}{4031} = 0.812$$

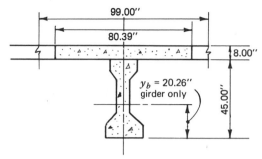

Figure 8.13   Composite section (positive moment region).

Reduced Width

$$0.812 \times 99.0$$

$$80.39$$

The composite section in which the slab is in compression is shown in Figure 8.13. The section properties for this positive moment section and the moment multipliers for stress are computed by using $I_T = \Sigma(I_0 + A_y^{-2})$.

Element	A	y_b	Ay_b	$\bar{y}$	A_y^{-2}	I_0	I_T
Slab	643	49.00	31507	13.38	115,113	3,430	118,543
Girder	560	20.26	11346	15.36	132,121	125,400	257,521
	1203		42853				376,064

$$y_b = 35.62 \text{ in.} \qquad S_b = 10{,}558 \text{ in.}^3$$

$$y_t = 9.38 \text{ in.} \qquad S_t = 40{,}092 \text{ in.}^3$$

$$y_{\text{slab}} = 17.38 \text{ in.} \qquad S_{\text{slab}} = 21{,}638 \text{ in.}^3$$

$$\frac{12{,}000}{S_b} = 1.137 \left(\frac{\text{lb-in.}}{\text{k-ft}} \times \frac{1}{\text{in.}^3} \right)$$

$$\frac{12{,}000}{S_t} = 0.299 \left(\frac{\text{lb-in.}}{\text{k-ft}} \times \frac{1}{\text{in.}^3} \right)$$

$$\frac{12{,}000}{S_{\text{slab}}} = 0.555 \times \overset{n}{0.812} = 0.450 \left(\frac{\text{lb-in.}}{\text{k-ft}} \times \frac{1}{\text{in.}^3} \right)$$

In the negative moment areas the deck slab is in tension; therefore the nominal deck reinforcing is considered as acting compositely with the

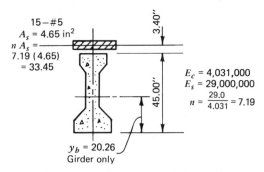

Figure 8.14   Composite section (negative moment region).

girder. This negative moment composite section is shown in Figure 8.14. The section properties for this negative composite moment section, with the moment multipliers for stress, are computed by using $I_T = \Sigma(I_0 + A_y^{-2})$.

Element	A	y_b	Ay_b	$\bar{y}$	A_y^{-2}	I_0	I_T
Reinforcement	33	48.40	1,597	26.57	23,297	—	23,297
Girder	560	20.26	11,346	1.57	1,380	125,400	126,780
	593		12,943				150,077

$$y_b = 21.83 \text{ in.} \qquad S_b = 6875 \text{ in.}^3$$

$$y_t = 23.17 \text{ in.} \qquad S_t = 6477 \text{ in.}^3$$

$$y_{st} = 26.57 \text{ in.} \qquad S_{st} = 5648 \text{ in.}^3$$

$$\frac{12,000}{S_b} = 1.746 \left(\frac{\text{lb-in.}}{\text{k-ft}} \times \frac{1}{\text{in.}^3} \right)$$

$$\frac{12,000}{S_t} = 1.853 \left(\frac{\text{lb-in.}}{\text{k-ft}} \times \frac{1}{\text{in.}^3} \right)$$

$$\frac{12,000}{S_{st}} = 2.124 \times 7.19 = 15.275 \left(\frac{\text{lb-in.}}{\text{k-ft}} \times \frac{1}{\text{in.}^3} \right)$$

Next, it is necessary to tabulate the moments due to the different increments of loading at the tenth points and in the negative moment region at the strand transfer point for stress computations (2.25 ft from the end of the beam or $0.958AB$ on the continuous span). Also in the negative moment region these moments should be calculated at the strand development point for capacity computations (9.0 ft from the end of the beam or $0.847AB$ on the continuous spans). These lengths are

Table 8.1   Girder Dead Load[a]

Point	x	$l - x$	$M = \dfrac{w}{2}(x)(l-x)$ (k-ft)
A			
0.1	6.12	54.04	+96
0.2	12.23	47.93	+171
0.3	18.35	41.81	+224
0.4	24.46	35.70	+255
0.5	30.58	29.58	+264
0.6	36.70	23.46	+251
0.7	42.81	17.35	+217
0.8	48.93	11.23	+160
0.9	55.04	5.12	+82
0.958	58.58	1.58	+27
0.847	51.83	8.33	+126

[a] $W = 583$ plf.

based on a generalization applied to the equations given in ref (3). The girder dead load is computed on the simple span and tabulated in Table 8.1. For an exterior girder the slab and diaphragm dead loads are computed by using a concrete weight equal to 150 pcf. (note that 6 lb-ft^{-1} is included as an approximation of the girder fillet).

Slab Dead Load (Exterior Girder)

$$W = \quad 0.667 \times 7.94 \text{ ft} \times 150 = 794$$
$$\tfrac{1}{2} \times 0.042 \times 3.15 \text{ ft} \times 150 = \quad 10 \quad \} \text{ sum} = 810 \text{ lb-ft}^{-1}$$
$$0.03 \times 1.33 \times 150 = \quad 6$$

Diaphragm Dead Load. $P = 3.17 \times 0.75 \times 3.83 \times 0.150 = 1.4$ k @ midspan. For the exterior girder the slab and diaphragm dead load moments are computed for the simple span and then tabulated and combined in Table 8.2. The slab and diaphragm dead loads are computed for the interior girder:

Slab Dead Load (Interior Girder)

$$W = 0.667 \times 8.25 \text{ ft} \times 150 = 825$$
$$0.03 \times 1.33 \text{ ft} \times 150 = \quad 6 \quad \} \text{ sum} = 831 \text{ lb-ft}^{-1}$$

Table 8.2

Point	M_{slab}	$M_{\text{diaphragm}}$	M_{S+D} (ext) (k-ft)
A			
0.1	+133	+4	+137
0.2	+238	+8	+246
0.3	+311	+13	+324
0.4	+354	+17	+371
0.5	+367	+21	+388
0.6	+349	+17	+366
0.7	+301	+13	+314
0.8	+222	+8	+230
0.9	+114	+4	+118
0.958	+38	+2	+40
0.847	+175	+6	+181

Diaphragm Dead Load (Interior Girder)

$$P = 2 \times 1.4 = 2.8 \text{ k @ midspan}$$

The slab and diaphragm dead load moments are then computed, tabulated, and summed for the interior girder in Table 8.3. The superimposed dead load for the exterior girder consists of a parapet dead load of 366 plf and 25 psf of wearing surface:

Superimposed DL (Exterior Girder)

$$W = 6.75 \times 25 = \left.\begin{matrix} 169 \\ 366 \end{matrix}\right\} \text{ sum} = 535 \text{ lb-ft}^{-1}$$

Table 8.3

Point	M_{slab}	$M_{\text{diaphragm}}$	M_{S+D} (int) (k-ft)
A			
0.1	+137	+8	+145
0.2	+244	+16	+260
0.3	+319	+26	+345
0.4	+363	+34	+397
0.5	+376	+42	+418
0.6	+358	+34	+392
0.7	+309	+26	+335
0.8	+228	+16	+244
0.9	+117	+8	+125
0.958	+38	+4	+42
0.847	+180	+12	+192

The superimposed dead load moments are taken from the tables in ref. (4) for two equal continuous spans and computed and tabulated in Table 8.4. For use in the foregoing tables, compute $Wl^2 = 0.535(61.16)^2 = 2001$ k-ft.

For the superimposed dead load on the interior girder the wearing surface attributed to that girder is the only item. For the interior girder the load and tabulated moments are listed in Table 8.5.

Superimposed DL (Interior Girder). $w = 8.25 \times 25 = 206$ (plf)

<table>
<tr><td colspan="2" align="center">Table 8.4</td><td colspan="2" align="center">Table 8.5</td></tr>
<tr><td>Point</td><td>M_{SDL} (ext) (k-ft)</td><td>Point</td><td>M_{SDL} (int) (k-ft)</td></tr>
<tr><td>A</td><td></td><td>A</td><td></td></tr>
<tr><td>0.1</td><td>+65</td><td>0.1</td><td>+25</td></tr>
<tr><td>0.2</td><td>+110</td><td>0.2</td><td>+42</td></tr>
<tr><td>0.3</td><td>+135</td><td>0.3</td><td>+52</td></tr>
<tr><td>0.4</td><td>+140</td><td>0.4</td><td>+54</td></tr>
<tr><td>0.5</td><td>+125</td><td>0.5</td><td>+48</td></tr>
<tr><td>0.6</td><td>+90</td><td>0.6</td><td>+35</td></tr>
<tr><td>0.7</td><td>+35</td><td>0.7</td><td>+13</td></tr>
<tr><td>0.8</td><td>-40</td><td>0.8</td><td>-15</td></tr>
<tr><td>0.9</td><td>-135</td><td>0.9</td><td>-52</td></tr>
<tr><td>0.958</td><td>-202</td><td>0.958</td><td>-78</td></tr>
<tr><td>B</td><td>-250</td><td>B</td><td>-96</td></tr>
<tr><td>0.847</td><td>-84</td><td>0.847</td><td>-32</td></tr>
</table>

The influence lines for two equal continuous spans can be found in many references. Listed in Table 8.6 are the ordinates and areas obtained from ref. (4). The tabulations shown are arranged in columns to read vertically to give influence lines and horizontally to give moment diagrams. The areas under the influence lines are also shown. These ordinates are plotted in Figure 8.15. The impact factor is $I = 50/(61.16 + 125) = 0.269$ according to AASHTO Article 1.2.12(c). The live load distribution is computed according to AASHTO Article 1.3.1, and for the exterior girder the simple beam reaction is computed as shown in Figure 8.16. The multiplier coefficient for combined impact and distribution becomes

$$\text{coefficient} = 1.426(1.269) = 1.809$$

AASHTO table 1.3.1(B) shows the live load distribution to the interior girder to be $S/5.5$, where S is girder spacing. The interior girder live load distribution is

$$DF = \frac{8.25}{5.5} = 1.500 \text{ lines of wheels}$$

Table 8.6   Influence Line Ordinates for Moments[a]

Load @	0.1	0.2	0.3	0.4	0.5	0.6	0.7	0.8	0.9	B
A										
0.1	+5.35	+4.59	+3.83	+3.06	+2.30	+1.54	+0.78	+0.01	-0.75	-1.52
0.2	+4.60	+9.20	+7.68	+6.16	+4.65	+3.13	+1.61	+0.10	-1.42	-2.94
0.3	+3.87	+7.73	+11.59	+9.34	+7.09	+4.84	+2.58	+0.33	-1.92	-4.18
0.4	+3.16	+6.31	+9.47	+12.62	+9.66	+6.70	+3.74	+0.78	-2.18	-5.14
0.5	+2.48	+4.97	+7.46	+9.94	+12.42	+8.79	+5.16	+1.53	-2.10	-5.74
0.6	+1.86	+3.72	+5.58	+7.44	+9.30	+11.16	+6.90	+2.64	-1.61	-5.87
0.7	+1.29	+2.58	+3.87	+5.16	+6.45	+7.74	+9.02	+4.20	-0.63	-5.46
0.8	+0.78	+1.57	+2.35	+3.13	+3.91	+4.70	+5.48	+6.26	+0.93	-4.40
0.9	+0.35	+0.70	+1.05	+1.40	+1.75	+2.10	+2.45	+2.80	+3.15	-2.62
B										
+ Area	+145	+252	+323	+355	+351	+309	+229	+112	+23	0
- Area	-24	-47	-70	-94	-117	-140	-164	-187	-275	-468
Σ Area	+121	+205	+253	+261	+234	+169	+65	-75	-252	-468

[a]Moments are in k-ft.

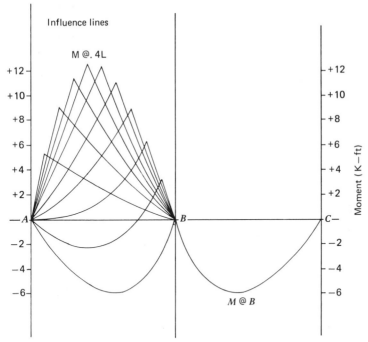

Figure 8.15   Influence line for moment.

The multiplier coefficient for combined live load plus impact on the interior girder is

$$\text{coefficient} = 1.500\,(1.269) = 1.904$$

Experience indicates that for shorter span ranges the HS 20 truck loading controls for positive moments and the HS 20 truck or lane controls for negative moments.

The following tabulations show the influence line ordinates for the various truck wheel loadings, the moment computed for one line of wheels, and the moment multiplied by the distribution plus impact coefficient to give the exterior and interior girder live load moments plus impact. The positive ordinates that give the maximum positive moment at the tenth points are selected first. The $0.958\,AB$ and

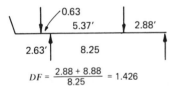

$$DF = \frac{2.88 + 8.88}{8.25} = 1.426$$

Figure 8.16   Exterior girder distribution factor.

Table 8.7

| | Positive Moment HS 20 Truck | | | One-Line Wheels | Distribution Plus Impact Coefficient | |
| | | | | | 1.809 (ext) | 1.904 (int) |
Point	16 k	16 k	4 K	(M_{LL})	(M_{LL+I})	(M_{LL+I})
A						
0.1	+5.35	+3.6	+2.0	+151	+273	+288
0.2	+9.20	+5.8	+3.1	+252	+456	+480
0.3	+11.59	+6.9	+3.0	+308	+557	+586
0.4	+12.62	+6.9	+5.0	+332	+601	+632
0.5	+12.42	+6.6	+5.8	+328	+593	+625
0.6	+11.16	+6.1	+4.1	+293	+530	+558
0.7	+9.02	+6.8	+3.1	+266	+481	+506
0.8	+6.26	+2.3	+0.6	+139	+251	+265
0.9	+3.15	—	—	+50	+90	+95
B						
0.958				+21	+38	+40
0.847				+97	+175	+185

0.847AB points are interpolated to obtain the maximum positive moment. This tabulation (in k-ft) is given in Table 8.7.

Note that the positive moment for this two-span bridge is zero at the support. For a three-span bridge positive moment can be developed at the support when the live loading is two spans away. This could result in a net positive moment because the negative dead load moment would be small and the result would be required reinforcement extending across the support between the girder ends.

The maximum negative truck loading is computed in a manner similar to that for positive moment. Note that only the ordinates for the moment at the interior support require influence line tabulation and all the other moments are interpolated down to zero at the exterior support. This is due to the fact that the same loading produces the maximum negative moment at all points. The maximum negative truck moments are tabulated (in k-ft) in Table 8.8.

The lane loadings are computed in a similar manner, except that areas are used for the uniform loads and ordinates for the concentrated loads. The maximum HS 20 lane-load-induced negative moments (k-ft) are shown in Table 8.9. Note that lane loading controls at point B and 0.958AB only and that HS truck loading controls elsewhere.

Now that all moments are computed the stresses can be evaluated for interior and exterior girders, top and bottom fiber, and the different increments of loading (Tables 8.10 and 8.11). Superimposed loads on the composite section are applied to the positive moment section (section with slab) and to the negative moment section (section composite

Table 8.8

| | Negative Moment HS 20 Truck | | | One-Line Wheels (M_{LL}) | Distribution Plus Impact Coefficient | |
| | | | | | 1.809 (ext) (M_{LL+L}) | 1.904 (int) (M_{LL+I}) |
Point	16 k	16 k	4 k			
A						
0.1				−20	−36	−38
0.2				−40	−72	−76
0.3				−60	−108	−114
0.4				−80	−145	−152
0.5				−100	−181	−190
0.6				−121	−219	−230
0.7				−141	−255	−268
0.8				−161	−291	−307
0.9				−181	−327	−345
B	−5.8	−5.8	−3.9	−201		
0.958				−193		
0.847				−170	−308	−324

with reinforcing). This is done for maximum positive and maximum negative live load plus impact. For each point we determine whether the slab is in tension or compression and the proper section is then selected by eliminating the stresses that would be applied to the wrong section. Stresses are computed by using the section multipliers shown at the top of the columns in Table 8.10.

Table 8.9

| | Negative Moment HS 20 Lane | | | One-Line Wheels (M_{LL}) | 1.809 (ext) (M_{LL+I}) | 1.904 (int) (M_{LL+I}) |
Point	0.32 k-ft	9 k	9 k			
A						
0.1				−13		
0.2				−26		
0.3				−38		
0.4				−51		
0.5				−64		
0.6				−77		
0.7				−89		
0.8	−187		−4.70	−102		
0.9	−275	−2.18	−5.28	−155		
B	−468	−5.87	−5.87	−255	−461	−486
0.958				−213	−385	−406
0.847				−127		

Table 8.10 Exterior Girder Stresses

Bottom Fiber

Point	Girder	Slab + Diaphragm	Positive Moment Section			Negative Moment Section		
	1.939		1.137			1.746		
			SDL	+LL + I	−LL + I	SDL	+LL + I	−LL + I
A								
0.1	−186	−265	−74	−310	+41			
0.2	−332	−477	−125	−518	+82			
0.3	−434	−628	−153	−633	+123			
0.4	−494	−719	−159	−683		−244		+253
0.5	−512	−752	−142	−674		−218		+316
0.6	−487	−710	−102	−603		−157		+382
0.7	−421	−609	−40	−547		−61		+445
0.8	−310	−446	+45	−285		+70		+508
0.9	−159	−229				+236	−157	+571
0.958	−52	−78				+353	−66	+672
B								

Top Fiber

Point	Girder	Slab + Diaphragm	Positive Moment Section			Negative Moment Section		
	2.367		0.299			1.853		
			SDL	+LL + I	−LL + I	SDL	+LL + I	−LL + I
A								
0.1	+227	+324	+19	+82	−11			
0.2	+405	+582	+33	+136	−22			
0.3	+530	+767	+40	+167	−32			
0.4	+604	+878	+42	+180		+259		−269
0.5	+625	+918	+37	+177		+232		−335
0.6	+594	+866	+27	+158		+167		−406
0.7	+514	+743	+10	+144		+65		−473
0.8	+379	+544	−12	+75		−74		−539
0.9	+194	+279				−250	+167	−606
0.958	+64	+95				−374	+70	−713
B								

Table 8.11 Interior Girder Stresses

Bottom Fiber

		1.939	Positive Moment Section 1.137			Negative Moment Section 1.746		
Point	Girder	Slab + Diaphragm	SDL	+LL + I	-LL + I	SDL	+LL + I	-LL + I
A								
0.1	-186	-281	-28	-327		-44		+66
0.2	-332	-504	-48	-546		-73		+133
0.3	-434	-669	-59	-666		-91		+199
0.4	-494	-770	-61	-719		-94		+265
0.5	-512	-811	-55	-711		-84		+332
0.6	-487	-760	-40	-634		-61		+402
0.7	-421	-650	-15	-575		-23		+468
0.8	-310	-473	+17	-301		+26		+536
0.9	-159	-242	+59	-108		+91		+602
0.958	-52	-81						+709
B						+136	-70	

Top Fiber

		2.367	Positive Moment Section 0.299			Negative Moment Section 1.853		
Point	Girder	Slab + Diaphragm	SDL	+LL + I	-LL + I	SDL	+LL + I	-LL + I
A								
0.1	+227	+343	+7	+86		+46		-70
0.2	+405	+615	+13	+144		+78		-141
0.3	+530	+817	+16	+175		+96		-211
0.4	+604	+940	+16	+189		+100		-282
0.5	+625	+989	+14	+187		+89		-352
0.6	+594	+928	+10	+167		+65		-426
0.7	+514	+793	+4	+151		+24		-497
0.8	+379	+578	-4	+79		-28		-569
0.9	+194	+296	-16	+28		-96		-639
0.958	+64	+99						-752
B						-145	+74	

The required prestress force is determined on the basis of the mid-span requirements; therefore the sum of the stresses due to externally applied loads at midspan is examined first.

Stresses @ 0.5

Exterior Girder

$$f_b = -512 - 752 - 142 - 674 = -2080 \text{ psi}$$

$$f_t = +625 + 918 + 37 + 177 = +1757 \text{ psi}$$

Interior Girder

$$f_b = -512 - 811 - 55 - 711 = -2089 \text{ psi}$$

$$f_t = +625 + 989 + 14 + 187 = +1815 \text{ psi}$$

Because the prestress force is placed as close to the bottom fiber as possible, it is reasonable to assume that the eccentricity of the prestress force will approximate 17.5 in.±. The allowable tensile stress on the bottom fiber = $4\sqrt{5000} = 424$ psi and the allowable compressive stress on the top fiber equals $0.4(5000) = 2000$ psi in accordance with AASHTO Article 1.6.6.

By using the allowables and assumed eccentricity a required prestress force and the approximate number of strands required can be determined. This is based on a working level after losses [45,000 psi per AASHTO Article 1.6.7(B)(2)] of 22.05 k per strand:

$$f_b = -424 = \frac{P}{560} \pm \frac{P(17.5)}{6190} - 2089$$

$$P = 360.9 \text{ k}$$

$$\text{number of strands required} = \frac{360.9}{22.05}$$

$$= 16.4 \text{ strands}$$

$$f_t = +2000 \text{ will not control}$$

Try 18 $\frac{1}{2}$-in. ϕ strands. By placing the 18 strands as close as possible to the bottom fiber a strand pattern can be selected with one bundle of bundled strands (Figure 8.17). The eccentricity is computed by locating the cgs in relation to the bottom fiber and subtracting this distance

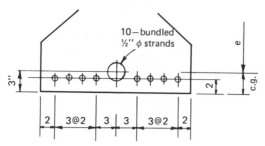

Figure 8.17   Strand pattern.

from the girder $\bar{y}_b$ value

$$cg = \frac{8(2) + 10(3)}{18}$$

$$= 2.56 \text{ in.}$$

$$e = 20.26 - 2.56$$

$$= 17.70 \text{ in.}$$

For $18\frac{1}{2}$-in ϕ strands the initial strand value equals $0.7f'_s$. The initial strand value minus a 19,000-psi allowance for immediate elastic shortening is used for initial stress computation. Again, the final strand value is the initial value minus 45,000 psi for all losses. Therefore the prestress forces to be used for design are

$$P_i = 18(28.91) = 520.4 \text{ k}$$

$$P_{i-ES} = 18(26.04) = 468.8 \text{ k}$$

$$E_s = 18(22.05) = 396.9 \text{ k}$$

On the basis of this prestress final force we compute the required eccentricity at $0.958AB$. The external induced stresses are summed:

Exterior Girder

$$f_b = -52 - 78 + 353 + 672 = +895 \text{ psi}$$

$$f_t = +64 + 95 - 374 - 713 = -928 \text{ psi}$$

Interior Girder

$$f_b = -52 - 81 + 136 + 709 = +712 \text{ psi}$$

$$f_t = +64 + 99 - 145 - 752 = -734 \text{ psi}$$

The allowable top fiber tension once again is $4\sqrt{5000}$, or 424 psi, and the allowable bottom fiber compression is 0.6(5000) = +3000 psi, based on AASHTO Article 1.6.12(C)(4). Using the final prestress force and these allowable stresses, we compute the required eccentricity:

$$f_b = +3000 = \frac{396.9}{560} + \frac{396.9(e)}{6190} + 895 \qquad \therefore e = 21.77 \text{ in.}$$

$$f_t = -424 = \frac{396.9}{560} - \frac{396.9(e)}{5069} - 928 \qquad \therefore e = +2.62 \text{ in.}$$

The strands are then laid out to deflect enough of them into the negative moment area to attain at least $e = +2.62$ in. at 0.958AB. For end spans the deflecting points should as a rule be 0.35 when enough strands are deflected toward the free end to provide acceptable stresses near the end bearing and 0.6 when deflecting toward the negative moment area. For interior spans of continuous units the deflecting points are usually at 0.40 and 0.60.

A good allowable for limiting the vertical uplift component of the initial force is 28 kips. If the uplift exceeds this value, it is recommended that the holddowns be staggered by 5 ft 0 in. This keeps the forces on the prestress bed within the capability of most prestressing plants.

These principles are applied to this problem and the strands are laid out as shown in Figure 8.18. The cgs, cgc, and eccentricities are plotted in Figure 8.19. The cgs slope and investigative points are located in Figure 8.19 in preparation for the computation of stresses due to prestress force. To compute the prestress stress at the investigative points the eccentricity at each point is first tabulated (Tables 8.12 and 8.13), Based on $P_f = 396.9$ k, $P_{i-ES} = 468.8$ k, the prestress values are computed. The initial minus elastic shortening (Int – ES) prestress stresses

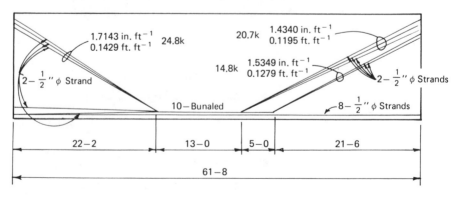

Figure 8.18   Girder elevation strand layout.

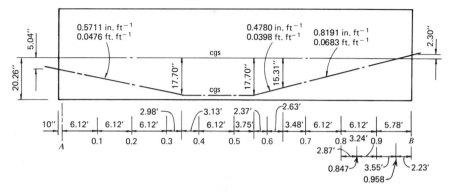

Figure 8.19   Girder layout: cgs profile.

are combined with the girder dead load stresses to give the initial stresses to be compared with the AASHTO allowables in Table 8.13.

According to AASHTO Article 1.6.6(B)(1) the allowable stresses are

$$\text{allowable compression} = 0.6(4500) = 2700 \text{ psi}$$

$$\text{allowable tension} = 200 \text{ psi w/o reinforcement}$$

$$\text{allowable tension} = 503 \text{ psi w/reinforcement}$$

Because the allowable for tension without reinforcement is exceeded, the required amount of reinforcing to replace the concrete tension

Table 8.12

Point	$\dfrac{P_f}{A}$	e	$\dfrac{P_f e}{S_b}$	$\dfrac{P_f e}{S_t}$	Final f_b (psi)	Final f_t (psi)	Int – ES f_b (psi)	Int – ES f_t (psi)
A	+709	+5.52	+354	−432	+1063	+277	+1256	+327
0.1	↑	+9.01	+578	−706	+1287	+3	+1520	+4
0.2		+12.51	+802	−980	+1511	−271	+1785	−320
0.3		+16.00	+1026	−1253	+1735	−544	+2049	−643
0.4		+17.70	+1135	−1387	+1844	−678	+2178	−801
0.5		+17.70	+1135	−1387	+1844	−678	+2178	−801
0.6		+16.57	+1062	−1297	+1771	−588	+2092	−695
0.7		+12.46	+799	−976	+1508	−267	+1781	−315
0.8		+7.45	+478	−583	+1178	+126	+1391	+149
0.9	↓	+2.43	+156	−190	+865	+519	+1022	+613
0.958	+709	−0.47	−30	+37	+679	+746	+802	+881
B								
0.847		+5.09						

Table 8.13   Initial Stresses

	Bottom Fiber			Top Fiber		
Point	Prestress	Girder	f_b (psi)	Prestress	Girder	f_t (psi)
A	+1256	-186	+1070	+327	+227	+554
0.1	+1520	-332	+1188	+4	+405	+409
0.2	+1785	-434	+1351	-320	+530	+210
0.3	+2049	-494	+1555	-643	+604	-39
0.4	+2178	-512	+1666	-801	+625	-176
0.5	+2178	-487	+1691	-801	+594	-207
0.6	+2092	-421	+1671	-695	+514	-181
0.7	+1781	-310	+1471	-315	+379	+64
0.8	+1391	-159	+1232	+149	+194	+343
0.9	+1022	-52	+970	+613	+64	+677
0.958 B	+802		+802	+881		+881

force is computed (Figure 8.20).   The summation of stresses showed that the interior girder controlled for positive moment; therefore all stresses are summed for the interior girder and compared with the allowable stresses in Table 8.14.   This summation is provided by using the loadings that produce maximum positive moment.

The exterior girder controls similarly for negative moment; therefore the sum of all stresses for the exterior girder is compared with the allowable stresses (Table 8.15).   This summation is applied to the loadings that produce maximum negative moment.

Now that the stresses have been checked the required ultimate moments are computed based on factored service load moments.   The superimposed dead load moment is not factored here because it is a moment that results from future loading.   The additional 0.3 (girder and slab) dead load is applied to the continuous unit, which is realistic for any future loadings.   These factored moments are listed in Table 8.16, for maximum positive (Figure 8.21) and maximum negative (Fig-

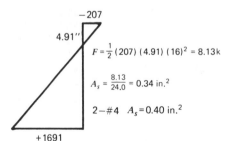

$$F = \frac{1}{2}(207)(4.91)(16)^2 = 8.13k$$

$$A_s = \frac{8.13}{24.0} = 0.34 \text{ in.}^2$$

$$2-\#4 \quad A_s = 0.40 \text{ in.}^2$$

Figure 8.20   Tensile reinforcement computation.

Table 8.14 Final Stresses: Interior Girder, Positive Moment

Bottom Fiber

	0.1	0.2	0.3	0.4	0.5	0.6	0.7	0.8	0.9	0.958
Prestress	+1287	+1511	+1735	+1844	+1844	+1771	+1508	+1178	+865	+679
Girder	-186	-332	-434	-494	-512	-487	-421	-310	-152	-52
Σ	+1101	+1179	+1301	+1350	+1332	+1284	+1087	+868	+713	+627
S + D	-281	-504	-669	-770	-811	-760	-650	-473	-242	-81
Σ	+820	+675	+632	+680	+521	+524	+437	+395	+471	+546
SDL	-28	-48	-59	-61	-55	-40	-15	+17	+59	+136
Σ	+792	+627	+573	+619	+466	+484	+422	+412	+530	+682
LL + I	-327	-546	-666	-719	-711	-634	-575	-301	-108	-70
Σ	+465	+81	-93	-100	-245	-150	-153	+111	+422	+612

Top Fiber

	0.1	0.2	0.3	0.4	0.5	0.6	0.7	0.8	0.9	0.958
Prestress	+3	-271	-544	-678	-678	-588	-267	+126	+519	+746
Girder	+227	+405	+530	+604	+625	+594	+514	+379	+194	+64
Σ	+230	+134	-14	-74	-53	+6	+247	+505	+713	+810
S + D	+343	+615	+817	+940	+989	+928	+793	+578	+296	+99
Σ	+573	+749	+803	+866	+936	+934	+1040	+1083	+1009	+909
SDL	+7	+13	+16	+16	+14	+10	+4	-4	-16	-145
Σ	+580	+762	+819	+882	+950	+944	+1044	+1079	+993	+764
LL + I	+86	+144	+175	+189	+187	+167	+151	+79	+28	+74
Σ	+666	+906	+994	+1071	+1137	+1111	+1195	+1158	+1021	+838

Table 8.15 Final Stresses: Exterior Girder, Negative Moment

Bottom Fiber

	0.1	0.2	0.3	0.4	0.5	0.6	0.7	0.8	0.9	0.958
Prestress Girder										
Σ	+1101	+1179	+1301	+1350	+1332	+1284	+1087	+868	+713	+627
$S+D$	-265	-477	-628	-719	-752	-710	-609	-446	-229	-78
Σ	+836	+702	+673	+631	+580	+574	+478	+422	+484	+549
SDL	-74	-125	-153	-244	-218	-157	-61	+70	+236	+353
Σ	+762	+577	+520	+387	+262	+417	+417	+492	+720	+902
$LL+I$	+41	+82	+123	+253	+316	+382	+445	+508	+571	+672
Σ	+803	+659	+643	+640	+578	+799	+862	+1000	+1291	+1574

Top Fiber

	0.1	0.2	0.3	0.4	0.5	0.6	0.7	0.8	0.9	0.958
Prestress Girder										
Σ	+230	+134	-14	-74	-53	+6	+247	+505	+713	+810
$S+D$	+324	+582	+767	+878	+918	+866	+743	+544	+279	+95
Σ	+554	+716	+753	+804	+865	+872	+990	+1049	+992	+905
SDL	+19	+33	+40	+259	+232	+167	+65	-74	-250	-374
Σ	+573	+749	+793	+1063	+1097	+1039	+1055	+975	+742	+531
$LL+I$	-11	-22	-32	-269	-335	-406	-473	+539	-606	-713
Σ	+562	+727	+761	+794	+762	+633	+582	+436	+136	-182

Table 8.16 Interior Girder

Point	Girder	Slab + Diaphragm	0.3 Girder + Slab[a]	SDL	$2.17 \times - LL + I$	$2.17 \times - LL + I$	Maximum Positive M_{ult}	Maximum Negative M_{ult}
A								
0.1	+96	+145	+51	+25	+625	-82	+942	
0.2	+171	+260	+86	+42	+1042	-165	+1601	
0.3	+224	+345	+107	+52	+1272	-247	+2000	
0.4	+255	+397	+111	+54	+1371	-330	+2188	
0.5	+264	+418	+99	+48	+1356	-412	+2185	
0.6	+251	+392	+72	+35	+1211	-499	+1961	
0.7	+217	+335	+27	+13	+1098	-582	+1690	-30
0.8	+160	+244	-31	-15	+575	-666	+979[b]	-308
0.847	+126	+192	-66	-32	+401	-703	+719[b]	-483
0.9	+82	+125	-107	-52	+206	-749	+413[b]	-701
0.958	+27	+42	-161	-78	+87	-881	+156[b]	-1051
B			-198	-96		-1055		-1349

[a] 0.3[girder + slab] = 0.3(583 + 831) = 424 lb·ft⁻¹.

[b] 0.3[girder + slab] and *SLD* are not included to be conservative.

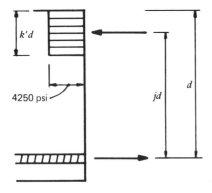

Figure 8.21   Positive ultimate moment stress diagram.

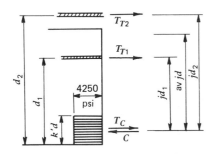

Figure 8.22   Negative ultimate stress diagram.

ure 8.22) ultimate moments. The values are given for the interior girder, which controls for the ultimate moment because of its live load effect. The negative moment at the interior support must be resisted by a reinforced concrete section.

Based on ultimate strength, the required negative moment reinforcing is

$$A_{s(required)} = \frac{1349(12)}{49(60)} = 5.51 \text{ in.}^2$$

The maximum service load moment at B is $M_B = -250 - 461 = -711$ k-ft for an exterior girder. Assuming that $j = 0.87$, the approximate required reinforcing for service load design is

$$A_{s(required)} = \frac{711(12)}{24(0.87)(49.69)} = 8.23 \text{ in.}^2$$

8-#9 $A_s = 8.00$ in.² is close. For 8-#9 $A_s = 8.00$ in.²; the capacity is computed according to Section 5 of AASHTO:

$$a = \frac{A_s f_y}{0.85 f'_c b} = \frac{800(60)}{0.85(3.3)(22)} = 7.78 \text{ in.}$$

$$\phi \, M_n = \phi \left[A_s f_y \left(d - \frac{a}{2} \right) \right]$$

$$\phi \, M_n = 0.90 \left[(8.00)(60) \left(49.69 - \frac{7.78}{2} \right) \right] \left(\frac{1}{12} \right)$$

$$= 1649 \text{ k-ft} > 1349 \text{ k-ft} \quad \text{OK}$$

Using 8–#9 we compute the actual stress:

$$p = \frac{8.00}{22(48.40)} = 0.0075$$

and from Appendix A

$$k = 0.306 \qquad j = 0.898$$

$$f_s = \frac{711(12)}{8.00(0.898)(48.40)}$$

$$= 24.5 \text{ ksi} > 24.0 \text{ ksi, by } 2.2\% \text{ which is close enough}$$

$$f_c = \frac{24.5(0.306)}{9(0.694)} = 1202 \text{ psi} < 1320 \text{ psi} \quad \text{OK}$$

To ensure failure in the steel AASHTO Article 1.6.10(A) limits the value of $p(f'_{su}/f'_c)$ to less than 0.30. The term f'_{su} is the stress in the reinforcement at failure and is given by an equation in AASHTO Article 1.6.9(C). The values of $p(f'_{su}/f'_c)$ are listed in Table 8.17 for positive moment, using the stress diagram in Figure 8.21. The term $p(f'_{su}/f'_c)$ is less than 0.30 in all cases; therefore the criteria are satisfied.

Table 8.17

Point	e	d	p	$0.5p \dfrac{f'_s}{f'_c}$	f'_{su}	$p \dfrac{f'_{su}}{f'_c}$
A						
0.1	+9.01	41.75	0.00082	0.022	264.0	0.043
0.2	+12.51	45.25	0.00076	0.020	264.5	0.040
0.3	+16.00	48.74	0.00070	0.019	264.9	0.037
0.4	+17.70	50.44	0.00068	0.018	265.0	
0.5	+17.70	50.44	0.00068	0.018	265.0	
0.6	+16.57	49.31	0.00070	0.019	264.9	
0.7	+12.46	45.20	0.00076	0.020	264.5	0.040
0.8	+7.45	40.19	0.00085	0.023	263.8	0.045
0.847	+5.09	37.82	0.00091	0.024	263.4	0.048

$b = 80.39$ in., $A_s = 18(0.1531) = 2.7558$ in.^{2}, $f'_s = 270{,}000$ psi, $f'_c = 5000$ psi.

Table 8.18

Point	T (kips)	$k'd$ (in.)	jd (in.)	$\dfrac{jd}{12}$ (ft)	M_n
A					
0.1	728	2.13	40.68	3.39	2468
0.2	729	2.13	44.18	3.68	2684
0.3	730	2.14	47.67	3.97	2900
0.4	730	2.14	49.37	4.11	3003 > 2188
0.5	730	2.14	49.37	4.11	3003 > 2185
0.6	730	2.14	48.24	4.02	2935
0.7	729	2.13	44.13	3.68	2681
0.8	727	2.13	39.12	3.26	2370
0.847	726	2.12	36.77	3.06	2224

[a]Moments are in k-ft.

Again using the stress diagram for positive moment (Figure 8.21) we compute the ultimate capacity as given in Table 8.18. The capacity exceeds the ultimate positive moment. Therefore the section is satisfactory.

For negative capacity moment computation some strands fall in the compression block area. Also the deck reinforcing offers some increased capacity as tensile reinforcement. These items are accounted for in Figure 8.22.

For concrete and prestressing steel the moduli of elasticity are

$$E_c = 4,031,000 \text{ psi}$$
$$E_s = 29,000,000 \text{ psi}$$

The concrete stress in the compression block is

$$f_c = 0.85(5000) = 4250 \text{ psi}$$

based on 28-day concrete strength, $f'_c = 5000$ psi.
The strain in the compression block is

$$E_c = \frac{4250}{4,031,000} = 0.0011 \text{ in.}^{-1}$$

After losses the stress in the prestressing steel is

$$f_s = 0.7(270 \text{ ksi}) - 45 \text{ ksi} = 144 \text{ ksi}$$

The resulting stress for the prestressing strands in the compression block is

$$f'_{su} = 144{,}000 - 0.0011(29{,}000{,}000) = 113{,}400 \text{ psi}$$

The total reinforcing ratio $p = f'_{su}/f'_c + P'(f_y/f'_c)$ must be less than 0.30 to ensure failure in steel.

The ratios $[p(f'_{su}/f'_c)]$ for the tensile prestressing strands is computed and tabulated at the points of critical negative moment listed in Table 8.19.

The ratio $p'(f'_y/f'_c)$ is computed for the conventional nominal deck reinforcing and combined with the values for tensile prestressing steel:

Point	d2	A'_s	$b = 22$ in. (p')	f'_y	$p' \dfrac{f'_y}{f'_c}$	$p \dfrac{f_{su}}{f'_c} + p' \dfrac{f'_y}{f'_c}$
0.7	48.40	4.65	0.0044	60	0.052	0.309
0.8	48.40	4.65	0.0044	60	0.052	0.212
0.847	48.40	4.65	0.0044	60	0.052	0.188

At the 0.7 point $[p(f'_{su}/f'_c) + p'(f'_y/f'_c)]$ is slightly greater than 0.30 but close enough to be considered acceptable. The negative moment capacities are computed, tabulated, and combined for the effects of both the prestressing and conventional reinforcing stresses, using the negative moment stress diagram, and assuming that $b = 22$ in. (Table 8.20). These capacities exceed the required M_u values and therefore are adequate.

The resisting moment of 8-#9 conventional reinforcing is combined with the development strength of prestressing steel and plotted against the ultimate moment envelope with cutoffs in the 8-#9 at 2-ft increments (Figure 8.23); these results show that adequate reinforcement has been supplied.

Next the shear requirements are examined according to the 1977 specifications. The girder and slab dead load shears are factored by adding 0.30 times the shears and applying these loads to the continuous

Table 8.19

Point	Number of Tensile Strands	d' (cgs to bottom)	A_s	$b = 22$ ft (p)	$0.5p \dfrac{f'_s}{f'_c}$	f'_{su}	$p \dfrac{f'_{su}}{f'_c}$
0.7	10	12.43	1.531	0.00560	0.151	229.2	0.257
0.8	10	21.46	1.531	0.00324	0.088	246.4	0.160
0.847	10	25.69	1.531	0.00271	0.073	250.3	0.136

Table 8.20

Point	T_{T1}	T_{T2}	T_c	C	A_c	$k'd$	
0.7	351	279	139	769	181	8.22	
0.8	377	279	139	795	187	8.50	
0.847	383	279	139	801	188	8.55	

Point	jd_1	$\dfrac{jd}{12}$	jd_2	$\dfrac{jd_2}{12}$	M_1	M_2	$\sum M_n$	
0.7	8.32	0.69	44.29	3.69	243	1030	1273	
0.8	17.21	1.43	44.15	3.68	541	1026	1567	OK
0.847	21.42	1.78	44.13	3.68	684	1026	1710	

beam. The superimposed dead load for the controlling interior girder is not factored, because it is a future load. The factored dead load shears for the controlling interior girder are listed in Table 8.21.

The factored live load shears are listed in Table 8.22.

Using v_c = 180 psi,

$$j = 0.8$$
$$d = 0.9 \times D$$

Then V_c = 180(7)(0.8)(0.9)(53.0) = 48.1 k.

⊓ #4 A_v = 0.40 jd = (0.8)(0.9)(53.0) = 38.2 in.

$$f_v = 60,000 \text{ psi}$$
$$\phi = 0.90$$

$$\text{stirrup spacing } s = \frac{\phi 2 A_v f_v jd}{V_u - V_c} = \frac{0.90(2)(0.40)(60)(38.2)}{V_u - V_c}$$
$$= \frac{1651}{V_u - V_c}$$

The dead load and $LL + I$ shears are tabulated and combined with the vertical component of the prestress shear in Table 8.23. The algebraic sum of these shears ($V_{DL} + V_{LL+I} + V_{prestress}$) = V_u is determined and the required stirrup spacing is computed at the tenth points.

The tie requirements according to AASHTO Article 1.6.14(D) are #4@1-10. The minimum shear reinforcing requirements of AASHTO

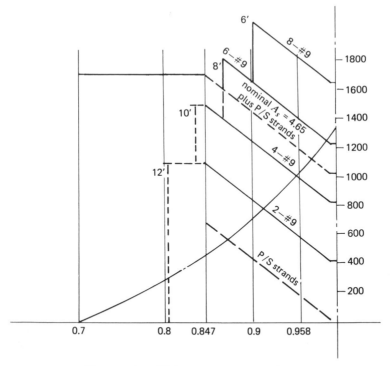

Figure 8.23 Ultimate moment envelope.

Article 1.6.13 are given as

$$s = \frac{60,000(0.40)}{100(7)} = 34.3 \text{ in.}$$

The ultimate shears are plotted against the resisting shears with variable spacing (Figure 8.24). The end block reinforcing is placed within $d/4$ according to AASHTO Article 1.6.15:

$$A_v = \frac{0.04(520.4)}{20.00} = 1.04 \text{ in.}^2$$

$$\frac{d}{4} = \frac{0.80(45)}{4} = 9.0 \text{ in.}$$

$$4 \text{ legs } \#5$$

$$A_s = 1.24 \text{ in.}^2 \text{ is sufficient}$$

Table 8.21

Point	DL (k)
AB	+54.4
0.1	+43.1
0.2	+31.8
0.3	+20.2
0.4	+9.5
0.5	−1.9
	−4.7
0.6	−16.0
0.7	−26.7
0.8	−38.3
0.9	−49.6
BA	−60.9

Table 8.22

Point	Ult. V_{LL+I} (k)
AB	+120.7
0.1	+102.9
0.2	+85.5
0.3	+69.0
0.4	+53.8
0.5	−62.5
0.6	−79.0
0.7	−94.6
0.8	−108.7
0.9	−122.0
BA	−133.0

Next, the shortening along the bottom girder fiber is computed to determine the additional length to be cast into the girders to give them the correct dimension at the bearings. The average initial stress along the bottom girder fiber is f_b = +1333 psi. The length of girder is about 703 in. The initial modulus is E_{ci} = 3,824,000 psi; therefore the initial shortening along the bottom fiber is

$$\text{initial } \Delta L = \frac{PL}{AE} = \frac{1333(703)}{3,824,000} = 0.245 \text{ in.}$$

Table 8.23

Point	V_{DL}	V_{LL+I}	Prestress V	V_u	V_c	$V_u - V_c$ (kips)	#4 $\sqcup$ s (in.)
AB	+54.4	+120.7	−18.9	+156.2	48.1	108.1	15.3
0.1	+43.1	+102.9	−18.9	+127.1	↑	79.0	20.9
0.2	+31.8	+85.5	−18.9	+98.4		50.3	32.8
0.3	+20.2	+69.0	−18.9	+70.3		22.2	
0.4	+9.5	+53.8		+63.3		15.2	
0.5	−1.9	−62.5		−64.4			
	−4.7			−67.2			
0.6	−16.0	−79.0	+15.8	−79.2		31.1	
0.7	−26.7	−94.6	+27.1	−94.2		46.1	35.8
0.8	−38.3	−108.7	+27.1	−119.9		71.8	23.0
0.9	−49.6	−122.0	+27.1	−144.5	↓	96.4	17.1
BA	−60.9	−133.0	+27.1	−166.8	48.1	118.7	13.9

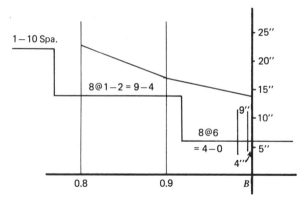

Figure 8.24   Ultimate shear versus stirrup spacing.

This is modified for an approximate creep before erection by assuming that the total creep equals $2\frac{1}{4}$ times the initial shortening and that 50% takes place before erection; therefore

$$\text{cast girders} = 0.245 + 1.25(\tfrac{1}{2})(0.245) = 0.398 \text{ in.}$$

longer than theoretical length ($\frac{3}{8}$ in.).

The approximate camber and deflections are computed to anticipate the required concrete fillet on top of the girder between the girder and the slab; therefore

$$\text{initial prestress } P_i = 520.4 \text{ k}$$

$$I = 125{,}400 \text{ in.}^4 = 6.0475 \text{ ft}^4$$

$$E_{ci} = 3{,}824{,}000 \text{ psi} = 550{,}700 \text{ ksf}$$

$$E_{cf} = 4{,}030{,}000 \text{ psi}$$

$$= 580{,}400 \text{ ksf}$$

The moment caused by the eccentric initial prestress force is shown in Figure 8.25. Assuming a conjugate beam with $EI = 1.0$, we sum the

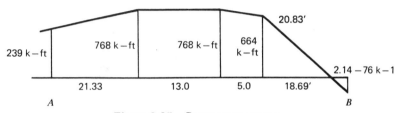

Figure 8.25   Prestress moment.

moments around B to compute the conjugate reaction at A:

$$\sum M_B = 0$$

239	$\times$ 21.33 =	5098	$\times$ 49.50 =	252,345
$\frac{1}{2} \times$ 529	$\times$ 21.33 =	5642	$\times$ 45.94 =	259,183
768	$\times$ 13.0 =	9984	$\times$ 32.33 =	322,783
664	$\times$ 5.0 =	3320	$\times$ 23.33 =	77,456
$\frac{1}{2} \times$ 104	$\times$ 5.0 =	260	$\times$ 24.17 =	6,284
$\frac{1}{2} \times$ 664	$\times$ 18.69 =	6205	$\times$ 14.60 =	90,594
$\frac{1}{2} \times (-76)$	$\times$ 2.14 =	-81	$\times$ 0.71 =	-58
		30,428		1,008,587

$$R_A = \frac{1,008,587}{60.16} = 16,765$$

The conjugate moment is a maximum where the conjugate shear equals zero. The moment is computed on the conjugate beam at this point to obtain the maximum conjugate moment and the resulting maximum camber.

	+16,765	$\times$ 29.18 =	+489,203
	-5098	$\times$ 18.51 =	$-94,364$
	-5642	$\times$ 14.96 =	$-84,404$
7.845 $\times$ 768 =	-6025	$\times$ 3.92 =	$-23,618$
	0		+286,817

Computation of the initial prestress camber is based on the actual EI:

$$\Delta_i^{\text{prestress}} = \frac{286,817}{6.0475(550,700)} = 0.0861 \text{ ft}$$

The girder dead load deflection, based on the initial modulus for the simple span, is

$$\Delta_{\text{girder } DL} = \frac{5(.583)(61.16)^4}{384(500,700)(6.0475)} = 0.0319 \text{ ft}$$

The slab and diaphragm deflections, based on the final modulus and the girder moment of inertia, are computed for a simple span:

$$\Delta_{\text{slab } DL} = \frac{5(0.820)(61.16)^4}{384(580,400)(6.0475)} = 0.0426 \text{ ft}$$

$$\Delta_{\text{diaphragm } DL} = \frac{2.1(61.16)^3}{48(580,400)(6.0475)} = 0.0029 \text{ ft}$$

Assume that creep equals $2\frac{1}{4}$ times the initial camber and that 50% creep takes place before erection. Then, before placing the slab and diaphragm, the girder should be cambered up by ($1\frac{1}{8}$ in.) as computed.
Before placing slab and diaphragm

$$\Delta = (0.0861 - 0.0319) + 1.25(\tfrac{1}{2})(0.0861 - 0.0319) = 0.0881(1\tfrac{1}{8} \text{ in.})$$

The deflection of the slab and diaphragm is

$$\Delta = 0.0426 + 0.0029 = 0.0455(\tfrac{1}{2} \text{ in.})$$

The net deflection up or camber is $\frac{5}{8}$ in., which should be accommodated by a $\frac{5}{8}$-in. fillet over the girder between the slab and girder at the bearings.
The horizontal shear is computed between the slab and diaphragm for a negative moment section:

$$\text{Ult } V_{BA} = 133.0 + 7.9 + 13.3 = 157.2 \text{ k}$$

$$I = \frac{22.00(14.81)^3}{3} + 72(33.59)^2 = 105,058 \text{ in.}^4$$

$$Q = 72(33.59) = 2418 \text{ in.}^3$$

$$v_n = \frac{157.2(2418)}{16(105,058)} = 226 \text{ psi}$$

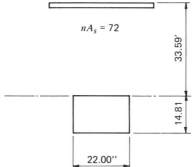

and for a positive moment section assume positive moment @ $0.8AB$:

$$V_{0.8AB} = 108.7 + 11.0 + 5.4 = 125.1 \text{ k}$$

$$I = 376{,}064 \text{ in.}^4$$

$$Q = 643(13.38) = 8603 \text{ in.}^3$$

$$v_n = \frac{125.1(8603)}{16(376{,}064)} = 179 \text{ psi}$$

Minimum ties are provided and the computed shears are less than the 300 psi allowable in AASHTO Article 1.6.14(C); therefore an artificially roughened surface is satisfactory in lieu of shear keys.

8.3   MULTIBEAM BRIDGES

8.3.1   General

A multibeam bridge is one where the tops of beams are connected to form the roadway riding surface. These beams also interact with one another to distribute loads. The interaction of the beams is developed by shear keys and transverse bolting, which, in effect, is light transverse posttensioning. Today most multibeam bridges use beams of pretensioned prestressing concrete. Reinforced concrete, however, is also a possibility for this construction.

Practical span lengths for this section vary according to the selection made. AASHTO-type prestressed slab units (see Appendix P) are reasonable for use on spans up to 50 ft. AASHTO-type prestressed box units (see Appendix Q) can be used on spans up to 100 ft. The slab and box units are the most likely to act according to the multibeam distribution criteria given in AASHTO Article 1.3.1(d) because of the large surface for interaction that results from transverse posttensioning.

Single- and multiple-tee sections are often used on multibeam superstructures. On this section the interacting surfaces often consist only of the connections between roadway deck slab and diaphragms. Therefore to ensure adequate interaction for a multibeam distribution it seems reasonable to provide diaphragms at close spacing, with transverse bolting (light posttensioning) through the diaphragms.

Protection of the roadway surface is desirable when using multibeam sections. Many sections have thin top decks for riding surfaces which, if allowed to deteriorate, would require replacement of the entire superstructure. Often an asphaltic wearing surface is used to remove transverse and/or longitudinal unevenness. These wearing surfaces can trap moisture at the top of the concrete deck which is a good reason for using a waterproof membrane as a protection system.

Some designers attempt to eliminate the camber "hump" caused by prestressing by the use of an asphaltic wearing surface. With large cambers this can result in a considerable asphaltic wearing surface thickness. For single-span bridges on low-speed roads this may not be necessary. For multiple-span bridges, however, we consider it essential that a leveling course be used to eliminate the "roller coaster" effect that would result from individual span camber.

Example 8.3. The prestressed slab superstructure shown in Figures 8.26 and 8.27 is designed with slab units 2 ft-2 in. deep × 4 ft-0 in. wide which are connected by grouted shear keys and transverse post-tensioning. These features ensure multibeam action in accordance with AASHTO Article 1.3.1(d).

The constants for the slab units are

$$A = 846 \text{ in.}^2 \qquad w = 881 \text{ lb-ft}^{-1}$$
$$I = 63,870 \text{ in.}^4 \qquad S = 4913 \text{ in.}^3$$

The slab dead load shear at the $\frac{1}{4}$ point of the span is

$$w = 881 \text{ lb-ft}^{-1}$$
$$V_{0.25} = (0.881)\left(\frac{60.67}{4}\right) = 13.4 \text{ k}$$

The slab dead load reaction is

$$\text{reaction} = (0.881)\left(\frac{62.0}{2}\right) = 27.3 \text{ k}$$

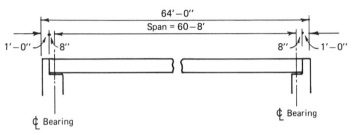

Figure 8.26 Elevation of bridge.

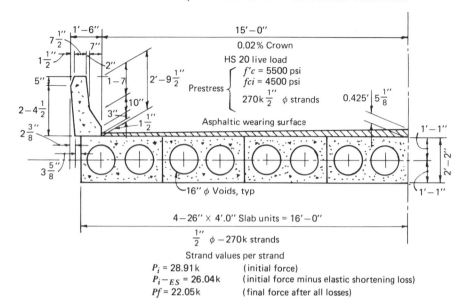

Figure 8.27   Half roadway section.

The slab dead load moment at midspan (0.5 point on span) and the top and bottom fiber stress are computed as

$$M_{0.5} = 0.881(60.67)^2(\tfrac{1}{8}) = 405 \text{ k-ft}$$

$$f_{SDL} = \frac{405(12)}{4913} = \pm 989 \text{ psi}$$

The parapet dead load is 394 lb-ft^{-1} where the weight of the parapet and the wearing surface (not constant depth) are to be considered as equally distributed to each slab unit.

The same shear, reaction, moment, and stress are computed for the parapet plus wearing surface dead load as for the slab dead load alone:

$$
\begin{aligned}
w = \text{parapet} \quad & 2(0.394) \\
\text{ws} \quad & 0.125 \text{ ft} \times 30.0 \text{ ft} \times 0.150 \\
\text{ws} \quad & \tfrac{1}{2} \times 0.30 \times 30.0 \times 1.50 \\[4pt]
& = 0.788 \text{ k-ft}^{-1} \\
& = 0.562 \text{ k-ft}^{-1} \left. \right\} \text{ Sum} = 2.025 \text{ k-ft}^{-1} \text{ divided by} \\
& = 0.675 \text{ k-ft}^{-1} \qquad\qquad 8 \text{ units} \\
& \qquad\qquad\qquad = 0.253 \text{ k-ft}^{-1} \text{ per unit.}
\end{aligned}
$$

$$V_{0.25} = (0.253)\left(\frac{60.67}{4}\right) = 3.8 \text{ k}$$

$$R = (0.253)\left(\frac{62.0}{2}\right) = 7.8 \text{ k}$$

$$M_{0.5} = 0.253(60.67)^2(\tfrac{1}{8}) = 116 \text{ k-ft}$$

$$f_{\text{ws+RDL}} = \frac{116(12)}{4913} = \pm 284 \text{ psi}$$

The impact factor according to AASHTO Article 1.2.2(c) is

$$I = \frac{50}{60.67 + 125} = 0.269$$

The distribution of lines of wheels for live load per slab unit is computed according to AASHTO Article 1.3.1(d)(1):

$$K = 0.8$$

$$C = 0.8\left(\frac{32}{60.67}\right) = 0.422$$

$$D = 5 + \frac{2}{10} + \left(3 - \frac{4}{7}\right)\left(1 - \frac{0.422}{3}\right)^2 = 6.994$$

$$S = \frac{12(2) + 9}{8} = 4.125$$

$$\frac{S}{D} = \frac{4.125}{6.994} = 0.590 \text{ lines of wheels per slab unit}$$

The coefficient that combines impact with live load distribution is

$$\text{coefficient} = 1.269(0.590) = 0.748$$

From AASHTO Appendix A, using

HS 20 live load

60.67-ft span

The maximum live load moment per lane = 818.5 k-ft. This maximum moment falls a small distance from the centerline of the span and is close to the maximum moment at midspan; therefore it is considered the midspan moment.

The live load plus impact moment per slab unit is

$$M_{0.5} = \tfrac{1}{2}(0.748)(818.5) = 306 \text{ k-ft}$$

The extreme fiber stress caused by this moment is

$$f_{LL+I} = \frac{306(12)}{4913} = \pm 747 \text{ psi}$$

Loading for maximum shear at intermediate points was explained in Example 5.1. The live load plus impact shear at the $\frac{1}{4}$ point is computed by placing the first 16-k wheel at the 0.25 point and the other 16-k wheel and 4-k wheel at 14-ft intervals behind the first wheel. The proportion of each wheel that produces shear at the nearest support is computed and summed to get the total shear. The maximum live load plus impact shear is

$$V_{0.25} = (0.748)(16)\left(0.75 + 0.519 + \frac{0.289}{4}\right) = 16.1 \text{ k}$$

On the basis of an allowable tensile stress $4\sqrt{f_c'} = 4\sqrt{5500} = 445$ psi and a maximum eccentricity of 5.81 in. the approximate required prestress force is computed. The maximum tensile stress at midspan and in the precompressed tensile zone due to applied loads is

$$f_b = -989 - 284 - 747 = -2020 \text{ psi}$$

Using these constraints, we can compute the required prestress force:

$$f_b = -445 = \frac{P}{846} + \frac{P(5.81)}{4913} - 2020$$

$$P = 666 \text{ k}$$

The number of strands required equals 666 divided by 22.05 = 30.2; 34 strands are used to provide the best pattern.

Thirty-four strands and a maximum e value of 5.81 in. are selected to fill the requirement for the strand layout in Figure 8.28. The actual eccentricity (e) of this section is

$$e = \frac{18(11) + 8(9) + 8(-11)}{34}$$

$$= 5.35 \text{ in. furnished} < 5.81 \text{ in. maximum} \quad \text{OK}$$

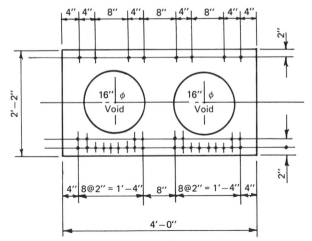

Figure 8.28 Strand layout.

The reason e has a maximum of 5.81 in. is explained in the following paragraphs.

The center of gravity of the prestressing strands must lie within the kern at the centerline of bearing. This means (per AASHTO specifications) that the top fiber in this location is in an area other than a precompressed tensile zone; therefore the allowable tensile stress is zero.

To maintain the top tensile stress at zero the prestress eccentricity (e) must be computed by using $f = P/A - P_e/S_t$:

$$f = 0 = \frac{P}{846} - \frac{P_e}{4913} = 0.001182 - 0.000204e = 0$$

$e = 5.81$ in. max (no tension allowed at bearings)

Most prestressing companies like to use lighter prestressing beds for this type of section and these beds are normally not designed to take uplift loads. Therefore only straight strand arrangements are used in slab units. With only straight strands the strands are placed at the top and bottom to keep within the desired eccentricity limit.

The prestress forces for this section are

$$P_i = 34(28.91) = 983 \text{ k (initial force)}$$

$$P_{i-ES} = 34(26.04)$$

$$= 885 \text{ k (initial force minus elastic shortening loss)}$$

$$P_f\ 34(22.05) = 750 \text{ k (final force after all losses)}$$

The extreme fiber stress due to prestress force equals $P/A \pm Pe/S$.

For the bottom fiber stress f_b

$$\text{final } f_b = \frac{750}{846} + \frac{750(5.35)}{4913} = 887 + 817 = +1704 \text{ psi}$$

For the top fiber stress f_t

$$\text{final } f_t = \frac{750}{846} - \frac{750(5.35)}{4913} = 887 - 817 = +70 \text{ psi}$$

These are the final stresses after all losses and are the total values at the support. They fall within the AASHTO limit.

For the initial and initial minus elastic shortening conditions the stresses due to prestress are prorated from those just computed:

initial	f_b = +2233 psi	
	f_t = +92 psi	
initial minus ES	f_b = +2011 psi	
	f_t = +83 psi	

The maximum final stresses at midspan due to all applied loads and prestress are compared with the allowables in AASHTO Article 1.6.6(b)(2).

Final Stresses @ 0.5

$$f_b = -989 - 284 - 747 + 1704 = -316 \text{ psi} < 445 \text{ psi}$$
$$f_t = +989 + 284 + 747 + 70 = +2090 \text{ psi} < 2200 \text{ psi}\quad \text{OK}$$

The maximum initial stresses are the algebraic sum of the slab unit dead load and the prestress, summed and compared with the allowables.

Initial Stresses @ 0.5

$$f_b = -989 + 2011 = +1022 \text{ psi} < 2700 \text{ psi}$$
$$f_t = +989 + 83 = +1072 \text{ psi} < 2700 \text{ psi}\quad \text{OK}$$

This completes the stress analysis; the ultimate moment capacity must be checked.

The required factored moment is $M_{\text{required}} = 1.3(405 + 116) + 2.17(306) = 1341$ k-ft.

For tensile bending strength use only the strands in the bottom of the section. Use $26\frac{1}{2}$-in. ϕ strands; $d = 23.33$ in. $A_s = 3.9806$ in.^{2}.
The reinforcement ratio is

$$p = \frac{5.9806}{48(23.38)} = 0.00355$$

The approximate steel stress at failure, according to AASHTO Article 1.6.9(c), equals

$$f'_{su} = f'_s \left[1 - 0.5p \frac{f'_s}{f'_c}\right]$$

$$f'_{su} = 270 \left[1 - 0.5(0.00355) \frac{270}{5.5}\right] = 246.5 \text{ ksi}$$

The capacity is based on $\phi = 1.0$ for factory-produced precast members and is computed as

$$\phi M_n = \phi \left[A_s f'_{su} d \left(1 - 0.6p \frac{f'_{su}}{f'_c}\right)\right]$$

therefore

$$\phi M_n = (3.9806)(246.5)(23.38)\left[1 - 0.6 \frac{(0.00355)(246.5)}{5.5}\right]\left(\frac{1}{12}\right)$$

$$= 1729 \text{ k-ft} > 1341 \text{ k-ft} \quad \text{OK}$$

If $[p(f'_{su}/f'_s)]$ is less than 0.30, the capacity analysis can be based on failure of the unit in tension in the prestressed reinforcement; therefore

$$\left(p \frac{f'_{su}}{f'_c}\right) = (0.00355)\left(\frac{246.5}{5.5}\right) = 0.16 < 0.30$$

This indicates that the capacity analysis for ϕM_n, which used the tensile strength of the reinforcement as the basis for strength, was correct. The capacity computation was based on a rectangular section. The depth of the compression block is

$$1.4 \, dp \frac{f'_{su}}{f'_c} = 1.4(23.38)(0.00355)\left(\frac{246.5}{5.5}\right) = 5.21 \text{ in.} > 5.0 \text{ in.}$$

which is close enough to call the section rectangular.

A number of problems have developed in the specifications for design of shear reinforcement that were put into the AASHTO specifications in 1980. The criteria used were selected for uniform loads applied to building floors and are not applicable to moving highway loadings. Therefore the following as suggested for shear design, based on recommendations made by the ACI-ASCE Committee 323(1). These requirements were the basis of the AASHTO specifications until 1979.

The required factored ultimate shear strength at the 0.25 point is (using $\phi = 0.90$ for prestress shear)

$$\frac{V_u @ 0.25}{\phi} = 1.44(13.4 + 3.8) + 2.41(16.1) = 63.6 \text{ k}$$

It is reasonable and safe to assume that the effective depth d is 80% of the total depth and that $j = 0.90$.

Using an ultimate concrete shear stress $v_c = 180$ psi, we compute the shear strength of the concrete:

$$V_c = v_c \, bjd = 0.180(48 - 2 \times 16)(0.90)(0.80)(26) = 53.9 \text{ k}$$

The spacing required is $s = A_v f_y jd/(V_u - V_c)$:

$$s = \frac{2(0.40)(60,000)(0.9)(0.8)(26)}{(63.6 - 53.9)(1000)} = 93 \text{ in.}$$

$$\min s = \frac{240}{48 - 2(160)} = 15.0 \text{ in. } (\tfrac{1}{2}D)$$

Use #4 @ 1'-3". ⌐⊐

End block reinforcing is provided to resist potential bursting stresses at the beam ends, where the pretensioned strands are developing the actual loads into the beams. This is done in accordance with AASHTO Article 1.6.15.

The reinforcement is placed in the distance $d/4$.

$$\frac{d}{4} = \frac{0.8(26)}{4} = 5.20 \text{ in.}$$

The required area of reinforcing in this spacing is

$$A_s = \frac{0.04(983)}{20} = 1.97 \text{ in.}^2$$

Use 2-#5 ⊡ (4 legs each).

$$A_s = 2.48 \text{ in.}^2$$

The anticipated elastic shortening is computed by using the stress and strain along the bottom fiber because it coincides with the center-to-center of the bearings.

The approximate average bottom fiber stress is av f_b^i = 1352 psi; the length of the girder L = 62 ft = 744 in.; the initial modulus of elasticity is E_{ci} = 57000$\sqrt{4500}$ = 3,824,000 psi. The initial elastic shortening Δ_i is

$$\Delta_i = \frac{1352(744)}{3,824,000} = 0.2630 \text{ in.}$$

Assume a creep factor equal to 2.25 times the initial elastic shortening and assume that half the creep beyond the elastic shortening will take place before girder placement. The additional girder length required to accommodate this shortening due to elasticity and creep is

$$\text{cast } \Delta_x \text{ longer} = 0.2630 + 1.25(\tfrac{1}{2})(0.2630) = 0.4274 \text{ in. } (\tfrac{3}{8} \text{ in.})$$

The camber is based on the initial force without reduction for elastic shortening. The elastic shortening could be deducted but it would not change the results significantly.

The prestress moment is constant over the full length of the slab unit because there are no deflected strands. This moment is

$$M = \frac{983(5.35)}{12} = 438 \text{ k-ft}$$

From the above

$$E_{ci} = 3,824,000 \text{ psi} = 550,656 \text{ ksf} \qquad I = 3.0802 \text{ ft}^4$$

A conjugate beam is loaded with a constant M/EI diagram. The mid-span simple-beam moment on the conjugate beam equals the midspan deflection. For a uniform load the midspan moment equals $\omega l^2/8$, where for this case $\omega = M/EI$. Therefore

$$\Delta \text{ prestress} = \frac{Ml^2}{EI8} = \frac{438(60.67)^2}{(550,656)(3.0802)(8)}$$

$$= 0.1188 \text{ (initial elastic deflection)}$$

The initial deflection due to the slab DL

$$\Delta \text{ slab} = \frac{5Ml^4}{384EI} = \frac{5(0.881)(60.67)^4}{384(550,656)(3.0802)} = 0.0916$$

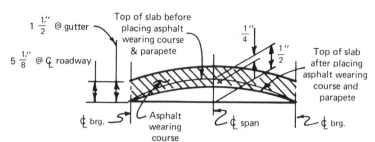

Figure 8.29   Camber diagram.

Assume the creep characteristics of beam shortening; then before placing any wearing surface or parapet the total deflection up is

$$\Delta \text{ before ws} + \text{parapet} = (0.1188 - 0.0916) + \tfrac{1}{2}(1.25)(0.1188 - 0.0916)$$
$$= 0.0442(\tfrac{1}{2} \text{ in.})$$

The deflection due to wearing surface plus parapet (based on $E_{cf} = 57000\sqrt{5500} = 4{,}277{,}000$ psi $= 608{,}700$ ksf) is

$$\Delta \text{ ws} + \text{parapet} = \frac{5(0.253)(60.67)^4}{384(608{,}700)(3.0802)} = 0.0238 \text{ ft } (\tfrac{1}{4} \text{ in.})$$

Figure 8.29 is the camber diagram.

BIBLIOGRAPHY

1. *AASHTO Standard Specifications for Highway Bridges, 1977*, with the 1978 to 1981 Interim Specifications, The American Association of State Highway and Transportation Officials.

2. Clifford L. Freyermuth, Design of Continuous Highway Bridges with Precast, Prestressed Concrete Girders, *J. Prestressed Concr. Inst.*, 14(2) (April 1969).

3. Talat Mostafa and Paul Zia, Development Length of Prestressing Strands, *J. Prestressed Concr. Inst.* (September/October 1977).

4. *Moments, Shears, and Reactions for Continuous Highway Bridges*, American Institute of Steel Construction, 1966.

5. Ned H. Burns and T. Y. Lin, *Design of Prestressed Concrete Structures*, Wiley, New York, 1981.

6. T. Y. Lin, *Prestressed Concrete Structures*, Wiley, New York, 1955.

7. James R. Libby, *Prestressed Concrete-Design and Construction*, Van Nostrand Reinhold, New York, 1961.

NINE

Segmental Box Bridges

The design and construction of segmental box girder bridges in the last decade has increased dramatically. The principles of design, design criteria, construction techniques, and case examples have been presented extensively in the Journal of the Prestressed Concrete Institute. With the kind permission of PCI much of the material contained in this chapter has been abstracted. More detailed design information can be obtained from PCI (see refs. 1–4).

Several techniques are used in the construction of segmental box girder bridges. Box girders can be cast-in-place or precast and can be built by falsework or by cantilever techniques. Other variations in construction methods such as span-by-span placement and incremental launching, are available. Cable-stayed bridges and arch structures are suitable for segmental construction as demonstrated on several occasions. Precast segmental box girder bridges of cantilever construction are the predominant segmental form; therefore most of the information presented here relates specifically to that design.

9.1 HISTORY[1,2]

9.1.1 General

The earliest known application of precast segmental bridge construction was a single-span county bridge built in 1952 in New York State. The bridge girders were divided longitudinally into three precast segments which were cast end-to-end. After curing the segments were transported to the job site where they were reassembled and posttensioned with cold joints.

The development in Europe of long-span prestressed concrete bridge construction techniques is well known. Of particular significance was the development in Germany of cast-in-place cantilever segmental construction by the firm of Dyckerhoff & Widmann. This technology was adapted and extended for use with precast segments in the Choisy-le-Roi Bridge over the Seine, south of Paris, in 1962. The Choisy-le-Roi Bridge was designed and built by Enterprises Campenon Bernard. Several other structures of the same type were built in due course. At the same time the techniques of precasting segments and placing them in the structure continued to be refined.

A major innovation in the construction of precast segmental bridges was the launching gantry, used for the first time on the Oleron Viaduct between 1964 and 1966. The launching gantry makes it possible to move segments over the completed part of a structure and to place them as cantilevers between successive piers. Use of a launching gantry resulted in the completion of the Oleron Viaduct at an average of 900 lin ft (270 m) of finished deck per month. Although the launching gantry is, in many cases, a useful means of erection the same can be accomplished with cranes and other means.

Experience with major precast segmental bridges in Europe advanced the refinement of the construction process. Improvements were made in precasting methods and in the design of erection equipment to permit the use of larger segments and longer spans and to accommodate horizontal curvature in roadway alignment.

This technique not only gained rapid acceptance in France but spread to other countries; for example, the Netherlands, Switzerland, and later Brazil and New Zealand adopted the method. Many other countries today are using the precast segmental techniques for various applications. The first in North America was a highway bridge over the Lievre River in Quebec. The Lievre River Bridge was built in 1967 and has a main span of 260 ft (79 m) and end spans of 130 ft (40 m). The Bear River bridge near Digby, Nova Scotia, contains six interior spans of 265 ft (81 m) and end spans of 203 ft (62 m). It was opened to traffic in December 1972.

The first U.S. precast segmental box girder bridge was built near Corpus Christi, Texas, and was opened to traffic in 1973. The Corpus Christi bridge has a central span of 200 ft (61 m) and end spans of 100 ft (30.5 m). Subsequent to the Corpus Christi bridge, precast segmental bridges were completed in Indiana and Colorado and another is now under construction in Illinois. A simple-span, precast segmental bridge was constructed at the Pennsylvania State University test track as a research project sponsored by the Federal Highway Administration and the Pennsylvania Department of Transportation. Numerous precast segmental bridges have been designed for other locations in the United States and Canada, and it is expected that this technique will see wide use in the years ahead.

9.1.2   Types of Precast Segmental Construction

Two main types of precast segmental bridge construction have developed which may be differentiated by the use of cast-in-place concrete or epoxy joints.

A number of precast segmental bridges have been built with cast-in-place joints 3–4 in. (76–102 mm) wide between segments. This procedure eliminates the need for match-casting and reduces the dimensional precision required in casting the segments; its major disadvantages, however, include the need for falsework to support the segments while the cast-in-place joint cures and substantial reduction in construction speed. Cast-in-place joints are not generally attractive and for this reason are not considered further in this text.

The prevailing system of precast segmental bridge construction uses an epoxy resin jointing material. The thickness of the joint is on the order of $\frac{1}{32}$ in. (0.8 mm). A perfect fit between the ends of adjacent segments is achieved by casting each segment against the end face of the preceding one (match-casting) and then erecting them in the same order in which they were cast. This chapter considers only design and construction techniques for bridges that use match-cast segments and an epoxy resin jointing material.

9.1.3   Advantages

The advantages of precast segmental construction are the following:

The economy of precast prestressed concrete construction is extended to a span range of 100–400 ft (30–120 m) and even longer spans may be economical in circumstances in which the use of heavy erection equipment is feasible.

Precast segments may be fabricated while the substructure is being built and rapid erection of the superstructure can be achieved.

The method makes use of repetitive industrialized manufacturing techniques with the inherent potential for achieving high-quality and high-strength concrete.

The need for falsework is eliminated and all erection may be accomplished from the completed portions of the bridge. These aspects may be particularly important for high-level crossings when it is necessary to minimize interference with the bridge environment or when heavy traffic must be maintained under the bridge during construction.

The structure geometry may be adapted to any horizontal or vertical curvature or any required roadway superelevation.

The effects of concrete shrinkage and creep may be substantially reduced during erection and in the completed structure because the segments will normally have matured to full design strength before erection.

Except for temperature and weather limitations related to mixing and placing epoxy, precast segmental construction is relatively insensitive to weather conditions.

The esthetic potential of concrete construction.

Enhanced durability of bridge decks by precompression of the concrete and elimination of cracking and by use of high-quality concrete produced under conditions that permit a high level of quality control.

9.1.4   Disadvantages

The disadvantages of precast segmental construction are the following:

The need for a high degree of geometry control during fabrication and erection of segments.

Temperature and weather limitations regarding mixing and placing epoxy joint material.

Lack of mild steel reinforcement across the joint and a limitation of tension stress.

9.2   PRECAST SECTIONS

9.2.1   General

Precast segmental bridge construction found initial acceptance for a span range of 160–350 ft (50–110 m). When the cantilever method of erection is used this span range is still the basic area of application.

In recent years the advantages of precast segmental construction have been extended to shorter span freeway overpasses in several European projects.

A procedure for precast segmental construction developed primarily for the span range of 100–160 ft (30–50 m) is the concept of progressive placement. With this procedure segments are placed continuously from one end of the deck to the other in successive cantilevers on the same side of the various piers rather than in a balanced cantilever at each pier.

9.2.2   Construction in North America

The first precast segmental bridge to be built in North America was the Lievre Bridge, located in Quebec, in 1967. The segmental Bear River Bridge, located in Nova Scotia, was constructed in 1972.

Table 9.1   Precast Segmental Concrete Bridges in North America.

Name and Location	Date of Construction	Method of Construction	Span Lengths [ft (m)]
Lievre River Notre Dame du Laus, Quebec	1967	Balanced cantilever	130-260-130 (39.6-79.2-39.6)
Bear River Digby, Nova Scotia	1972	Balanced cantilever	203.75-6 @ 265-203.75 (62.1-6 @ 80.77-62.1)
JFK Memorial Causeway Corpus Christi, Texas	1973	Balanced cantilever	100-200-100 (30.5-61-30.5)
Muscatuck River U.S. 50 North Vernon, Indiana	1975	Balanced cantilever	95-190-95 (29-58-29)
Sugar Creek, State Route 1620 Parke County, Indiana	1976	Balanced cantilever	90.5-180.5-90.5 (27.6-55-27.6)
Vail Pass, I-70 West of Denver, Colorado (four bridges)	1977	Balanced cantilever	134-200-200-134 (40.8-61-61-40.8) 134-200-200-145 (40.8-61-61-44) 151-155-210-210-154 (46-47.2-64-64-47) 153-210-210-154 (46.6-64-64-47)
Penn DOT Test Track Bridge Penn State University, State College, Pennsylvania	1977	On falsework	124 (37.8)
Turkey Run State Park Parke County, Indiana	1977	Balanced cantilever	180-180 (54.9-54.9)
Pasco-Kennewick, Columbia River between Pasco and Kennewick, Washington (cable-stay spans)	1978	Balanced cantilever	406.5-981-406.5 (124-299-124)
Wabash River U.S. 136 Covington, Indiana	1978	Incremental launching	93.5-4 @ 187-93.5 (28.5-4 @ 57-28.5)
Kishwaukee River Winnebago Co. near Rockford, Illinois (dual structure)	1979	Incremental launching	170-3 @ 250-170 (51.8-3 @ 76.2-51.8)
Islington Avenue Extension Toronto, Ontario	1979	Incremental launching	2 @ 161-200-5 @ 272 (2 @ 49-61-5 @ 83)
Kentucky River Frankfort, Kentucky (dual structure)	1979	Balanced cantilever	228.5-320-228.5 (69.6-97.5-69.6)
Long Key, Florida (contract let late 1978)	—	Span-by-span	113-101 @ 118-113 (34.4-101 @ 36-34.4)
Linn Cove, Blue Ridge Parkway North Carolina (contract let late 1978)	—	Progressive placing	98.5-163-4 @ 180-163-98.5 (30-49.7-4 @ 54.9-49.7-30)
Zilwaukee, Michigan (dual structure) (bids opened late 1978)	—	Balanced cantilever	26 north bound spans, total length 8087.5 (2465) 25 south bound spans, total length 8057.5 (2456)

In 1972 the JFK Memorial Causeway Bridge, located in Corpus Christi, Texas, was the first precast prestressed segmental bridge to be built in the United States. Many other bridges have been constructed, as listed in Table 9.1.[2]

Summary of Precast Segmental Concrete Bridges in the United States and Canada With Cross Sections

Note: for metric dimensions
   1 ft. = 0.3048 m
   1 in. = 25.4 mm

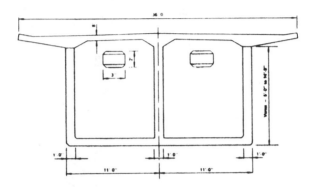

Lievre River Bridge, Quebec
Spans: 130 feet — 260 feet — 120 feet
Bridge Length: 520 feet
Segment Length: 9 feet 6 inches

Figure 9.1   Lievre River bridges.

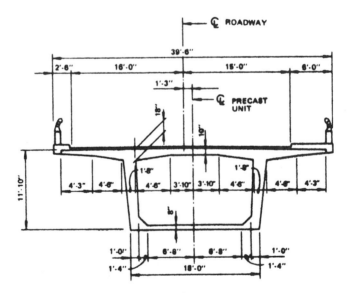

Bear River Bridge, Nova Scotia
End Spans: 203 feet 9 inches
Interior Spans: 265 feet
Bridge Length: 1997.50 feet
Segment Length: 14 feet 2 inches

Figure 9.2   Bear River Bridge.

9.2.3   Typical Sections

A catalogue of the general cross sections of many of the bridges listed in Table 9.1 is contained in ref. (1). These illustrations have been reproduced and are presented in figures 9.1–9.16.

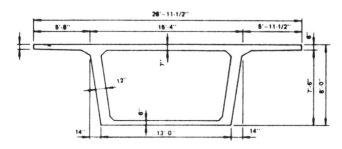

Corpus Christi, Texas
Spans: 100 feet − 200 feet − 100 feet
Bridge Length: 400 feet
Two Segments Wide
Segment Length: 10 feet

Figure 9.3   Corpus Christi Bridge.

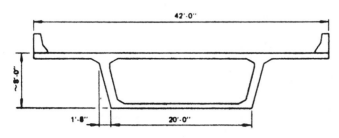

Vail Pass, Colorado
End Spans: 160 feet
Main Spans: 210 feet
Segment Length: 7 feet 4 inches

Figure 9.4   Vail Pass bridges.

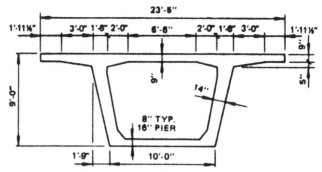

North Vernon, Indiana
Over Muscatatuck River
Spans: 95 feet — 190 feet — 95 feet
Bridge Length: 380 feet
Segment Length: 8 feet

Figure 9.5 · Muscatatuck River Bridge.

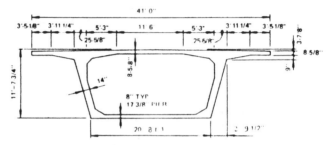

Kishwaukee River Bridge, Illinois
End Spans: 170 feet
Interior Spans: 250 feet
Northbound: 1090 feet
Southbound: 1090 feet
Segment Length: 7 feet 0-5/8 inches

Figure 9.6   Kishwaukee River Bridge.

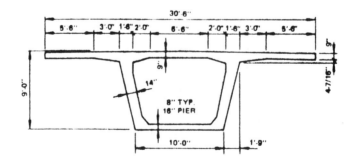

Parke County, Indiana
Bridge Length: 276 feet
Spans: 90 feet — 180 feet — 90 feet
Segment Length: 8 feet

Figure 9.7   Parke Bridge.

428

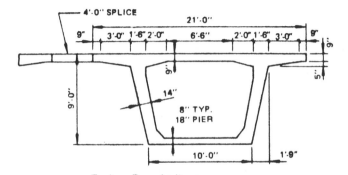

Turkey Run, Indiana
Bridge Length: 322 feet
Spans: 180 feet – 180 feet
Segment Length: 8 feet

Figure 9.8   Turkey Run Bridge.

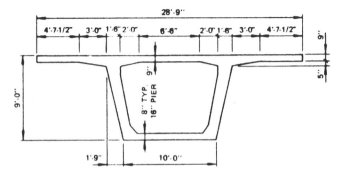

Pike County, Kentucky
Bridge Length: 372 feet
Spans: 93.5 feet – 185 feet – 93.5 feet
Segment Length: 7 feet 10 inches

Figure 9.9   Pike Bridge.

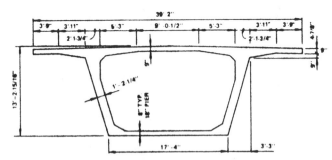

Lake Oahe Crossing Missouri River,
   North Dakota
Bridge Length: 3020 feet
Spans: 179 feet – 10 @ 265 feet – 179 feet
Segment Length: 8 feet 4 inches

Figure 9.10   Missouri River Bridge.

429

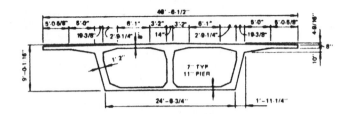

Scottdale Bridge, Michigan
Bridge Length: 407 feet
Spans: 97 feet — 206.5 feet — 97 feet
Segment Length: 8 feet

Figure 9.11   Scottdale Bridge.

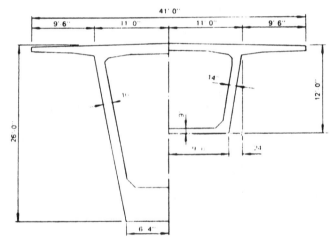

Illinois River, Illinois
Eastbound: 3329.5 feet
Westbound: 3203.5 feet
Approach Spans: 175 feet — 230 feet
Main Spans: 390 feet — 550 feet — 390 feet
Segment Length: 10 feet

Figure 9.12   Illinois River Bridge.

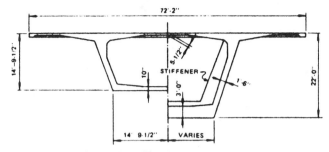

Zilwaukee, Michigan
Bridge Over Saginaw River
Bridge Length: 8000 feet
Two segments Wide
Spans: Variable 155–392 feet
Segment Length: 8 feet to 12 feet

Figure 9.13   Saginaw River Bridge.

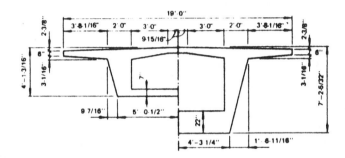

St. Louis Missouri
Bridge Length: 405 feet
Spans: 100 feet – 200 feet – 100 feet
Segment Length: 9 feet 4 inches

Figure 9.14   St. Louis Bridge.

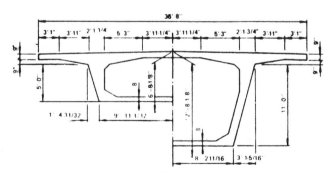

Akron Bridge, Ohio
Westbound: 3660 feet
Eastbound: 3646 feet
Spans: Variable 100 to 290 feet
Segment Length: 6, 7 and 8 feet

Figure 9.15   Akron Bridge.

431

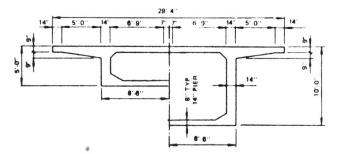

Lake County Ramps, Indiana
Bridge Length: 6240 feet
(Ramps plus Mainline)
Spans: Variable 100 to 315 feet
Segment Length: 7 feet 9 inches

Figure 9.16   Lake County ramp structure.

9.3 DESIGN CRITERIA

9.3.1 General Segment Dimensions

The principal segment dimensions are top slab width W, construction depth D, width of bottom slab B, web spacing s, and segment length L, as shown in Figure 9.17.

In most cases the segment width W is selected equal to the width of the bridge. When the bridge width exceeds about 40 ft (12 m) or when it is necessary to minimize segment weight or size the structure width can be divided into a multiple of the segment width (Figure 9.18). The transverse connection of the top slabs may be accomplished by transverse posttensioning, which extends through all the boxes and the cast-in-place joint(s).

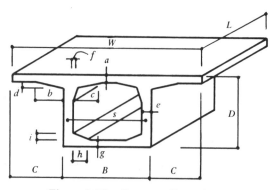

Figure 9.17   Segment dimensions.

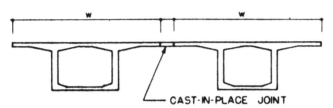

Figure 9.18 Superstructure with parallel segments and cast-in-place joints.

As an alternative to the use of multiple boxes for structures wider than 40 ft (12 m) single boxes with multiple webs have been used for widths up to 70 ft (21 m). For intermediate widths single box sections may be used with integral transverse floor beams under the roadway slab or boxed cantilevers.

The construction depth D is determined by the spans. Most European bridges have been built with span/depth ratios of 18-20. Ratios of 20-30, however, are considered feasible and structurally satisfactory. Deflection tests on the model of the Corpus Christi Bridge which had a span/depth ratio of 25 resulted in a deflection of only $L/3200$ which is only 25% of the deflection permitted in steel structures in the United States. Span/depth ratios for end spans are usually somewhat lower than those for interior spans. The shallower depth structures require more high strength posttensioning materials. Variable depth structures become appropriate for spans in excess of 250-300 ft (75-90 m). In this case the span/depth ratios have normally been selected as 18-20 at support and 40-50 at midspan.

When webs are vertical the bottom slab width B follows from the width W and the structurally acceptable length of the cantilever. Sloping webs present no problems when the box girder depth is constant but do require significant form adjustments for production of variable depth segments due to the variation in bottom slab width. A narrow bottom slab is desirable to reduce segment weight because the bottom slab area is usually a factor for structural consideration only in the negative moment area adjacent to piers.

The segment length L has a pronounced effect on the economy of a bridge. The selection of the segment length determines the total number of segments that must be produced and erected. Because most of the cost of production and erection is fixed per unit and only a small share is variable, economy is achieved by using the smallest number of segments consistent with transportation requirements and the capacity of the erection equipment; and because the cost of handling and erection increases with L it is necessary to make a study of the total in-place economy of various segment lengths to determine the most economical value. When segments must be transported over highways the weight and size limitations usually determine the value of L.

The spacing of webs s is normally determined purely on structural criteria. In principle, any web spacing can be used if all pertinent structural aspects are thoroughly investigated—if necessary by more sophisticated structural analysis techniques. The need for analysis is greatly reduced when the web spacing selected allows ordinary beam theory to be applied for longitudinal moments. The beam theory may be used when the depth of the section is equal to or greater than $\frac{1}{30}$ of the span and when the width W, divided by the number of webs, is not more than $7\frac{1}{2}$% of the span length. For the sections shown in Figures 9.17 and 9.18 the slab cantilever C is about one-fourth W. For box sections with more than two webs the slab cantilever dimension should be selected to provide reasonable requirements for the depth of the top slab and the amount of transverse top slab reinforcement.

9.3.2   Detail Segment Dimensions

The concrete dimensions of top slab, webs, bottom slab, and haunches are determined by structural considerations and by numerous practical factors related to the production of the segments.

The top slab thickness (a in Figure 9.17) is usually 7–10 in. (175–250 mm). It is necessary to consider the following structural factors when selecting the top slab thickness:

1. Bending moments in the transverse direction caused by slab dead load, permanent loads, and live load.
2. Compression zone requirements for longitudinal bending moments must be considered when determining top slab thickness in structures with spans of 350 ft (110 m) or more.
3. Local bending stresses due to wheel loads applied directly over epoxy joints.
4. Local anchorage bearing and splitting stresses for transverse posttensioning (when used) require a minimum thickness of about $8\frac{1}{2}$ in. (216 mm) for tendon forces of 100–120 k (445–534 kN).

In addition to the above structural considerations, the top slab thickness must accommodate four layers of transverse and longitudinal mild steel reinforcement, transverse and longitudinal tendons, and a minimum concrete cover of 2 in. (51 mm) on top and 1 in. (25 mm) on the bottom.

The dimensions of haunches b, c, and d in Figure 9.17 are determined by the transverse bending moments and by the space required for the anchorages of the longitudinal posttensioning tendons. It is generally necessary to accommodate at least two layers of longitudinal tendons. A concrete depth of 14 in. (356 mm) is required at anchorages of longitudinal strand tendons. A depth of 10 in. (254 mm) may suffice for bar tendons. Although it is essential to provide adequate space in the top slab and haunch thicknesses for these considerations, it should also be kept in mind that the top slab is the heaviest part of the box girder, and from this standpoint it is desirable to keep the dimensions as small as practicable.

The web thickness e must be 14 in. (356 mm) or more to provide room for the anchorage hardware of 12-strand tendons, a size frequently selected. Minimum anchorage space requirements for bar tendons is about 10 in. (254 mm). The 14-in. (356-mm) width may also be desirable or necessary to accommodate the bursting and splitting force from anchorages for 12-strand tendons. This thickness may be reduced when tendons are anchored in ribs or anchor blocks. Thicknesses as small as 8 in. (203 mm) have been used with strand tendons when webs were vertically prestressed. When shear forces near supports are reduced by upward shear from the posttensioning tendons, and segment depth is within specified limits, the shear stress requirements for highway bridges are generally met when the total width of webs amounts to 7 or 8% of the bridge width. The principal tensile stresses that result from a combination of vertical shear stresses and compressive stresses reach a maximum value at the intersection of the top slab and web. An effort should be made to keep these principal stresses within allowable limits to avoid the use of additional reinforcement for this purpose. This requires widening the webs f (Figure 9.17).

The web is a stiff element in the box section which provides substantial moment restraint to the top slab; consequently transverse moments at the junction of the web and top slab are high. Increased concrete thickness, obtained by widening the web f (Figure 9.17) reduces the amount of reinforcement required. Particular attention should be given to lapping the reinforcement in this area to avoid discontinuity in areas of high moment.

A different situation exists in positive and negative moment areas related to the required bottom slab thickness g. The structural significance of the bottom slab in the positive moment area relates only to the bottom slab contribution to the section properties. As a result the bottom slab thickness is reduced in positive moment areas to the minimum required to carry the slab dead load and the space needed for reinforcement and concrete cover. Space for one layer of tendons, mild steel reinforcement, and concrete cover require a minimum bottom slab thickness of about 7 in. (178 mm). In the negative moment area the bottom slab thickness is controlled by high compressive stresses. Thickening of the bottom slab near piers is nearly always required to keep the compressive stresses within the allowable limits. The bottom slab thickening for this purpose should be reduced to the minimum thickness required in the shortest distance possible to facilitate manufacture of the segments.

9.3.3   Tentative Design and Construction Specifications

In 1975 the PCI Bridge Committee prepared tentative design and construction specifications and an accompanying commentary in the form of a proposed addition to the AASHTO Standard Specifications for Highway Bridges. These documents were presented to the AASHTO Committee on Bridges and Structures for evaluation and published by the Prestressed Concrete Institute [*J. Prestressed Concr. Inst.* (July-August 1975)] to encourage comments and discussion.

The PCI Bridge Committee evaluated the comments received on the 1975 tentative specifications, as well as new information on design and construction of precast segmental box girder bridges, and prepared the following version of the design and construction specifications for consideration by the AASHTO Subcommittee on Bridges at the 1977 regional meetings. The specification proposals presented in this section represent the recommendations of the PCI Bridge Committee and may be modified before final adoption as AASHTO Standard Specifications for Highway Bridges.

Precast Segmental Box Girders

(A)   General

Except as otherwise noted in this section, the provisions of AASHTO Section 6—Prestressed Concrete shall apply to the analysis and design of precast segmental box girder bridges. Deck slabs without transverse post-tensioning shall be designed under the applicable provisions of AASHTO Section 5—Concrete Design.

Elastic analysis and beam theory may be used in the design of precast segmental box girder structures. For box girders of unusual proportions, methods of analysis which consider shear lag shall be used to determine stresses in the cross section due to longitudinal bending.

(B)   Design of Superstructure

(1)   Flexure

The transverse design of precast segments for flexure shall consider the segment as a rigid box frame. Top slabs shall be analyzed as variable depth sections considering the fillets between the top and webs. Wheel loads shall be positioned to provide maximum moments, and elastic analysis shall be used to determine the effective longitudinal distribution of wheel loads for each load location. Transverse post-tensioning of top slabs is generally recommended.

In the analysis of precast segmental box girder bridges, no tension shall be permitted at the top of any joint between segments during any stage of erection or service loading. The allowable stresses at the bottom of the joint shall be as specified in the AASHTO Code, Article 1.6.6(B)(2).

(2) Shear

(a)   Reinforced keys shall be provided in segment webs to transfer erection shear. Possible reverse shearing stresses in the shear keys shall be investigated, particularly in segments near a pier. At time of erection, the shear stress carried by the shear key shall not exceed $2\sqrt{f'_c}$.

(b)   Design of web reinforcement for precast segmental box girder bridges shall be in accordance with the provisions of AASHTO Article 1.6.13.

(3) Torsion

In the design of the cross section, consideration shall be given to the increase in web shear resulting from eccentric loading or geometry of structure.

(4) Deflections

Deflection calculations shall consider dead load, live load, prestressing, erection loads, concrete creep and shrinkage, and steel relaxation.

Deflections shall be calculated prior to manufacture of segments, based on the anticipated production and erection schedules. Calculated deflections shall be used as a guide against which erected deflection measurements are checked.

(5) Details

(a)   Epoxy bonding agents for match-cast joints shall be thermosetting 100 percent solid compositions that do not contain solvent or any non-reactive organic ingredient except for pigments required for coloring. Epoxy bonding agents shall be of two components, a resin and a hardener. The two components shall be distinctly pigmented, so that mixing produces a third color similar to the concrete in the segments to be joined, and shall be packaged in pre-proportioned, labeled, ready-to-use containers.

Epoxy bonding agents shall be formulated to provide application temperature ranges which will permit erection of match-cast segments at substrate temperatures from 40°F (5°C) to 115°F (46°C). If two surfaces to be bonded have different substrate temperatures, the adhesive applicable at the lower temperature shall be used.

If a project would require or benefit from erection at concrete substrate temperatures lower than 40°F, the temperature of the concrete to a depth of approximately 3 in. (76 mm) should be elevated to at least 40°F to insure effective wetting of the surface by the epoxy compound and adequate curing of the epoxy compound in a reasonable length of time. An artificial environment will have to be provided to accomplish this elevation in temperature and should be created by an enclosure heated by circulating warm air or by radiant heaters. In any event, localized heating shall be avoided and the heat shall be provided in a manner that prevents surface temperatures greater than 110°F (43°C) during the epoxy hardening period. Direct flame jetting of concrete surfaces shall be prohibited.

Epoxy bonding agents shall be insensitive to damp conditions during application and, after curing, shall exhibit high bonding strength to cured concrete, good water resistivity, low creep characteristics and tensile strength greater than the concrete.

In addition, the epoxy bonding agents shall function as a lubricant during the joining of the match-cast segments being joined, as a durable, watertight bond at the joint. See AASHTO Article 2.4.33(M) for epoxy bonding agent specifications.

(b)   AASHTO Articles 1.6.24(C) and 1.6.24(F) relating to flange thickness and diaphragms shall not apply to precast segmental box girders.

(C)   Design of Substructure

In addition to the usual substructure design considerations, unbalanced cantilever moments due to segment weights and erection loads shall be accommodated in pier design or with auxiliary struts.   Erection equipment which can eliminate these unbalanced moments may be used.

COMMENTARY

Precast Segmental Box Girders

(A)   General

Material strengths and allowable stresses need be no different from other prestressed concrete bridges; therefore, current limits in AASHTO Standard Specifications for Highway Bridges should apply.   However, higher strength concrete has advantages and should be used when available.   Higher strength concrete has more durability, not only because of the mix design but also because of the greater quality control required to produce it.

(B)   Design of Superstructure

Influence surfaces for design of constant and variable depth deck slabs have been published (see References 5 and 6).

The following limitations are recommended:

1.   When beam theory is used, single cell boxes should be no more than 40 ft. (12 m) wide, including cantilevers.   For bridges wider than 40 ft., multiple box cross sections or multiple cell boxes are usually used.   Single cell boxes of width greater than 40 ft. can be used if carefully analyzed for shear lag to determine the portion of cross section capable of handling longitudinal moment.

2.   For maximum economy, the span-to-depth ratio for constant depth structures should be 18 to 20.   However, span-to-depth ratios of 20 to 30 have been used when required for clearances or aesthetics. The shallower depths require the use of more high strength post-tensioning steel which may cause congested cross sections. Variable depth structures usually have span-to-depth ratios of 18 to 20 at the supports and 40 to 50 at midspan.

3.   Width-to-depth ratios should also be considered.   A shallow box girder that is too wide begins to behave as a slab.   No criteria have been established, but when the width-to-depth ratio is greater than six, considering the total width of the section including slab cantilevers, it is recommended that the designers use multiple cell boxes or carefully analyze the cross section.

4.   Proper fillets should be used in the cross section to allow stress transfer

around the box perimeter and to provide ample room for the large number of tendons.

5.   Diaphragms should be considered. These are usually required only at piers, abutments, and expansion joints.

6.   The thickened bottom slab in pier segments, when required for stresses, should taper down or step down to the minimum midspan segment bottom slab thickness in as short a distance as is practical.

7.   Web thicknesses should be chosen for production ease. If post-tensioning anchorages are located in the webs, web thickness may be governed by the anchorage requirements.

8.   Permanent access holes into the box section should be limited in size to the minimum functional dimension and should be located near points of minimum stress.

(C)   Design of Substructure

Unbalanced cantilever moments occur during erection only and are usually greater in magnitude than service load moments. Wind loads in combination with erection loads could develop critical stresses and, thus, wind loads should be considered in accordance with Article 1.2.22.

9.4   DESIGN EXAMPLE

To show a complete example of this type of design would be quite extensive.   Therefore it is intended to show only that portion of an analysis of a precast segmental structure which handles the major loadings on the structure—the most significant effects to be evaluated. Moments caused by creep and temperature differential would normally be investigated even though they are not included in this example. Also, a transverse analysis that uses the influence surfaces from refs. (5) and (6) would need to be done. In the transverse direction any effects of applied loadings should be distributed around the entire box section. In addition, a transverse analysis should include an investigation of shear flow, as described in Chapter 7.

Example 9.1.   The precast segmental bridge to be designed is shown in elevation in Figure 9.19 and in section in Figure 9.20. To eliminate rounding of numbers the spans are modified under "Span arrangement" in Figure 9.21. The section dimensions and section properties are listed in Table 9.2.

Order of Erection

Step 1. The segmental cantilevers are erected from each pier.

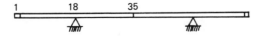

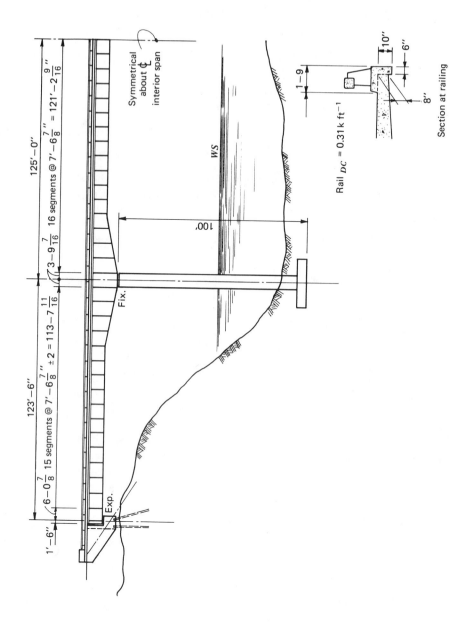

Figure 9.19 Bridge elevation: Example 9.1 with Section at railing.

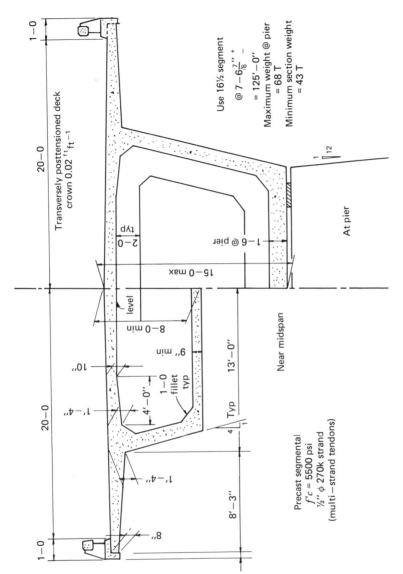

Figure 9.20 Typical Section: Example 9.1.

Transversely posttensioned deck
crown 0.02 ^{ft}ft^{-1}

Use 16½ segment
@ 7–6$\frac{7}{8}$'' +
= 125'–0''   –
Maximum weight @ pier
= 68 T
Minimum section weight
= 43 T

At pier

1
12

20–0

20–0

1–0

1–0

2–0 typ

1–6 @ pier

15–0 max

level

8–0 min

9'' min

13'–0''

10''

1'–4''

4'–0''

1–0
fillet
typ

1'–4''

8''

8'–3''

4
1 Typ

Near midspan

Precast segmental
f''_c = 5500 psi
½'' ϕ 270k strand
(multi–strand tendons)

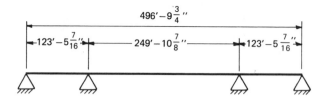

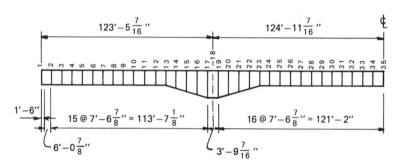

Figure 9.21   Span arrangement.

Table 9.2

Joint Numbers	1 ~ 13 23 ~ 35	13 ~ 14 22 ~ 23	14 ~ 15 21 ~ 22	15 ~ 16 20 ~ 21	16 ~ 17 19 ~ 20	17 ~ 18 18 ~ 19
A (ft^2)	76.3	82.4	88.2	93.8	99.2	104.4
c_t (ft)	2.92	3.62	4.34	5.08	5.83	6.58
I (ft^4)	727.5	1129.3	1635.2	2248.0	2974.5	3816.3
Section Z_t (ft^3)	249.1	312.0	376.8	442.6	509.7	580.0
Modulus Z_b (ft^3)	143.2	195.4	253.1	315.7	382.8	453.2
Bottom thickness (ft)	0.75	0.90	1.05	1.20	1.35	1.5
Average height (ft)	8.00	9.40	10.80	12.20	13.60	15.00

Step 2.   The end supports are erected.

Step 3.   The continuity of midspan is made.

Posttensioning Details

Fourteen $\frac{1}{2}$-in. ϕ 270 k strand tendons except Group 3(b).
Ultimate strength = 578 k per tendon except Group 3(b).
Tendons are stressed initially to 70% of their ultimate strength.

Assume final tendon forces after losses are 60% of ultimate. Assume the following layout for permanent tendons:

Group 1: 38 tendons, 19 in each web (cantilever construction).
Group 2: two tendons, one in each web (tail span continuity).
Group 3(A): eight tendons, four in each web (center span continuity).
Group 3(B): four tendons, located in top flange (seven strands each, center span continuity).

The layout of these tendons is shown in Figure 9.22. The location of these tendon groups is shown in the box section in the following sectional drawings (Figure 9.23). Note the numbering of the tendons for groups 1 and 3a.

Design Loadings and Properties of Material

$f_c' = 5500$ psi.
No tension is allowed for combinations of all loadings.
Cracking safety under 110% of DL and 125% of LL.
Ultimate load capacity of 175% of DL and 225% of LL (more conservative assumption than PCI recommendations).
Final tendon forces are 60% of ultimate.

The design is carried out for loading by the following:

Dead load during construction.
Initial prestress.
Superimposed permanent load.
Live loads.
Temperature differential.*
Creep under box girder dead load.*
Creep under posttensioning.*
Loss of prestress.

*These factors are not considered as part of this example.

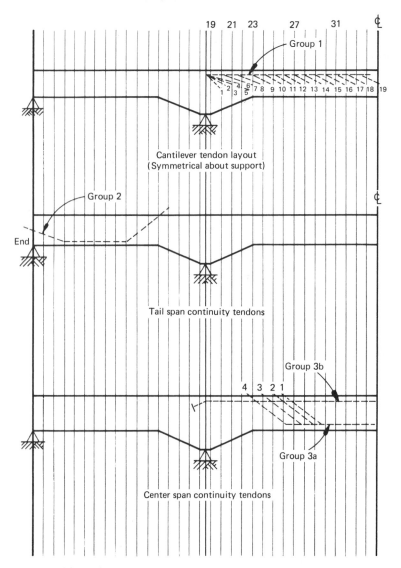

Figure 9.22   Tendon layout: longitudinal section.

Step 1. Free cantilever plus initial cantilever Group 1 posttensioning. The loads to be analyzed are the following:

1.   Dead load of box girders.
2.   Group 1 cantilever posttensioning.

For the dead load of the box girder assume a unit weight of concrete of 150 lb-ft^{-3}; then $WDL = 150 \times (area)$ unit in k-ft^{-1}.

Joint No.	1 ~ 13	13 ~ 14	14 ~ 15	15 ~ 16	16 ~ 17	17 ~ 18	
	23 ~ 35	22 ~ 23	21 ~ 22	20 ~ 21	19 ~ 20	18 ~ 19	
Area	76.3	82.4	88.2	93.8	99.2	104.4	ft^2
WDL	11.45	12.36	13.23	14.07	14.88	15.66	lb-ft^{-1}

The tendon location and the cgs locations are shown in Figure 9.24 for the sections at joints 18, 21, 23, and 27 (Group I tendons).

The cantilever moment is $MDL = \frac{1}{2} w_{DL} \times l^2$ and the segment length = 7.573 ft. The cantilever moments are as follows:

Joint 31. $11.45 \times \frac{1}{2} (7.573 \times 4)^2 = 5253$ k-ft.

Joint 27. $11.45 \times \frac{1}{2} (7.573 \times 8)^2 = 21013$ k-ft.

Joint 23. $11.45 \times \frac{1}{2} (7.573 \times 12)^2 = 47280$ k-ft.

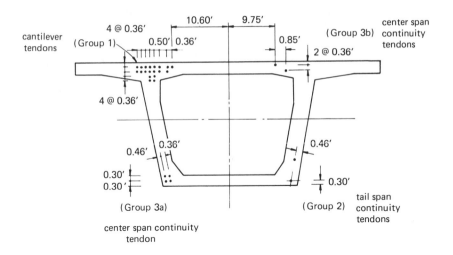

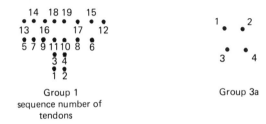

Figure 9.23   Tendon layout: transverse section.

Joint 21. $11.45 \times \frac{1}{2}(7.573 \times 14)^2 + (12.36 - 11.45) \times 7.573 \times$
$(7.573 \times 1.5) + (13.23 - 11.45) \times 7.573 \times (7.573 \times 0.5) = 64482$ k-ft.

Joint 18. $11.45 \times \frac{1}{2}(7.573 \times 16.5)^2 + (12.36 - 11.45) \times 7.573^2 \times$
$4 + (13.23 - 11.45) \times 7.573^2 \times 3 + (14.07 - 11.45) \times 7.573^2 \times 2 +$
$(14.88 - 11.45) \times 7.573^2 \times 1 + (15.66 - 11.45) \times 7.573^2 \times \frac{1}{4} =$
90430 k-ft.

The final tendon forces are 60% of the ultimate strength $578 \times 0.60 =$
346.5 k. The total force at each joint is therefore $346.5 \times$ the number
of tendons:

Joint No.	18	21	23	27	31	35
No. of tendons	38	30	24	16	8	0
Force (kips)	13,167	10,395	8,316	5,544	2,772	0

The eccentricity from the centroid of the groups of tendons to the

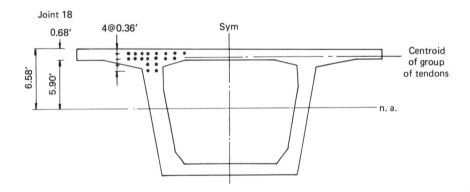

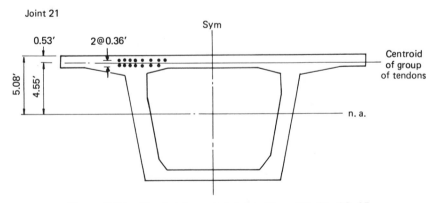

Figure 9.24   Tendon layout: joint sections 18, 21, 23, 27.

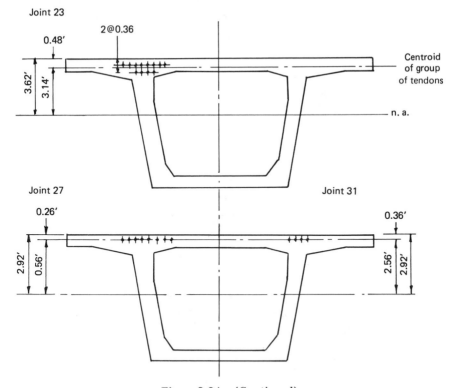

Figure 9.24   *(Continued)*

neutral axis is computed as follows:

No. of tendons   distance to top fiber

Joint 18. $-(8 \times 0.36 + 7 \times 0.36 \times 2 + 2 \times 0.36 \times 3 + 2 \times 0.36 \times 4)/19 + 6.58 = 5.90$ ft.

Joint 21. $-(8 \times 0.36 + 7 \times 0.36 \times 2)/15 + 5.08 = 4.55$ ft.

Joint 23. $-(8 \times 0.36 + 4 \times 0.36 \times 2)/12 + 3.62 = 3.14$ ft.

Joint 27. $-(8 \times 0.36)/8 + 2.92 = 2.56$ ft.

Joint 31. $2.92 - 0.36 = 2.56$ ft.

Use linear interpolation to calculate the eccentricity of joints other than 18, 21, 23, 27, 31:

Joint 19. $(5.90 - 4.55) \times \frac{2}{3} + 4.55 = 5.45$ ft.

Joint 20. $(5.90 - 4.55) \times \frac{1}{3} + 4.55 = 5.00$ ft.

Joint 22. $(3.14 + 4.55)/2 = 3.85$ ft.

Joint 24. $(3.14 - 2.56)\frac{3}{4} + 2.56 = 3.00$ ft.

Joint 25. $(3.14 - 2.56)\frac{2}{4} + 2.56 = 2.85$ ft.

Joint 26. $(3.14 - 2.56)/4 + 2.56 = 2.71$ ft.

Table 9.3

Joint	18	19	20	21	22	23	24	25	26	
e	5.90	5.45	5.00	4.55	3.85	3.14	3.00	2.85	2.71	(ft)
joint	27	28	29	30	31					
e	2.56	2.56	2.56	2.56	2.56	(ft)				
joint	31	32	33	34	35					
e	2.56	1.92	1.28	0.64	0	(ft)				

The eccentricities are summarized in Table 9.3.

The moment due to prestress = $F_p \times e$ (bending moment due to eccentricity) is computed in Table 9.4(a).

The prestress stresses are converted to equivalent moments for purposes of analysis by establishing the following relationship:

$$\frac{F_p}{A} + \frac{M_p}{Z} = \text{prestress stress}$$

$$\frac{F_p(Z)}{A} + M_p = (\text{prestress stress}) \times (Z)$$

where $F_p(Z)/A$ = moment of resistance. The moment of resistance is computed for the compressive stress on the top and bottom fibers;

$$M_{\text{top}} = \frac{F_p}{A} Z_t \quad \text{(moment of resistance)}$$

$$M_{\text{bottom}} = \frac{F_p}{A} Z_b \quad \text{(moment of resistance)}$$

The moment of resistance is listed in Table 9.4(b).

Table 9.4(a)

Joint	18	21	23	27	31	
e	5.90	4.55	3.14	2.56	2.56	(ft)
F_p	13167	10395	8316	5544	2772	(k)
M_p	77685	47300	26112	14193	7096	(k-ft)

Table 9.4(b)

Joint	18	21	23	27	31
F_p =	13167	10395	8316	5544	2772 k
Z_t/A =	5.556	4.719	3.786	3.265	3.265 ft
M_{top} =	73150	49049	31488	18101	9050 k-ft
Z_b/A =	4.341	3.366	2.371	1.877	1.877 ft
M_{bottom} =	-57157	-34986	-19720	-10406	-5202 k-ft

The top and bottom fiber moments are combined for prestress, that is, the M_p values are combined with $(F_p/A)\,Z$ to give the total prestress effect (Table 9.5). This is done first for the top fiber:

$$M_p + M_{\text{top}} = \text{(i)}$$

then for the bottom fiber:

$$M_p + M_{\text{bottom}} = \text{(ii)}$$

therefore the final moments are M_{DL} + (i) or M_{DL} + (ii) (both are compressive stress). The allowable compressive stress = $0.55 f_c'$ = 0.55×5500 = 3025 psi and the allowable moment = $M_{\text{allowable}}$ = $(0.55 f_c')\,Z$.

The prestress moments are then combined with the dead load moments and the summation is compared with an allowable moment by using $0.55 f_c'$ as an upper limit (Table 9.6). These moments are plotted to show the adequacy of the design for Step 1; that is, free cantilever plus group I posttensioning (Figure 9.25).

Table 9.5

Joint	18	21	23	27	31
M_p	77,685	47,300	26,112	14,193	7,096
M_{top}	73,150	49,049	31,488	18,101	9,050
$M_p + M_{\text{top}}$ (i)	150,835	96,349	57,600	32,294	16,146 compression (k-ft)
M_{bottom}	-57,157	-34,986	-19,720	-10,406	-5,202
$M_p + M_{\text{bottom}}$ (ii)	20,528	12,314	6,492	3,787	1,894 tension (k-ft)

Table 9.6

Joint	18	21	23	27	31
M_{DL}	-90,430	-64,482	-47,280	-21,013	-5253 k-ft
$M_p + M_{\text{top}} + M_{DL}$	60,405	31,867	10,320	11,281	10,893 compression
$M_p + M_{\text{bottom}} + M_{DL}$	-69,902	-52,168	-40,888	-17,226	-3359 compression
Z_t	580	442.6	312	249.1	249.1
$M_{\text{allowable}}$ (top)	252,648	192,797	135,907	108,508	108,508 OK
Z_b	453.2	315.7	195.4	143.2	143.2
$M_{\text{allowable}}$ (bottom)	197,414	137,519	85,116	62,378	62,378 OK

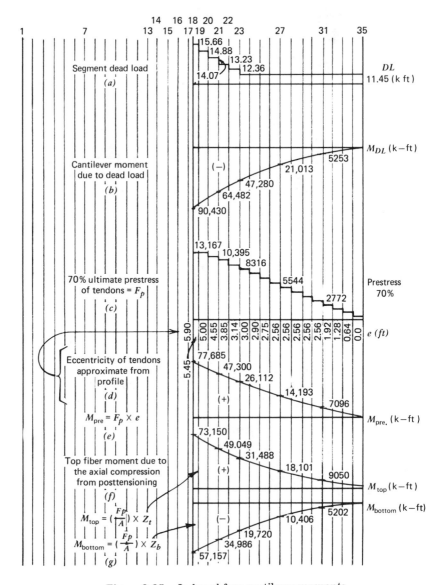

Figure 9.25 Induced free cantilever moments.

450

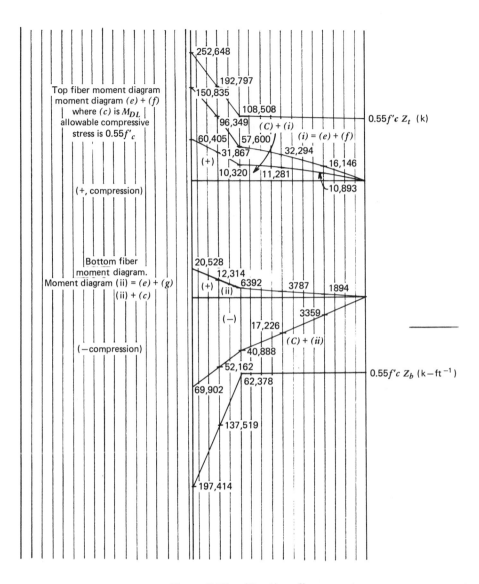

Figure 9.25   (Continued)

Step 2. Completion of end span plus continuity Group 2 post-tensioning. For analysis purposes the changes with respect to Step 1 are the following:

1. end support is added.
2. continuity Group 2 post-tensioning is installed (Figure 9.26).

Referring to Figure 9.27

   Diagram (*a*)   The dead load is the same as that in Step 1.

   Diagram (*b*)   The box dead load moment diagram due to the introduction of the end support is the same as that in Step (1).

   Diagram (*c*)   Prestress force = $-578 \times 0.60 \times 2 = -693$ k; the average box height = 8 ft (1–13) joint.

$$c_t = 2.92 \text{ ft}$$

$$c_b = 5.08 \text{ ft}$$

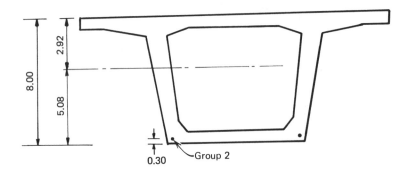

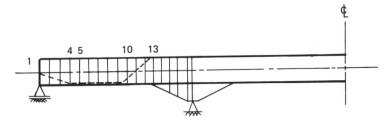

Figure 9.26   Posttensioning: Group 2.

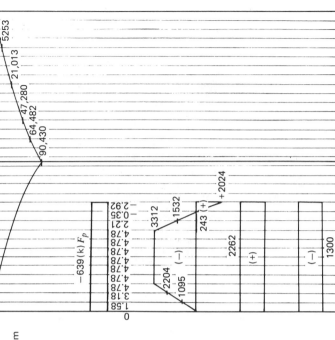

structure configuration of Step 2

(a)

because of no reaction on the left support
moment diagram is the same as cantilever beam

(b)

forces and eccentricities of tendons Group 2

(c)

Prestress bending moment $M_p = F_p e$

(d)

$$M_{top} = \frac{F_p}{A}(Z_t)$$

(e)

$$M\,bottom = \frac{F_b}{A}(Z_b)$$

(f)

Figure 9.27 Induced end span moments.

453

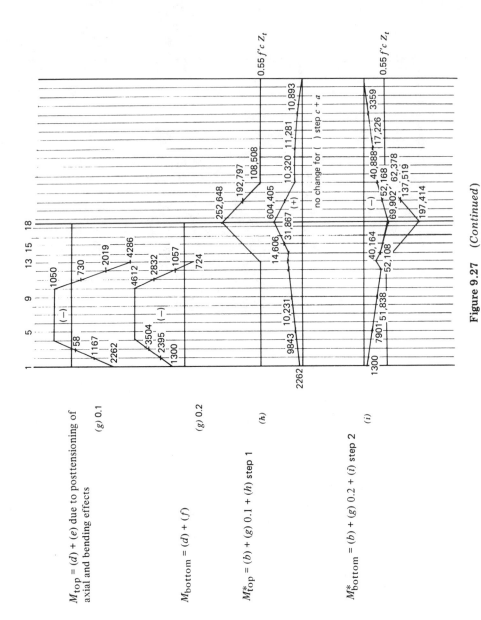

Figure 9.27 (*Continued*)

The eccentricity is estimated from the profile of the tendons (linear interpolation):

$$\text{maximum } e = 5.08 - 0.30 \text{ ft} = 4.78 \text{ ft}$$

$$\text{joint } 11 - \frac{(4.78 + 2.92)}{3} + 4.78 = 2.21 \text{ ft}$$

$$\text{joint } 12 - (4.78 + 2.92)\tfrac{2}{3} + 4.78 = -0.35 \text{ ft}$$

The prestress moment (M_p) and the prestress moment of resistance $F_p Z/A$ are listed in Table 9.7. The dead load effects are then combined with the Group 1 and Group 2 posttensioning effects, tabulated in Table 9.8, and shown in Figure 9.27.

Table 9.7

Joint No.	1	2	3	4 ~ 10	11	12	13
Eccentricity	0	1.58	3.18	4.78	2.21	-0.35	-2.92 ft
Diagram (d) M_p	0	-1095	-2204	-3312	-1532	+243	+2024 k-ft
Z_t/A	3.265	3.265	3.265	3.265	3.265	3.265	3.265 ft
Diagram (e)							
$M_{\text{top}} = F_p(Z_t/A)$	+2262	+2262	+2262	+2262	+2262	+2262	+2262 k-ft
Z_b/A	1.877	1.877	1.877	1.877	1.877	1.877	1.877 ft
Diagram (f)							
$M_{\text{bottom}} = F_b(Z_b/A)$	-1300	-1300	-1300	-1300	-1300	-1300	-1300 k-ft
Diagram (g)							
(1) $M_p + M_{\text{top}}$	2262	1167	58	-1050	730	2019	4286 k-ft
(2) $M_p + M_{\text{bottom}}$	-1300	-2395	-3504	-4612	-2832	-1057	+724 k-ft

Table 9.8

Joint No.	1	5	9	13	15	18
M_{DL}	0	-5253	-21013	-47280	-64482	-90430
Step 2						
$M_p + M_{\text{top}}$	2262	-1050	-1050	4286	0	0
Step 1						
$M_p + M_{\text{top}}$	0	16146	32294	57600	96349	150833
M_{top}^* [a]	2262	9843	10231	14606	31867	6040
Step 2						
$M_p + M_{\text{bottom}}$	-1300	-4612	-4612	+724	0	0
Step 1						
$M_p + M_{\text{bottom}}$	0	1894	3787	6392	12314	20528
M_{bottom}^* [a]	-1300	-7971	-21838	-40164	-52168	-69902

[a] M_{top}^*, M_{bottom}^* are equivalent top or bottom moments due to the combination of dead load and prestress for Steps 1 and 2.

Note. The moments M_{top}^* or M_{bottom}^* from Joints 15-35 are the same as in Step 1 because there is no prestress effective between Joints 15-35 (end cable only).

Step 3. Completion of center span. The tendon location for the Group 3a and 3b tendons is given in Figure 9.28.

The tendon centroid is shown in Figure 9.29 for joints 30–35 and 18–21.

(1) Normally there is a cast-in-place closure pour at midspan that requires the calculation of an additional bending moment due to the weight of this closure pour. In this example we assume that the effects of this weight are already included in the moments computed.

Tendon forces [group 3a and 3b] are computed:

$$
\left.
\begin{array}{ll}
\text{Group 3a} & 8 \times 578 \times 0.6 = 2772 \text{ k} \\[2mm]
\text{Group 3b} & 4 \times \left(\dfrac{578}{2}\right) \times 0.6 \\[3mm]
& = 693 \text{ k } (\because \text{ seven strands only})
\end{array}
\right\} \text{ sum} = 3465 \text{ k}
$$

Joint No.	18	21	26	27	31	35
F_p	-693	-693	-3456	-3456	-3465	-3465

Next, eccentricities are computed by assuming that the area per tendon is equal to one. For joints 30–35 the moments are summed about the top fiber.

Joints 30–35

$$
\frac{(0.5 \times 0.36 \times 0.5 \times 0.72 + 2 \times 7.7 + 2 \times 7.34)}{5} = 6.124 \text{ ft}
$$

Then

$$
e = 6.124 - C_t = 6.124 - 2.92 = 3.20 \text{ ft}
$$

$$
\text{Joint 18, 19} \quad e = 6.58 - 0.54 = 6.04 \text{ ft}
$$

$$
20 \quad e = 5.83 - 0.54 = 5.29 \text{ ft}
$$

$$
21 \quad e = 5.08 - 0.54 = 4.54 \text{ ft}
$$

$$
\text{Joint 22} \quad e = \frac{(+4.43 + 3.20)}{9} - 4.54 = -3.68 \text{ ft}
$$

$$
\text{Joint 23} \quad e = 0.86 \times 2 - 4.54 = -2.82 \text{ ft}
$$

$$
\text{Joint 24} \quad e = 0.86 \times 3 - 4.54 = -1.96 \text{ ft}
$$

$$
\text{Joint 25} \quad e = 0.86 \times 4 - 4.54 = -1.1 \text{ ft}
$$

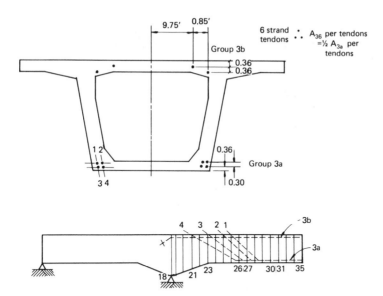

Figure 9.28 Tendon layout: center span, Group 3.

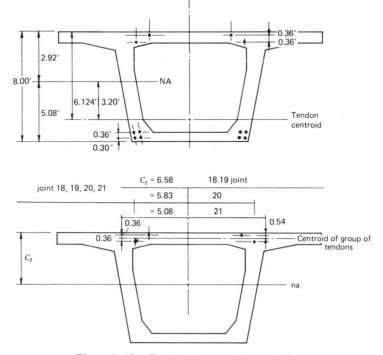

Figure 9.29 Tendon layout: joints 30 ∼ 35.

457

Joint 26 $e = 0.86 \times 5 - 4.54 = -0.24$ ft

Joint 27 $e = 0.86 \times 6 - 4.54 = +0.62$ ft

Joint 28 $e = 0.86 \times 7 - 4.54 = 1.48$ ft

Joint 29 $e = 0.86 \times 8 - 4.54 = 2.34$ ft

The resulting M_p for Joints 18–35 are listed in Table 9.9.

Using linear interpolation, we estimate the e values from Joints 22–30:

The rotation angle at the support is computed by the continuous conjugate beam method, assuming a constant I as an approximate solution. The secondary moment M_s is computed to provide equal slopes across the support. Continuity conditions that develop from this are:

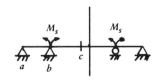

$$V_R = V_{\hat{L}}$$

$$(76610 - 28946 - 16.5M_s) = 5.5M_s$$

$$M_s = 2167 \text{ k-ft}$$

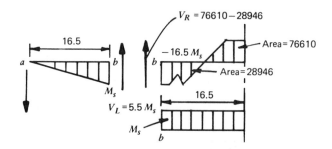

The secondary moment (M_s) can be added to the primary moment (M_p), as listed in Table 9.9, for all the joints; this results in the final moments $(M_{p \text{ cont.}})$ described in Table 9.10. Also listed are the moments at the top and bottom that are due to the prestressing force, where $M_p + M_{\text{bot}}$ or $M_p + M_{\text{top}}$ are the equivalent top or bottom fiber moment to account for the prestress. These moments are then combined with the moments computed with the other prestressing operations, as given in Table 9.11. The results in Tables 9.9–9.11 are shown in Figure 9.30.

Table 9.9

Joint	18	19	20	21	22	23	24	25	26	27	28	29	30	35	
F_p =	-695	-695	-695	-695	-1249	-1802	-2356	-2910	-3465	-3465	-3465	-3465	-3465	-3465	k
e =	-6.04	-6.04	-5.29	-4.54	-3.68	-2.82	-1.96	-1.10	-0.24	0.62	1.48	2.34	3.20	3.20	ft
$M_p = F_p e$ =	4200	4200	3676	3155	4596	5082	4618	3201	832	-2148	-5128	-8108	-11088	-11088	k-ft
Average per segment	4200	3938	3416	3876	4839	4850	3910	2017	-1316	-3638	-6618	-9598	-11088		
Length	0.5	1.0	1.0	1.0	1.0	1.0	1.0	1.0	1.0	1.0	1.0	1.0	5.0		
Area under moment curve	2100	3938	3416	3876	4839	4850	3910	2017	-1316	-3638	-6618	-9598	-55440		

$\sum$ Area (+) = 28946.

$\sum$ Area (−) = 76610.

where M_p is the bending moment due to posttensioning for the hinged end.

459

Table 9.10

Joint	18	19	20	21	22	23	24	25	26	27	28	29	30	31	35
M_p cont.	6367	6367	5843	5322	6763	7249	6785	5368	2999	1900	-2961	-5941	-8921	-8921	-8921
F_p	-695	-695	-695	-695	-1249	-1802	-2356	-2910	-3465	-3465	-3465	-3465	-3465	-3465	-3465
Z_t/A	5.556	5.556	5.138	4.719	4.272	3.786	3.265	3.265	3.265	3.265	3.265	3.265	3.265	3.265	3.265
$M_{top} = (Z_t/A)Z_p$	3861	3861	3571	3280	5336	6822	7692	9501	11313	11313	11313	11313	11313	11313	11313
Z_b/A	4.341	4.341	3.859	3.366	2.870	2.371	1.877	1.877	1.877	1.877	1.877	1.877	1.877	1.877	1.877
$M_{bot} = (Z_b/A)F_p$	-3017	-3017	-2682	-2339	-3585	-4273	-4422	-5462	-6504	-6504	-6504	-6504	-6504	-6504	-6504
$M_p + M_{top} = M^*_{top}$	10228	10228	9414	8602	12099	14071	14477	14869	14312	13213	8352	5372	2392	2392	2392
$M_p + M_{bot} = M^*_{bot}$	3350	3350	3161	2983	3178	2976	2363	-94	-3505	-4604	-9465	-12445	-15425	-15425	-15425

Table 9.11

	1	5	9	13	15	18	21	23	27	31	35
Step 2. M^*_{top}	2262	9843	10231	14606	31867	60405	31867	10320	11281	10893	0 (incl. dead ld.)
Step 3. M^*_{top}	0	525	1051	1576	1839	2167 10228	8602	14071	13213	2392	2392
M^*_{top} (g)	2262	10368	11282	16182	33706	62572 70633	40469	24391	24494	13285	2392 k-ft
Step 2. M^*_{bot}	-1300	-7971	-21838	-40164	-52168	-69902	-52168	-40888	-17236	-3359	0
Step 3. M^*_{bot}	0	525	1051	1576	1839	2167 3350	2983	2976	-4604	-15425	-15425
$M^*_{bot}(h)$	-1300	-7446	-20787	-38588	-50329	-67735 -66552	-49185	-37912	-21840	-18784	-15425 k-ft

461

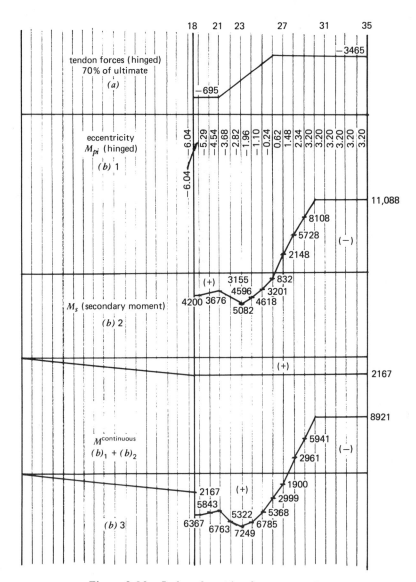

Figure 9.30 Induced cast-in-place moments.

462

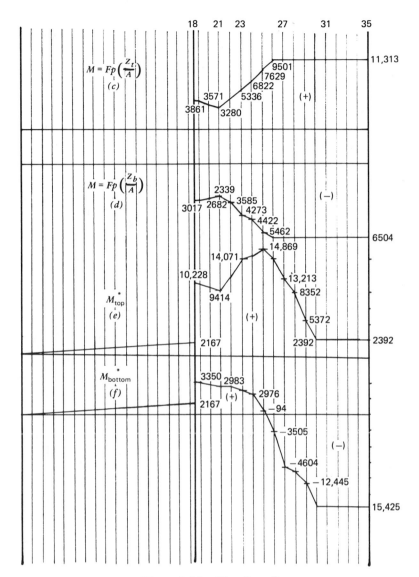

Figure 9.30 *(Continued)*

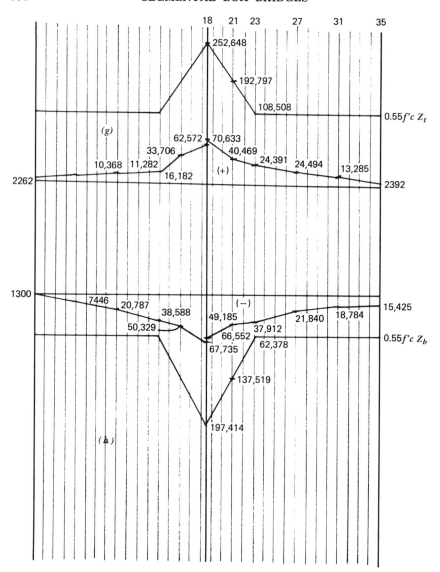

Figure 9.30   (*Continued*)

Step 4. Apply the superimposed dead load @ 0.32 k-ft^{-1}. Using conventional methods (e.g., stiffness method, neglect warping, and shear lag) the moment on this continuous bridge due to a dead load of 0.32 k-ft^{-1} is computed, as shown in Table 9.12, where this moment is combined with the computed moments. Again these values are plotted against an initial allowable of 0.55f'_c and a final allowable of 0.4f'_c (Figure 9.31).

Table 9.12

Joint	1	5	9	13	15	28	21	23	27	31	35
Diag. (a) M_{DL} (0.32 k·ft)	0	48	-189	-714	-1083	-1646	-964	-499	216	644	787
Diag. (g) Step 3 M_{top}^3	2262	10368	11282	16182	33706	62572 / 70633	40469	24391	24494	13285	2392
Diag. (b) M_{top}^4	2262	10416	11394	15468	32623	60926 / 68987	39552	23892	24710	13929	3179 k·ft
Step M_{bot}^3	-1300	-7446	-20787	-38588	-50329	-67735 / -66552	-49185	-37912	-21840	-18784	-15425
Diag. (c) M_{bot}^4	-1300	-7398	-20976	-39302	-51412	-69381 / -68198	-50149	-38411	-21624	-18140	-14638 k·ft

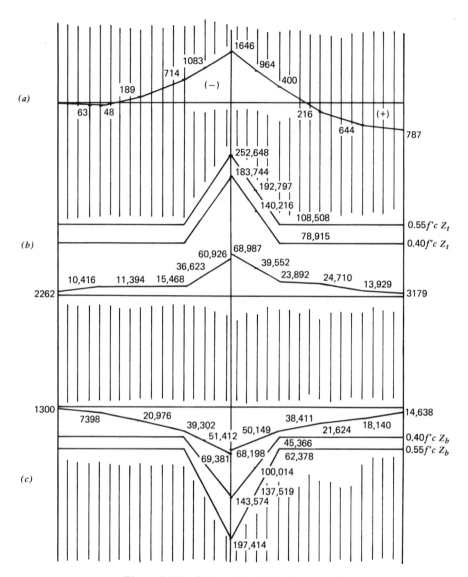

Figure 9.31   Induced dead load moments.

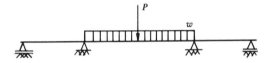

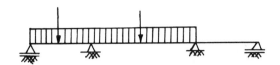

Figure 9.32 Live loading locations.

Step 5. Application of live load. HS 20-44 loading: $P = 18$ k, $w = 640$ lb-ft^{-1}, which are placed as shown in Figure 9.32. Using these loading combinations to get the maximum moments (negative or positive) at the critical points and other variations for maximum moments at the intermediate points, we obtain the moments shown in Table 9.13. In this tabulation the live load plus impact moment is combined with the preceding M_{top} for maximum positive moment and for the maximum negative moment with the preceding M_{bottom}. These values are then plotted against the final allowable of $0.4f'_c$ as shown in Figure 9.33.

Creep on structures with significant spans should also be included with an additional M_{top} and M_{bottom} which result from this action on the continuous unit.

Review of the M_{top} and M_{bottom} diagrams indicates that the design falls well within acceptable limits. Therefore the influence of these items is not significant.

BIBLIOGRAPHY

1. *Precast Segmental Box Girder Bridge Manual*, Prestressed Concrete Institute, 20 N. Wacker Drive, Chicago, Illinois, 1978.
2. W. Podolny, An Overview of Precast Prestressed Segmental Bridges, *J. Prestressed Concr. Inst.*, **24**(1), 56–87 (January–February 1979).

Table 9.13

Joint	1	5	9	13	15	18	21	23	27	31	35
Max. positive	0	2932	3747	2449	1324	462	577	711	2596	5078	6167
M^4_{top} Step 4	2262	10416	11394	15468	32623	60926 68987	39552	23892	24710	13929	3179
M^5_{top}	2262	13348	15141	17917	33947	61388 69449	40129	24603	27306	19007	9346
Max. negative	0	-2323	-4649	-6975	-8137	-10506	-6351	-3618	-736	-866	-995
M^*_{bot} Step 4	-1300	-7398	-20976	-39302	-51412	-69381 -68198	-50149	-38411	-21624	-18140	-14638
M^5_{bot}	-1300	-9721	-25625	-46277	-59549	-79887 -78704	-56500	-42029	-22360	-19006	-15633

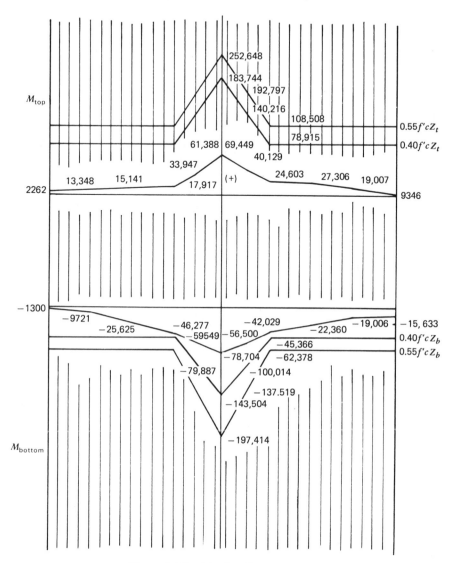

Figure 9.33 Live loading moments.

469

3. Jean Muller, Ten Years of Experience in Precast Segmental Construction, *J. Prestressed Concr. Inst.*, **20**(1), 28-61 (January–February 1975).

4. PCI Committee on Segmental Construction, Recommended Practices for Segmental Construction in Prestressed Concrete, *J. Prestressed Concr. Inst.*, **20**(2), 22-41 (March–April 1975).

5. Helmut Homberg, *Fahrbahnplatten mit Verandlicher Dicke*, Springer-Verlag, New York, 1968.

6. Helmut Homberg and Walter Ropers, *Fahrbahnplatten mit Verandlicher Dicke*, Springer-Verlag, New York, 1965.

7. James R. Libby, Long Span Precast, Prestressed Girder Bridges, *J. Prestressed Concr. Inst.*, **16**(4) 80-98 (July–August 1971).

8. T. Y. Lin, *Design of Prestressed Concrete Structures*, 2nd ed., Wiley, New York, 1963.

9. *Post-Tensioning Manual*, Post-Tensioning Institute, Glenview, Illinois, 1976.

10. *AASHTO Standard Specifications for Highway Bridges, 1977* with the 1978 to 1981 Interim Specifications, The American Association of State Highway and Transportation Officials.

TEN

Substructure Design

The substructure of a bridge is one of its most significant components in regard to design and layout and greatly influences its economy.

Substructure design is significant because of the many possible load variations that can exercise control. Ordinarily most loadings and their combinations, as described in Chapter III, are important only in the design of the substructure. In addition to the seven or more group-type loadings or combinations that could be applied are three or four variations within the groups. The substructure design process therefore is the most demanding in bridge construction.

The bridge layout is designated by the proper spacing of the substructure units and is often the most significant factor in producing the most economical design. Placement and location of the substructure also will influence the aesthetics of the structure. Therefore its layout must be considered an integral part of the total bridge design.

Generally the substructure must be designed to accommodate adequate resistance to vertical and lateral loads and be designed to prevent any serious settlement that might cause distress to any of the portions of the superstructure.

In the design of the substructure units the details of three basic components are presented: (1) bearings, (2) piers, and (3) abutments.

Bearings

In general, steel roller or rocker-type bearings are not practical for concrete bridges. The most common type for high loads is the "pot bearing" which can be provided with a pinned or expansion condition. It is common also on shorter concrete bridge spans to use elastomeric bearing pads for pinned or expansion conditions. The design of these bearings is discussed in Section 10.2.

Piers

Piers may have many different configurations (see Figure 10.1). Normally they are required to withstand not only high vertical loads but large lateral loads as well because of the various possible loading effects mentioned in the AASHTO specifications. When cast-in-place concrete structures are used it is common practice to cast the pier and super-

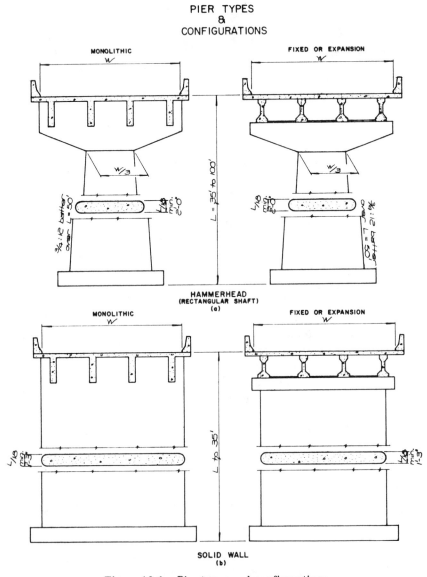

Figure 10.1   Pier types and configurations.

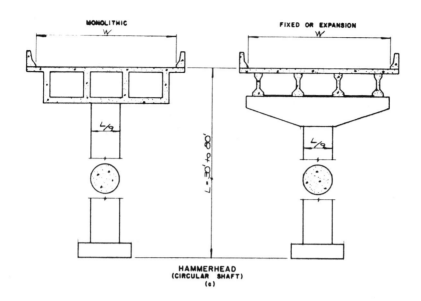

HAMMERHEAD
(CIRCULAR SHAFT)
(c)

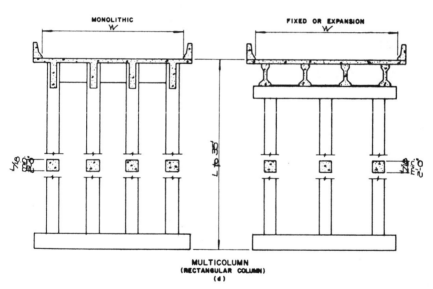

MULTICOLUMN
(RECTANGULAR COLUMN)
(d)

Figure 10.1   (*Continued*)

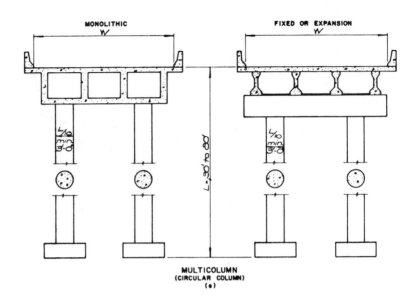

MULTICOLUMN
(CIRCULAR COLUMN)
(e)

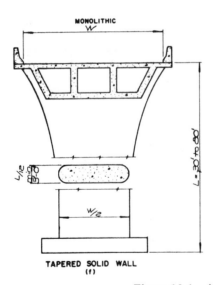

TAPERED SOLID WALL
(f)

Figure 10.1   (*Continued*)

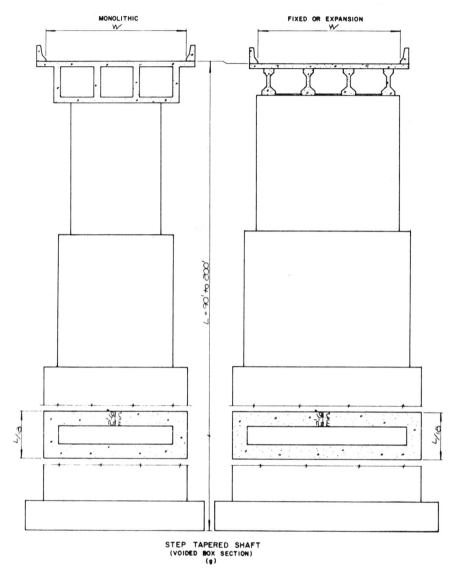

STEP TAPERED SHAFT
(VOIDED BOX SECTION)
(g)

Figure 10.1   (Continued)

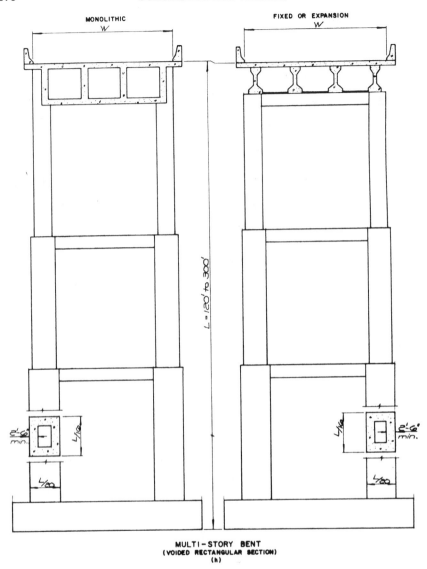

MULTI - STORY BENT
(VOIDED RECTANGULAR SECTION)
(h)

Figure 10.1 (*Continued*)

structure monolithically with adequate reinforcement to make framed corners. These efficient rigid frames have a number of redundancies that give them added reserve strength not accounted for in the design calculations. Precast concrete superstructures are normally placed on bearings that sit on pier caps in which the bearings can transmit longitudinal and transverse lateral loads.

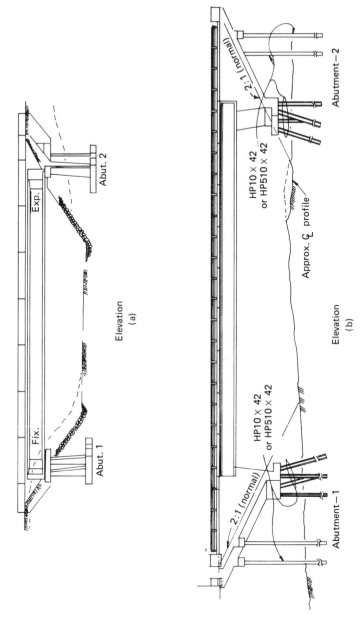

Figure 10.2 Abutment types. (a) Spill-through abutment; (b) closed abutment (partially retained embankment).

477

Abutments

Abutments can be spill-through or closed (Figure 10.2). The spill-through abutment generally has a substantial berm to help restrain embankment settlement at the approach of the structure. Approach embankment settlement can also be accommodated by approach slabs to eliminate bumps at the bridge ends. Closed abutments partially or completely retain the approach embankment from spilling under the span. Bridges of several spans require expansion at the abutments. Therefore they are not usually required to resist the longitudinal forces that develop. In general, the load applied to abutments are the vertical reactions, the lateral earth pressure and an accompanying surcharge effect. Sometimes the friction or shear force produced by expansion bearings is significant and worthy of consideration and can have an additive effect on the earth pressure and surcharge forces.

10.1   FOUNDATION INVESTIGATIONS

The performance of a bridge structure is obviously dependent on the foundation materials and type of foundation provided for the structure. Therefore it is important that a good foundation report, complete with analyses of tests and boring logs and recommendations for various foundation supports, be supplied on every project. The most significant design changes that occur during construction result from poor foundation investigations. Significant sums of money can be invested in foundations, particularly when pile or drilled shaft-type foundations are used. Therefore it is imperative that a careful study be made of the foundation required and the type that would be most feasible for any given structure. AASHTO has provided a guide to the type of information that should be obtained in a normal foundation investigation.[31] It is also appropriate that the foundations engineer report the various alternates that are possible for the structural engineer to consider in his preliminary layout.

10.2   BEARINGS

10.2.1   General

Bridge bearings serve several purposes, first of which is to transmit vertical loads from the superstructure to the substructure. These bearings must also transmit lateral forces, longitudinal or transverse in direction, between the superstructure and substructure.

Bearings may be classed as fixed (sometimes called pinned) and will transmit longitudinal and transverse lateral loads. Bearings that are classed as expansion generally transmit only friction or longitudinal shear force from the movement of the bearing during longitudinal expansion and contraction. Expansion bearings often transmit lateral thrust in transverse direction between the superstructure and substructure.

Numerous types of bearing are available. Some years back steel bearings made of steel plates, rollers, or rockers and sometimes steel pins were quite common. Today they are not generally used in concrete bridge construction. On some rare occasions they may be practical, but elastomeric bearing pads, teflon coated or with no special surface coating, are the normal choice. This is particularly true in shorter span bridges; for longer span bridges in which reactions could be 300 k or greater high-load "pot bearings" are a practical consideration.

The pot bearings shown in Figure 10.3 are made of elastomeric material confined within a ring. This material is capable of resisting high

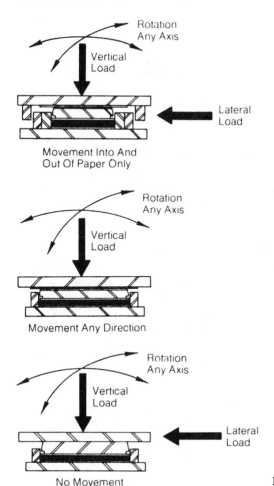

Figure 10.3　"Pot" bearings.

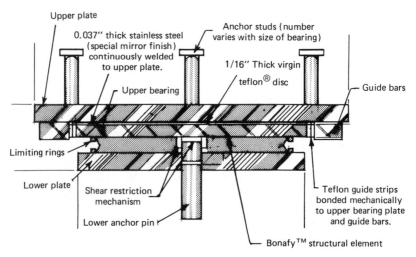

Figure 10.4   Wabo Fyfe bearing.

loads and significant rotations due to the floating characteristic of the
confined material.   The fluid-type action distributes loads equally on
the base plate even with rotations of 1:50.   A TFE surface moving
against a stainless steel surface or a TFE surface against another TFE
surface can provide a high-load bearing surface with low friction values.
Keeper bars or pins can restrain lateral movement in one or two directions.   Different manufacturers provide variations in detail; therefore
it is important that the designer seek out manufacturers' data before
working out a detailed design.   An interesting variation is the propri-
etary Wabo-Fyfe bearing shown in Figure 10.4 which uses a cast poly-
urethane elastomeric element.

TFE sliding surfaces of self-lubricating polytetrafluoroethylene are
designed to slide on a hard smooth mating surface at a low friction
value, which is usually stainless steel and sometimes another TFE surface.   AASHTO requires that a bearing with a TFE surface provide for
a rotation of not less than 0.015 rad to prevent excessive local stresses
on the sliding surface. It is also important that the mating surface cover
the TFE surface completely in all operating positions.   Holes or slots
should not be used on any sliding surfaces.

According to AASHTO, the following TFE minimum and maximum
thicknesses are required:

Unfilled or filled TFE	$\frac{1}{32}$ in. (0.8 mm) minimum to $\frac{3}{32}$ in. (2.4 mm) maximum
Fabric containing TFE fibers	$\frac{1}{80}$ in. (0.03 mm) minimum to $\frac{1}{8}$ in. (3.2 mm) maximum

| Interlocked bronze and filled TFE structures | $\frac{1}{32}$ in. (0.8 mm) minimum to $\frac{1}{8}$ in. (3.2 mm) maximum |
| TFE-perforated metal composite | $\frac{1}{16}$ in. (1.6 mm) minimum to $\frac{1}{8}$ in. (3.2 mm) maximum |

Bearing manufacturers generally assign a value for the coefficient of friction of TFE surfaces. AASHTO supplies alternate values for this coefficient of friction:

Material	Bearing Pressure		
	500 psi (3.447 MPa)	2000 psi (13.790 MPa)	3500 psi (24.132 MPa)
Unfilled TFE, fabric containing TFE fibers, TFE-perforated metal composite	0.08	0.06	0.04
Filled TFE	0.12	0.10	0.08
Interlocked bronze and filled TFE structures	0.10	0.17	0.05

AASHTO also limits bearing pressures on TFE surfaces to:

Filled TFE	3500 psi (24.132 MPa)
Unfilled TFE (recessed)	3500 psi (24.132 MPa)
Unfilled TFE (not recessed)	2000 psi (13.790 MPa)
Fabric-containing TFE fibers	3500 psi (24.132 MPa)
Interlocking bronze and filled TFE structures	6000 psi (41.369 MPa)
TFE-perforated metal composite	5000 psi (34.474 MPa)

10.1.2   Elastomeric Bearings

AASHTO Article 1.12 contains specifications for the design of elastomeric pads. These pads are available with durometer hardness values that vary between 50 and 70 hardness; 70 hardness material should not be used in laminated pads. Most agencies are now specifying a hardness value of 55 ± 5. AASHTO Article 2.25 lists the physical properties for this bearing.

Pads are laminated to keep the compressive strain and the bulging that results within a specified limit of $0.07\,t$, where t is the total thickness for pads with no laminations or the lamination thickness in laminated pads. Laminated pads must consist of alternate laminations of elastomer and metal or elastomer and fabric separators bonded together. Metal laminations must be covered at the edge with $\frac{1}{8}$ in. of elastomer. Pads 1 in. thick or less can be laminated and pads more than 1 in. thick must be laminated.

The following definitions are applicable to the design of elastomeric pads

L = length of a rectangular bearing parallel to the direction of translation.

W = width of a rectangular bearing perpendicular to the direction of translation.

R = radius of a circular bearing.

t = average thickness of a plain bearing or the thickness of any layer of elastomer in a laminated bearing (including the top and bottom layers).

T = total effective elastomer thickness (summation of t's).

S = shape factor (the area of the loaded face divided by the side area free to bulge).

$S = LW/2t(L + W)$ for rectangular bearings, $S = R/2t$ for circular bearings.

AASHTO provides the following limits in bearing sizes:

Plain bearings	minimum $L = 5T$
	minimum $W = 5T$
	minimum $R = 3T$
Laminated bearings	minimum $L = 2T$
	minimum $W = 2T$
	minimum $R = 2T$

Other AASHTO requirements to be observed in the bearing design are the following:

1. For expansion pads the total movement in both longitudinal directions should not exceed $0.5T$.
2. Maximum dead load plus live load pressure should not exceed 800 psi.
3. Maximum dead load pressure should not exceed 500 psi.
4. Dead load pressure should not be less than 200 psi without positive attachment (such as cementing) to the top surface or the top and bottom surfaces.
5. Compressive strain must be less than $0.07\ t$.
6. Nonparallel load surfaces that produce an offset greater than $0.06\ t$ over the bearing length shall not be permitted.
7. If tapered bearings are used, the maximum taper shall be $\frac{5}{8}$ in.-ft^{-1}.

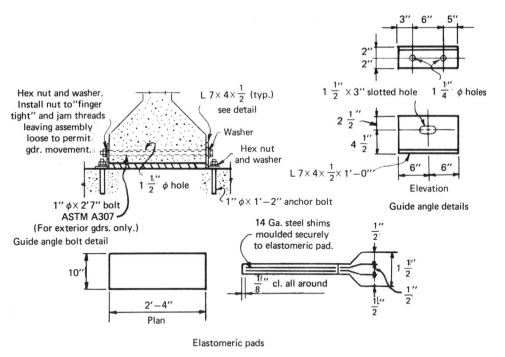

Hex nut and washer. Install nut to "finger tight" and jam threads leaving assembly loose to permit gdr. movement.

L 7× 4× $\frac{1}{2}$ (typ.) see detail

Washer

Hex nut and washer

$1\frac{1}{2}''$ φ hole

1″ φ × 2′7″ bolt ASTM A307 (For exterior gdrs. only.)

Guide angle bolt detail

$1\frac{1}{2}'' \times 3''$ slotted hole $1\frac{1}{4}''$ φ holes

L 7× 4× $\frac{1}{2}$× 1′−0″

1″ φ× 1′−2″ anchor bolt

Elevation

Guide angle details

14 Ga. steel shims moulded securely to elastomeric pad.

$\frac{1}{8}''$ cl. all around

10″

2′−4″

Plan

Elastomeric pads

Figure 10.5 Expansion end details.

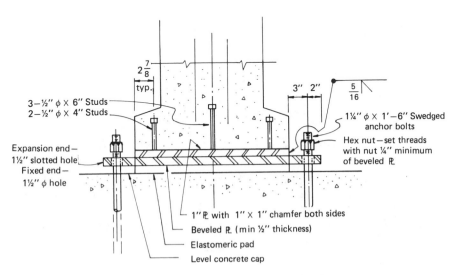

$2\frac{7}{8}$
typ.

3″ 2″

$\frac{5}{16}$

3−½″ φ X 6″ Studs
2−½″ φ X 4″ Studs

1¼″ φ X 1′−6″ Swedged anchor bolts

Hex nut−set threads with nut ¼″ minimum of beveled ℝ

Expansion end−
1½″ slotted hole
Fixed end−
1½″ φ hole

1″ ℝ with 1″ X 1″ chamfer both sides

Beveled ℝ (min ½″ thickness)

Elastomeric pad

Level concrete cap

Figure 10.6 Beveled sole plate detail.

483

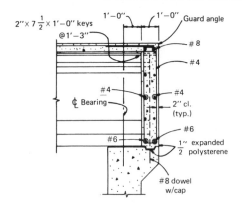

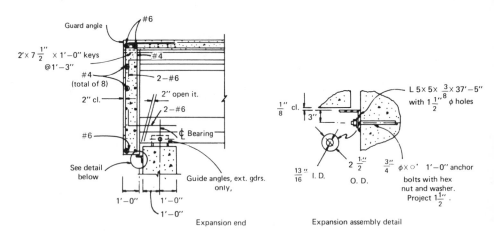

Figure 10.7   Details at expansion and fixed ends.

Figures 10.5 to 10.7 illustrate typical details that can be used for bearings and connections at fixed and expansion locations.

In general the pad design process is as follows:

1. Select a pad thickness based on the temperature motion requirements.

2. Proportion the pad area based on allowable pressures. With precast members make the bearing width equal to the beam width.

3. Compute actual pressures and shape factor.

4. Select durometer hardness and read the compressive strain from the charts in Figures 10.8–10.11. Compare with the allowable strain.

5. Check for nonparallel load surfaces.

6. Compute the shear forces applied to the substructure.

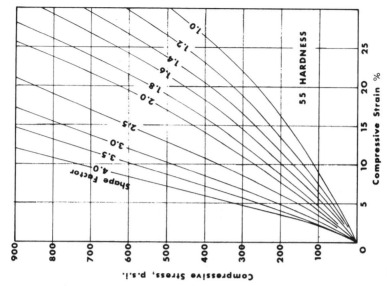

Figure 10.9 Compressive strain versus compressive stress (55 hardness).

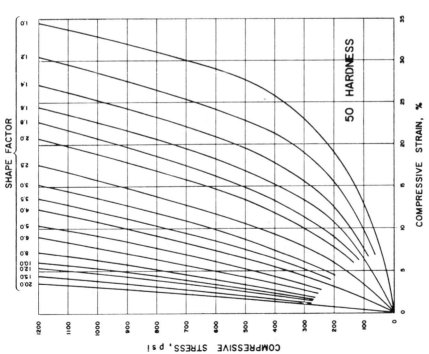

Figure 10.8 Compressive strain versus compressive stress (50 hardness).

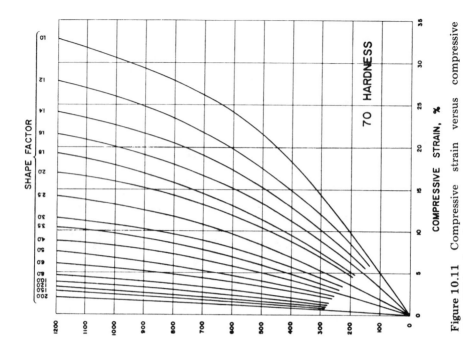

Figure 10.11 Compressive strain versus compressive stress (70 hardness).

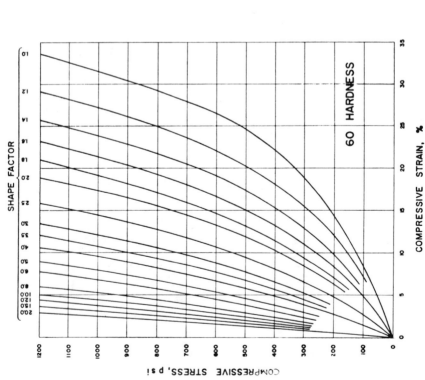

Figure 10.10 Compressive strain versus compressive stress (60 hardness).

An example of this design process follows:

Example 10.1 (Figure 10.12)

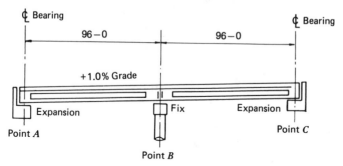

Figure 10.12 Bridge girder layout.

Reactions	@ A	@ B
Girder, slabs diaphragm	94.0	188.0
SDL	14.4	48.0
End diaphragm	14.2	7.6
Total *DL*	122.6 k	243.6 k
LL (no impact)	95.5	106.9
DL and *LL*	218.1 k	350.5 k

Design Bearing Pads

Elastomeric Pad @ Abutment.   Assume a cold climate:

$$\Delta \text{ temperature} = 0.000006 \times 96.0 \times (35 + 45°)$$

$$= 0.0461 \text{ ft} = 0.553 \text{ in.}$$

$$\Delta \text{ temperature} \leqslant \frac{T}{0.5}$$

$$0.553 \leqslant \frac{T}{0.5}$$

$$T \leqslant 1.106 \text{ in.}$$

Try a $1\frac{1}{2}$-in. pad with two laminations ($\frac{5}{8}$ in. each) or a $1\frac{1}{2}$-in. pad with three laminations ($\frac{1}{2}$ in. each).   In some areas it is common to use only $\frac{1}{2}$ in. laminations. Therefore select $T = 1\frac{1}{2}$ in. $t = \frac{1}{2}$ in.

$$A_{required} = \frac{122.6}{0.500} = 245.2 \text{ in.}^2 \text{ for } (DL)$$

$$A_{required} = \frac{218.1}{0.800} = 272.6 \text{ in.}^2 \text{ for } (DL + LL)$$

Try a 10×28 in. pad; $A = 280$ in.^{2}. The bottom flange width $= 28$ in. and thus the 28-in. pad width:

$$f_{DL} = \frac{122.6}{280} = 438 \text{ psi} \begin{array}{l} < 500 \text{ psi} \\ > 200 \text{ psi} \end{array}$$

$$f_{DL+LL} = \frac{218.1}{280} = 779 \text{ psi} < 800 \text{ psi}$$

$$S = \frac{10(28)}{2(0.5)(28 + 10)} = 7.36$$

The end compressive strain $= 5\%$ from Figure 10.10, assuming 60 hardness. Check nonparallel surfaces (bridge grade $= +1.0\%$), rotation @ A:

prestress $\theta_A = +0.0074$

girder DL $\theta_A = \frac{\omega l^3}{24EI} = -\frac{(1.055)(94.75)^3}{24(583,200)(25.13)} = -0.0026$

slab DL $\theta_A = \frac{\omega l^3}{24EI} = -\frac{0.858(94.75)^3}{24(619,200)(25.13)} = -0.0020$

Neglect diaphragm and SDL:

net $\theta_A = +0.0100 + 0.0074 - 0.0026 - 0.0020 = +0.0128$

offset $= +0.0128(10) = 0.1280$ in.

allowable offset $= 0.06(1.50) = 0.0900$ in.

Therefore provide a beveled sole plate as shown in Figure 10.6.

Elastomer Pads @ Pier.   Provide two pads; $t = \frac{1}{2}$ in. (fixed condition):

$$A_{required} = \frac{243.6}{0.500} = 487.2 \text{ in.}^2 \text{ for } DL$$

$$A_{required} = \frac{350.5}{0.800} = 438.1 \text{ in.}^2 \text{ for } DL + LL$$

Try 2 – 9 × 28 in. pads; $A = 504$ in.^{2}

$$f_{DL} = \frac{243.6}{504} = 483 \text{ psi} \begin{array}{l} < 500 \text{ psi} \\ > 200 \text{ psi} \end{array}$$

$$f_{DL+LL} = \frac{350.5}{504} = 695 \text{ psi} < 800 \text{ psi}$$

$$S = \frac{9 \times 28}{2(0.5)(9 + 28)} = 6.81$$

Compressive strain = 4.5 < 7% from Figure 10.10, assuming 60 hardness. Check nonparallel load surfaces (side span AB) (bridge grade = +1.0%):

prestress	$\theta_B = -0.0055$
girder DL	$\theta_B = +0.0026$
slab DL	$\theta_B = +0.0020$

Neglect diaphragm + SDL:

net θ_B = +0.0100 – 0.0055 + 0.0026 + 0.0020 = +0.0091

offset = +0.0091(10) = 0.0910 in.

allowable offset = 0.06(0.5) = 0.0300 in.

Therefore provide a beveled sole plate.

Abutment Design Shear Force from Pad

$$\text{shear force} = \frac{\text{modulus} \times \text{area} \times \text{movement}}{\text{pad thickness}}$$

60 hardness

0°F temperature

From Figure 10.13, modulus = 180. Movement,

$$\Delta = 0.000006 \times 45° \times 96.0$$

$$= 0.0259 \text{ ft (temperature drop)}$$

$$= 0.0259 \text{ ft}$$

creep + shrinkage = 0.0279 ft

Sum   0.0538 ft = 0.6458 in.

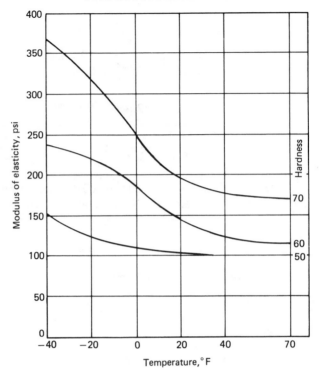

Figure 10.13 Shear modulus.

$$\text{area} = 10(28) = 280 \text{ in.}^2$$

$$T = 1.50 \text{ in.}$$

$$\text{shear force} = \frac{180(280)(0.6458)}{1.5} = 21.7 \text{ k}$$

$$\text{max force before slip} = 20\% \, DL$$

$$= 0.20(122.6) = 24.5 \text{ k}$$

Therefore design the abutment for 21.7 k shear force per pad.

10.3 BRIDGE PIER DESIGN

10.3.1 General

Today it is common practice to build bridge piers with reinforced concrete. Economy and aesthetics are probably the prime reasons. Other innovations, such as segmental and posttensioned cast-in-place piers, have been used. Most designers, however, favor the various reinforced concrete pier designs.

It is reasonable to build piers of reinforced concrete, for, by using different configurations (Figure 10.1) most practical conditions can be accommodated. Piers can be monolithic with cast-in-place superstructures. This is an efficient design because the rigid frame incorporates a substructure stiffness that reduces superstructure moments. In addition, monolithic construction reduces the large substructure moments that result from a free-standing cantilever pier because a pier with moment restraints, top and bottom, will have an inflection point at which the direction of bending will reverse itself. Other piers have fixed or expansion connections to the superstructure.

A number of pier configurations are shown in Figure 10.1. The wall pier is practical for low river heights; the hammer head pier is economical for high river heights; and the multiple column bent is practical on wider structures when river debris collection is not a problem. The hammer head pier and multiple column bent require caps to transfer the superstructure load to the substructure unit. In monolithic construction pier caps are required to transfer moment to the columns by torsional action. On wide caps, when more than one girder is framed into them, this action can become a difficult design problem. Therefore it is best to limit torsional action by framing only one or two girders into the pier. Another common pier is the pile bent, which is normally used on low-level, short-span structures.

For high piers one should consider a battered shaft on a hammer head pier or the stepped shaft as shown in Figures 10.1a and 10.1g. These variable moment-of-inertia piers offer some complications in analysis because of the difficulty of computing critical buckling loads; however, they are efficient and economical. Another option for high piers is the multistory frame shown in Figure 10.1h, which is efficient and economical for heights of more than 100 ft.

A pier shaft configuration is usually circular or rectangular, but more exotic configurations are now prevalent because of today's demand for more aesthetics, as shown in Figure 10.14 for the Linn Cove Viaduct.

Rectangular shafts are often modified with rounded ends to improve stream flow characteristics. Reinforced concrete rectangular shafts are more efficient than circular shafts when designed to carry moment; therefore they work well in monolithic rigid frames. Circular shafts, when used in rigid frames, tend to attract moment because of their stiffness and then have difficulty in developing adequate strength to carry it. Circular shafts have some advantages, such as ease in forming and the added confining strength that results from spirals incorporated in the section. The close spacing of spiral reinforcing also provides excellent buckling strength characteristics to the main reinforcing, a definite advantage under heavy earthquake loading.

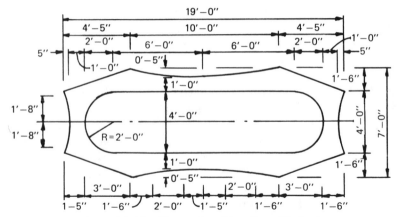

Figure 10.14 Pier section Linn Cove Viaduct.

10.3.2 Design Specifications

Current bridge design codes generally contain criteria developed for use in building design.[3] Although for many reasons they are satisfactory, it is important that the bridge designer realize that the requirements in the design of bridge piers vary in the design of building columns. Most bridge piers have low (P/P_0) ratios and high bending moments and often the bending moment controls their design. Other differences such as massiveness and shape are typical of bridge piers. Bridge pier shapes are frequently dictated by aesthetic requirements.

AASHTO Section 1.5 for the design of reinforced concrete offers two basic design procedures: service load and load factor. When column design is considered the specifications given under service load design state that the column design must be based on capacity. The code requires that the service loads and moments may not exceed 35% of the capacity of the column section with $\phi = 1.0$. This service load procedure also accounts for moment magnification and slenderness, as outlined in the load factor design section of this chapter. AASHTO Section 1.5 on load factor design follows the more traditional ultimate design procedures for columns in which $\phi = 0.70$ for tied columns and $\phi = 0.75$ for spiral columns. The specifications also allow for a variation from the column ϕ value to the flexural value of $\phi = 0.90$ for pure bending as the axial load strength decreases from $0.10f_c'A_g$ or P_b, whichever is smaller, to pure bending with no axial load. Most bridge pier designs fall within this range; thus modification for ϕ should be taken into account. It can be shown that the service load procedures provided for in the AASHTO specifications are extremely conservative and somewhat unrealistic. Therefore it is recommended that only the load factor portion of this section be used for the design of reinforced concrete columns.

Although the AASHTO design specifications no longer provide for service load stress level computations, some consideration should be given to computing stresses at the service load. The specifications do not contain special serviceability criteria for piers. It is desirable, however, that these criteria be applied to piers in water or buried in the ground, where they might be exposed to cyclic freezing and thawing. As a matter of judgment it is reasonable to assume a maximum permissible crack width of 0.006 in.; the formula for crack-width determination at the level of reinforcing is $w = 0.115\sqrt[4]{A}\,(f_s)(1/10^6)$,[27] where A = the area of concrete surrounding one bar in square inches $(3.0 < A < 50.0 \text{ in.}^2)$.

10.3.3   Pier Moments and Loads

When computing pier loads and moments the designer is forced to make a number of assumptions and approximations based on engineering judgment.

The AASHTO specifications require that first-order loads and moments be computed on the basis of elastic analysis, whereas stiffness assumptions are to be reasonable and consistent. AASHTO offers no guidance to what is reasonable and consistent, but the commentary on the Ontario Bridge Code suggests an uncracked cross section, neglecting all reinforcement, or a transformed cracked section throughout. The most correct representation of the ultimate state would require moment–curvature relationships. Reasonable results, however, can be obtained if EI values are assumed to be equal to $E_c I_g (0.2 + 1.2 \rho_t E_s / E_c)$ for piers and to $0.5 E_c I_g$ for superstructure stiffness, where I_g = gross moment of interia, $\rho_t = A_s/A_g$, E_s = 29,000,000 psi, and E_c = 57,000 $\sqrt{f_c'}$ (psi).

The specifications allow for no redistribution of moments, but one important assumption to be made when loads and moments are computed involves the degree of fixity for foundation conditions. Minimum information is available, however, for the determination of end fixity. Reference 4 contains some information that relates pier stiffness to foundation stiffness (Table 10.1). In general, the following is recommended for foundation fixity in lieu of a sophisticated analysis.

1. Piers on multiple rows of piles are 100% fixed at the connections to the piles.

2. Piers on a single row of piles are pinned at the pile connections to the footing.

3. Piers on spread footings with an allowable foundation pressure* of more than 3–6 TSF = 30% fixed at the bottom of the footing.

4. Piers on spread footings with an allowable foundation pressure* of more than 6–9 TSF = 40% fixed at the bottom of the footing.

*Allowable foundation pressure at a service-load level.

Table 10.1   Foundation Fixity Parameters[a]

G_B	
1.5	Footing on rock anchored
3.0	Footing on rock not anchored
5.0	Footing on soil
1.0	Footing on piles (add 10 ft (3.05 m) to the effective length)

$$G_B = \frac{EI/L \text{ columns}}{EI/L \text{ members resisting column bending @ } B \text{ end of column}}$$

[a]Taken from ref. 4.

5. Piers on spread footings with an allowable foundation pressure*
    of more than 9 TSF (competent rock) = 100% fixed at the bot-
    tom of the footing.

Many structures are detailed to eliminate deck joints. One technique
removes abutment backwalls and compacts the approach embankments
against the superstructure end beams. These embankments offer minimal restraint against sidesway; thus sidesway and all second-order effects should be included in bridge pier design. The AASHTO specifications[1] specify that an accurate secondary analysis be used to account
for "the influence of axial loads and variable moment of inertia on
member stiffness and fixed end moments, the effect of deflections on
moments and forces, and the effects of the duration of loads." These
secondary effects can be approximated by the moment magnifier
method.

The moment magnifier equation can be derived by examining Figure 10.15. It is important to note that this equation is intended only to
magnify pier moments that result from lateral loads, in spite of which
most designers magnify the total pier moment conservatively.

By examining Figure 10.15 we obtain the primary deflection Δ_1
caused by the end moments M_1:

$$\Delta_1 = \frac{M_1 L^2}{8EI}$$

The secondary deflection Δ_2, caused by the axial load influence, gives

$$\Delta_2 \simeq \frac{0.10P(\Delta_1 + \Delta_2)(L^2)}{EI}$$

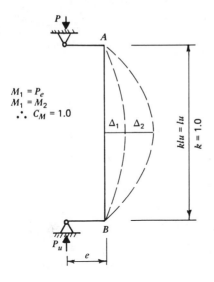

$M_1 = P_e$
$M_1 = M_2$
$\therefore\ C_M = 1.0$

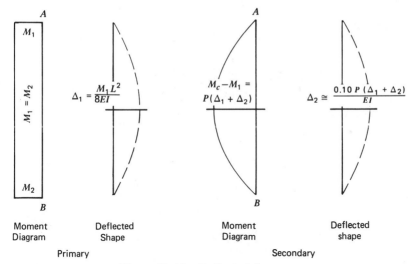

Figure 10.15 Deflected shapes.

However, the critical Euler load is

$$P_{cr} = \frac{\pi^2 EI}{L^2}$$

or

$$\frac{L^2}{EI} = \frac{\pi^2}{P_{cr}}$$

Substitution of this term into the Δ_1 and Δ_2 equations yields

$$\Delta_1 = \frac{M_1}{8}\left(\frac{\pi^2}{P_{cr}}\right)$$

and

$$\Delta_2 \simeq 0.10P(\Delta_1 + \Delta_2)\left(\frac{\pi^2}{P_{cr}}\right)$$

Summing Δ_1 and Δ_2 gives

$$\Delta_1 + \Delta_2 = \left[\frac{M_1}{8} + P(\Delta_1 + \Delta_2)(0.10)\right]\left(\frac{\pi^2}{P_{cr}}\right)$$

However,

$$M_c = M_1 + P(\Delta_1 + \Delta_2)$$

Therefore

$$M_c = M_1 + P\left(\frac{\pi^2}{P_{cr}}\right)\left[\frac{M_1}{8} + P(\Delta_1 + \Delta_2)(0.10)\right]$$

Multiplying gives

$$M_c = M_1 + \left(\frac{P}{P_{cr}}\right)[M_1 + P(\Delta_1 + \Delta_2)]$$

or

$$M_c = M_1 + \left(\frac{P}{P_{cr}}\right)[M_1 + M_c - M_1]$$

Thus

$$M_c = \frac{M_1}{1 - P/P_{cr}}$$

By letting $M_c = \delta M$, where δ is the moment magnifier factor,

$$\delta = \frac{1}{1 - (P/P_{cr})}$$

The specifications introduce ϕ to give

$$\delta = \frac{1}{1 - (P/\phi P_{cr})}$$

The moment magnifier procedure, as first derived, assumes columns with pinned ends, single curvature, equal end moments, and no sidesway. Therefore each pier design must be modified by an effective length factor k to correct it to an equivalent pinned end column with single curvature and no sidesway. The value of k is used to determine reasonable column dimensions for stability and to compute the critical buckling load $P_{cr} = \pi^2 EI/(klu)^2$. The variation in effective length factors can be large as the end conditions change and as the degree of bracing against sidesway varies.

Effective length factors can be determined from Figure 10.16 or for framed structures from the Jackson-Moreland Charts[7] shown in Figure 10.17. These charts use the parameter

$$G_x = \frac{\sum EI/L \text{ for columns}}{\sum EI/L \text{ for members resisting column bending @ } x \text{ end of column}}$$

Effective length factors, K						
	(a)	(b)	(c)	(d)	(e)	(f)
Buckled shape of column is shown by dashed line						
Theoretical K value	0.5	0.7	1.0	1.0	2.0	2.0
Design value of K when ideal conditions are approximated	0.65	0.80	1.2	1.0	2.1	2.0
End condition code	Rotation fixed Translation fixed					
	Rotation free Translation fixed					
	Rotation fixed Translation free					
	Rotation free Translation free					

Figure 10.16 Effective length factors, K.

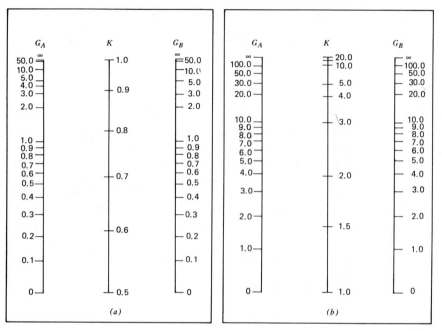

Figure 10.17   Alignment charts for effective length of columns in continuous frames: (*a*) sideways prevented; (*b*) sideways permitted.

The same parameter can be applied in the equations suggested by Cranston[8] for braced columns, where

$$k = 0.7 + 0.05(G_A + G_B) \leqslant 1.0$$
$$k = 0.85 + 0.05(G_{\min}) \leqslant 1.0$$

(use the smaller of the two values) and for unbraced columns, as suggested by Furlong[9] for

$$G_{av} < 2 \qquad k = \frac{20 - G_{av}}{20} \sqrt{1 + G_{av}}$$

and for

$$G_{av} \geqslant 2 \qquad k = 0.9\sqrt{1 + G_{av}}$$

Table 10.1 lists the value *G* to be used for various foundation conditions.

Another important influence on the effective length factor *k* is bracing against sidesway. Presently *k* can be determined only for braced and unbraced conditions; therefore the unbraced condition for bridge pier design is recommended. A braced condition should be considered

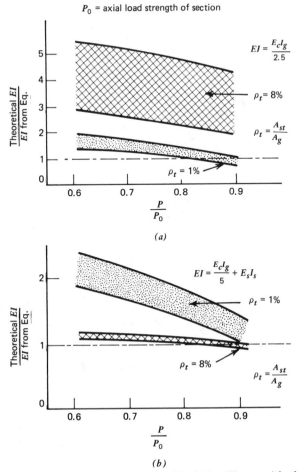

P_0 = axial load strength of section

(a)

(b)

Figure 10.18 Comparison of equations for flexural stiffness with theoretical values from moment-curvature diagrams.

only in a case similar to a row of columns in a bent strutted by a large drift wall with considerable stiffness against lateral movement in the transverse direction.

In the computation of the critical buckling load the formulas for EI were developed using the data shown in Figure 10.18. These curves, which are plots of the ratio of the theoretical to the specification EI for different ratios and percentages of reinforcement, show that $EI = E_c I_s/2.5$ is extremely inaccurate for high percentages of reinforcing and that $EI = E_c I_g/5 + E_s I_s$ is somewhat inaccurate for low percentages.

Therefore it is recommended that $EI = E_c I_g/2.5$ be used for the design of piers with 2% reinforcement or less and $EI = E_c I_g/5 + E_s I_s$ in the design of piers with reinforcement of more than 2%. Again these values are to be applied only in the computation of critical buckling

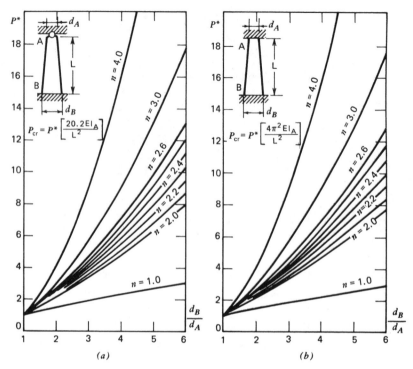

Figure 10.19   Critical buckling loads for tapered shafts: (a) Case 3 column, fixed-pinned ends; (b) Case 4 column, fixed ends.

loads. These specification formulas also give buckling loads for a constant moment of inertia. If the columns are tapered, the critical buckling loads can be computed by referring to the charts shown in Figure 10.19 and Table 10.2. Also, Newmark[11] has developed a numerical procedure for computing buckling loads for columns and piers with a variable moment of inertia.

When a series of piers with a variable stiffness is attached to a continuous superstructure it is appropriate to evaluate the buckling strength of the entire continuous unit. Reference 5 illustrates the procedure, which involves transmitting lateral thrust through the superstructure to provide additional lateral strength to the more limber piers.

The formulas for EI contain the expression $1 + \beta_D$ which accounts for the effect of creep due to sustained loads. The value of $1 + \beta_D$ decreases EI and thereby increases the moment magnification. The specifications define β_D as the ratio of maximum dead load moment to the maximum design total load moment (always positive), which means that β_D is the ratio of dead load moment to the sum of all moments neglecting signs.

Table 10.2   Shape Factors[a]

	Solid, rectangular section Constant width b Varying depth d Buckling about horizontal axis	$n = 3$
	Solid, rectangular section Constant width b Varying depth d Buckling about vertical axis	$n = 1$
	Solid, circular section Varying diameter d	$n = 4$
	Solid, square section Varying dimension d	$n = 4$

[a]Critical buckling loads for tapered shafts taken from Ref. 10.

10.3.4   Section Analysis

In the analysis of a pier section, subjected to the most critical loading conditions, computation of stresses for serviceability and capacity for maximum loads is involved. Modern aesthetics often require that an analysis be made for irregular and nonsymmetrical shapes. In addition to vertical reinforcement, it is essential to evaluate the requirements for lateral reinforcing (hoops, ties, and spirals).

In general, bridge pier sections are governed by moment and not load. Therefore the most critical loading of all the many possible loadings on a pier will be the one that produces the largest moment about the weakest pier axis. When two loads produce similar moments, with variations in load, the load producing the maximum eccentricity will normally control.

The serviceability requirements in the AASHTO specifications are not practical for use in bridge piers and columns in that special crack width control at the reinforcement for service level loadings is not considered. This control is particularly necessary in piers exposed to moisture and freezing and thawing. For piers positioned in water or buried in the ground a limiting maximum crack width of 0.006 in. is recommended. The crack width at the level of the reinforcement can be computed from the following:

$$w = \frac{0.115 \sqrt[4]{A}}{10^6} (f_s) \text{ (Ref. 27)}$$

where A = area of concrete surrounding one bar in square inches (3.0 $<$
    $A < 50.0$ in.^{2}),
   f_s = steel stress at a working level in psi.

Serviceability requirements require that all piers and columns be analyzed for stress. The stresses due to direct load and bending can be evaluated by using the procedure given in the ACI handbook.[32] It is recommended, however, when using this handbook, that the formulas instead of the tables be used to provide more accurate results and minimal work.

Other means of computing stresses in piers include methods by Goodall[33] and Hu.[34] Examples 10.2 and 10.3 illustrate Hu's technique.[34] These methods do not account for double compression in the reinforcing steel; therefore results will be conservative in relation to stress. A method for the stress analysis of circular columns, which does include double compression in the reinforcement, has been developed by Toprac[35] and is illustrated in Examples 10.4 and 10.5.

The axial load capacity can be computed by considering a uniform strain between the extreme concrete fiber and reinforcing steel location at which the maximum allowable concrete strain is 0.3%. Because most columns fall in an area below the balance point on the interaction curve, it is reasonable to compute two steel strains, one at the balance point of 0.207% and the other at about 0.5%. Then, by connecting those two points on an interaction diagram a reasonable approximation can be made of the axial load capacity of most bridge piers and columns. Computation of the biaxial load capacity requires only the evaluation of the two uniaxial capacities and the rotation of a curve between the points, as described and illustrated in the examples that follow.

The AASHTO design specifications indicate that biaxial bending and direct load should be analyzed at ultimate load, based on stress and strain compatibility, or, as an alternate, by one of the following approximate formulas:

$$\frac{1}{P_{nxy}} = \frac{1}{P_{nx}} + \frac{1}{P_{ny}} - \frac{1}{P_0}$$

where $P_u > 0.10 f_c' A_g$ or

$$\frac{M_{ux}}{\phi M_{nx}} + \frac{M_{uy}}{\phi M_{ny}} \leqslant 1.0$$

where $P_u < 0.10 f_c' A_g$. In general, the ultimate strength analysis of bridge piers (P_u) is less than $0.10 f_c' A_g$; however, on occasion the service load (35% ultimate criteria) will cause the value P_u to be greater than $0.10 f_c' A_g$, which would then require use of the first equation (reciprocal load equation). It is possible to simulate the stress and strain compatibility analysis closely by using the PCA Computer Program[13] that employs the equation

$$\left(\frac{M_{ux}}{\phi M_{nx}}\right)^{\log 0.5/\log \beta} + \left(\frac{M_{uy}}{\phi M_{ny}}\right)^{\log 0.5/\log \beta} = 1$$

This equation is based on the work of Parme.[25] According to Furlong,[12] however, it is reasonable to use the elliptic equation

$$\left(\frac{M_{ux}}{\phi M_{nx}}\right)^2 + \left(\frac{M_{uy}}{\phi M_{ny}}\right)^2 \leqslant 1$$

(a simulation of the interaction plane produced by a stress and strain compatibility analysis; Figure 10.20).

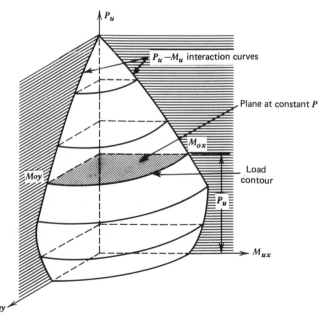

Figure 10.20 Load contours for constant P_u on failure surface.

It should be noted that two inaccuracies occur in the biaxial bending and direct load analysis:

1. An approximation of the true shape of the compression block with a rectangular block over a portion of the compressive area tends to underestimate the capacity.

2. Experimental data show that in some cases a strain value of 0.4% is reasonable as a limiting value before the concrete deteriorates.

These inaccuracies are on the conservative side and are accepted by most designers.

The ultimate strength analysis of irregular and unsymmetrical shapes involves problems similar to those encountered in ultimate biaxial bending analysis; however, minimal research has been conducted in pier analysis. Marin[16] has performed some mathematical modeling of this type of section (Figure 10.21), but experimental tests are required to verify his work. The designer should therefore be cautious when designing this type of section. In particular adequate shear and lateral reinforcement across thin wall portions must be examined.

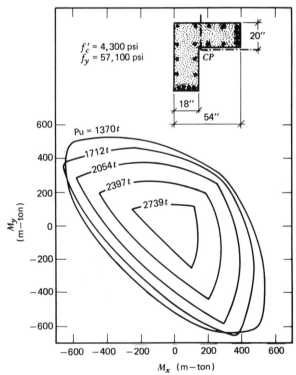

Figure 10.21   Isoload diagram for L cross section. Developed by Marin.[16]

There is considerable variation in thought in regard to the requirements for hoops, ties, and spirals. It is known that lateral reinforcement does little to enhance structural performance before initial concrete deterioration, after which the action of lateral reinforcement is significant. Appropriate details and design can be the difference between catastrophic failure and safety. Experience has shown that spirals perform better than hoops and ties. This is probably due to the closer spacing provided by spirals and to the better confining capabilities of the circular hoop tension they develop. The historic $1'-0''$ hoop spacing is not adequate to prevent vertical reinforcing from buckling when bond is lost due to concrete deterioration. Studies[14,15] have shown that column ductility can be greatly improved by using closely spaced hoops, ties, and spirals in potential plastic hinge areas. Ductility in these areas can serve as a means of energy dissipation under extreme dynamic earthquake loadings. Ductility is measured by what is called the ductility factor (μ), which is defined as the ratio of the maximum displacement under the design earthquake to the theoretical yield displacement. According to studies performed in New Zealand, the design value of μ can be reasonably set at 6.

Ductility requirements should be applied to Zone 3 earthquake areas. The following requirement from the draft of the New Zealand Concrete Design Code, seems, however, to be the most reasonable.

In potential plastic hinge regions, the maximum center-to-center spacing of transverse reinforcement should not be less than the larger of one-fifth of the smaller column dimension (column diameter for circular columns), or six longitudinal bar diameters, or 7.88 in. (200 mm). In circular columns ρ_s should not be less than the smaller of

$$\rho_s = 0.45 \left(\frac{A_g}{A_c} - 1\right) \frac{f'_c}{f_{yh}} \left(0.5 + 1.25 \frac{P_e}{\phi f'_c A_g}\right)$$

or

$$\rho_s = 0.12 \frac{f'_c}{f_{yh}} \left(0.5 + 1.25 \frac{P_e}{\phi f'_c A_g}\right)$$

where P_e should be less than either $\phi 0.7 f'_c A_g$ or $\phi 0.7 P_0$.

In rectangular columns, the total area of hoops and supplementary cross ties in each of the principal directions of the cross section within spacing s_h should be not less than the smaller of

$$A_{sh} = 0.3 s_h h'' \left(\frac{A_g}{A_c} - 1\right) \frac{f'_c}{f_{yh}} \left(0.5 + 1.25 \frac{P_e}{\phi f'_c A_g}\right)$$

or

$$A_{sh} = 0.12 s_h h'' \frac{f'_c}{f_{yh}} \left(0.5 + 12.5 \frac{P_e}{\phi f'_c A_g}\right)$$

where P_e should be less than either $\phi 0.7 f'_c A_g$ or $\phi 0.7 P_0$.

The term P_e in these equations is the maximum design compressive load due to gravity and seismic loading acting on a column during an earthquake. Because AASHTO does not provide for vertical seismic loading, it is reasonable to ignore the vertical seismic effect. It will only add a few percent to P_e and the equations are not that accurate; h'' is the dimension of the concrete core at right angles to the hoops being investigated.

Also in rectangular shafts the center-to-center spacing of tied bars should not exceed the larger of one-third of the column cross-section dimension in that direction, or 7.88 in. (200 mm).

10.3.5 Design Examples

The pier for the bridge in Figures 3.9 and 3.11 is analyzed by service load and load factor design procedures. The design load for this bridge is HS 20 and $\beta_L = 1.67$ for an overload. It is possible that this bridge will occasionally have to carry a special overload vehicle. Therefore we compare the overload vehicle (using $\beta_L = 1.0$) with HS 20 to determine which controls. It is reasonable to check the overload vehicle only for capacity, even under service load design.

Example 10.2. Load factor design. First the loadings must be grouped and tabulated. Group loadings are established in accordance with AASHTO 1.2.22. Cases of live load are for bending in opposite directions longitudinally. Direct loads and moments are tabulated at the top and bottom of the pier shaft, for HS 20, and for the special overload vehicle. Comparisons are made of these tabulations and the most critical loadings are selected for analysis. Computations of all the loadings are given in Example 3.1.

First these tabulations are made for the loads and moments by using the HS 20 live load at the top of the pier (Table 10.3), which is point BD in examples 5.1 and 5.2; β_D is used as specified in AASHTO 1.2.22 when the dead load reaction is tabulated; that is, β_D equals 0.75 to reduce the reaction and 1.0 for moments. This produces a maximum eccentricity, which is a conservative approach.

The same tabulation is then made for the loads and moments at the bottom of the pier shaft, again using the HS 20 live load (Table 10.4).

The same two tabulations are made for the overload vehicle (Tables 10.5 and 10.6).

In the next step in the analysis we compute the section capacity. Because biaxial bending is the most critical, it is necessary to compute the bending capacity about each axis and to rotate a curve between the two axes. It can be assumed that the requirements for capacity are below the balance point on the interaction curve. Therefore the capacity is

Table 10.3   Loads and Moments @ BD Top Pier (HS 20)

Group	Loading	Case IB P	M_L	M_T	Case II P	M_L	M_T
I	$\beta_D D$	226.4	+430		226.4	+322	
	1.67 $(L+I)$	97.5	+578	391	150.3	−364	601
	Σ	323.9	+1008	391	376.7	−42	601
	1.3 × Σ	421.1	+1310	508	489.7	−55	781
II	$\beta_D D$	226.4	+430		+226.4	+322	
	$W + OT$			94			94
	Σ	226.4	+430	94	+226.4	+322	94
	1.3 × Σ	294.3	+559	122	294.3	+419	122
III	$\beta_D D$	226.4	+430		+226.4	+322	
	$L + I$	58.4	+346	234	90.0	−218	360
	0.3W						
	0.30T			28			28
	WL						
	Σ	284.8	+776	262	+316.4	+104	388
	1.3 × Σ	370.2	+1009	341	+411.3	+135	504
IV	$\beta_D D$	226.4	+430		+226.4	+322	
	$L + I$	58.4	+346	234	90.0	−218	360
	$S + T$		+227			−556	
	Σ	284.8	+1003	234	316.4	−452	360
	1.3 × Σ	370.2	+1304	304	411.3	−588	468
V	Σ Group II	226.4	+430	94	+226.4	+322	94
	$S + T$		+227			−556	
	Σ	226.4	+657	94	+226.4	−234	94
	1.25 × Σ	283.0	+821	118	+283.0	−292	118
VI	Σ Group III	284.8	+776	262	+316.4	+104	388
	$S + T$		+227			−556	
	Σ	284.8	+1003	262	+316.4	−452	388
	1.25 × Σ	356.0	+1254	328	+395.5	−565	485
VII	$\beta_D D$	226.4	+430		226.4	+322	
	EQ			0			
	Σ			Neglect			
	1.3 × Σ						

Table 10.4   Loads and Moments @ DB Bottom Pier (HS 20)

Group	Loading	Case IB P	Case IB M_L	Case IB M_T	Case II P	Case II M_L	Case II M_T
I	$\beta_D D$	254.9	+149		254.9	+112	
	$1.67\,L + I$	97.5	+194	391	150.0	-125	360
	Σ	352.4	+343	391	404.9	-13	360
	$1.3 \times \Sigma$	458.1	+446	508	526.4	-17	468
II	$\beta_D D$	254.9	+149				
	W			317			
	OT			94			
	Σ	254.9	+149	411			
	$1.3 \times \Sigma$	331.4	+194	534			
III	$\beta_D D$	254.9	+149		254.9	+112	
	$L + I$	58.4	+116	234	90.0	-75	360
	βW			95			95
	$0.3\,OT$			28			28
	WL			170			170
	Σ	313.3	+265	527	344.9	+37	653
	$1.3 \times \Sigma$	407.3	+344	685	448.4	+48	849
IV	$\beta_D D$	254.9	+149		254.9	+112	
	$L + I$	58.4	+116	234	90.0	-75	360
	$S + T$		+229			-560	
	Σ	313.3	+494	234	344.9	-223	360
	$1.3 \times \Sigma$	407.3	+642	304	448.4	-290	468
V	Group II	254.9	+149	411			
	$S + T$		+229				
	Σ	254.9	+378	411			
	$1.25 \times \Sigma$	318.6	+472	514			
VI	Group III	313.3	+265	527	344.9	+37	653
	$S + T$		+229			-560	
	Σ	313.3	+494	527	344.9	-523	653
	$1.25 \times \Sigma$	391.6	+617	659	431.1	-654	816
VII	$\beta_D D$	254.9	+149				
	EQ			1161			
	Σ	254.9	+149	1161			
	$1.3 \times \Sigma$	331.4	+194	1509			

508

Table 10.5   Loads and Moments @ *BD* Top Pier (Overload Vehicle)

	LL Case	Case V*B*			Case VI		
Group	Loading	P	M_L	M_T	P	M_L	M_T
I	$\beta_D D$	226.4	+430		226.4	+322	
	$L + I$	102.6	+652	290	135.5	-499	384
	Σ	329.0	+1082	290	361.9	-177	384
	$1.3 \times \Sigma$	427.7	+1407	377	470.5	-230	499
II	$\beta_D D$	226.4	+430				
	$W + OT$			94			
	Σ	226.4	+430	94			
	$1.3 \times \Sigma$	294.3	+559	122			
III	$\beta_D D$	226.4	+430		226.4	+322	
	$L + I/1.67$	61.4	+390	174	81.1	-299	230
	$0.3W$						
	$0.3OT$			28			
	WL						
	Σ	287.8	+820	202	307.5	+23	230
	$1.3 \times \Sigma$	374.1	+1066	263	399.8	+30	299
IV	$\beta_D D$	226.4	+430		226.4	+322	
	$L + I/1.67$	161.4	+390	174	81.1	-299	230
	$S + T$		+227			-556	
	Σ	287.8	+1047	174	307.5	-533	230
	$1.3 \times \Sigma$	374.1	+1361	226	399.8	-692	299
V	Σ Group III	226.4	+430	94			
	$S + T$		+227				
	Σ	226.4	+657	91			
	$1.25 \times \Sigma$	283.0	+821	118			
VI	Σ Group III	287.8	+820	202	307.5	+23	230
	$S + T$		+227			-556	
	Σ	287.8	+1047	202	307.5	-533	230
	$1.25 \times \Sigma$	359.8	+1309	252	384.4	-666	288
VII	$\beta_D D$						
	EQ						

Table 10.6   Loads and Moments @ DB Bottom Pier (Overload Vehicle)

	LL Case	Case VB			Case VI		
Group	Loading	P	M_L	M_T	P	M_L	M_T
I	$\beta_D D$	254.9	+149		254.9	+112	
	$LL + I$	102.6	+222	290	135.5	−172	384
	Σ	357.5	+371	290	390.4	−60	384
	$1.3 \times \Sigma$	464.8	+482	377	507.5	−78	499
II	$\beta_D D$	254.9	+149				
	$W + OT$			411			
	Σ	254.9	+149	411			
	$1.3 \times \Sigma$	331.4	+194	534			
III	$\beta_D D$	254.9	+149		254.9	+112	
	$LL + I/1.67$	61.4	+133	174	81.1	−103	230
	$0.3W$			95			95
	$0.3OT$			28			28
	WL			170			170
	Σ	316.3	+282	467	336.0	+9	523
	$1.3 \times \Sigma$	411.2	+367	607	436.8	+12	680
IV	$\beta_D D$	254.9	+149		254.9	+112	
	$L + I/1.67$	61.4	+133	174	81.1	−103	230
	$S + T$		+229			−560	
	Σ	316.3	+511	174	336.0	−551	230
	$1.3 \times \Sigma$	411.2	+664	226	436.8	−716	299
V	Σ Group II	254.9	+149	411			
	$S + T$		−560				
	Σ	254.9	−411	411			
	$1.25 \times \Sigma$	318.6	−514	514			
VI	Σ Group III	316.3	+282	467	336.0	+9	523
	$S + T$		+229			−560	
	Σ	316.3	+511	467	336.0	−551	523
	$1.25 \times \Sigma$	395.4	+639	584	420.0	−689	654
VII	$\beta_D D$	254.9	+149				
	EQ			1161			
	Σ	254.9	+149	1161			
	$1.3 \times \Sigma$	331.4	+194	1509			

computed as referenced about each axis for a compressive concrete strain of 0.003 and a steel compressive strain of 0.00207 at the balance point. The direct load and moment capacity for a concrete compressive strain of 0.003 and a steel compressive strain 0.005 which are close to a level at which the direct load is small or almost zero are computed next. The capacity interaction curve can be approximated by a straight line between these two points. This procedure is done as follows: first draw the reinforcing in the column section and plot the strain diagrams in a position projected from the column section (Figure 10.22).

Using the strain diagram with a maximum tensile steel strain of 0.00207, we compute the capacity for direct load and bending about the Y-Y axis. The reinforcing is located in relation to the centroid of

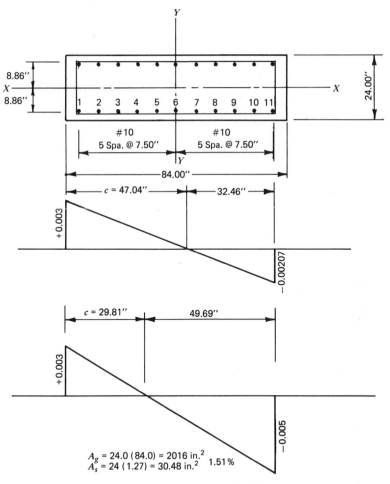

Figure 10.22 Pier section analysis: axis YY (transverse moment).

Table 10.7 Pier Capacity Analysis $\epsilon_{s_-} = 0.00207$ (Axis Y-Y)

Reinforcing Location	Distance R/F to Centroid	Reinforcing Area A_s	Strain @ R/F	Stress in R/F	Force in R/F	Moment (k-ft)
1	37.50	3.81	+0.00271	+60.00	+228.6	714
2	30.00	2.54	+0.00223	+60.00	+152.4	381
3	22.50	2.54	+0.00176	+51.04	+129.6	243
4	15.00	2.54	+0.00128	+37.12	+94.3	118
5	7.50	2.54	+0.00080	+23.20	+58.9	37
6	0	2.54	+0.00032	+9.28	+23.6	0
7	7.50	2.54	-0.00016	-4.64	-11.8	7
8	15.00	2.54	-0.00064	-18.56	-47.1	59
9	22.50	2.54	-0.00111	-32.19	-81.8	153
10	30.00	2.54	-0.00159	-46.11	-117.1	293
11	37.50	3.81	-0.00207	-60.00	-228.6	714
					+687.4 −486.4	2719

$c = 47.04$ $\beta_1 = 0.85$

$\beta_1 c = 39.98$

$A_c = 960$

$C = 0.85(3250)(960) = 2652$ k

$M_c = 2652[42.0 - (39.98)/2](\frac{1}{12}) = 4864$ k-ft

$\left.\begin{array}{l} \phi P_b = 2652 + 687 - 486 = 2853 \text{ k} \\ \phi M_b = 2719 + 4864 = 7583 \text{ k-ft} \end{array}\right\}$ $\phi = 1.0$

$\left.\begin{array}{l} \phi P_b = 2568 \text{ k} \\ \phi M_b = 6825 \text{ k-ft} \end{array}\right\}$ $\phi = 0.90$

$\left.\begin{array}{l} \phi P_b = 1997 \text{ k} \\ \phi M_b = 5308 \text{ k-ft} \end{array}\right\}$ $\phi = 0.70$

Table 10.8 Pier Capacity Analysis $\epsilon_s = 0.005$ (Axis Y-Y)

Reinforcing Location	Distance R/F to Centroid	Reinforcing Area A_s	Strain @ R/F	Stress in R/F	Force in R/F	Moment (k-ft)
1	37.50	3.81	+0.00255	+60.00	+228.6	714
2	30.00	2.54	+0.00179	+51.91	+131.9	330
3	22.50	2.54	+0.00104	+30.16	+76.6	144
4	15.00	2.54	+0.00028	+8.12	+20.6	26
5	7.50	2.54	-0.00047	-13.63	-34.6	-22
6	0	2.54	-0.00123	-35.67	-90.6	0
7	7.50	2.54	-0.00198	-57.42	-145.8	91
8	15.00	2.54	-0.00274	-60.00	-152.4	190
9	22.50	2.54	-0.00349	-60.00	-152.4	286
10	30.00	2.54	-0.00425	-60.00	-152.4	381
11	37.50	3.81	-0.00500	-60.00	-228.6	714
					+457.7 -956.8	2854

$c = 29.81$

$\beta_1 c = 0.85(29.81) = 25.34$ in.

$A_c = 25.34(24.0) = 608$ in.^{2}

$C = 0.85(3250)(608) = 1680$ k

$M_c = 1680[42.0 - (25.34)/2](\frac{1}{12}) = 4106$ k·ft

$\phi P_n = +457.7 - 956.8 + 1680 = 1181$ k $\left.\right\}\ \phi = 1.0$

$\phi M_n = 2854 + 4106 = 6960$ k·ft

$\phi P_n = 1063$ k $\left.\right\}\ \phi = 0.90$
$\phi M_n = 6264$ k·ft

$\phi P_n = 827$ k $\left.\right\}\ \phi = 0.70$
$\phi M_n = 4872$ k·ft

513

the section.   The strains at each reinforcing bar are taken from the
strain diagram and are tabulated as + for compression and - for tension.
The stresses in the reinforcing are computed from the strain by using
E_s = 29,000 ksi and f_y = 60 ksi as a maximum stress. The force in the
reinforcing is the stress times the area A_s. The moments due to the
reinforcing are then taken around the centroid. The rectangular concrete compression block is computed according to AASHTO and the
moment of the compression block force is taken around the centroid.
All these forces and moments are added to give the balanced load and
moment P_b and M_b. These values are listed for the various capacity reduction factors ϕ (Table 10.7).

The procedure on Table 10.7 is used for an intermediate point with
0.005 as the maximum tensile strain value in the reinforcing; the concrete strain as 0.003 maximum (Table 10.8).

The values just computed are plotted in Figure 10.23, with the most
critical cases obtained from the table of ultimate loads and moments.
These values are shown for bending around the Y-Y axis.

The capacity is computed for bending about the X-X axis as shown
in Tables 10.9 and 10.10 and Figure 10.24.

These values are plotted again, in addition to the ϕP_0 term, which is
the direct load capacity according to AASHTO (eq. 6-9) shown in
Figure 10.25.

Next the critical loading is selected and compared with a simulated

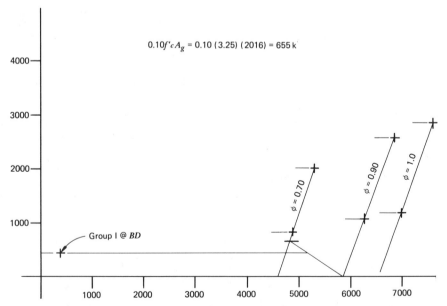

Figure 10.23   Pier section analysis: axis Y-Y (transverse moment).

Table 10.9 Pier Capacity Analysis $\epsilon_s = 0.005$ (Axis x-x)

Reinforcing Location	Distance R/F to Centroid	Reinforcing Area A_s	Strain @ R/F	Stress in R/F	Force in R/F	Moment (k-ft)
1	8.86	13.97	+0.00180	+52.20	+729.2	538
2	0	2.54	-0.00160	-46.40	-117.9	0
3	8.86	13.97	-0.00500	-60.00	-838.2	619
					+729.2 -956.1	1157

$c = 7.82$

$\beta_1 c = 0.85(7.82) = 6.65$

$A_c = 6.65(84.0) = 558$

$C = 558(3.25)(0.85) = 1542$ k

$M_c = 1542[12.0 - (6.65)/2](\frac{1}{12}) = 1115$ k-ft

$\phi P_n = 1542 + 729.2 - 956.1 = 1315$ k $\Big\}$ $\phi = 1.0$

$\phi M_n = 1115 + 1157 = 2272$ k-ft

$\phi P_n = 1184$ k $\Big\}$ $\phi = 0.90$
$\phi M_n = 2045$ k-ft

$\phi P_n = 920$ k $\Big\}$ $\phi = 0.70$
$\phi M_n = 1590$ k-ft

Table 10.10 Pier Capacity Analysis $\epsilon_s = 0.00207$ (Axis X–X)

Reinforcing Location	Distance R/F to Centroid	Reinforcing Area A_s	Strain @ R/F	Stress in R/F	Force in R/F	Moment (k-ft)
1	8.86	13.97	+0.00224	+60.00	+838.2	619
2	0	2.54	+0.00008	+2.32	+5.9	0
3	8.86	13.97	−0.00207	−60.00	−838.2	619
					+844.1 −838.2	1238

$c = 12.34$ in.

$\beta_1 c = 0.85(12.34) = 10.49$

$A_c = 84.0(10.49) = 881$

$C = 881(3.25)(0.85) = 2434$ k

$M_c = 2434[12.0 - (10.49)/2](\frac{1}{12}) = 1370$ k-ft

$\phi P_b = 2434 + 844.1 - 838.2 = 2440$ k $\Big\}$ $\phi = 1.0$

$\phi M_b = 1370 + 1238 = 2608$ k-ft

$\phi P_b = 2196$ k $\Big\}$ $\phi = 0.90$
$\phi M_b = 2347$ k-ft

$\phi P_b = 1708$ k $\Big\}$ $\phi = 0.70$
$\phi M_b = 1826$ k-ft

516

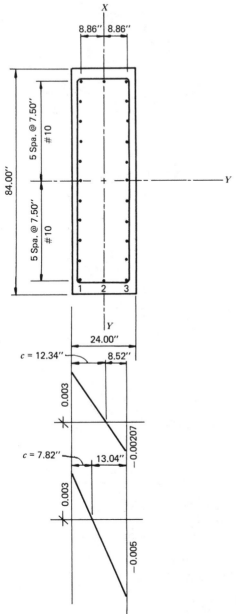

Figure 10.24 Pier section analysis: axis X-X (longitudinal moment).

interaction plane. Two formulas for the interaction plane are used for comparison: the elliptic equation developed by Furlong[12] and the AASHTO expression. The most critical is the Group I loading at the top of the pier:

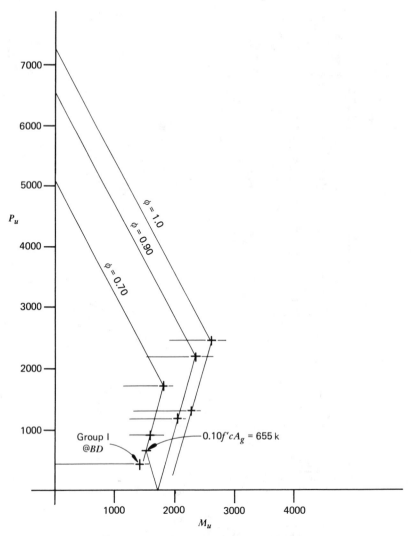

Figure 10.25 Pier section capacity: axis X-X (longitudinal moment.

$$\phi P_0 = [(0.85)(f'_c)(A_g - A_{st}) + A_{st}f_y]\ \phi;$$

$$\phi P_0 = [(0.85)(3.25)(2016 - 30.48) + 30.48(60)]$$

$$\phi = 7314\ k \qquad \phi = 1.0,$$

$$= 6582\ k \qquad \phi = 0.90,$$

$$= 5120\ k \qquad \phi = 0.70.$$

Pier Capacity Analysis @ Top Pier (BD)

Biaxial Bending

Group I - Case VB

P_u = 427.7 k

M_u^{x-x} = 1407 k-ft

M_u^{y-y} = 377 k-ft

Moment magnification occurs only in the transverse direction because the column is braced against sidesway (longitudinally). For this sidesway the specifications indicate that in the design the minimum value of e equals $\frac{1}{10}$ the depth: min e = 8.4 in.; M_u^{yy} would be 299 k-ft. Therefore minimum eccentricity does not control.

Next (EI) stiffness is computed for the critical buckling loads. Moment magnification is in the transverse direction only (braced longitudinally):

min e = 8.4 in.

M_u^{yy} = 299 k-ft does not control

E_c = 3,249,000 psi   for creep β_D = 0

I_g = 24.0(84.0)^3/12 = 1,185,408 in.^{4}

E_s = 29,000,000 psi

I_s = 4(1.27)($\overline{7.5}^2$ + $\overline{15.0}^2$

$\qquad$ + $\overline{22.5}^2$ + $\overline{30.0}^2$) = 8572 $\left.\begin{array}{l}\\\\\\\end{array}\right\}$ sum = 19,288 in.^{4}

$\qquad$ = 6(1.27)($\overline{37.5}^2$) = 10,716

$EI = \dfrac{E_c I_g/5 + E_s I_s}{1 + \beta_D}$

$\quad = \dfrac{[3,249,000(1,185,408)]/5 + 29,000,000(19,288)}{1.0}$

$\quad = 1.3296 \times 10^{12}$

The critical buckling load is then computed; use k = 2.1:

$P_{cr} = \dfrac{\pi^2 EI}{(klu)^2} = \dfrac{\pi^2(1.3296 \times 10^{12})}{[(2.1)(23.75)(12)]^2} = 36,636$ k

$\phi = 0.769$, as established by the capacity charts. Because the column is in the unbraced condition, $C_m = 1.0$. According to AASHTO (eq. 6-15) the magnification is

$$\delta = \frac{C_m}{1 - P_u/\phi P_{cr}} = \frac{1.0}{1 - 427.7/[0.769(36{,}636)]} = 1.015$$

Therefore the following values are plotted on the preceding capacity curves; use $P_u = 427.7$ k:

$$M_u^{xx} = 1407 \text{ k-ft}$$

$$M_u^{yy} = 1.015(377) = 383 \text{ k-ft}$$

Scaling from the capacity curves gives

$$\frac{M_{ux}}{\phi M_{ny}} = 0.89 \text{ from plot}$$

$$\frac{M_{uy}}{\phi M_{ny}} = 0.07 \text{ from plot}$$

The biaxial capacity is checked by using two procedures:

1. According to Furlong[12]

$$\sqrt{\left(\frac{M_{ux}}{\phi M_{nx}}\right)^2 + \left(\frac{M_{uy}}{\phi M_{ny}}\right)^2} \leqslant 1$$

$$\sqrt{(0.89)^2 + (0.07)^2} = 0.89 < 1.0 \quad \text{OK}$$

2. According to AASHTO

$$\frac{M_{ux}}{\phi M_{nx}} + \frac{M_{uy}}{\phi M_{ny}} \leqslant 1$$

$$0.89 + 0.07 = 0.96 \leqslant 1.0 \quad \text{OK}$$

The ultimate requirements at the bottom of the pier are less critical than at the top; therefore the pier design is satisfactory.

Serviceability requirements are checked next. It is assumed that the crack width at the top of the pier does not exceed 0.008 in. and that at the bottom of the pier it does not exceed 0.006 in. The computations are based on the equation developed by Kaar et al.[27] To obtain this crack width it is necessary to compute the service load level stress in the reinforcing steel at the top and bottom of the column. This stress, which results from biaxial bending, is computed by using the procedure and charts developed by Hu.[34] These charts are shown in Figure 10.26.

Analysis of Direct Load and Biaxial Bending by the Lu-Shien Hu Method

The stresses for a given column section are determined as follows:

1. Section constants are computed.
2. Load constants are computed.

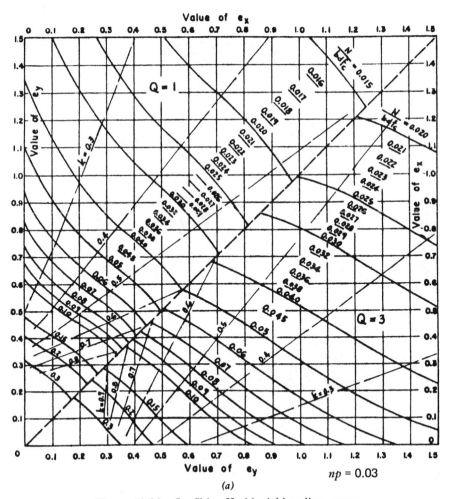

Figure 10.26 Lu-Shien Hu biaxial bending curves.

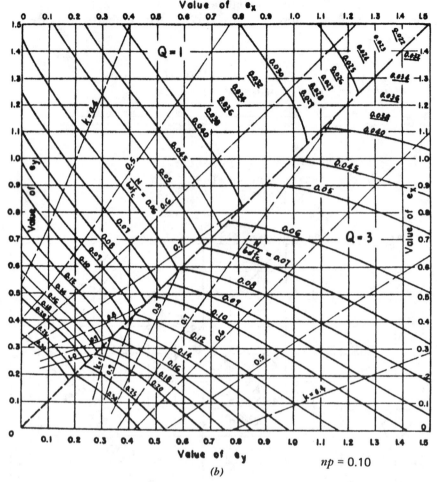

Figure 10.26   (*Continued*)

3. The value Q is computed from Sectional constants and Load constants.

4. The values of N/bdf_c and k are obtained from the attached charts.

5. The concrete stress is obtained from N/bdf_c. The steel stress is obtained from the relationship:

$$f_s = nf_c \left(1 - \frac{1 - a_x}{h} - \frac{1 - a_y}{k}\right)$$

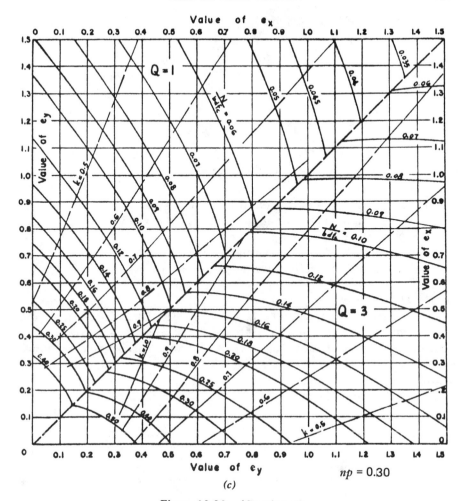

Figure 10.26 (Continued)

1. The sectional constants are as follows:

A_c = gross area of section

A_s = area of steel

I_x^s = moment of inertia of all steel about the principal axis x–x (nontransformed section).

I_y^s = moment of inertia of all steel about the principal axis y–y (nontransformed section).

I_x^c = moment of inertia of gross section about the principal axis x–x.

I_y^c = moment of inertia of gross section about the principal axis y-y.

$$np = \frac{nA_s}{A_c} \qquad nq_x = \frac{nI_x^s}{I_x^c} \qquad nq_y = \frac{nI_y^s}{I_y^c}$$

(n has usual meaning)

2.   The load constants are as follows:

$$e_x = \frac{M_x}{Nd} \qquad e_y = \frac{M_y}{Nb} \qquad m = \frac{e_x}{e_y}$$

where M_x = bending moment about principal axis x-x,
      M_y = bending moment about principal axis y-y,
      N = total direct load,
      d = length of section normal to x-x axis,
      b = length of section normal to y-y axis.

3. $Q = nq_y + m^2 nq_x/np$
4.   Interpolate between charts shown in Fig. 10.26 for both np and Q as computed. *Caution.* Note that the e_x and e_y axes are not consistent on the charts. The value of Q can be kept within the chart range by the proper selection of x and y axes. The chart values of np are within the practical range.
5.   In the computation of f_s

$$h = \frac{k}{m}$$

a_x and a_y = clearance to ₵ of bars in the x and y directions, respectively, expressed as a fraction of the section dimension in the same direction

This procedure is used in the following computations for stresses of the column top (Figure 10.27).

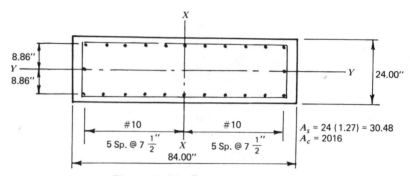

Figure 10.27   Pier reinforcement.

Pier Serviceability @ Top Pier

Service Load Stresses @ Col top: Hu Method

Group I $\qquad P = 301.8 + 58.4 = 360.2$ k

$M_{yy} = +430 + 346 = 776$ k-ft

$M_{xx} = 234$ k-ft

$$I_x^s = 4(1.27)(7.50)^2 = 286$$
$$4(1.27)(15.00)^2 = 1{,}143$$
$$4(1.27)(22.50)^2 = 2{,}572 \left.\right\} \text{ sum} = 19{,}289 \text{ in.}^4$$
$$4(1.27)(30.00)^2 = 4{,}572$$
$$6(1.27)(37.50)^2 = 10{,}716$$

$$I_x^c = 24.0(84.0)^3/12 = 1{,}185{,}408 \text{ in.}^4$$

$$I_y^s = 2(11)(1.27)(8.86)^2 = 2{,}193 \text{ in.}^4$$

$$I_y^c = \frac{84.0(24.0)^3}{12} = 96{,}768 \text{ in.}^4$$

$$np = \frac{9(30.48)}{2016} = 0.1361$$

$$nq_x = \frac{9(19{,}289)}{1{,}185{,}408} = 0.1464$$

$$nq_y = \frac{9(2193)}{96{,}768} = 0.2040$$

$$e_x = \frac{234}{(360.2)(7.00)} = 0.0928$$

$$e_y = \frac{776}{(360.2)(2.00)} = 1.077$$

$$m = \frac{0.0928}{1.077} = 0.086$$

$$Q = \frac{nq_y + m^2 nq_x}{np}$$

$$= \frac{0.2040 + (0.086)^2(0.1464)}{0.1361} = 1.51$$

	$np = 0.10$		$np = 0.30$	
	$\dfrac{N}{bdf'_c}$	k	$\dfrac{N}{bdf'_c}$	k
$Q = 1$	0.072	0.35	0.115	0.45
$Q = 3$	0.115	0.38	0.205	0.52
$Q = 1.51$	0.083	0.36	0.138	0.47

$$np = 0.1361$$

$$\frac{N}{bdf'_c} = 0.093$$

$$k = 0.38$$

$$f_c = \frac{N}{bd(0.093)} = \frac{360,200}{(24.0)(84.0)(0.093)} = 1921 \text{ psi}$$

$$h = \frac{k}{m} = \frac{0.38}{0.086} = 4.42$$

$$a_x = \frac{4.5}{84.0} = 0.0536$$

$$a_y = \frac{3.14}{24.0} = 0.1308$$

$$f_s = 9(1921)\left(1 - \frac{1 - 0.0536}{442} - \frac{1 - 0.1308}{0.38}\right) = 26.0 \text{ ksi max}$$

On the basis of this stress the crack width is computed:

Top of Column

$$A = 2(3.14)(7.5) = 47.1$$
$$\omega = 0.115 \sqrt[4]{A} \; f_s \times 10^{-6}$$
$$= 0.115 \sqrt[4]{47.1} \; (26.0) \times 10^{-3}$$
$$= 0.0078 \text{ in.} < 0.008 \text{ in.} \quad \text{OK}$$

Using the preceding computations and reversing the axis, we compute the maximum stress at the bottom of the column:

Pier Serviceability @ Bottom Pier

Group VII $P = 339.9 \text{ k}$

$M_{xx} = 149 \text{ k-ft}$ (longitudinal)

$M_{yy} = 1161 \text{ k-ft}$ (transverse)

$$np = 0.1361$$

$$nq_x = 0.2040$$

$$nq_y = 0.1464$$

$$e_x = \frac{149}{(339.9)(2.00)} = 0.2192$$

$$e_y = \frac{1161}{(339.9)(7.00)} = 0.4880$$

$$n = \frac{0.2192}{0.4880} = 0.4491$$

$$Q = \frac{nq_y + m^2 nq_x}{np}$$

$$= \frac{0.1464 + (0.4491)^2(0.2040)}{0.1361} = 1.38$$

	$np = 0.10$		$np = 0.30$	
	$\dfrac{N}{bdf'_c}$	k	$\dfrac{N}{bdf'_c}$	k
$Q = 1$	0.125	0.62	0.178	0.72
$Q = 3$	0.175	0.71	0.280	0.85
$Q = 1.38$	0.134	0.64	0.197	0.74

$$np = 0.1361$$

$$\frac{N}{bdf'_c} = 0.145$$

$$k = 0.66$$

$$f_c = \frac{N}{bd(0.145)} = \frac{339{,}900}{(24.0)(84.0)(0.145)} = 1163 \text{ psi}$$

$$h = \frac{k}{m} = \frac{0.66}{0.4491} = 1.47$$

$$a_x = 0.1308$$

$$a_y = 0.0536$$

$$f_s = 9(1163)\left(1 - \frac{1 - 0.1308}{1.47} - \frac{1 - 0.0536}{0.66}\right) = 10.7 \text{ ksi}$$

Based on this stress, the crack width is computed:

Bottom of Column

$$A = 47.1$$

$$\omega = 0.115\sqrt[4]{47.1}\,(10.7) \times 10^{-3}$$

$$= 0.0032 < 0.006 \text{ in.} \quad \text{OK}$$

Now the footing can be designed. It is generally the case that the allowable footing pressures are given at a service load level. Therefore the footing is proportioned for a size on service load. First the service level loads and moments are tabulated for the footing and the actual pressures are computed. These pressures are then compared with the allowable pressures for sizing the footing. Footing pressures should be computed by using elastic analysis for the foundation. If the load is outside the kern, the pressures can be computed by using the chart in Figure 10.28. On the basis of the computed pressures the footing should be checked for serviceability. If the footing reinforcement has not yet been selected, this will have to be done after the capacity analysis. The serviceability requirements selected are similar to those for the bottom of the pier shaft; that is, 0.006 in. as a maximum crack width. Crack widths are again computed with the Kaar formula.[27]

The footing loads and moments listed for a service load level (Table 10.11) include a tabulation of longitudinal and transverse moments. Again the tabulation is in the order of group loadings contained in AASHTO Article 1.2.22. Each group loading is developed for three possible live loadings and, for comparison, summed up and divided by the allowable overstress. The loads and moments were calculated in Example 3.1.

The footing is assumed to be 6 × 12 ft in plan. Therefore the following are the elastic properties:

$$A = 72 \text{ ft}^2$$

$$S_L = \frac{12(6)^2}{6} = 72 \text{ ft}^3$$

$$S_T = \frac{6(12)^2}{6} = 144 \text{ ft}^3$$

In the first case to be considered the resultant is outside the kern which gives the pressure obtained by using Figure 10.28. This is the Group I loading with the maximum longitudinal moment.

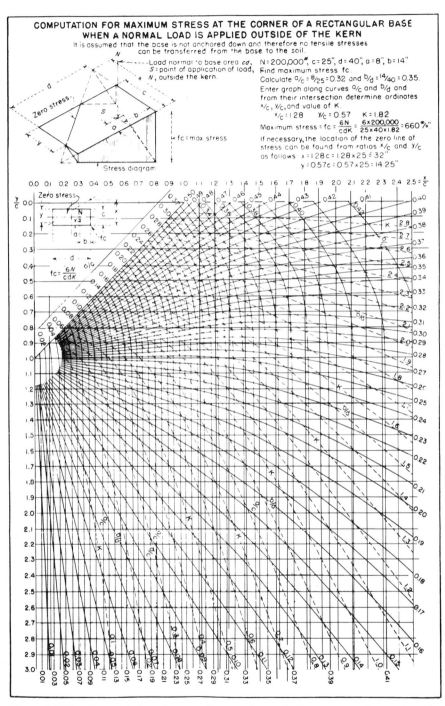

Figure 10.28 Computation for maximum stress at the corner of a rectangular base when a normal load is applied outside the kern.

Table 10.11   Pier Footing Design Service Load Level.
Loads and Moments @ DB (Footing)

LL Case		Case IB			Case II			Case IIA		
Group	Loading	P	M_L	M_T	P	M_L	M_T	P	M_L	M_T
I	D	377.6	+216		377.6	+216		418.6	+216	
	LL	46.2	+136	185	69.2	−84	277	76.0	−72	304
	Σ	423.8	+352	185	446.8	+132	277	494.6	+144	304
II	D	377.6	+216					418.6	+216	
	W			351						351
	OT			94						94
	Σ	377.6	+216	445				418.6	+216	445
	÷ 1.25	302.1	+173	356				334.9	+173	356
III	Group I	423.8	+352	185	446.8	+132	277	494.6	+144	304
	0.3W			105			105			105
	0.3OT			28			28			28
	WL			184			184			184
	Σ	423.8	+352	502	446.8	+132	594	494.6	+144	621
	÷ 1.25	339.0	+282	402	357.4	+106	475	395.7	+115	497
IV	Group I	423.8	+352	185	446.8	+132	277	494.6	+144	304
	$S + T$		+283			−691			−691	
	Σ	423.8	+635	185	446.8	−559	277	494.6	−547	304
	÷ 1.25	339.0	+508	148	357.4	−447	222	395.7	−438	243
V	Group II	377.6	+216	445				418.6	+216	445
	$S + T$		+283						+283	
	Σ	377.6	+499	445				418.6	+499	445
	÷ 1.40	269.7	+356	318				299.0	+356	318
VI	Group III	423.8	+352	502	446.8	+132	594	494.6	+144	621
	$S + T$		+283			−691			−691	
	Σ	423.8	+635	502	446.8	−559	594	494.6	−547	621
	÷ 1.40	302.7	+454	359	319.1	−399	424	353.3	−391	444
VII	D	377.6	+216					418.6	+216	
	EQ			1282						1282
	Σ	377.6	+216	1282				418.6	+216	1282
	÷ 1.33	283.9	+162	964				314.7	+162	964

Group I: Case IB

$$P = 423.8 \text{ k} \qquad M_L = 352 \text{ k-ft} \qquad M_T = 185 \text{ k-ft}$$

To apply Figure 10.28 the following constants are computed:

$$c = 6.00 \qquad d = 12.00$$

$$a = 2.17 \qquad b = 5.56$$

$$\frac{a}{c} = 0.361 \qquad \frac{b}{d} = 0.464$$

From the curves shown in Figure 10.28

$$\frac{x}{c} = 1.23 \qquad \frac{y}{c} = 0.03 \qquad K = 2.88$$

$$x = 7.38 \text{ ft} \qquad y = 0.18 \text{ ft}$$

The maximum pressure is

$$f = \frac{6M}{cdK}$$

$$f = \frac{6(423.8)}{(6.0)(12.0)(2.88)} = 12.26 \text{ ksf}$$

For the maximum reaction case in Group I the resultant falls within the kern:

Group I: Case IIA

$$P = 494.6 \text{ k} \qquad M_L = 144 \text{ k-ft} \qquad M_T = 304 \text{ k-ft}$$

The pressures are computed by using

$$f = \frac{N}{A} \pm \frac{M_L}{S_L} \pm \frac{M_T}{S_T}$$

$$f = \frac{494.6}{72} \pm \frac{144}{72} \pm \frac{304}{144}$$

$$= 6.87 \pm 2.00 \pm 2.11 = 10.98 \text{ ksf max}$$

Group IV, Case IB, is investigated by using Figure 10.28:

Group IV, Case IB

$$P = 339.0 \text{ k} \qquad M_L = 508 \text{ k-ft} \qquad M_T = 148 \text{ k-ft}$$

$$c = 6.00 \qquad d = 12.00$$

$$a = 1.50 \qquad b = 5.56$$

$$\frac{a}{c} = 0.250 \qquad \frac{b}{d} = 0.464$$

$$\frac{x}{c} = 0.82 \qquad \frac{y}{c} = 0.33 \qquad K = 2.00$$

$$f = \frac{6(339.0)}{6.0(12.0)(2.00)} = 14.12 \text{ ksf}$$

$$x = 4.92 \text{ ft} \qquad y = 1.98 \text{ ft}$$

Group VII, Case IIA, is done in the same manner.

Group VII: Case IIA

$$P = 283.9 \text{ k} \qquad M_L = +162 \text{ k-ft} \qquad M_T = 964 \text{ k-ft}$$

$$d = 6.00 \qquad c = 12.00$$

$$b = 2.43 \qquad a = 2.60$$

$$\frac{b}{d} = 0.405 \qquad \frac{a}{c} = 0.217$$

$$\frac{x}{c} = 0.79 \qquad \frac{y}{c} = 0.56 \qquad K = 1.47$$

$$f = \frac{6(283.9)}{12.0(6.0)(1.47)} = 16.09 \text{ ksi} \longrightarrow 8.0 \text{ TSF max}$$

$$x = 9.48 \text{ ft} \qquad y = 6.72 \text{ ft}$$

This case produces the maximum foundation pressure 8.0 TSF which is considered equal to the allowable.

Next a diagram is drawn (Figure 10.29) of the most significant loadings as far as bending and shear in the footing are concerned. These are Group I, Case I*B*, and Group VII, Case II*A*.

From visual observation of these diagrams it can be seen that Group VII, Case II*A*, controls bending and shear. Serviceability criteria are examined for crack control at a service load level. Assumption of a footing 2.5 ft thick and 6.5 ft of earth over the footing develops a uniform downward overburden load.

Serviceability

Group VII: Case IIA

$$\text{Overburden:} \quad \omega = 2.5 \times 0.150 = 0.38$$
$$6.5 \times 0.120 = \underline{0.78}$$
$$1.16 \text{ ksf} \div 1.33 = 0.87 \text{ ksf}$$

Note that the total overburden load is divided by 1.33, the allowable overstress for Group VII loadings. It places the overburden at the same level as the pressures shown.

In the transverse direction at the face of the column the shear and bending around the edges of the columns are computed by cantilever action.

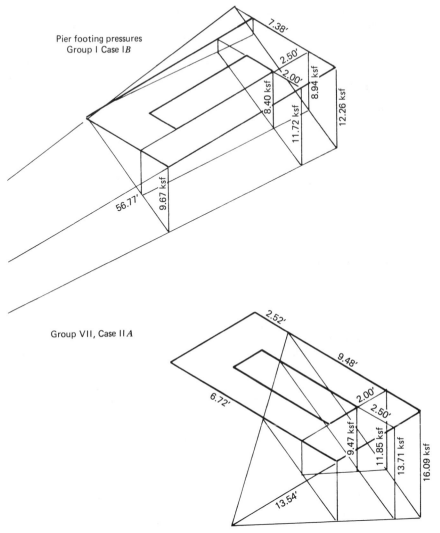

Figure 10.29 Pier footing pressures Group 1, Case IB; Group VII, Case IIA.

Transverse

	V (k-ft^{-1})	M (k-ft ft^{-1})
$9.47 \times 2.50 =$	$+23.7 \times 1.25 =$	$+29.6$
$\frac{1}{2} \times 4.24 \times 2.50 =$	$+5.3 \times 1.67 =$	$+8.9$
$-0.87 \times 2.50 =$	$-2.2 \times 1.25 =$	-2.7
	$+26.8$ k	$+35.8$ k-ft

The reinforcing used, based on the capacity analysis and the minimum reinforcing requirements, is

#7 @ 9 A_s = 0.80 in.^{2}

The effective depth d and the reinforcement ratio p are computed and k and j are taken from Appendix A:

$$d = 30 - 3.0 - 0.44 = 26.56$$

$$p = \frac{0.80}{12(26.56)} = 0.0025$$

$$k = 0.191$$

$$j = 0.936$$

The stress in the reinforcing is computed as

$$f_s = \frac{M}{A_s jd}$$

$$f_s = \frac{35.8(12)}{0.80(0.936)(26.50)} = 21.6 \text{ ksi}$$

The crack width is computed by using the formula

$$w = 0.115\sqrt{A}\,f_s\left(\frac{1}{10^{-6}}\right)$$

where dc = 2.00 + 0.44 = 2.44 in.
　　　A = (2.44)(2)(9.0) = 43.9
　　　w = 0.115 $\sqrt[4]{43.9}$ (21,600)(10^{-6})
　　　 = 0.0064 > 0.006   close enough

Next the response of the footing in the longitudinal direction is examined by again computing by cantilever action the shear and moments per foot at the corner of the column shaft.

Longitudinal	V (k-ft^{-1})	M (k-ft ft^{-1})
9.47 × 2.00 = 18.9 × 1.00 =		18.9
$\frac{1}{2}$ × 2.38 × 2.00 = 2.4 × 1.33 =		3.2
−0.87 × 2.00 = −1.7 × 1.00 =		−1.7
	19.6 k	20.4 k-ft

Assuming the minimum reinforcing controls, we use #6 @ 9 in., A_s = 0.59 in.^{2}. Then, as done in the tranverse direction, the crack width is computed. Use min R/F regm't #6 @ 9 in., A_s = 0.59 in.^{2}.

$$d = 30 - 3.0 - 0.88 - 0.38 = 25.74$$

$$p = \frac{0.59}{12(25.74)} = 0.00191 \qquad k = 0.169 \qquad j = 943$$

$$f_s = \frac{20.4(12)}{0.59(0.943)(25.74)} = 17.1 \text{ ksi}$$

$$dc = 2.0 + 0.88 + 0.38 = 3.26$$

$$A = 3.26(2)(9.0) = 58.7 \qquad \text{use } 50.0$$

$$w = 0.115 \sqrt[4]{50.0}(17,100)(10^{-6}) = 0.0052 < 0.006 \text{ in.} \quad \text{OK}$$

The loads and moments must be tabulated again for capacity analysis. These actions are the required ultimate values for section analysis of the footing. Loads and moments must be tabulated for the various group loadings shown in AASHTO and for three cases of live load, which would include bending in each direction and maximum live load reaction. These ultimate loads and moments should also be tabulated for the HS 20 loading and special overload vehicle, after which the most critical cases are again selected, pressure diagrams are drawn, and section capacity requirements are checked. It should be noted that the punching shear considerations are not critical on bridge pier footings and only in unusual circumstances would they be checked. Beam bending can be checked by computing the moment per foot around the critical face of the pier in each direction.

First the ultimate loads and moments are tabulated for the HS 20 vehicle, as given in Table 10.12, and computed in Example 3.1; β_D = 0.75 for P in Cases IB and II, in which maximum eccentricity is being considered. Elsewhere β_D = 1.0.

The same tabulation is made for the overload vehicle in Table 10.13.

For critical ultimate loads and moments the footing pressures are computed as for the service load level. For Group I, Case VB, the resultant is outside the kern; therefore use Figure 10.28.

Footing Pressures @ DB (Footing) (Ultimate)

Group I: Case VB

$$P = 473.7 \text{ k} \qquad M_L = 616 \text{ k-ft} \qquad M_T = 299 \text{ k-ft}$$

$$c = 6.00 \qquad d = 12.00$$

$$a = 1.70 \qquad b = 5.37$$

$$\frac{a}{c} = 0.283 \qquad \frac{b}{d} = 0.447$$

$$\frac{x}{c} = 0.98 \qquad \frac{y}{c} = 0.30 \qquad K = 2.17$$

$$x = 5.88 \qquad y = 1.80$$

$$f = \frac{6(473.7)}{(6.0)(12.0)(2.17)} = 18.19 \text{ ksf}$$

Table 10.12 Pier Footing Design (HS 20): Loads and Moments @ DB (Footing)

	LL Case	Case IB			Case II			Case IIA		
Group	Loading	P	M_L	M_T	P	M_L	M_T	P	M_L	M_T
I	$\beta_D D$	283.2	+216		283.2	+216		418.6	+216	
	1.67LL	77.2	+227	309	115.6	−140	463	126.9	−120	508
	Σ	360.4	+443	309	398.8	+76	463	545.5	+96	508
	1.3 × Σ	468.5	+576	402	518.4	+99	602	709.2	+125	660
II	$\beta_D D$	283.2	+216					418.6	+216	
	$W + OT$			445						445
	Σ	283.2	+216	445				418.6	+216	445
	1.3 × Σ	368.2	+281	578				544.2	+281	578
III	$\beta_D D$	283.2	+216		283.2	+216		418.6	+216	
	LL	46.2	+136	185	69.2	−84	277	76.0	−72	304
	0.3W			105			105			105
	0.3OT			28			28			28
	WL			184			184			184
	Σ	329.4	+352	502	352.4	+132	594	494.6	+144	621
	1.3 × Σ	428.2	+458	653	458.1	+172	772	643.0	+187	807
IV	$\beta_D D$	283.2	+216		283.2	+216		418.6	+216	
	LL	46.2	+136	185	69.2	−84	277	76.0	−72	304
	$S + T$		+283			−691			−691	
	Σ	329.4	+826	185	352.4	−559	277	494.6	−547	304
	1.3 × Σ	428.2	+1073	240	458.1	−727	360	643.0	−711	395
V	Σ Group II	283.2	+216	445				418.6	+216	445
	$S + T$		+283						+283	
	Σ	283.2	+499	445				418.6	+499	445
	1.25 × Σ	354.0	+624	556				523.2	+624	556
VI	Σ Group III	329.4	+352	502	352.4	+132	594	494.6	+144	621
	$S + T$		+283			−691			−691	
	Σ	329.4	+635	502	352.4	−559	594	494.6	−547	621
	1.25 × Σ	411.8	+794	628	440.5	−699	742	618.2	−684	776
VIII	$\beta_D D$	283.2	+216					418.6	+216	
	EQ			1282						1282
	Σ	283.2	+216	1282				418.6	+216	1282
	1.3 × Σ	368.2	+281	1667				544.2	+281	1667

Table 10.13 Pier Footing Design, Overload Vehicle $\beta_L = 1.0$: Loads and Moments @ DB (Footing)

LL Case		Case VB			Case VI			Case VIA		
Group	Loading	P	M_L	M_T	P	M_L	M_T	P	M_L	M_T
I	$\beta_D D$	283.2	+216		283.2	+216		418.6	+216	
	LL	81.2	+258	230	104.2	−192	295	144.2	−192	409
	Σ	364.4	+474	230	387.4	+24	295	562.8	+24	409
	1.3 × Σ	473.7	+616	299	503.6	+31	383	731.6	+31	532
II	$\beta_D D$	283.2	+216					418.6	+216	
	$W + OT$			445						445
	Σ	283.2	+216	445				418.6	+216	445
	1.3 × Σ	368.2	+281	578				544.2	+281	578
III	$\beta_D D$	283.2	+216		283.2	+216		418.6	+216	
	LL/1.67	48.6	+154	138	62.4	−115	177	86.3	−115	245
	0.3W			105			105			105
	0.3OT			28			28			28
	WL			184			184			184
	Σ	331.8	+370	455	345.6	+101	494	504.9	+101	562
	1.3 × Σ	431.3	+481	592	449.3	+131	642	656.4	+131	731
IV	$\beta_D D$	283.2	+216		283.2	+216		418.6	+216	
	LL/1.67	48.6	+154	138	62.4	−115	177	86.3	−115	245
	$S + T$		+283			−691			−691	
	Σ	331.8	+653	138	345.6	−590	177	504.9	−590	
	1.3 × Σ	431.3	+849	179	449.3	−767	230	656.4	−767	318
V	Σ Group II	283.2	+216	445				418.6	+216	445
	$S + T$		+283						+283	
	Σ	283.2	+499	445				418.6	+499	445
	1.25 × Σ	354.0	+624	556				523.2	+624	556
VI	Σ Group III	331.8	+370	445	345.6	+101	494	504.9	+101	562
	$S + T$		+283			−691			−691	
	Σ	331.8	+653	455	345.6	−590	494	504.9	−590	562
	1.25 × Σ	414.8	+816	569	432.0	−737	617	631.1	−738	702
VIII	$\beta_D D$	283.2	+216					418.6	+216	
	EQ			1282						1282
	Σ	283.2	+216	1282				418.6	+216	1282
	1.3 × Σ	368.2	+281	1667				544.2	+681	1667

The pressure is computed for Group VI, Case VB, for which the resultant is outside the kern.

Group VI: Case VB

$$P = 414.8 \qquad M_L = +816 \text{ k-ft} \qquad M_T = +569 \text{ k-ft}$$

$$c = 6.00 \qquad\qquad d = 12.00$$

$$a = 1.03 \qquad\qquad b = 4.63$$

$$\frac{a}{c} = 0.172 \qquad\qquad \frac{b}{d} = 0.386$$

$$\frac{x}{c} = 0.64 \qquad \frac{y}{c} = 0.69 \qquad K = 1.12$$

$$x = 3.84 \qquad y = 4.14$$

$$f = \frac{6(414.8)}{(6.0)(12.0)(1.12)} = 30.86 \text{ ksf}$$

In the same manner the pressure is computed for Group VII, Case V*B*.

Group VII: Case VB

$$P = 368.2 \text{ k} \qquad M_L = +281 \text{ k-ft} \qquad M_T = 1667 \text{ k-ft}$$

$$d = 6.00 \qquad c = 12.00$$

$$b = 2.24 \qquad a = 1.47$$

$$\frac{b}{d} = 0.373 \qquad \frac{a}{c} = 0.122$$

$$\frac{x}{c} = 0.47 \qquad \frac{y}{c} = 0.79 \qquad K = 0.78$$

$$x = 5.64 \qquad y = 9.48$$

$$f = \frac{6(368.2)}{(6.0)(12.0)(0.78)} = 39.33 \text{ ksf}$$

Group VII, Case V*B* is selected as the most critical for bending the footing; therefore the pressure diagram for this loading can be drawn as in Figure 10.30. The computed overburden is factored for the overload condition:

$$\text{Overburden:} \qquad \omega = 1.16 \times \overset{\beta_D}{0.75} \times \overset{\gamma}{1.3}$$

$$= 1.13 \text{ k-ft}^{-1} \qquad \text{Group VII}$$

The loads and moments per foot are computed by cantilever action at the corner of the column shaft in the transverse direction.

Transverse	V (k-ft^{-1})	M (k-ft ft^{-1})
$14.65 \times 2.5 =$	$+36.6 \times 1.25 =$	$+45.8$
$\frac{1}{2} \times 17.43 \times 2.5 =$	$+21.8 \times 1.67 =$	$+36.4$
$-1.13 \times 2.5 =$	$-2.8 \times 1.25 =$	-3.5
	$+55.6$ k	$+78.7$ k-ft

The reinforcing already selected is checked for capacity.

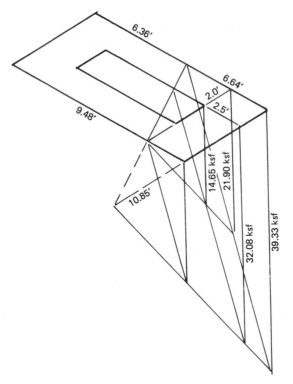

Figure 10.30 Pressure diagram: Group VII, Case V*B*.

#7 @ 9 in.

$$A_s = 0.80 \text{ in.}^2$$

$$a = \frac{0.80(60)}{0.85(3.25)(12)} = 1.44 \text{ in.}$$

$$\phi M_n = \phi\left[A_s f_y \left(d - \frac{a}{2}\right)\right]$$

$$= 0.90[(0.80)(60)(26.56 - 0.72)](\tfrac{1}{12})$$

$$= 93.0 \text{ k-ft} > 78.7 \text{ k-ft} \quad \text{OK}$$

Moments are then computed and capacity is checked in the longitudinal direction:

Longitudinal	V (k-ft^{-1})	M (k-ft ft^{-1})
14.65 × 2.00 = +29.3 × 1.00 =		+29.3
$\frac{1}{2}$ × 7.25 × 2.00 = +7.2 × 1.33 =		+9.6
−1.13 × 2.00 = −2.3 × 1.00 =		−2.3
+34.2 k		+36.6 k-ft

#6 @ 9 $A_s = 0.59$ in.^{2}

$$d = 30 - 3.0 - 0.88 - 0.38 = 25.74 \text{ in.}$$

$$a = \frac{0.59(60)}{0.85(3.25)(12)} = 1.06 \text{ in.}$$

$$\phi M_n = 0.90[(0.59)(60)(25.74 - 0.53)](\tfrac{1}{12})$$

$$= 66.9 \text{ k-ft} > 36.6 \text{ k-ft} \quad \text{OK}$$

The minimum reinforcing bar requirements of AASHTO Article 1.5.7 are checked. This requirement says that adequate reinforcing must be provided to develop a moment capacity of 1.2 times the cracking moment. The cracking moment is

$$M_{cr} = 7.5\sqrt{3250}\left[\frac{12(30)^2}{6}\right]\left(\frac{1}{12}\right) = 64 \text{ k-ft}$$

The reinforcing area required for a capacity equal to 1.2 times the cracking moment is

$$\text{approx } A_{s \text{ (required)}} = \frac{64(12)(1.2)}{(26.5 - 1.0)(60)}\left(\frac{1}{0.90}\right) = 0.60 \text{ in.}^2 \quad \text{OK}$$

The maximum reinforcing requirements of AASHTO Article 1.5.32(A)(1) are checked, where 75% of the balanced reinforcing ratio is

$$0.75\rho_b = \left[\frac{0.85\beta_1 f'_c}{f_y}\left(\frac{87{,}000}{87{,}000 + f_y}\right)\right]0.75$$

$$= \left[\frac{0.85(0.85)(3.25)}{60}\left(\frac{87{,}000}{87{,}000 + 60{,}000}\right)\right]0.75$$

$$= 0.0174$$

The reinforcing ratio is

$$\rho_{max} = \frac{0.80}{(12)(26.56)} = 0.0025 < 0.0174 \quad \text{OK}$$

Shear should be checked at d distance. If, however, the designer studies the pressure diagram, he will see that the shear is negligible at a d distance; therefore we make no further investigation.

Top reinforcement requirements, based on the potential for cracking under ultimate load and on a stress level check at a service load level for plain concrete, are then checked by using the allowables according to the AASHTO specifications.

Top R/F

$$\phi M_{cr} = 0.90(64.0) = 57.6 \text{ k-ft}$$

$$M_u = 1.3 \times 2.7 = 3.5 \text{ k-ft}$$

Serviceability: Check Stress

$$M = 2.7 \text{ k-ft}$$

$$S = \frac{12(30)^2}{6} = 1800 \text{ in.}^2$$

$$f = \frac{2.7(12)}{1800} = 18 \text{ psi} < 90 \text{ psi} = 0.21(715)\sqrt{3250}$$

No *R/F* required; nominal #5 @ 1–3 each way will be supplied.

Example 10.3. Service load design. This example relates to the bridge in Example 10.2, but the design is by service load. It is still reasonable to compute service load stresses and to check the stress levels according to the allowables. In addition, the pier section should be checked for AASHTO criteria because the service loads and moments should not exceed 35% of the capacity of the section. First, the service loads and moments are tabulated according to the AASHTO Article 1.2.22, Group Loadings, for two cases of live load with maximum bending in each direction. These loads are tabulated for the top and bottom of the pier shaft in Tables 10.14 and 10.15. The stresses are then checked by the Hu method and compared with the working level allowables. Note that the two group loadings checked yield similar results for somewhat different loadings. First, the loads and moments are tabulated for a service load level and each group is summed; the sum is then divided by the appropriate overstress factor. Therefore the group loadings can be compared.

Group I, Case I*B*, and Group IV, Case I*B*, appear to be the most critical loadings because they produce the largest moments about the weakest axis. These are at point *BD* at the top of the column. Using the Hu method described in Example 10.2, the stresses are computed and compared with the allowables. First this is performed for Group I, Case I*B*. Note that this section is larger than that required by the load factor design (Figure 10.31).

Table 10.14   Pier Design, Service Load:
Loads and Moments @ Column Top (*BD*)

LL Case		Case I*B*			Case II		
Group	Loading	P	M_L	M_T	P	M_L	M_T
I	D	301.8	+430		301.8	+430	
	$L + I$	58.4	+346	234	90.0	−218	360
	Σ	360.2	+776	234	391.8	+212	360
II	D	301.8	+430				
	W						
	OT			94			
	Σ	301.8	+430	94			
	÷ 1.25	241.4	+344	75			
III	Group I	360.2	+776	234	391.8	+212	360
	0.3W						
	0.3OT			28			28
	WL						
	Σ	360.2	+776	262	391.8	+212	388
	÷ 1.25	288.2	+621	210	313.4	+170	310
IV	Group I	360.2	+776	234	391.8	+212	360
	$S + T$		+227			−556	
	Σ	360.2	+1003	234	391.8	−344	360
	÷ 1.25	288.2	+802	187	313.4	−275	288
V	Group II	301.8	+430	94			
	$S + T$		+227				
	Σ	301.8	+657	94			
	÷ 1.40	215.6	+469	67			
VI	Group III	360.2	+776	262	391.8	+212	388
	$S + T$		+227			−556	
	Σ	360.2	+1003	262	391.8	−344	388
	÷ 1.40	257.3	+716	187	279.9	−246	277
VII	D	301.8	+430				
	EQ			0			
	Σ		neglect				

| LL Case | | Case IB | | | Case II | | |
Group	Loading	P	M_L	M_T	P	M_L	M_T
I	D	339.9	+149		339.9	+149	
	$L + I$	58.4	+116	234	90.0	−75	360
	Σ	398.3	+265	234	429.9	+74	360
II	D	339.9	+149				
	W			317			
	OT			94			
	Σ	339.9	+149	411			
	÷ 1.25	271.9	+119	329			
III	Group I	398.3	+265	234	429.9	+74	360
	0.3W			95			95
	0.3OT			28			28
	WL			170			170
	Σ	398.3	+265	527	429.9	+74	653
	÷ 1.25	318.6	+212	422	343.9	+59	522
IV	Group I	398.3	+265	234	429.9	+74	360
	$S + T$		+229			−560	
	Σ	398.3	+494	234	429.9	−486	360
	÷ 1.25	318.6	+395	187	343.9	−389	288
V	Group II	339.9	+149	411			
	$S + T$		−560				
	Σ	339.9	−411	411			
	÷ 1.40	242.8	−294	294			
VI	Group III	398.3	+265	527	429.9	+74	653
	$S + T$		+229			−560	
	Σ	398.3	+494	527	429.9	−486	653
	÷ 1.40	284.5	+353	376	307.1	−347	466
VII	D	339.9	+149				
	EQ			1161			
	Σ	339.9	+149	1161			
	÷ 1.33	255.6	+112	873			

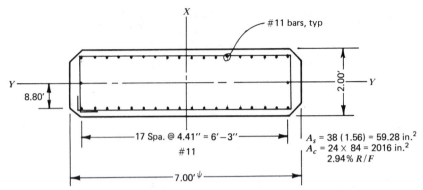

Figure 10.31   Pier reinforcement, column BD.

Pier Section Analysis @ Column Top (BD)

Service Load: Stresses—Lu Shien Hu Method

Group I: $P = 360.2$ k

$M_{yy} = 776$ k-ft

$M_{xx} = 234$ k-ft

$$I_x^s = 4(1.56)(\overline{2.21}^2) = 30$$
$$4(1.56)(\overline{6.62}^2) = 273$$
$$4(1.56)(\overline{11.03}^2) = 759$$
$$4(1.56)(\overline{15.44}^2) = 1488$$
$$4(1.56)(\overline{19.85}^2) = 2459$$
$$4(1.56)(\overline{24.26}^2) = 3673$$
$$4(1.56)(\overline{28.68}^2) = 5133$$
$$4(1.56)(\overline{33.09}^2) = 6832$$
$$6(1.56)(\overline{37.50}^2) = 13162$$

sum = 33,809 in.^{2}

$$I_x^c = \frac{24.0(84.0)^3}{12} = 1,185,408 \text{ in.}^4$$

$$I_y^s = 36(1.56)(8.80)^2 = 4349 \text{ in.}^4$$

$$I_y^c = \frac{84.0(24.0)^3}{12} = 96,768 \text{ in.}^4$$

$$np = \frac{9(59.28)}{2016} = 0.2646$$

$$nq_x = \frac{9(33,809)}{1,185,000} = 0.2568$$

$$nq_y = \frac{9(4349)}{96,768} = 0.4045$$

$$e_x = \frac{234}{360.2(7.00)} = 0.093$$

$$e_y = \frac{776}{360.2(2.00)} = 1.077$$

$$m = \frac{0.093}{1.077} = 0.086$$

$$Q = \frac{nq_y + m^2 nq_x}{np}$$

$$= \frac{0.4045 + (0.086)^2(0.2568)}{0.2646} = 1.54$$

	$np = 0.10$		$np = 0.30$	
	$\dfrac{N}{bdf'_c}$	k	$\dfrac{N}{bdf'_c}$	k
$Q = 1.0$	0.072	0.35	0.115	0.45
$Q = 3.0$	0.115	0.38	0.205	0.52
$Q = 1.54$	0.084	0.36	0.141	0.47

$$np = 0.2646$$

$$\frac{N}{bdf'_c} = 0.131$$

$$k = 0.45$$

$$f_c = \frac{360.2(1000)}{24.0(84.0)(0.131)} = 1364 \text{ psi} > 0.4f'_c$$

$$= 1300 \text{ psi, } 4.9\% \text{ overstress, close enough to say the section is OK}$$

$$h = \frac{k}{m} = \frac{0.45}{0.086} = 5.23$$

$$a_x = \frac{4.5}{84.0} = 0.0536$$

$$a_y = \frac{3.20}{24.0} = 0.1333$$

$$f_s = 9(1364)\left(1 - \frac{1 - 0.0536}{5.23} - \frac{1 - 0.1333}{0.45}\right) = 13.6 < 24 \text{ ksi}$$

Next we check Group IV, Case IB:

Group IV: $P = 288.2$ k

$M_{xx} = 802$ k-ft

$M_{yy} = 187$ k-ft

$$e_x = \frac{187}{288.2(7.00)} = 0.0927$$

$$e_y = \frac{802}{288.2(2.00)} = 1.3914$$

$$m = \frac{0.0927}{1.3914} = 0.0666$$

$$Q = \frac{0.4045 + (0.0666)^2(0.2568)}{0.2646} = 1.53$$

	$np = 0.10$		$np = 0.30$	
	$\dfrac{N}{bdf'_c}$	k	$\dfrac{N}{bdf'_c}$	k
$Q = 1$	0.057	0.31	0.091	0.41
$Q = 3$	0.092	0.35	0.166	0.47
$Q = 1.53$	0.067	0.32	0.113	0.43

$$np = 0.2646$$

$$\frac{N}{bdf'_c} = 0.105$$

$$k = 0.41$$

$$f'_c = \frac{288.2(1000)}{(24.0)(84.0)(0.105)^2} = 1361 \text{ psi} > 0.4f'_c$$

$= 1300$ psi, 4.7% overstress, close enough to say section is OK

$$h = \frac{k}{m} = \frac{0.41}{0.0666} = 6.76$$

$a_x = 0.0536$

$a_y = 0.1333$

$$f_s = 9(1361)\left(1 - \frac{1 - 0.0536}{6.16} - \frac{1 - 0.1333}{0.41}\right) = 15.5 < 24 \text{ ksi}$$

The development length of #11 bars with a hook would require a $3'\text{-}6''$ footing (Appendix F), but because the stress in the steel is not up to the allowable there is no need to develop the full strength. The strength required can be proportioned to the strength supplied by #11s and the $2'\text{-}6''$ thickness is then adequate. This is performed for tension strength:

Tension.   The maximum $f_s = 15.5$ ksi and the allowable $f_s = 24.0$ ksi; the development length $l_d = 65$ in.; the reduced l_d required $= 15.5/24 \times l_d$ (65 in.) $= 42$ in.

According to Appendix F, $E = 38$ in. with a hook for full development strength. The hook strength is equivalent to $65 - 38 = 27$ in. of development length. The reduced requirement for E is $42 - 27 = 15$ in. The footing depth should then be $15 + 3$ clear $= 18$ in. However, the $2'\text{-}6''$ footing used is conservative.

Next the capacities are checked for the AASHTO criteria which limit the service loads and moments to 35% of the section capacity, using $\phi = 1.0$.   AASHTO further specifies that these service load moments must be magnified for slenderness according to the moment magnification procedures provided. Capacities are computed by procedures similar to those in load factor design but for an increased section required by service load. These capacities are shown about the x–x axis in two diagrams. Figure 10.32 shows the capacity curve below the balance point for scaling moment ratios. Figure 10.33 shows the entire ca-

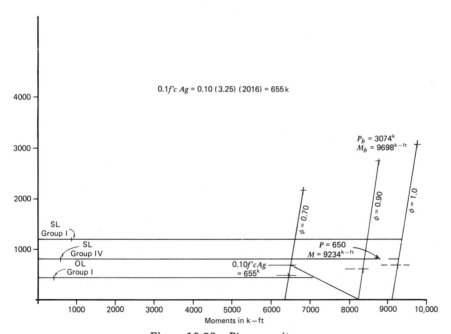

Figure 10.32   Pier capacity.

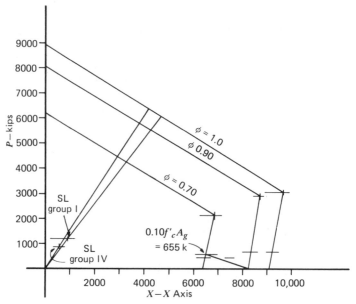

Figure 10.33   Pier capacity:

$$\phi P_0 = [(0.85)(3.25)(2016 - 59.28) + 59.28(60)] = 8962 \text{ k}$$

$$\phi = 1.0;$$

$$\phi P_0 = 8066 \text{ k},$$

$$\phi = 0.9;$$

$$\phi P_0 = 6274 \text{ k};$$

$$\phi = 0.7.$$

pacity curve about the x-x axis for establishing values for the Bressler reciprocal load equation. Computations for the different loads on the curves are shown after the capacity curve about the y-y axis (Figure 10.34). Note that they also have capacity levels at $\phi = 0.70$ and $\phi = 0.90$ for evaluation of overload capacity. The overload point is given in Example 10.2.

The capacity curve for the moment about the y-y axis, shown in Figure 10.34, is used for scaling the moment ratios and selecting values for the Bressler reciprocal load equation. Note that the curve also has capacity levels at $\phi = 0.70$ and $\phi = 0.90$ for the evaluation of the overload capacity. The Group I overload point is the same as the loading examined in Example 10.2.

As mentioned earlier, the overload moment plotted is taken from the loadings in Example 10.2. For the service load maxima the loadings in Group I and IV, which were checked for stress, are used. First the mo-

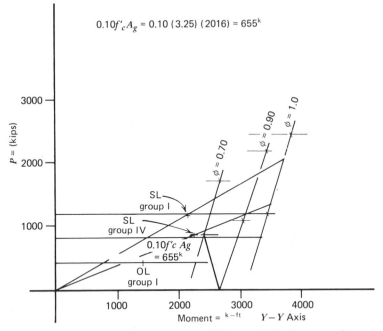

Figure 10.34 Pier capacity axis YY (longitudinal moment).

ment magnification about the x-x axis is computed. Consider the Group I loading in which the service level load and moment are increased to the required ultimate level;

$$\text{service load} = 35\% \text{ ultimate criteria}$$

Group I @ BD.   Axis x-x Transverse Moment

$$P_u = \frac{360.2}{0.35} = 1201 \text{ k}$$

$$M_u = \frac{234}{0.35} = 669 \text{ k-ft}$$

The moment should be checked for minimum eccentricity of 10%.

$$\min e = 0.10(84 \text{ in.}) = 8.40 \text{ in.}$$

$$M_u = 1201 \left(\frac{8.40}{12}\right) = 841 \text{ k-ft}$$

which controls. The larger formula for stiffness in computing critical buckling is used. The values needed for this formula are

$$E_c = 57,000 \sqrt{3250} = 3,249,000 \text{ psi}$$

$$I_g = 1,185,408 \text{ in.}^4$$

$$E_s = 29,000,000 \text{ psi}$$

$$I_s = 33,809 \text{ in.}^4$$

$$\beta_D = \frac{0}{234} = 0$$

The stiffness EI is then computed:

$$EI = \frac{(E_c I_g)/5 + E_s I_s}{1 + \beta_D}$$

$$= \left[\frac{[(3,249)(1,185,408)]/5 + (29,000)(33,809)}{1.0} \right] \times 10^3$$

$$= 1.7507 \times 10^{12}$$

From Figure 10.15 $k = 2.1$. The critical buckling load is

$$P_{cr} = \frac{\pi^2 EI}{(klu)^2} = \frac{\pi^2 (1.7507 \times 10^{12})}{[(2.1)(23.75)(12)]^2} = 48,238 \text{ k}$$

and the magnification is

$$\delta = \frac{C_m}{1 - P_u/\phi P_{cr}} = \frac{1.0}{1 - 1201/[1.0(48,233)]} = 1.026$$

with $\phi = 1.0$. Use $P_u = 1201$ k:

$$M_u = 841 \times 1.026 = 863 \text{ k-ft about } x\text{-}x \text{ axis}$$

The same computations are made for the Group IV loading about the x-x axis:

Group IV @ BD.   Axis x-x Transverse Moment

$$P_u = \frac{288.2}{0.35} = 823 \text{ k} \qquad \min e = 8.4 \text{ in.}$$

$$M_u = \frac{187}{0.35} = 534 \text{ k-ft} \qquad M_u = 576 \text{ k-ft} \leftarrow \text{use}$$

$$EI = 1.7507 \times 10^{12}$$

because $\beta_D = 0$

$$k = 2.1$$

$$P_{cr} = 48,238$$

$$\delta = \frac{1.0}{1 - 823/[48,238(1.0)]} = 1.017$$

Use $P_u = 823$ k:

$$M_u = 576 \times 1.017 = 586 \text{ k-ft}$$

The pier can be considered braced at the abutment in the longitudinal direction. Therefore a magnification $\delta = 1.0$ is used. The ultimate load and moments about the y-y axis for Groups I and IV are as follows:

Group I @ Top Column.   Axis y-y

$$P_u = \frac{360.2}{0.35} = 1201 \text{ k}$$

$$M_u = \frac{776}{0.35} = 2217 \text{ k-ft}$$

$$\delta = 1.0$$

Use $P_u = 1201$ k:

$M_u = 2217$ k-ft

Group IV @ Top Column.   Axis y-y

$$P_u = \frac{288.2}{0.35} = 823 \text{ k}$$

$$M_u = \frac{802}{0.35} = 2291 \text{ k-ft}$$

Use $\delta = 1.0$; P_u for both group loadings exceeds $0.10 f'_c A_g$. Therefore, according to the AASHTO specifications, the biaxial bending capacity, using the Bressler reciprocal load equation, should be evaluated.

$$\frac{1}{P_{nxy}} = \frac{1}{P_{nx}} + \frac{1}{P_{ny}} - \frac{1}{P_0}$$

or

$$P_{nxy} = \frac{1}{(1/P_{nx}) + (1/P_{ny}) - (1/P_0)}$$

P_{nx} and P_{ny} are taken graphically from the interaction curves plotted. First this equation is solved for the Group I loadings:

$$P_u = 1201 \text{ k}$$

$$M_u = 863 \text{ k-ft (about } x\text{-}x \text{ axis)}$$

$$M_u = 2217 \text{ k-ft (about } y\text{-}y \text{ axis)}$$

$$P_{nx} = 6100 \text{ k}$$

$$P_{ny} = 2050 \text{ k}$$

$$P_0 = 8962 \text{ k}$$

$$P_{nxy} = \frac{1}{(1/6100) + (1/2050) - (1/8962)} = 1851 > 1201 \text{ k} \quad \text{OK}$$

For Group IV loadings

$$P_u = 823 \text{ k}$$

$$M_u = 586 \text{ k-ft (about } x\text{-}x \text{ axis)}$$

$$M_u = 2291 \text{ k-ft (about } y\text{-}y \text{ axis)}$$

$$P_{nx} = 6400 \text{ k}$$

$$P_{ny} = 1300 \text{ k}$$

$$P_0 = 8962 \text{ k}$$

$$P_{nxy} = \frac{1}{(1/6400) + (1/1300) - (1/8962)} = 1229 > 823 \text{ k} \quad \text{OK}$$

When an overload vehicle is specified it should be checked for capacity only according to the load factor principles in Example 10.7. The bottom of the shaft in a framed structure such as this will often have less reinforcing than the section at the top of the shaft. Therefore it is practical to stop half the required top shaft reinforcing at midheight and to provide alternate bars at the bottom of the shaft extending into the footing. Therefore the section shown in Figure 10.35 will be adequate for the bottom shaft section. An analysis similar to that for the top column section would indicate this condition. The footing loads and moments are tabulated as in Example 10.2 for the service load design of pier footings. The footing is proportioned on the basis of allowable pressure. The footing pressure computations and drawings of the pressure diagrams are as given in Example 10.2. The footing section is then designed for stress for the Group VII, Case IIA, loading. The overburden is the weight of 2.5 ft of footing and 6.5 ft of earth over the footing.

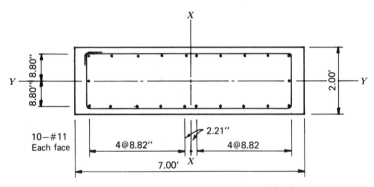

Figure 10.35 Section at bottom of shaft.

Overburden

$$w = 2.5 \times 0.150 = 0.38$$
$$6.5 \times 0.120 = \underline{0.78}$$
$$1.16 \text{ ksf} \div 1.33 = 0.87 \text{ ksf} \quad \text{Group VII}$$

Moments from the overburden and foundation pressure are summed as cantilever action at the corner of the pier shaft. First the response is evaluated in the transverse direction:

Transverse

	V (k-ft^{-1})	M (k-ft ft^{-1})
$9.47 \times 2.50 =$	$+23.7 \times 1.25 =$	$+29.6$
$\frac{1}{2} \times 4.24 \times 2.50 =$	$+5.3 \times 1.67 =$	$+8.9$
$-0.87 \times 2.50 =$	$\underline{-2.2} \times 1.25 =$	$\underline{-2.7}$
	$+26.8$ k	$+35.8$ k-ft

The effective depth d is approximately

$$d = 30 - 3.0 - 0.50 \pm = 26.5 \text{ in.} \pm$$

Shear is computed at a d distance from support.

Shear @ d Distance

$$13.21 \times (2.50 - 2.21) = \quad 3.9$$
$$\frac{1}{2} \times 0.50 \times (2.50 - 2.21) = \quad 0.1$$
$$-0.87 \times (2.50 - 2.21) = \underline{-0.3}$$
$$3.7 \text{ k}$$

The shear is insignificant; therefore it need not be considered further. The approximate d required is

$$d_{(\text{required})} = \sqrt{\frac{35.8(12)}{(0.190)(12)}} = 13.7 < 26.5 \text{ in.} \pm$$

The approximate area of reinforcing required is

$$A_{s\ (\text{required})} = \frac{35.8(12)}{24(0.87)(26.5)} = 0.78 \text{ in.}^2$$

Try #7 @ 9 in.

$$A_s = 0.80 \text{ in.}^2$$

$$d = 30 - 3.0 - 0.44 = 26.56$$

From Appendix A the k and j values are computed as

$$p = \frac{0.80}{12(26.56)} = 0.0025 \qquad k = 0.191 \qquad j = 0.936$$

The stresses are computed and compared with the allowables:

$$f_s = \frac{35.8(12)}{0.80(0.936)(26.56)} = 21.6 < 24.0 \text{ ksi}$$

$$f'_c = \frac{21.6(0.191)}{9(0.809)} = 567 \text{ psi} < 1300 \text{ psi} \quad \text{OK}$$

The minimum reinforcing requirements were computed in Example 10.2 as $A_s = 0.56$ in.^{2}. Next the response in the longitudinal direction is examined. The shears and moments are

Longitudinal

	V (k-ft^{-1})	M (k-ft ft^{-1})
$9.47 \times 2.00 =$	$18.9 \times 1.00 =$	18.9
$\frac{1}{2} \times 2.38 \times 2.00 =$	$2.4 \times 1.33 =$	3.2
$-0.87 \times 2.00 =$	$-1.7 \times 1.00 =$	-1.7
	19.6 k	20.4 k-ft

The approximate area of steel required is

$$\text{approx } A_{s\ (\text{required})} = \frac{20.4(12)}{24(0.87)(25.5)} = 0.46 \text{ in.}^2$$

The minimum reinforcement requirements control; therefore use #6 @ 9 in.

$$A_s = 0.59 \text{ in.}^2$$

The requirement for top reinforcement is investigated by examining the stress in the plain concrete with only the overburden load. The moment = 2.7 k-ft.

The plain concrete bending stress is

$$S = \frac{12(30)^2}{6} = 1800 \text{ in.}^2$$

$$f = \frac{2.7(12)}{1800} = 18 \text{ psi} < 90 \text{ psi} = 0.21(7.5)\sqrt{3250} \text{ allowable}$$

OK as plain concrete but nominal #5 @ 1–3 each way in top footing is used. By observation peripheral or punching shear is insignificant. The footing is examined for overload capacity in Example 10.2.

Examples 10.4 and 10.5. Figures 3.13 and 3.14 show details for a bridge with a three-column bent, for which one of the columns is analyzed by service load and load factor design procedures. Computations for the cap design indicate that the interior column of the three-column bent will control. Therefore only the interior column is examined. The superstructure for this bridge is a longitudinally reinforced concrete slab and the columns are monolithic with the slab. The design loading for the bridge is H 15 and the concrete strength is $f'_c = 4000$ psi, $f_c = 1600$ psi, $n = 8$. The reinforcing steel yield stress is $f_y = 60,000$ psi. For a working stress design $f_s = 24,000$ psi. The columns are circular and the calculations show some of the differences between circular shaft and rectangular shaft design given in Examples 10.2 and 10.3. The footings for these circular shafts are supported on foundation piling. Therefore these examples are contrasted with Examples 10.2 and 10.3, for they exhibit a different type of foundation. The method given herein is somewhat abbreviated in that certain group loadings are eliminated because they cannot control the design. The superstructure design for this bridge is described in Examples 4.1 and 4.2.

Example 10.4. Load factor design.   The loads must be tabulated as in the load factor design in Example 10.1. Because in this example the design is for an H 15 live load, only Group IA load factors for live load are used. Observation indicates that the only groups worth considering are I, IV, and VII. Group I should be examined because live load may produce significant moments at the tops of columns. Group IV should be examined because the columns are relatively short and moderately stiff and significant moments could be produced due to changes in the length of the span. Group VII is generally significant for most designs because it includes earthquake loadings. Observation of the data shows that only the moments computed in a longitudinal direction are given. No transverse moments are computed because forces in the longitudinal direction are most critical. If longitudinal and transverse moments develop on a circular shaft, they would be resolved in a moment for analysis. The moments are tabulated at the top and bottom of columns for the three groups mentioned (Table 10.16). These groups use the γ and

Table 10.16   Bent No. 1: Column Loads and Moments (Interior Column)

LL Case		Case III			Case IV			Case VI		
Group	Loading	M	M_L	M_T	M	M_L	M_T	M	M_L	M_T
@ Column *BE*										
I	$\beta_D D$	115	− 26		115	− 26		115	− 26	
	2.20L + I	88	+185		75	− 262		176	− 57	
	Σ	203	+159		190	− 288		291	− 83	
	× 1.3	264	+207		247	− 374		378	−108	
IV	$\beta_D D$	115	− 26		115	− 26		115	− 26	
	L + I	40	+84		34	− 119		80	− 26	
	S + T		+60			− 24			− 24	
	Σ	155	+118		149	− 169		195	− 76	
	× 1.3	202	+153		194	− 220		254	− 99	
VII	$\beta_D D$	115	− 26							
	EQ		− 163							
	Σ	115	− 189							
	× 1.3	150	− 246							
@ Column *EB*										
I	$\beta_D D$	128	− 12		128	− 12		128	− 12	
	2.20L + I	88	+20		75	− 88		176	− 37	
	Σ	216	+8		203	− 100		304	− 49	
	× 1.3	281	+10		264	− 130		395	− 64	
IV	$\beta_D D$	128	− 12		128	− 12		128	− 12	
	L + I	40	+9		34	− 40		80	− 17	
	S + T		+67			− 27			− 27	
	Σ	168	+64		162	− 79		208	− 56	
	× 1.3	218	+83		211	− 103		270	− 73	
VII	$\beta_D D$	128	− 12							
	EQ		− 162							
	Σ	128	− 174							
	× 1.3	166	− 226							

β factors given in AASHTO Article 1.2.22; β_D is assumed to be 0.75 for direct load and 1.0 for moment. The loads and moments in the following tabulations were computed in Example 3.2.

In the next step in the analysis we calculate the section capacity like the computation for the rectangular shaft, except that the properties of segments of a circle must be included. These properties can be taken from Figure 10.36. Because it is obvious that bending will control the design, we can assume that the critical loadings will fall below the balance point and below 10% of the squash load $(0.10f'_c A_g)$ on the interaction curve. Therefore we compute the capacity about each axis for a concrete compressive strain of 0.003 and a steel compressive strain of 0.00207 at the balance point. We also compute another direct load and moment capacity for a concrete compressive strain of 0.003 and a steel compressive strain of 0.005, which again is close to a level where the direct load is small or almost zero. We can then approximate the necessary portion of capacity interaction curve by a straight line between these two points (Figure 10.37). This computation is as follows:

First a half-column is drawn in section with the reinforcing spaced around the face. The two strain possibilities already mentioned are

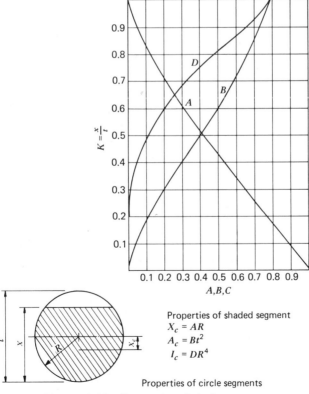

Properties of shaded segment
$X_c = AR$
$A_c = Bt^2$
$I_c = DR^4$

Properties of circle segments

Figure 10.36   Properties of circle segments.

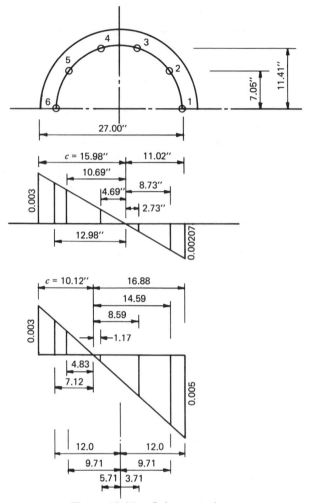

Figure 10.37   Column strains.

projected from the half section and necessary dimensions are added. The capacity is computed for the strain condition at the balance point of 0.00207 steel strain and 0.003 concrete strain. First the reinforcing is located in relation to the section centroid and the strain at each bar is computed on the basis of its location in the assumed strain diagram. The strain in the reinforcing is then converted to stress by multiplying by E_s = 29,000 ksi using the yield strength of 60 ksi as a maximum stress. These stresses are converted to forces by multiplying by the area of the reinforcing bars, and the moments for each bar are computed by multiplying the bar force by its distance from the centroid. These results are listed in Table 10.17. The size of the compression block is computed by using β_1 = 0.85, according to AASHTO, and the area of

Table 10.17 Maximum $\epsilon_s = 0.00207$

Reinforcing Location	Distance R/F to Centroid	Reinforcing Area A_s	Strain @ R/F	Stress in R/F (kips)	Force in R/F (kips)	Moment (k-ft)
1	12.00	0.79	-0.00207	-60.00	-47.4	47
2	9.71	1.58	-0.00164	-47.56	-75.1	61
3	3.71	1.58	-0.00051	-14.79	-23.4	7
4	3.71	1.58	+0.00088	+25.52	+40.3	12
5	9.71	1.58	+0.00201	+58.29	+92.1	75
6	12.00	0.79	+0.00244	+60.00	+47.4	47
					+179.8 -145.9	249

$c = 15.98$ in.

$\beta_1 c = 0.85(15.98) = 13.58$ in.

$k = \dfrac{\beta_1 c}{t} = \dfrac{13.58}{0.30} = 0.453$

$B = 0.345$ (Figure 10.36) $A_c = 0.345(30)^2 = 310$ in.^{2}

$C = 0.85(4000)(310) = 1054$ k

$A = 0.475$ (Figure 10.36) $x_c = 0.475(15.0) = 7.12$ in.

for C $M_c = 1054\left(\dfrac{7.12}{12}\right) = 626$ k-ft

$P_b = +179.8 - 145.9 + 1054 = 1088$ k $\left.\begin{array}{l}\ \end{array}\right\}\ \phi = 1.0$

$M_b = 249 + 626 = 875$ k-ft

$P_b = 979$ k $\left.\begin{array}{l}\ \end{array}\right\}\ \phi = 0.90$
$M_b = 787$ k-ft

$P_b = 816$ k $\left.\begin{array}{l}\ \end{array}\right\}\ \phi = 0.75$
$M_b = 656$ k-ft

the compression block is computed by using Figure 10.36 from which the centroid of the compression block is also taken. The force in the compression block and its moment about the centroid are then computed. All forces and moments are added to give the capacity values P_b and M_b and are shown for different levels of the capacity reduction factor ϕ.

The same procedure is used to compute the capacity values of P and M for the condition in which the concrete strain is 0.003 and the reinforcing steel strain is 0.005 (Table 10.18).

It is now essential to compute any moment magnification that might be applied to the different loadings for this column. Group IV, Case IV, and Group VII at the top of the column are computed. These are the two groups deemed by observation to be the most critical. This structure should be considered unbraced in both longitudinal and transverse directions. Computations for effective length factors are based on the unbraced condition and use of the Jackson Moreland Charts in Figure 10.17. The Group I, Case IV, loading case is examined first. The ultimate load and moment to be considered are the following:

Group I: Case IV @ BE

$$P_u = 247 \text{ k}$$

$$M_u = 374 \text{ k-ft}$$

According to AASHTO Article 1.5.23(D), the modulus of elasticity is

$$E_c = 57,000 \sqrt{4000} = 3,605,000 \text{ psi}$$

For the 30 in. diameter shaft the moment of inertia is

$$I_g = 0.049087(30)^4 = 39,760 \text{ in.}^4$$

The creep ratio β_D is the ratio of the dead load moment to the total moment:

$$\beta_D = \frac{26}{288} = 0.0903$$

Using AASHTO eq. 6.18, we compute the EI stiffness for critical buckling:

$$EI = \frac{E_c I_g / 2.5}{1 + \beta_D}$$

$$EI = \frac{3,605,000(39,760)}{2.5(1.0903)} = 5.259 \times 10^{10} \text{ psi}$$

Table 10.18 Maximum $\epsilon_s = 0.005$

Reinforcing Location	Distance R/F to Centroid	Reinforcing Area A_s	Strain @ R/F	Stress in R/F (kips)	Force in R/F (kips)	Moment (k-ft)
1	12.00	0.79	-0.00500	-60.00	-47.4	47
2	9.71	1.58	-0.00432	-60.00	-94.8	77
3	3.71	1.58	-0.00254	-60.00	-94.8	24
4	3.71	1.58	-0.00035	-10.15	-16.0	-5
5	9.71	1.58	+0.00143	+41.47	+65.5	53
6	12.00	0.79	+0.00211	+60.00	+47.4	47
					+112.9 -253.0	243

$c = 10.12$ in.

$\beta_1 c = 0.85(10.12) = 8.60$ in.

$k = \dfrac{\beta_1 c}{t} = \dfrac{8.60}{30} = 0.287$

$B = 0.182$ (Table 10.36) $A_c = 0.182(30)^2 = 164$ in.^{2}

$C = 0.85(4000)(164) = 558$ k

$A = 0.665$ (Figure 10.36) $x_c = 0.665(15.0) = 9.98$ in.

for C $M_c = 558\left(\dfrac{9.98}{12}\right) = 464$ k-ft

$\left.\begin{array}{l} P_n = +112.9 - 253.0 + 558 = 418 \text{ k} \\ M_n = 243 + 464 = 707 \text{ k-ft} \end{array}\right\}$ $\phi = 1.0$

$\left.\begin{array}{l} P_n = 376 \text{ k} \\ M_n = 636 \text{ k-ft} \end{array}\right\}$ $\phi = 0.90$

$\left.\begin{array}{l} P_n = 314 \text{ k} \\ M_n = 530 \text{ k-ft} \end{array}\right\}$ $\phi = 0.75$

From the Jackson Moreland Charts in Figure 10.17 the top and bottom stiffness parameters are computed as

$$G_{top} = \frac{0.971}{0.990 + 0.947} = 0.501$$

$$G_{bottom} = 1.0 \text{ (from Table 10.1)}$$

We then obtain the effective length factor k:

$$k = 1.2$$

The critical buckling load P_c is computed as

$$P_c = \frac{\pi^2 EI}{(klu)^2} = \frac{\pi^2(5.259 \times 10^{10})}{[(1.2)(26.19)(12)]^2} = 3649 \text{ k}$$

The moment magnification in AASHTO eq. 6-15 is therefore

$$\delta = \frac{C_m}{1 - P_u/\phi P_c} = \frac{1.0}{1 - 247/[(0.75)(3649)]} = 1.099$$

Then for the required ultimate column capacity use

$$P_u = 247 \text{ k}$$

$$M_u = 1.099(374) = 411 \text{ k-ft}$$

With the same procedure the required ultimate capacity for the Group VII loading is computed:

Group VII @ BE

$$P_u = 150 \text{ k}$$

$$M_u = 246 \text{ k-ft}$$

$$\beta_D = \frac{26}{189} = 0.138$$

$$EI = 5.259 \times 10^{10} \times \frac{1.0903}{1.138} = 5.040 \times 10^{10}$$

$$P_c = \frac{\pi^2(5.040 \times 10^{10})}{[(1.2)(26.19)(12)]^2} = 3498 \text{ k}$$

$$\delta = \frac{1.0}{1 - 150/[(0.75)(3498)]} = 1.060 \text{ (the required magnification)}$$

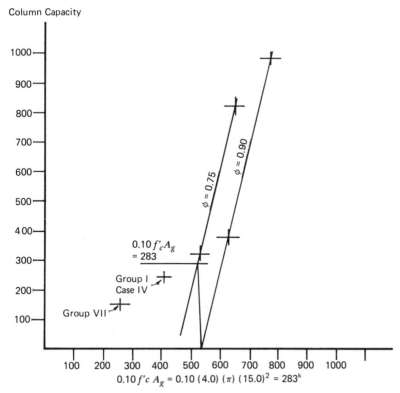

Column Capacity

$0.10 f'c A_g = 0.10 (4.0) (\pi) (15.0)^2 = 283^k$

Figure 10.38   Column capacity.

Use P_u = 150 k:

$$M_u = 1.060(246) = 261 \text{ k-ft}$$

The magnified moments and loads are then plotted against the computed capacity (Figure 10.38).

It is now necessary to compute the serviceability requirements. In the first step in this procedure we tabulate the service load level loads and moments for the top and bottom of the column and then select the most critical loadings. The steel stress is computed by using the procedure derived by Toprac.[35] The steel stress is converted to a crack width according to the Kaar formula and compared with the allowable of 0.006 in. The method for computing stresses, outlined by Toprac,[35] is basically an elastic analysis for a transformed reinforced concrete section. This method applies the properties for circle segments shown in Figure 10.36 and uses double compression in the reinforcing steel. The terminology for this method is defined in Figure 10.39. Groups I, IV, and VII are tabulated for a service load level (Tables 10.19 and 10.20) according to the groups in AASHTO Article

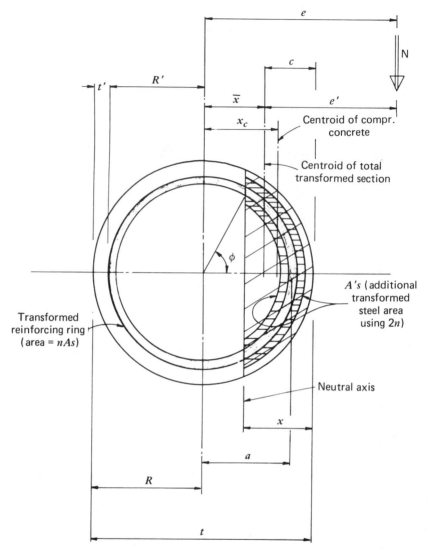

Figure 10.39   Notations for stress analysis of circular column.

1.2.22.   The loads and moments for each group loading are summed and divided by the allowable overstress that provides for all loadings to be at 100% of the allowable stress; therefore the loadings are easily compared.

The proposed section is analyzed for stress by the method developed by Toprac.[35]   As stated earlier, this elastic analysis uses a transformed section and $2n$ for the compressive reinforcement.   Refer to Figure 10.36 for the properties of the circle segments and Figure 10.39 for the definition of the terms in this analysis.   This method involves iteration

Table 10.19 Bent No. 1: Column Loads and Moments (Interior Column). Service Load Level

Group	Loading	Case III			Case IV			Case VI		
		N	M_L	M_T	N	M_L	M_T	N	M_L	M_T
@ BE										
I	D	153	-26		153	-26		153	-26	
	L + I	40	+84		34	-119		80	-26	
	Σ	193	+58		187	-145		233	-52	
IV	Group I	193	+58		187	-145		233	-52	
	S + T		+60			-24			-24	
	Σ	193	+118		187	-169		233	-76	
	÷ 1.25	154	+94		150	-135		186	-61	
VII	D	153	-26							
	EQ		-163							
	Σ	153	-189							
	÷ 1.33	115	-142							

of the assumed locations for the neutral axis. The proper approach is to assume a neutral axis location and compute a neutral axis location as a check of the assumption. If they do not compare, the next trial should be made by assuming that the neutral axis is approximately where it was computed to be during the preceding trial.

Table 10.20

Group	Loading	Case III			Case IV			Case VI		
		N	M_L	M_T	N	M_L	M_T	N	M_L	M_T
@ Column EB										
I	D	170	-12		170	-12		170	-12	
	L + I	40	+9		34	-40		80	-17	
	Σ	210	-3		204	-52		250	-29	
IV	Group I	210	-3		204	-52		250	-29	
	S + T		+67			-27			-27	
	Σ	210	+64		204	-79		250	-56	
	÷ 1.25	168	+51		163	-63		200	-45	
VII	D	170	-12							
	EQ		-162							
	Σ	170	-174							
	÷ 1.33	128	-131							

Group VII is picked for a stress analysis in which the notation in Figure 10.39 is applied. Assuming a reinforcing area equal to 1% of the column gross area, we compute the column reinforcing as

$$1\% \, R/F \qquad A_s = 0.01(\pi)(15)^2 = 7.07 \text{ in.}^2$$

$$10 - \#8 \, A_s = 7.90 \text{ in.}^2$$

This area is assumed to be a ring of reinforcing with $A_s = 7.90$ in.^{2}, as shown in Figure 10.37. Again, using Figure 10.37, the critical data are

Group VII	$N = 115$ k	$M = 142$ k-ft	$e = 14.82$ in.
	$t = 30.0$ in.	$t' = 3.00$ in.	$d = 27.00$ in.
	$A_g = 707$ in.^{2}	$A_s = 7.90$ in.^{2}	$p = 0.0112$
	$n = 8$	$R = 15.0$ in.	$R' = 12.0$ in.

Assume that $x = 12.60$ in. For Figure 10.23

$$k = \frac{x}{t} = \frac{12.60}{30.0} = 0.420$$

Next the area of concrete in compression is computed by using Figure 10.36:

$$B = 0.31 \qquad A_c = 0.31(30)^2 = 279 \text{ in.}^2$$

The transformed reinforcing area for the ring is nA_s:

$$nA_s = 8(7.90) = 63.20 \text{ in.}^2$$

From Figure 10.39 the angle ϕ to the compressive reinforcement with $\cos \phi = 1 - a/R'$ is computed as

$$a = 12.60 - 3.0 = 9.60 \text{ in.} \qquad R' = 12.0 \text{ in.}$$

$$\cos \phi = 1 - \frac{9.60}{12.0} = 0.200$$

$$\phi = 78.5°$$

The angular fraction of a half-circle $\lambda = \phi/180 = 78.5/180° = 0.436$. The area of the additional compressive reinforcement, which results from the use of $2nA_s'$, is computed as

$$A_s' = \lambda(nA_s) = 0.436(63.20) = 27.55 \text{ in.}^2$$

The total transformed steel area is then

$$A_{ST} = nA_s + A_s' = 63.20 + 27.55 = 90.75 \text{ in.}^2$$

The total transformed area is

$$A_T = A_c + A_{ST} = 279 + 91 = 370 \text{ in.}^2$$

Then

$$\frac{N}{A_T} = \frac{115}{370} = 311 \text{ psi}$$

Using Figure 10.36, we obtain the centroid of the compressed concrete:

$$\text{coefficient } A = 0.51$$
$$x_c = 0.51(15.0) = 7.65 \text{ in.}$$

and the centroid of the compressive reinforcing A_s' is computed by using

$$x' = 57.3R' \frac{\sin \phi}{\phi}$$

$$x' = 57.3(12.0) \frac{\sin 78.5}{78.5} = 8.58 \text{ in.}$$

The centroid of the entire transformed section is computed by summing moments around the column axis:

$$A_c = \quad 279 \ \times 7.65 = 2134$$
$$A_b = 27.55 \times 8.58 = \underline{\ 236}$$
$$2370 \div A_T \ (370 \text{ in.}^2) = 6.41 \text{ in.} = \bar{x}$$

Other dimensions are located with respect to the transformed section centroid:

$$x_c - \bar{x} = 7.65 - 6.41 = 1.24 \text{ in.}$$
$$x' - \bar{x} = 8.58 - 6.41 = 2.17 \text{ in.}$$
$$c = 15.0 - 6.41 = 8.59 \text{ in.}$$
$$e' = 14.82 - 6.41 = 8.41 \text{ in.}$$

Next compute the moment of inertia of the compressed concrete by using Figure 10.36:

$$\text{coefficient } D = 0.062$$

$$I_c = 0.062(15)^4 = 3139 \text{ in.}^4$$

The moment of inertia of the reinforcing ring nA_s is computed by using

$$I_s = \frac{A_s(2R')^2}{8}$$

$$I_s = \frac{63.20(24.0)^2}{8} = 4550 \text{ in.}^4$$

Because of its eccentric location, the additional moment of inertia of the compressed concrete is

$$A_c(x_c - \bar{x})^2 = 279(1.24)^2 = 429 \text{ in.}^4$$

and because of its eccentric location the additional moment of inertia of the reinforcing ring is

$$nA_s\bar{x}^2 = 63.2(6.41)^2 = 2597 \text{ in.}^4$$

The moment of inertia of the additional compressive reinforcing area, using $2n$, is negligible about its own axis but is computed for its eccentric location:

$$A_s'(x' - \bar{x})^2 = 27.55(2.17)^2 = 130 \text{ in.}^4$$

The total moment of inertia I_T equals $I_c + I_s + A_c(x_c - \bar{x})^2 + nA_s\bar{x}^2 + A_s'(x' - \bar{x})^2 = 10{,}845 \text{ in.}^4$. The values of stress $Ne'c/I_T$ and $Ne'(R' + \bar{x})/I_T$ are computed as follows:

$$\frac{Ne'c}{I_T} = \frac{115(8.41)(8.59)}{10{,}845} = 766 \text{ psi}$$

$$\frac{Ne'(R' + \bar{x})}{I_T} = \frac{115(8.41)(18.41)}{10{,}845} = 1642 \text{ psi}$$

The maximum concrete stress equals

$$f_c = \frac{N}{A_T} + \frac{Ne'c}{I_T} = 311 + 766 = 1077 \text{ psi}$$

The maximum tensile stress in the reinforcing equals

$$\frac{f_s}{n} = \frac{N}{A_T} - \frac{Ne'(R' + \bar{x})}{I_T} = 311 - 1642 = -1331 \text{ psi}$$

This yields $f_s = 8(-1331) = -10.6$ ksi.

Now the x value is examined. Any large deviation from our assumption would mean that another iteration of these computations is needed:

$$x = d \times \frac{f_c}{f_c + f_s/n} = 27.0 \times \frac{1077}{2408} = 12.08$$

This approximates the assumption of $x = 12.60$ in. Therefore the stress computations are reasonably accurate.

The crack width is computed and compared with the allowable of 0.006 in. The 10-#8 are at 9.42 in. oc:

$$d_c = 3.0 \text{ in.}$$

$$A = 2(d_c)(\text{cc of bars})$$

$$= 2(3.0)(9.42) = 56.5 \text{ in.}^2$$

$$\omega = 0.115 \sqrt[4]{A}(f_s)(10^{-3})$$

$$= 0.115 \sqrt[4]{56.5}(10.6)(10^{-3})$$

$$= 0.0033 < 0.006 \text{ allowable \quad OK}$$

It is apparent from the design computations that the column is somewhat overdesigned. A 2'-6" diameter column, however, is a desirable minimum size and less than the 1% reinforcing used in these calculations is not appropriate.

The footing is designed to consider the pile foundation. It is generally assumed that the foundation will react in an elastic manner; therefore the pile loads are computed by elastic analysis. Pile loads are shown as an allowable at the service load level; therefore the footing is sized and the number of piles selected is based on a service load level. Initially the service level loads and moments for the footing are tabulated and the actual pile loads computed are compared with the allowables. Under normal conditions uplift probably should not be allowed on piles. Under some rather unusual conditions, however, such as earthquake loadings, it might be appropriate to allow a small amount of uplift. When uplift is allowed in the design, the piles should have some additional anchor in the footing. Using the computed pile loads, we check the footing for serviceability requirements by using the

Table 10.21 Service Load Level: Footing Loads and Moments @ *EB* (Interior Column)

Group	Loading	Case III			Case IV			Case VI		
		N	M_L	M_T	N	M_L	M_T	N	M_L	M_T
I	D	199	−16		199	−16		199	−16	
	L	31	+15		26	−44		61	−17	
	Σ	230	−1		225	−60		260	−33	
IV	Group I	230	−1		225	−60		260	−33	
	$S + T$		+80			−33			−33	
	Σ	230	+79		225	−93		260	−66	
	÷ 1.25	184	+63		180	−74		208	−53	
VII	D	199	−16							
	EQ		−197							
	Σ	199	−213							
	÷ 1.33	150	−160							

normal 0.006 in. as a maximum crack width and then compare it against crack widths computed by the Kaar formula. After the pile loads are computed at a service load level and serviceability is checked the footing is checked for capacity by retabulating the loads as factored values and computing new pile loads. We then apply these pile loads to the footing and compute the maximum applied shears and moments. These shears and moments should then be checked against capacity. First the service load level reactions and moments are tabulated (Table 10.21). The assumed pile group is

The moment of inertia of this pile group in each direction is

$$I = 4(2.0)^2 = 16$$

Group I, Case VI, and Group VII are selected as critical and are checked for maximum pile load:

$$P = \frac{N}{4 \text{ piles}} \pm \frac{M \, (2.0 \text{ ft})}{I_{\text{pile group}}}$$

Group I, Case VI

$$N = 260 \text{ k}$$

$$M_L = 33 \text{ k-ft}$$

$$P = \frac{260}{4} \pm \frac{33(2.0)}{16} = 65.0 \pm 4.1 = +69.1 \text{ k} \qquad \textit{34.5 tons max}$$

Group VII

$$N = 150 \text{ k}$$

$$M_L = 160 \text{ k-ft}$$

$$P = \frac{150}{4} \pm \frac{160(2.0)}{16} = 37.5 \pm 20.0 = 57.5 \text{ k}$$

The 34.5-ton maximum pile load is within allowables. To check serviceability the bending around a square section equal in area to the circular 30-in. diameter column is computed:

2'-6" ϕ Column

$$\text{equiv sq} = \sqrt{\pi (1.25)^2} = 2.22 \text{ ft}$$

The footing is assumed to be 6'-6" × 6'-6" × 2'-6" thick. The overburden on the footing consists of 2.5 ft of concrete and 3.0 ft of earth over the footing (120 pcf):

Overburden

$$2.5 \times 0.15 = 0.38$$
$$3.0 \times 0.12 = \underline{0.36}$$
$$0.74 \text{ ksf}$$

Half the width of a footing is designed to incorporate one pile load. For half the footing the overburden is 0.74 ksf × 3.25 = 2.40 k-ft⁻¹. A group I loading controls the design; therefore no adjustment is necessary in the overburden to account for overstress.

The loads on the footing are shown in the following diagram:

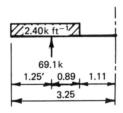

Moments are summed around the face of the column.

V (k-ft^{-1})	M (k-ft ft^{-1})
$+69.1 \times 0.89 =$ +61.5	
$2.40 \times 2.14 =$ $-5.1 \times 1.07 =$ -5.5	
$+64.0$ k	$+56.0$ k-ft

Assume 3-#8 reinforcing bars with each pile placed longitudinally and transversely:

$$3\text{-}\#8 \text{ each}$$

$$A_s = 2.37 \text{ in.}^2$$

k and j are found by using Appendix A:

$$d = 30 - 3.0 - 1.0 - 0.5 = 25.5 \text{ in. (min)}$$

$$p = \frac{2.37}{39(25.5)} = 0.0024$$

therefore $k = 0.178$, $j = 0.941$. The steel stress is computed by

$$f_s = \frac{M}{A_s jd}$$

$$f_s = \frac{56.0(12)}{2.37(0.941)(25.5)} = 11.8 \text{ ksi}$$

The crack width is checked as before.
    Assume a 2-in. cover for crack width:

$$dc = 2.0 + 1.0 + 0.5 = 3.50 \text{ in.}$$

$$A = 2(3.50)(39.0) \div 3 = 91.0 \qquad \text{use } 50.0 \text{ max}$$

$$\omega = 0.115 \sqrt[4]{50.0}(11.8)(10^{-3}) = 0.0036 < 0.006 \quad \text{OK}$$

To check the footing capacity the loads and moments at an ultimate level are now tabulated (Table 10.22), where β_D is assumed to be equal to 1.0.
    Using the equation

$$P = \frac{N}{4 \text{ piles}} \pm \frac{M \,(2.0 \text{ ft})}{I_{\text{pile group}}}$$

we compute the ultimate pile loads for the same two groups previously considered:

Group I, Case VI

$$N = 446 \text{ k}$$

$$M_L = -69 \text{ k-ft}$$

$$P = \frac{446}{4} \pm \frac{69(2.0)}{16} = 111.5 \pm 8.6 = 120.1 \text{ k}$$

Group VII

$$N = 259 \text{ k}$$

$$M_L = 277 \text{ k-ft}$$

$$P = \frac{259}{4} \pm \frac{277(2.0)}{16} = 64.8 \pm 34.6 = 99.4 \text{ k}$$

The loading diagram is

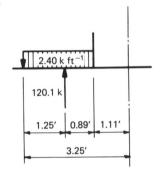

Table 10.22   Footing *EB*: Loads and Moments.   Capacity Required
(Interior Column)

Group	Loading	Case III			Case IV			Case VI		
		N	M_L	M_T	N	M_L	M_T	N	M_L	M_T
I	$\beta_D D$	199	-16		199	-16		199	-16	
	$2.20LL$	68	$+33$		57	-97		134	-37	
	Σ	267	$+17$		256	-113		343	-53	
	$\times 1.3$	347	$+22$		333	-147		446	-69	
IV	$\beta_D D$	199	-16		199	-16		199	-16	
	LL	31	$+15$		26	-44		61	-17	
	$S + T$		$+80$			-33			-33	
	Σ	230	$+79$		225	-93		260	-66	
	$\times 1.3$	299	$+103$		292	-121		338	-86	
VII	$\beta_D D$	199	-16							
	EQ		-197							
	Σ	199	-213							
	$\times 1.3$	259	-277							

Moments are summed around the face of the column:

	V (k-ft^{-1})	M (k-ft ft^{-1})
	$120.1 \times 0.89 =$ 106.9	
$2.40 \times 2.14 =$	$-5.1 \times 1.07 =$	-5.5
	115.0 k	101.4 k-ft

The footing ultimate moment for half the width is 101.4 k-ft. The capacity is then computed:

$$3\text{-}\#8 \qquad A_s = 2.37 \text{ in.}^2$$

$$a = \frac{A_s f_y}{0.85 f'_c b} = \frac{2.37(60)}{0.85(4.0)(39)} = 1.07 \text{ in.}$$

$$\phi M_n = \phi\left[A_s f_y \left(d - \frac{a}{2}\right)\right]$$

$$\phi M_n = 0.90[(2.37)(60)(25.5 - 0.54)](\tfrac{1}{12})$$

$$= 266 > 101.4 \text{ k-ft (required)}$$

Shear distance d from the face of the column is outside the piles. Therefore the shear is examined for the pile load:

$$V_u = 120.1 - 2.40(1.25) = 117.1 \text{ k}$$

$$\phi = 0.85$$

$$v_u = \frac{V}{\phi bd} = \frac{117.1}{0.85(39)(25.5)} = 138 \text{ psi} > 126 \text{ psi} = 2\sqrt{f'_c}$$

This is close; therefore footing is OK as designed.

Example 10.5. Service load design. The column and footing derived by load factor design in Example 10.4 is now computed by service load design. The tabulation of loads and moments in group loadings for a service load level was given in Example 10.4. The computations for stress were discussed in the same example as part of the serviceability computations. Because Example 10.4 assumed minimum desirable sections, the same sections are used in this example. From Example 10.4 the column stresses computed with 10-#8 are

$$f_c = 1077 \text{ psi} < 0.4 f'_c = 1600 \text{ psi}$$

$$f_s = 10.6 \text{ ksi} < 24.0 \text{ ksi}$$

Using the data from that example we obtain the compressive stresses in the reinforcing:

$$f_s' = \left[f_c' \times \left(\frac{x - t'}{x} \right) \right] 2(n)$$

$$f_s' = \left(1077 \times \frac{9.60}{12.60} \right) 2(8) = 13.1 \text{ ksi} < 24.0 \text{ ksi}$$

All stresses are within allowables.

Next, according to AASHTO Article 1.5.28, it is necessary to check to see that the critical loading does not exceed 35% of the capacity of the section, assuming a capacity reduction factor equal to 1.0. The section capacity is computed exactly as in Example 10.4. The critical service loads and moments are then divided by 0.35 to compare them with the capacity at $\phi = 1.0$. Before making this comparison, however, the moments must be magnified because, as previously determined, the structure is unbraced in the longitudinal and transverse directions. The magnified moments and loads are then checked against the capacity, again assuming $\phi = 1.0$. For Group VII loading the ultimate load and moment are

Group VII @ BE

$$P_u = \frac{115}{0.35} = 329 \text{ k}$$

$$M_u = \frac{142}{0.35} = 406 \text{ k-ft (longitudinal)}$$

EI is computed according to AASHTO eq. 6.18:

$$EI = \frac{E_c I_g / 2.5}{1 + \beta_D}$$

$$E_c = 57000 \sqrt{4000} = 3,605,000 \text{ psi}$$

$$I_g = 0.049087(30)^4 = 39,760 \text{ in.}^4$$

$$\beta_D = \frac{26}{189} = 0.138$$

$$EI = \frac{3,605,000(39,760)}{2.5(1.138)} = 5.038 \times 10^{10}$$

From Example 10.4 the effective length factor k is 1.2. The critical buckling load is

$$P_c = \frac{\pi^2 EI}{(klu)^2} = \frac{\pi^2(5.038 \times 10^{10})}{[(1.2)(26.19)(12)]^2} = 3496 \text{ k}$$

The magnification is

$$\delta = \frac{C_m}{1 - P_u/\phi P_c} = \frac{1.0}{1 - 329/[1.0(3496)]} = 1.104$$

The ultimate load and moment to be checked against capacity are

$$P_u = 329 \text{ k}$$

$$M_u = 406(1.104) = 448 \text{ k-ft}$$

Using the capacity computed in Example 10.4, we plot the capacity against the ultimate required in Figure 10.40.

Example 10.4 presented the tabulations of group loadings for a ser-

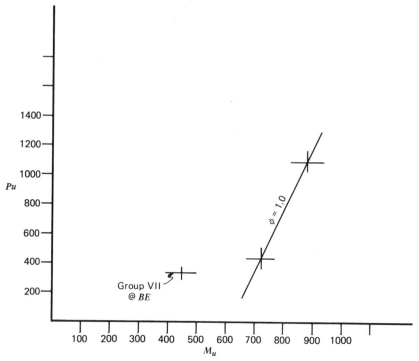

Figure 10.40   Pier capacity.

vice level loading for footing design. The pile loads and footing moment computations are also contained in the serviceability computations of Example 10.4. From the same example the stresses in the footing are

$$f_s = 11.8 < 24.0 \text{ ksi}$$

$$f_c = \frac{11.8(0.178)}{8(0.822)} = 319 < 1600 \text{ psi} \quad \text{OK}$$

The shear at the pile load is

$$v_L = \frac{V}{bd} = \frac{64.0}{39(26.0)} = 63 \text{ psi} > 0.95 \sqrt{f_c'} = 60 \text{ psi}$$

This is close enough to say that the footing is satisfactory as designed.

10.4 ABUTMENTS

10.4.1 General

Abutments can be defined as the support structure at the end of a bridge, where it connects with the approach roadway. There are numerous configurations and various types of abutment. Some can become extremely complicated and require considerable detail. Many have exterior faces treated architecturally and masonry facing is often used on abutment walls. Sometimes, the designer will build a bridge with cantilever end spans that eliminate the design and construction of complicated abutments.

Abutments fall into two basic categories: the spill-through abutment in Figure 10.2(a) and the closed abutment in Figure 10.2(b). The closed design prevents the approach embankment from spilling out in front of the abutment. The partially closed abutment shown in Figure 10.2 allows a limited amount of the embankment to spill out. A bridge with a spill-through abutment requires additional length to cover the embankment slopes. However, the abutments would be considerably cheaper, and the cost comparison between a longer bridge with a spill-through abutment and a shorter bridge with a large closed abutment would be about equal. Deflections on abutments are critical because joints can close and highly undesirable forces can be induced into the superstructure. Therefore it is important that abutment deflection be controlled and abutments be built with a high degree of stiffness. Closed abutments of more than 20 ft in height should be counterforted to provide adequate stiffness. An alternative for the control of deflections in large abutments provides a concrete deck slab over the wing walls parallel to the roadway. The embankment is then allowed to spill down inside the abutment and much of the high earth pressure on the backwalls is eliminated. This design is often called a cellular abutment. Caution should be exercised during abutment design to check the stability of the abutment structure when loaded with full earth pressure plus live load surcharge without superstructure dead load reaction. If an unreasonably low factor of safety against overturning is attained, backfilling should be limited until the superstructure dead load is in place.

In some abutment designs it is advisable to check heel pressures with a maximum reaction and one-half the normal earth pressure. An experienced designer can look over his design problem and determine whether this additional check is necessary.

BIBLIOGRAPHY

1. *AASHTO Standard Specifications for Highway Bridges*, with 1977–1981 Interim Specifications. The American Association of State Highway and Transportation Officials.
2. *Ontario Highway Bridge Design Code*, 1979, Vols. 1 and 2. Ontario, Ministry of Transportation and Communications.
3. *ACI Building Code Requirements for Reinforced Concrete (ACI 318-77)* and Commentary, American Concrete Institute.
4. *Strength and Serviceability Criteria, Reinforced Concrete Bridge Members, Ultimate Design*, FHWA, October 1969.
5. Joseph A. Yura, The Effective Length of Columns in Unbraced Frames, *AISC Eng. J.* (April 1974).
6. *Guide to Stability Design Criteria for Metal Structures*, Structural Stability Research Council, Wiley, New York, 1976.
7. *Guide to Design Criteria for Metal Compression Members*, Column Research Council Engineering Foundation, 1960.
8. W. B. Cranston, *Analysis and Design of Reinforced Concrete Columns*, Research Report 20, Paper 41.020, Cement and Concrete Association, London, 1972.
9. R. W. Furlong, Column Slenderness and Charts for Design, *J. Am. Concr. Inst.*, **68**(1) 9–17 (January 1971).
10. J. M. Gere and W. O. Carter, Critical Buckling Loads of Tapered Columns, *ASCE Struct. J.* (February 1962).
11. N. M. Newmark, Numerical Procedure for Computing Deflections, Moments, and Buckling Loads, ASCE Proceedings (May 1942).
12. R. W. Furlong, Concrete Columns under Biaxially Eccentric Thrust, *J. Am. Concr. Inst.* (October 1979).
13. *Notes on Load Factor Design for Reinforced Concrete Bridge Structures with Design Applications*, Portland Cement Association.
14. R. T. Potangaroa, M. J. N. Priestley, and R. Park, *Ductility of Spirally Reinforced Concrete Columns Under Seismic Loading*, Research Report 79-8, Department of Civil Engineering, University of Canterbury, February 1979.

15. W. D. Gill, R. Park, and M. J. N. Priestley, *Ductility of Rectangular Reinforced Concrete Columns with Axial Load*, Research Report 79-1, Department of Civil Engineering, University of Canterbury, February 1979.

16. J. Marin, Design Aids for L-Shaped Reinforced Concrete Shapes, *J. Am. Concr. Inst.* (November 1979).

17. J. G. MacGregor, U. H. Oelhafen, and S. E. Hage, A Re-examination of the *EI* Value for Slender Columns, ACI Publication SP-50-1, *Reinforced Concrete Columns*, 1975.

18. A. Singh, R. E-Wei Hu, and R. D. Cousineau, Lateral Load Capacity of Piles in Sand and Normally Consolidated Clay, *ASCE Civil Eng.* (August 1971).

19. Peter Kocsis, The Equivalent Length of a Pile or Caisson in Soil, *ASCE Civil Eng.* (December 1976).

20. B. R. Wood, D. Beaulieu, and P. F. Adams, Column Design by P Delta Method, ST 2, *ASCE J. Constr. Div.* (February 1976).

21. B. R. Wood, D. Beaulieu, and P. F. Adams, Further Aspects of Design by P-Delta Method, ST 3, *ASCE J. Constr. Div.* (March 1976).

22. J. G. MacGregor and S. E. Hage, Stability Analysis and Design of Concrete Frames, ST 10, *ASCE J. Constr. Div.* (October 1977).

23. Joseph E. Bowles, *Foundation Analysis and Design*, McGraw-Hill, New York, 1977.

24. Boris Bressler, Design Criteria for Reinforced Columns under Axial Load and Biaxial Bending, *J. Am. Concr. Inst.* (November 1960).

25. A. L. Parme, J. M. Nieves, A. Gouwens, Capacity of Reinforced Rectangular Columns Subject to Biaxial Bending, *J. Am. Concr. Inst.* (September 1966).

26. R. Park and T. Paulay, *Reinforced Concrete Structures*, Wiley, New York, 1975.

27. P. H. Kaar and A. H. Mattock, High Strength Bars as Concrete Reinforcement, Part 4, *Control of Cracking*, PCA Development Department Bulletin D59 (also Journal of PCA Research and Development Laboratories, 51(1) 15 (January 1963).

28. P. Gergely and L. A. Lutz, Maximum Crack Width in Reinforced Flexural Members, *Causes, Mechanism and Control of Cracking in Concrete*, SP-20, American Concrete Institute, 1968, pp. 87–117.

29. N. M. Newmark and E. Rosenblueth, *Fundamentals of Earthquake Engineering*, Prentice Hall, Englewood Cliffs, New Jersey, 1971.

30. Phil M. Ferguson, *Reinforced Concrete Fundamentals*, 4th ed., Wiley, New York, 1979.

31. *AASHTO Manual on Foundation Investigations*, American Association of State Highway and Transportation Officials.

32. *Reinforced Concrete Design Handbook, Working Stress Method*, SP-3, 3rd ed, American Concrete Institute, 1980.

33. David M. Goodall, Analysis of Rectangular Reinforced Concrete Sections Subjected to Direct Stress and Binding in Two Directions, *Public Roads Magazine* (June 1948).

34. Lu-Shien Hu, Eccentric Bending in Two Directions of Rectangular Concrete Columns, *J. Am. Concr. Inst.* (May 1955).

35. A. Anthony Toprac, Circular Reinforced Concrete Columns Subjected to Direct Stress and Bending, *Public Roads Magazine* (December 1959).

Appendixes

Appendix A   Values of k and j for Single Reinforcement
(Rectangular Sections)

	$N = 6$		$N = 7$		$N = 8$		$N = 9$		$N = 10$	
P	k	j	k	j	k	j	k	j	k	j
0.0010	0.104	0.965	0.112	0.963	0.119	0.960	0.126	0.958	0.132	0.956
0.0012	0.113	0.962	0.121	0.960	0.129	0.957	0.137	0.954	0.143	0.952
0.0014	0.121	0.960	0.131	0.956	0.139	0.954	0.147	0.951	0.154	0.949
0.0016	0.129	0.957	0.139	0.954	0.148	0.951	0.156	0.948	0.164	0.945
0.0018	0.137	0.954	0.147	0.951	0.156	0.948	0.165	0.945	0.173	0.942
0.0020	0.143	0.952	0.154	0.949	0.164	0.945	0.173	0.942	0.181	0.940
0.0022	0.150	0.950	0.161	0.946	0.171	0.943	0.180	0.940	0.189	0.937
0.0024	0.156	0.948	0.167	0.944	0.178	0.941	0.188	0.937	0.196	0.935
0.0026	0.162	0.946	0.173	0.942	0.184	0.939	0.194	0.935	0.204	0.932
0.0028	0.167	0.944	0.179	0.940	0.190	0.937	0.201	0.933	0.210	0.930
0.0030	0.173	0.942	0.185	0.938	0.196	0.935	0.207	0.931	0.217	0.928
0.0032	0.178	0.941	0.190	0.937	0.202	0.933	0.213	0.929	0.223	0.926
0.0034	0.183	0.939	0.196	0.935	0.208	0.931	0.219	0.927	0.229	0.924
0.0036	0.187	0.938	0.201	0.933	0.213	0.929	0.224	0.925	0.235	0.922
0.0038	0.192	0.936	0.206	0.931	0.218	0.927	0.230	0.923	0.240	0.920
0.0040	0.196	0.935	0.210	0.930	0.223	0.926	0.235	0.922	0.246	0.918
0.0042	0.201	0.933	0.215	0.928	0.228	0.924	0.240	0.920	0.251	0.916
0.0044	0.205	0.932	0.219	0.927	0.232	0.923	0.245	0.918	0.256	0.915
0.0046	0.209	0.930	0.224	0.925	0.237	0.921	0.249	0.917	0.261	0.913
0.0048	0.213	0.929	0.228	0.924	0.241	0.920	0.254	0.915	0.266	0.911
0.0050	0.217	0.928	0.232	0.923	0.246	0.918	0.259	0.914	0.270	0.910
0.0052	0.221	0.926	0.236	0.921	0.250	0.917	0.263	0.912	0.275	0.908
0.0054	0.224	0.925	0.240	0.920	0.254	0.915	0.267	0.911	0.279	0.907
0.0056	0.228	0.924	0.244	0.919	0.258	0.914	0.271	0.910	0.283	0.906
0.0058	0.231	0.923	0.247	0.918	0.262	0.913	0.275	0.908	0.287	0.904
0.0060	0.235	0.922	0.251	0.916	0.266	0.911	0.279	0.907	0.292	0.903

Appendix A   (*Continued*)

P	N = 6		N = 7		N = 8		N = 9		N = 10	
	k	j	k	j	k	j	k	j	k	j
0.0062	0.238	0.921	0.254	0.915	0.269	0.910	0.283	0.906	0.296	0.901
0.0064	0.241	0.920	0.258	0.914	0.273	0.909	0.287	0.904	0.299	0.900
0.0066	0.245	0.918	0.261	0.913	0.276	0.908	0.290	0.903	0.303	0.899
0.0068	0.248	0.917	0.265	0.912	0.280	0.907	0.294	0.902	0.307	0.898
0.0070	0.251	0.916	0.268	0.911	0.283	0.906	0.298	0.901	0.311	0.896
0.0072	0.254	0.915	0.271	0.910	0.287	0.904	0.301	0.900	0.314	0.895
0.0074	0.257	0.914	0.274	0.909	0.290	0.903	0.305	0.898	0.318	0.894
0.0076	0.260	0.913	0.277	0.908	0.293	0.902	0.308	0.897	0.321	0.893
0.0078	0.263	0.912	0.280	0.907	0.296	0.901	0.311	0.896	0.325	0.892
0.0080	0.266	0.911	0.283	0.906	0.299	0.900	0.314	0.895	0.328	0.891
0.0082	0.268	0.911	0.286	0.905	0.303	0.899	0.318	0.894	0.331	0.890
0.0084	0.271	0.910	0.289	0.904	0.306	0.898	0.321	0.893	0.334	0.889
0.0086	0.274	0.909	0.292	0.903	0.308	0.897	0.324	0.892	0.338	0.887
0.0088	0.276	0.908	0.295	0.902	0.311	0.896	0.327	0.891	0.341	0.886
0.0090	0.279	0.907	0.298	0.901	0.314	0.895	0.330	0.890	0.344	0.885
0.0092	0.282	0.906	0.300	0.900	0.317	0.894	0.333	0.889	0.347	0.884
0.0094	0.284	0.905	0.303	0.899	0.320	0.893	0.335	0.888	0.350	0.883
0.0096	0.287	0.904	0.306	0.898	0.323	0.892	0.338	0.887	0.353	0.882
0.0098	0.289	0.904	0.308	0.897	0.325	0.892	0.341	0.886	0.355	0.882
0.0100	0.292	0.903	0.311	0.896	0.328	0.891	0.344	0.885	0.358	0.881
0.0102	0.294	0.902	0.313	0.896	0.331	0.890	0.347	0.884	0.361	0.880
0.0104	0.296	0.901	0.316	0.895	0.333	0.889	0.349	0.884	0.364	0.879
0.0106	0.299	0.900	0.318	0.894	0.336	0.888	0.352	0.883	0.366	0.878
0.0108	0.301	0.900	0.321	0.893	0.338	0.887	0.354	0.882	0.369	0.877
0.0110	0.303	0.899	0.323	0.892	0.341	0.886	0.357	0.881	0.372	0.876
0.0112	0.306	0.898	0.325	0.892	0.343	0.886	0.359	0.880	0.374	0.875
0.0114	0.308	0.897	0.328	0.891	0.346	0.885	0.362	0.879	0.377	0.874
0.0116	0.310	0.897	0.330	0.890	0.348	0.884	0.364	0.879	0.379	0.874
0.0118	0.312	0.896	0.332	0.889	0.350	0.883	0.367	0.878	0.382	0.873
0.0120	0.314	0.895	0.334	0.889	0.353	0.882	0.369	0.877	0.384	0.872
0.0122	0.316	0.895	0.337	0.888	0.355	0.882	0.372	0.876	0.387	0.871
0.0124	0.318	0.894	0.339	0.887	0.357	0.881	0.374	0.875	0.389	0.870
0.0126	0.321	0.893	0.341	0.886	0.359	0.880	0.376	0.875	0.392	0.869
0.0128	0.323	0.892	0.343	0.886	0.362	0.879	0.379	0.874	0.394	0.869
0.0130	0.325	0.892	0.345	0.885	0.364	0.879	0.381	0.873	0.396	0.868
0.0132	0.327	0.891	0.347	0.884	0.366	0.878	0.383	0.872	0.398	0.867
0.0134	0.329	0.890	0.349	0.884	0.368	0.877	0.385	0.872	0.401	0.866
0.0136	0.331	0.890	0.351	0.883	0.370	0.877	0.387	0.871	0.403	0.866
0.0138	0.332	0.889	0.353	0.882	0.372	0.876	0.390	0.870	0.405	0.865
0.0140	0.334	0.889	0.355	0.882	0.374	0.875	0.392	0.869	0.407	0.864

Appendix A (*Continued*)

P	N = 6 k	N = 6 j	N = 7 k	N = 7 j	N = 8 k	N = 8 j	N = 9 k	N = 9 j	N = 10 k	N = 10 j
0.0142	0.336	0.888	0.357	0.881	0.376	0.875	0.394	0.869	0.410	0.863
0.0144	0.338	0.887	0.359	0.880	0.378	0.874	0.396	0.868	0.412	0.863
0.0146	0.340	0.887	0.361	0.880	0.380	0.873	0.398	0.867	0.414	0.862
0.0148	0.342	0.886	0.363	0.879	0.382	0.873	0.400	0.867	0.416	0.861
0.0150	0.344	0.885	0.365	0.878	0.384	0.872	0.402	0.866	0.418	0.861
0.0152	0.346	0.885	0.367	0.878	0.386	0.871	0.404	0.865	0.420	0.860
0.0154	0.347	0.884	0.369	0.877	0.388	0.871	0.406	0.865	0.422	0.859
0.0156	0.349	0.884	0.371	0.876	0.390	0.870	0.408	0.864	0.424	0.859
0.0158	0.351	0.883	0.373	0.876	0.392	0.869	0.410	0.863	0.426	0.858
0.0160	0.353	0.882	0.374	0.875	0.394	0.869	0.412	0.863	0.428	0.857
0.0162	0.354	0.882	0.376	0.875	0.396	0.868	0.414	0.862	0.430	0.857
0.0164	0.356	0.881	0.378	0.874	0.398	0.867	0.415	0.862	0.432	0.856
0.0166	0.358	0.881	0.380	0.873	0.399	0.867	0.417	0.861	0.434	0.855
0.0168	0.359	0.880	0.381	0.873	0.401	0.866	0.419	0.860	0.436	0.855
0.0170	0.361	0.880	0.383	0.872	0.403	0.866	0.421	0.860	0.437	0.854
0.0172	0.363	0.879	0.385	0.872	0.405	0.865	0.423	0.859	0.439	0.854
0.0174	0.364	0.879	0.387	0.871	0.406	0.865	0.425	0.858	0.441	0.853
0.0176	0.366	0.878	0.388	0.871	0.408	0.864	0.426	0.858	0.443	0.852
0.0178	0.368	0.877	0.390	0.870	0.410	0.863	0.428	0.857	0.445	0.852
0.0180	0.369	0.877	0.392	0.869	0.412	0.863	0.430	0.857	0.446	0.851
0.0182	0.371	0.876	0.393	0.869	0.413	0.862	0.432	0.856	0.448	0.851
0.0184	0.372	0.876	0.395	0.868	0.415	0.862	0.433	0.856	0.450	0.850
0.0186	0.374	0.875	0.396	0.868	0.417	0.861	0.435	0.855	0.452	0.849
0.0188	0.375	0.875	0.398	0.867	0.418	0.861	0.437	0.854	0.453	0.849
0.0190	0.377	0.874	0.400	0.867	0.420	0.860	0.438	0.854	0.455	0.848
0.0192	0.378	0.874	0.401	0.866	0.422	0.859	0.440	0.853	0.457	0.848
0.0194	0.380	0.873	0.403	0.866	0.423	0.859	0.442	0.853	0.458	0.847
0.0196	0.381	0.873	0.404	0.865	0.425	0.858	0.443	0.852	0.460	0.847
0.0198	0.383	0.872	0.406	0.865	0.426	0.858	0.445	0.852	0.462	0.846
0.0200	0.384	0.872	0.407	0.864	0.428	0.857	0.446	0.851	0.463	0.846

Appendix B   Tee Beams *k* and *j* Values

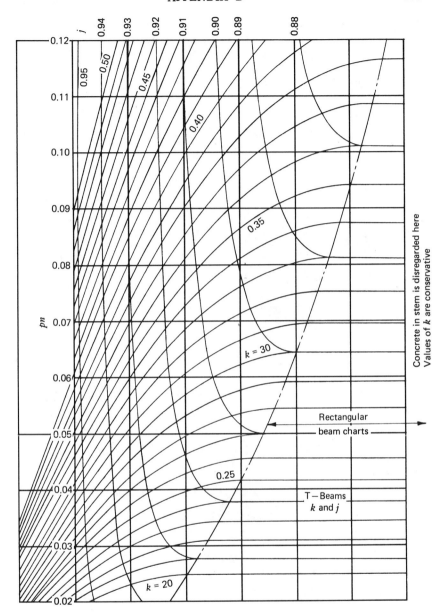

Appendix C   Minimum Spacing of Reinforcement (c-to-c) (AASHTO Article 1.5.4A

Size	ϕ	2.5ϕ	$1.50 + \phi$	$2.25 + \phi$	Minimum Spacing[a] (1 in. ϕ agg)	(1$\frac{1}{2}$ in. ϕ agg)
#4	0.500	1.25	2.00	2.75	2	$2\frac{3}{4}$
#5	0.625	1.56	2.13	2.88	$2\frac{1}{8}$	$2\frac{7}{8}$
#6	0.750	1.88	2.25	3.00	$2\frac{1}{4}$	3
#7	0.875	2.19	2.38	3.12	$2\frac{3}{8}$	$3\frac{1}{8}$
#8	1.000	2.50	2.50	3.25	$2\frac{1}{2}$	$3\frac{1}{4}$
#9	1.128	2.82	2.63	3.38	$2\frac{7}{8}$	$3\frac{3}{8}$
#10	1.270	3.17	2.77	3.52	$3\frac{1}{4}$	$3\frac{1}{2}$
#11	1.410	3.52	2.91	3.67	$3\frac{1}{2}$	$3\frac{3}{4}$
#14	1.692	4.22	3.19	3.94	$4\frac{1}{4}$	$4\frac{1}{4}$
#18	2.256	5.65	3.76	4.51	$5\frac{3}{4}$	$5\frac{3}{4}$

[a]$2.5 \times \phi$ or 1.5 agg + ϕ or 1$\frac{1}{2}$ in. + ϕ.

Appendix D   Tension Development Length ($f_c' = 3300$ psi)

	Grade 40		Grade 60	
	Other	Top	Other	Top
#2	12	12	12	13
#4	12	12	12	17
#5	12	14	15	21
#6	12	17	18	26
#7	17	23	25	35
#8	22	31	33	46
#9	28	39	42	58
#10	35	50	53	74
#11	43	61	65	91
#14	59	83	89	124
#18	77	107	115	161

Source: AASHTO Article 1.5.14.

Appendix E   Lap Splice Lengths ($f'_c = 3300$ psi)

| | Grade 40 | | | | | | | | Grade 60 | | | | | | | |
| | Top | | | | Other | | | | Top | | | | Other | | | |
	A	B	C	D	A	B	C	D	A	B	C	D	A	B	C	D
#3	12	12	14	17	12	12	12	12	13	16	21	25	12	12	15	18
#4	12	15	19	22	12	12	14	16	17	22	29	34	12	16	20	24
#5	14	18	24	28	12	13	17	20	21	27	36	42	15	20	26	30
#6	17	22	29	34	12	16	21	25	26	33	44	51	18	24	31	37
#7	23	30	40	47	17	22	28	33	35	46	60	70	25	33	43	50
#8	31	40	52	62	22	29	37	44	46	60	79	92	33	43	56	66
#9	39	51	66	78	28	36	47	56	58	76	99	117	42	54	71	84
#10	50	64	84	99	35	46	60	71	74	97	126	149	53	69	90	106
#11	61	79	103	122	43	56	74	87	91	119	155	182	65	85	111	130

Source: AASHTO Article 1.5.22(B).

Appendix F Equivalent Length (l_e) and Embedment Length (E) for Hooks (f'_c = 3300 psi)

	Grade 40						Grade 60							
	No Enclosure			Enclosed			No Enclosure				Enclosed			
		Top	Other		Top	Other	Top		Other		Top		Other	
	l_e	E	E	l_e	E	E	l_e	E	l_e	E	l_e	E	l_e	E
#3	3	11	11	4	10	10	5	10	5	9	6	9	6	8
#4	4	10	10	5	9	9	6	13	6	8	8	11	8	6
#5	5	12	10	7	10	8	8	16	8	10	10	14	10	8
#6	6	14	9	8	12	7	8	21	10	11	10	19	12	9
#7	9	16	12	11	16	9	9	30	13	16	11	28	17	12
#8	11	24	15	15	20	11	11	39	17	20	15	35	22	15
#9	14	31	20	19	26	15	14	50	22	26	19	45	28	20
#10	18	38	23	24	32	17	18	62	24	35	24	56	32	27
#11	22	46	28	29	39	21	22	76	26	46	29	69	34	38
#14	28	65	41	36	67	33	28	106	28	71	36	98	36	63
#18	24	97	67	31	90	60	24	151	24	105	31	144	31	98

Source: AASHTO Article 1.5.17.

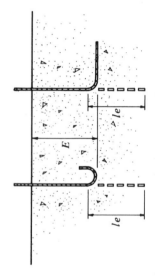

Appendix G   Standard Hooks

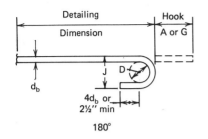

180°

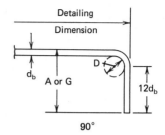

90°

		180° Hooks		90° Hooks
Bar Size	Finished Bend Diameter D (in.)	A or G (in.)	I (in.)	A or G (in.)
#3	$2\frac{1}{4}$	5	3	6
#4	3	6	4	8
#5	$3\frac{3}{4}$	7	5	10
#6	$4\frac{1}{2}$	8	6	1-0
#7	$5\frac{1}{4}$	10	7	1-2
#8	6	11	8	1-4
#9	$9\frac{1}{2}$	1-3	$1-1\frac{3}{4}$	1-7
#10	$10\frac{3}{4}$	1-5	$1-1\frac{1}{4}$	1-10
#11	12	1-7	$1-2\frac{3}{4}$	2-0
#14	$18\frac{1}{4}$	2-3	$1-9\frac{3}{4}$	2-7
#18	24	3-0	$2-4\frac{1}{2}$	3-5

Recommended End Hooks, All Grades

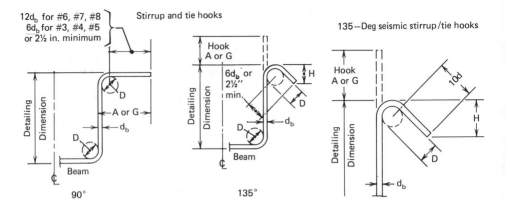

		Stirrup and Tie Hook Dimensions (in.)[a]			Seismic 135° Stirrup-Tie Hook Dimensions (in.)[a]	
		90° Hook	135° Hook		135° Hook	
Bar Size	D (in.)	A or G	A or G	H (Approx.)	A or G	H (Approx.)
#3	$1\frac{1}{2}$	4	4	$2\frac{1}{2}$	5	$3\frac{1}{2}$
#4	2	$4\frac{1}{2}$	$4\frac{1}{2}$	3	$6\frac{1}{2}$	$4\frac{1}{2}$
#5	$2\frac{1}{2}$	6	$5\frac{1}{2}$	$3\frac{3}{4}$	8	$5\frac{1}{2}$
#6	$4\frac{1}{2}$	1-0	$7\frac{3}{4}$	$4\frac{1}{2}$	$10\frac{3}{4}$	$6\frac{1}{2}$
#7	$5\frac{1}{4}$	1-2	9	$5\frac{1}{4}$	$1\text{-}0\frac{1}{2}$	$7\frac{3}{4}$
#8	6	1-4	$10\frac{1}{4}$	6	$1\text{-}2\frac{1}{4}$	9

[a]Grades 40, 50, and 60.

Appendix H Reinforcing Bar Properties

Weight (lb ft^{-1})	Diameter (in.)	Size	Number	Area (in.^{2})	Perimeter (in.)
0.167	0.250	$\frac{1}{4}$	#2	0.05	0.786
0.376	0.375	$\frac{3}{8}$	#3	0.11	1.178
0.668	0.500	$\frac{1}{2}$	#4	0.20	1.571
1.043	0.625	$\frac{5}{8}$	#5	0.31	1.963
1.502	0.750	$\frac{3}{4}$	#6	0.44	2.356
2.044	0.875	$\frac{7}{8}$	#7	0.60	2.749
2.670	1.000	1	#8	0.79	3.142
3.400	1.128	1	#9	1.00	3.544
4.303	1.270	$1\frac{1}{8}$	#10	1.27	3.990
5.313	1.410	$1\frac{1}{4}$	#11	1.56	4.430
7.650	1.692	$1\frac{1}{2}$	#14	2.25	5.316
13.600	2.256	2	#18	4.00	7.088

Appendix I Modulus of Elasticity for Concrete

f_c'	psi	ksf
3000	3,122,019	449,571
3300	3,274,401	471,514
3500	3,372,165	485,592
4000	3,604,997	519,120
4500	3,823,676	550,609
5000	4,030,508	580,393
5500	4,227,233	608,722
6000	4,415,201	635,789
6500	4,595,487	661,750
7000	4,768,962	686,731

Source: Equation $E_c = 57,000 \sqrt{f_c'}$.

Appendix J   Frame Constants and Fixed End Moments for Parabolic Haunched Girders and Slabs

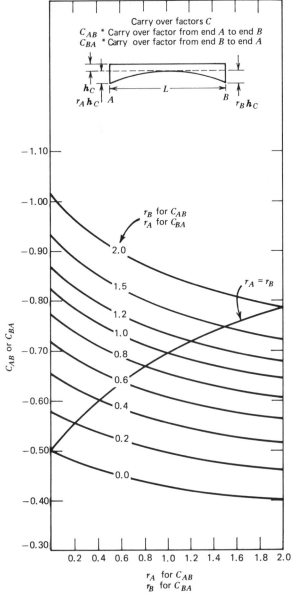

When using these charts for T-beams and box girders $r = \sqrt[3]{(I_{max}/I_{min})} - 1$. Taken from *Continuous Concrete Bridges*, P.C.A.

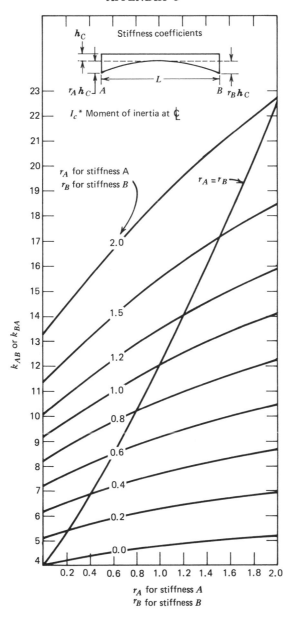

Stiffness coefficients

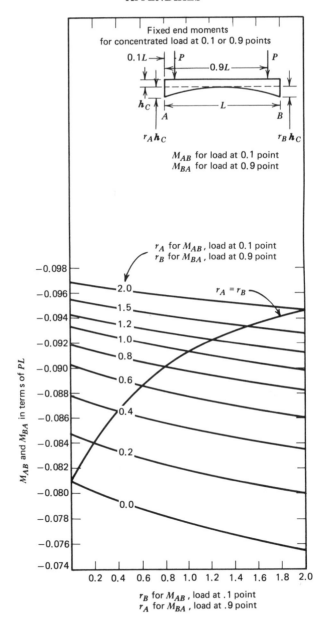

Fixed end moments
for concentrated load at 0.1 or 0.9 points

M_{AB} for load at 0.1 point
M_{BA} for load at 0.9 point

r_A for M_{AB}, load at 0.1 point
r_B for M_{BA}, load at 0.9 point

$r_A = r_B$

M_{AB} and M_{BA} in terms of PL

r_B for M_{AB}, load at .1 point
r_A for M_{BA}, load at .9 point

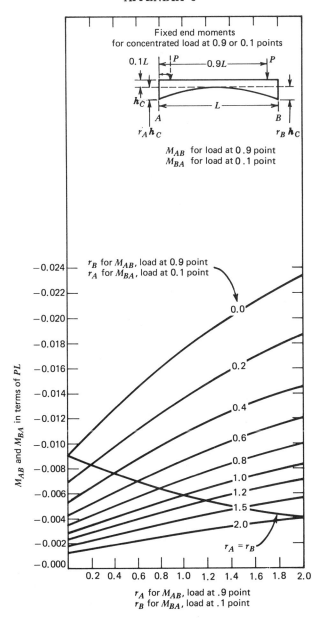

Fixed end moments
for concentrated load at 0.9 or 0.1 points

M_{AB} for load at 0.9 point
M_{BA} for load at 0.1 point

M_{AB} and M_{BA} in terms of PL

r_B for M_{AB}, load at 0.9 point
r_A for M_{BA}, load at 0.1 point

r_A for M_{AB}, load at .9 point
r_B for M_{BA}, load at .1 point

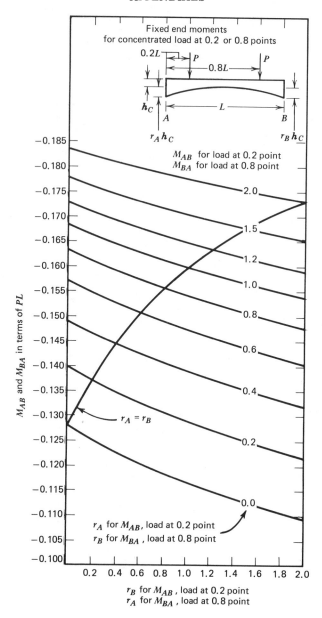

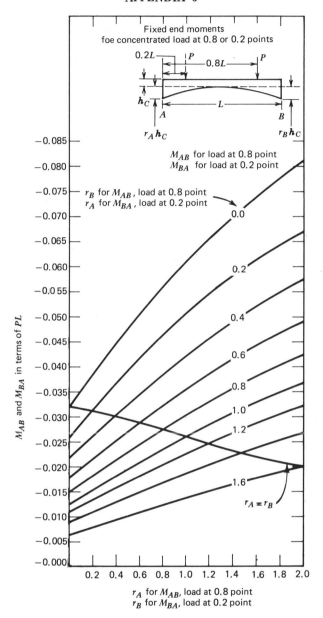

Fixed end moments
foe concentrated load at 0.8 or 0.2 points

M_{AB} and M_{BA} in terms of PL

M_{AB} for load at 0.8 point
M_{BA} for load at 0.2 point

r_B for M_{AB}, load at 0.8 point
r_A for M_{BA}, load at 0.2 point

r_A for M_{AB}, load at 0.8 point
r_B for M_{BA}, load at 0.2 point

$r_A = r_B$

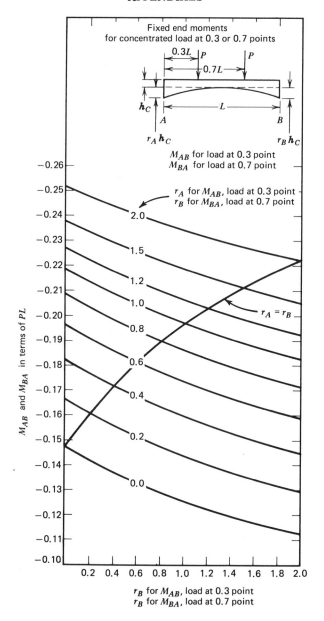

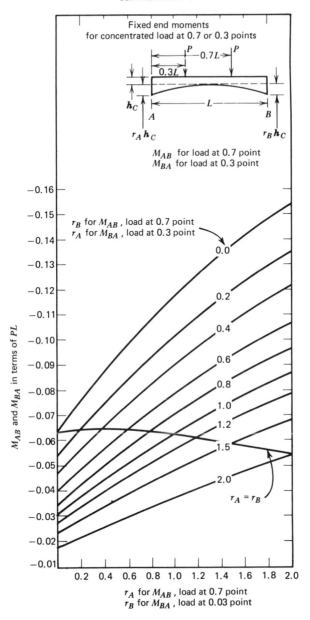

Fixed end moments
for concentrated load at 0.7 or 0.3 points

M_{AB} for load at 0.7 point
M_{BA} for load at 0.3 point

r_B for M_{AB}, load at 0.7 point
r_A for M_{BA}, load at 0.3 point

M_{AB} and M_{BA} in terms of PL

r_A for M_{AB}, load at 0.7 point
r_B for M_{BA}, load at 0.03 point

$r_A = r_B$

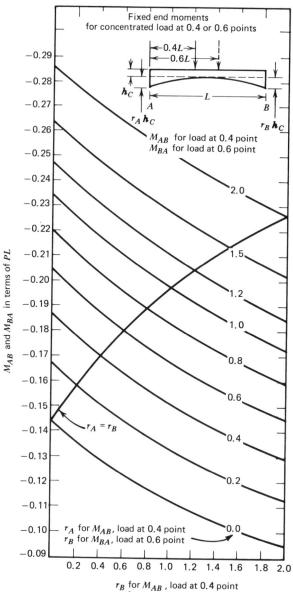

M_{AB} and M_{BA} in terms of PL

Fixed end moments
for concentrated load at 0.6 or 0.4 points

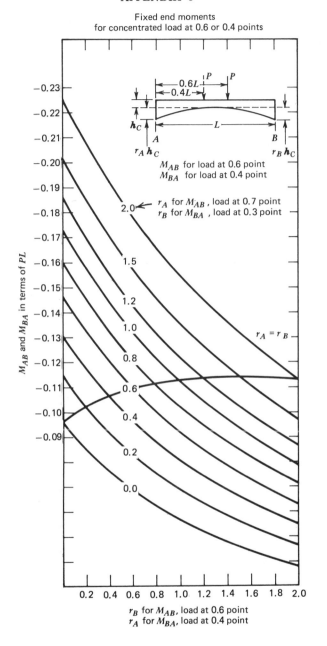

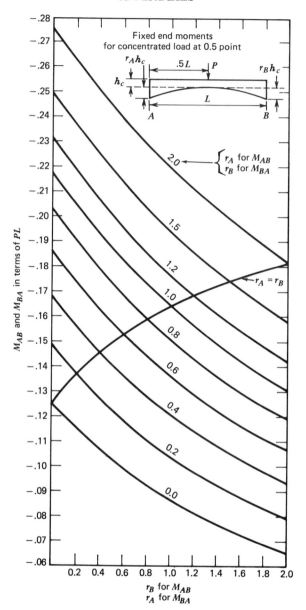

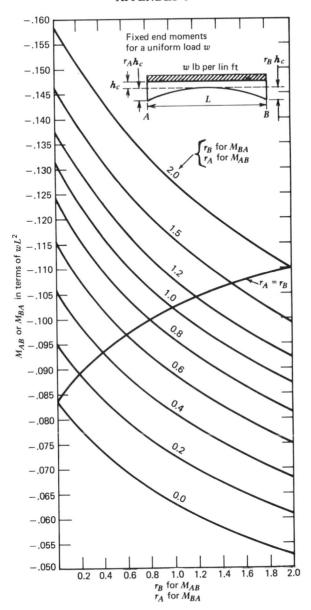

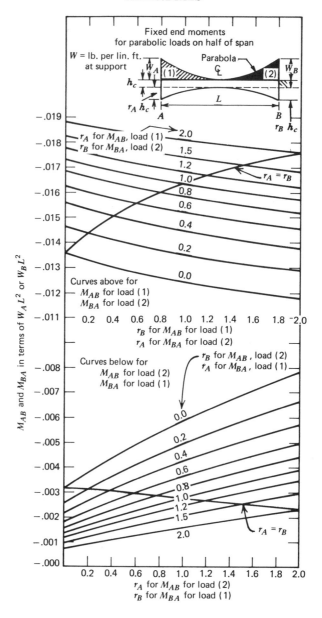

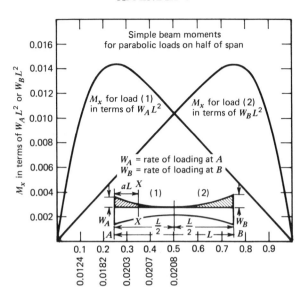

Appendix K   Frame Constants and Fixed End Moments for Parabolic and Straight Haunches of Varying Length

SYMMETRICAL MEMBERS WITH STRAIGHT HAUNCHES

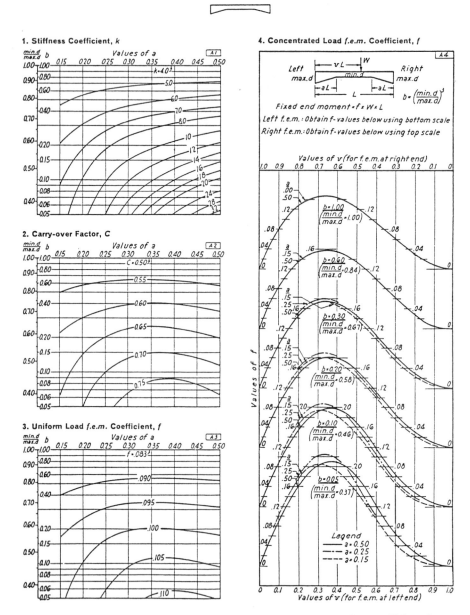

1. Stiffness Coefficient, k

2. Carry-over Factor, C

3. Uniform Load f.e.m. Coefficient, f

4. Concentrated Load f.e.m. Coefficient, f

Fixed end moment $= f \times W \times L$

Left f.e.m.: Obtain f-values below using bottom scale
Right f.e.m.: Obtain f-values below using top scale

Values of v (for f.e.m. at right end)

Values of v (for f.e.m. at left end)

$b = \left(\frac{min.d}{max.d}\right)^{3}$

Legend
—— $a = 0.50$
— — $a = 0.25$
---- $a = 0.15$

When using these charts for T-beams and box girders $b = I_{min}/I_{max}$. Taken from *Concrete Members with Variable Moments of Inertia*, PCA-ST 103.

SYMMETRICAL MEMBERS WITH PARABOLIC HAUNCHES

1. Stiffness Coefficient, k

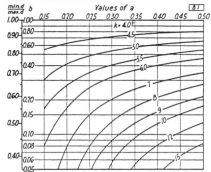

2. Carry-over Factor, C

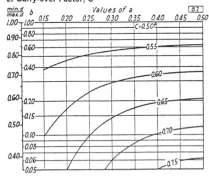

3. Uniform Load f.e.m. Coefficient, f

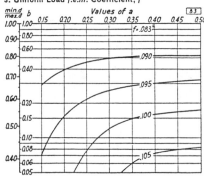

4. Concentrated Load f.e.m. Coefficient, f

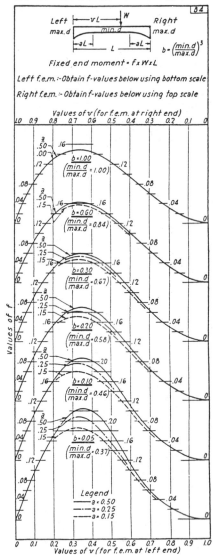

UNSYMMETRICAL MEMBERS WITH STRAIGHT HAUNCH AT ONE END
Coefficients at Haunched End

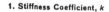

1. Stiffness Coefficient, k

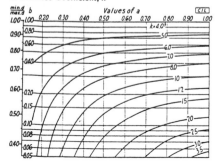

2. Carry-over Factor, C

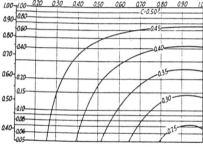

3. Uniform Load f.e.m. Coefficient, f

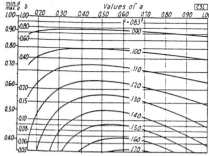

4. Concentrated Load f.e.m. Coefficient, f

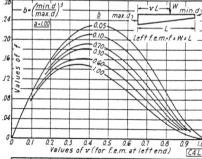

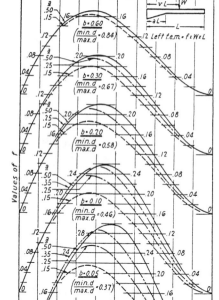

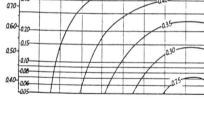

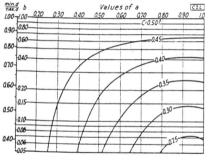

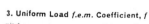

UNSYMMETRICAL MEMBERS WITH STRAIGHT HAUNCH AT ONE END

Coefficients at Small End

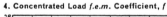

1. Stiffness Coefficient, k

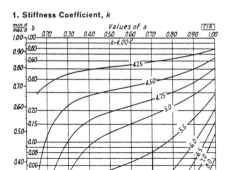

2. Carry-over Factor, C

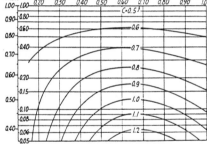

3. Uniform Load f.e.m. Coefficient, f

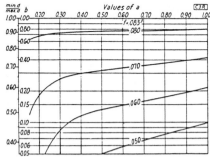

4. Concentrated Load f.e.m. Coefficient, f

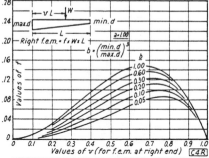

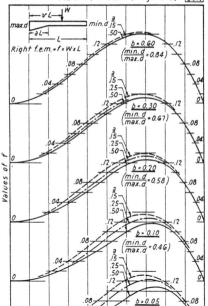

UNSYMMETRICAL MEMBERS WITH PARABOLIC HAUNCH AT ONE END

Coefficients at Haunched End

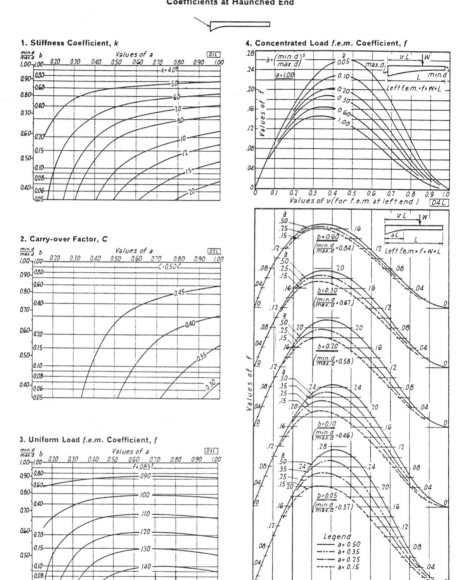

UNSYMMETRICAL MEMBERS WITH PARABOLIC HAUNCH AT ONE END

Coefficients at Small End

1. Stiffness Coefficient, k

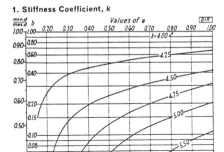

4. Concentrated Load f.e.m. Coefficient, f

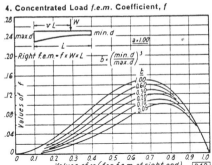

2. Carry-over Factor, C

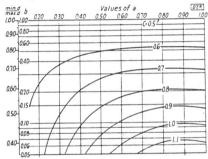

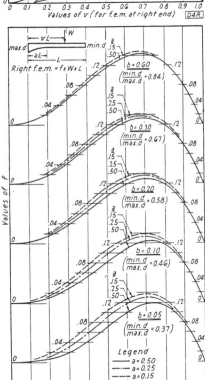

3. Uniform Load f.e.m. Coefficient, f

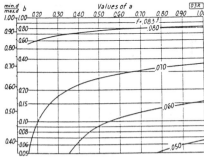

Appendix L   Frame Constants and Fixed End Moments for Prismatic Haunched Members

Prismatic Haunch at One End

Right Haunch		Carry-Over Factors		Stiffness Factors		Unif. Load F.E.M. Coef. $\times wL^2$		Concentrated Load b			
								0.1		0.3	
a_B	τ_B	C_{AB}	C_{BA}	k_{AB}	k_{BA}	M_{AB}	M_{BA}	M_{AB}	M_{BA}	M_{AB}	M_{BA}
						$r_A = 0$		$a_A = 0$			
0.1	0.4	0.593	0.491	4.24	5.12	0.0749	0.1016	0.0799	0.0113	0.1397	0.0788
	0.6	0.615	0.490	4.30	5.40	0.0727	0.1062	0.0797	0.0119	0.1378	0.0828
	1.0	0.639	0.488	4.37	5.72	0.0703	0.1114	0.0794	0.0125	0.1358	0.0873
	1.5	0.652	0.487	4.40	5.89	0.0690	0.1143	0.0792	0.0129	0.1346	0.0898
	2.0	0.658	0.487	4.42	5.97	0.0684	0.1156	0.0791	0.0131	0.1341	0.0910
0.2	0.4	0.677	0.469	4.42	6.37	0.0705	0.1126	0.0791	0.0134	0.1345	0.0925
	0.6	0.730	0.463	4.56	7.18	0.0664	0.1225	0.0785	0.0149	0.1302	0.1025
	1.0	0.793	0.458	4.74	8.22	0.0610	0.1353	0.0777	0.0168	0.1248	0.1154
	1.5	0.831	0.455	4.86	8.88	0.0576	0.1434	0.0772	0.0180	0.1214	0.1235
	2.0	0.849	0.453	4.91	9.20	0.0559	0.1473	0.0769	0.0186	0.1197	0.1276
0.3	0.4	0.741	0.439	4.52	7.63	0.0698	0.1155	0.0787	0.0149	0.1319	0.1013
	0.6	0.831	0.427	4.75	9.24	0.0642	0.1296	0.0777	0.0175	0.1255	0.1182
	1.0	0.954	0.415	5.09	11.69	0.0559	0.1511	0.0762	0.0215	0.1158	0.1440
	1.5	1.036	0.409	5.34	13.53	0.0497	0.1673	0.0751	0.0245	0.1085	0.1633
	2.0	1.078	0.407	5.48	14.54	0.0464	0.1762	0.0745	0.0262	0.1045	0.1740
0.4	0.4	0.774	0.405	4.55	8.70	0.0703	0.1117	0.0786	0.0156	0.1315	0.1035
	0.6	0.901	0.386	4.83	11.28	0.0646	0.1269	0.0774	0.0192	0.1240	0.1254
	1.0	1.102	0.367	5.33	16.03	0.0549	0.1548	0.0752	0.0257	0.1105	0.1658
	1.5	1.260	0.357	5.79	20.46	0.0462	0.1807	0.0732	0.0319	0.0982	0.2035
	2.0	1.349	0.352	6.09	23.32	0.0407	0.1975	0.0719	0.0358	0.0903	0.2278
0.5	0.4	0.768	0.371	4.56	9.45	0.0700	0.1048	0.0786	0.0154	0.1312	0.0993
	0.6	0.919	0.343	4.84	12.94	0.0651	0.1176	0.0774	0.0193	0.1240	0.1218
	1.0	1.200	0.316	5.42	20.61	0.0561	0.1451	0.0749	0.0280	0.1096	0.1709
	1.5	1.470	0.301	6.10	29.74	0.0466	0.1777	0.0720	0.0384	0.0934	0.2290
	2.0	1.647	0.295	6.63	37.04	0.0393	0.2036	0.0698	0.0466	0.0807	0.2755
0.6	0.4	0.726	0.341	4.62	9.84	0.0675	0.0986	0.0782	0.0146	0.1280	0.0916
	0.6	0.872	0.305	4.88	13.97	0.0630	0.1072	0.0771	0.0183	0.1214	0.1096
	1.0	1.196	0.267	5.43	24.35	0.0560	0.1277	0.0748	0.0274	0.1092	0.1537
	1.5	1.588	0.247	6.18	39.79	0.0482	0.1572	0.0718	0.0408	0.0939	0.2183
	2.0	1.905	0.237	6.92	55.51	0.0412	0.1870	0.0688	0.0544	0.0792	0.2839
0.7	0.4	0.657	0.321	4.86	9.96	0.0631	0.0954	0.0770	0.0138	0.1175	0.0846
	0.6	0.770	0.275	5.14	14.39	0.0580	0.1006	0.0758	0.0167	0.1097	0.0955
	1.0	1.056	0.224	5.62	26.45	0.0516	0.1122	0.0738	0.0243	0.0992	0.1213
	1.5	1.491	0.196	6.24	47.48	0.0463	0.1304	0.0714	0.0371	0.0890	0.1633
	2.0	1.944	0.183	6.95	73.85	0.0417	0.1523	0.0687	0.0530	0.0793	0.2149
0.8	0.4	0.583	0.319	5.46	9.97	0.0585	0.0951	0.0741	0.0137	0.1040	0.0837
	0.6	0.645	0.263	5.89	14.44	0.0516	0.0990	0.0721	0.0160	0.0921	0.0907
	1.0	0.818	0.196	6.47	27.06	0.0435	0.1053	0.0696	0.0211	0.0781	0.1025
	1.5	1.128	0.155	6.98	50.85	0.0385	0.1130	0.0676	0.0296	0.0692	0.1175
	2.0	1.533	0.135	7.47	84.60	0.0355	0.1222	0.0658	0.0412	0.0638	0.1357
0.9	0.4	0.524	0.356	6.87	10.10	0.0604	0.0948	0.0674	0.0157	0.1031	0.0835
	0.6	0.542	0.295	7.95	14.58	0.0497	0.0991	0.0623	0.0184	0.0866	0.0913
	1.0	0.594	0.206	9.44	27.16	0.0372	0.1052	0.0553	0.0226	0.0642	0.1023
	1.5	0.695	0.142	10.48	51.25	0.0289	0.1098	0.0506	0.0266	0.0492	0.1105
	2.0	0.842	0.107	11.07	86.80	0.0245	0.1147	0.0481	0.0306	0.0414	0.1159

Note: All carry-over factors are negative and all stiffness factors are positive. All fixed-end moment coefficients are negative except where plus sign is shown.

F.E.M.—Coef. × PL								Moment, M at $b = (1 - a_B)$		Haunch Load	
		b						F.E.M. Coef. × M		F.E.M. Coef. × $W_B L^2$	
0.5		0.7		0.9		$1 - a_B$					
M_{AB}	M_{BA}	M_{AB}	M_{BA}	M_{AB}	M_{BA}	M_{AB}	M_{BA}	M_{AB}	M_{BA}	M_{AB}	M_{BA}
					$a_A = 0$						
0.1110	0.1553	0.0478	0.1798	0.0042	0.0911	0.0042	0.0911	0.0793	0.8275	0.0001	0.0047
0.1074	0.1630	0.0439	0.1881	0.0029	0.0937	0.0029	0.0937	0.0561	0.8780	0.0001	0.0048
0.1035	0.1716	0.0396	0.1974	0.0016	0.0966	0.0016	0.0966	0.0304	0.9339	0.0001	0.0049
0.1012	0.1764	0.0373	0.2026	0.0008	0.0982	0.0008	0.0982	0.0161	0.9651	0.0000	0.0049
0.1002	0.1786	0.0361	0.2050	0.0005	0.0990	0.0005	0.0990	0.0094	0.9795	0.0000	0.0050
0.1020	0.1788	0.0409	0.1975	0.0050	0.0890	0.0182	0.1581	0.1640	0.6037	0.0013	0.0171
0.0942	0.1972	0.0335	0.2148	0.0037	0.0917	0.0137	0.1684	0.1241	0.7006	0.0010	0.0178
0.0843	0.2207	0.0242	0.2368	0.0022	0.0951	0.0080	0.1815	0.0728	0.8245	0.0006	0.0187
0.0781	0.2355	0.0182	0.2507	0.0012	0.0973	0.0044	0.1897	0.0403	0.9029	0.0003	0.0193
0.0750	0.2429	0.0153	0.2576	0.0007	0.0984	0.0026	0.1939	0.0242	0.9418	0.0002	0.0196
0.0987	0.1899	0.0420	0.1929	0.0056	0.0868	0.0420	0.1929	0.2371	0.3457	0.0045	0.0338
0.0877	0.2185	0.0338	0.2130	0.0045	0.0893	0.0338	0.2130	0.1935	0.4682	0.0036	0.0359
0.0711	0.2621	0.0217	0.2436	0.0028	0.0930	0.0217	0.2436	0.1261	0.6548	0.0023	0.0391
0.0587	0.2948	0.0128	0.2665	0.0017	0.0959	0.0128	0.2665	0.0750	0.7952	0.0014	0.0415
0.0520	0.3129	0.0080	0.2792	0.0010	0.0974	0.0080	0.2792	0.0467	0.8725	0.0008	0.0448
0.0992	0.1855	0.0445	0.1773	0.0059	0.0849	0.0713	0.1938	0.2780	0.0876	0.0106	0.0509
0.0875	0.2182	0.0377	0.1932	0.0049	0.0869	0.0611	0.2204	0.2456	0.2035	0.0089	0.0547
0.0671	0.2780	0.0267	0.2222	0.0034	0.0904	0.0438	0.2689	0.1817	0.4177	0.0063	0.0616
0.0485	0.3339	0.0173	0.2491	0.0022	0.0938	0.0284	0.3142	0.1198	0.6183	0.0037	0.0679
0.0367	0.3699	0.0113	0.2664	0.0014	0.0959	0.0187	0.3434	0.0793	0.7479	0.0027	0.0720
0.0983	0.1679	0.0442	0.1663	0.0059	0.0836	0.0983	0.1679	0.2710	+0.1319	0.0189	0.0556
0.0884	0.1935	0.0386	0.1769	0.0051	0.0849	0.0884	0.1935	0.2593	+0.0493	0.0167	0.0702
0.0706	0.2486	0.0299	0.1993	0.0038	0.0877	0.0706	0.2486	0.2203	0.1356	0.0131	0.0802
0.0516	0.3137	0.0215	0.2255	0.0027	0.0909	0.0516	0.3137	0.1663	0.3579	0.0094	0.0918
0.0370	0.3655	0.0153	0.2463	0.0019	0.0934	0.0370	0.3655	0.1209	0.5361	0.0067	0.1011
0923	0.1519	0.0419	0.1603	0.0056	0.0829	0.1154	0.1276	0.2103	+0.2862	0.0283	0.0769
0.0835	0.1664	0.0368	0.1666	0.0048	0.0837	0.1068	0.1463	0.2221	+0.2453	0.0254	0.0813
0.0705	0.1999	0.0299	0.1804	0.0038	0.0854	0.0926	0.1910	0.2190	+0.1321	0.0212	0.0913
0.0572	0.2478	0.0237	0.1997	0.0030	0.0878	0.0762	0.2559	0.1926	0.0433	0.0171	0.1055
0.0455	0.2960	0.0186	0.2189	0.0023	0.0901	0.0611	0.3215	0.1589	0.2243	0.0136	0.1197
0.0844	0.1461	0.0392	0.1582	0.0053	0.0827	0.1175	0.0846	0.0959	+0.3666	0.0372	0.0854
0.0745	0.1543	0.0335	0.1621	0.0045	0.0832	0.1097	0.0955	0.1322	+0.3615	0.0330	0.0890
0.0626	0.1710	0.0269	0.1694	0.0035	0.0841	0.0992	0.1213	0.1655	+0.3228	0.0280	0.0965
0.0537	0.1959	0.0223	0.1796	0.0028	0.0854	0.0890	0.1633	0.1731	+0.2367	0.0241	0.1076
0.0468	0.2255	0.0191	0.1915	0.0024	0.0869	0.0793	0.2149	0.1646	+0.1219	0.0210	0.1210
0.0793	0.1456	0.0380	0.1580	0.0053	0.0826	0.1023	0.0461	+0.0804	+0.3734	0.0452	0.0917
0.0667	0.1520	0.0311	0.1614	0.0043	0.0831	0.0950	0.0517	+0.0150	+0.3956	0.0388	0.0951
0.0521	0.1615	0.0232	0.1660	0.0031	0.0838	0.0863	0.0628	0.0588	+0.4118	0.0314	0.1004
0.0432	0.1715	0.0184	0.1705	0.0024	0.0844	0.0802	0.0793	0.0990	+0.4009	0.0268	0.1064
0.0384	0.1824	0.0159	0.1750	0.0020	0.0849	0.0759	0.1009	0.1150	+0.3684	0.0242	0.1133
0.0844	0.1439	0.0418	0.1568	0.0059	0.0824	0.0674	0.0157	+0.3652	+0.2913	0.0550	0.0942
0.0691	0.1510	0.0339	0.1605	0.0048	0.0830	0.0623	0.0184	+0.2658	+0.3364	0.0460	0.0985
0.0484	0.1609	0.0231	0.1656	0.0032	0.0837	0.0553	0.0226	+0.1311	+0.3969	0.0337	0.1044
0.0346	0.1680	0.0159	0.1692	0.0021	0.0842	0.0505	0.0266	+0.0410	+0.4351	0.0255	0.1089
0.0274	0.1723	0.0121	0.1714	0.0016	0.0845	0.0481	0.0306	0.0049	+0.4515	0.0213	0.1117

Prismatic Haunch at Both Ends

Note: All carry-over factors and fixed-end moment coefficients are negative and all stiffness factors are positive.

a	r	Carry-Over Factors c	Stiffness Factors k	Unif. Load F.E.M. Coef. × wL² M	Concentrated Load F.E.M.—Coef. × PL b										Haunch Load, Both Haunches F.E.M. Coef. × WL² M
					0.1		0.3		0.5		0.7		0.9		
					M_{AB}	M_{BA}	M_{AB}	M_{BA}	M_{AB}	M_{BA}	M_{AB}	M_{BA}	M_{AB}	M_{BA}	
0.1	0.4	0.583	5.49	0.0921	0.0905	0.0053	0.1727	0.0606	0.1396	0.1396	0.0606	0.1727	0.0053	0.0905	0.0049
	0.6	0.603	5.93	0.0940	0.0932	0.0040	0.1796	0.0589	0.1428	0.1428	0.0589	0.1796	0.0040	0.0932	0.0049
	1.0	0.624	6.45	0.0961	0.0962	0.0023	0.1873	0.0566	0.1462	0.1462	0.0566	0.1873	0.0023	0.0962	0.0050
	1.5	0.636	6.75	0.0972	0.0980	0.0013	0.1918	0.0551	0.1480	0.1480	0.0551	0.1918	0.0013	0.0980	0.0050
	2.0	0.641	6.90	0.0976	0.0988	0.0008	0.1939	0.0543	0.1489	0.1489	0.0543	0.1939	0.0008	0.0988	0.0050
0.2	0.4	0.634	7.32	0.0970	0.0874	0.0079	0.1852	0.0623	0.1506	0.1506	0.0623	0.1852	0.0079	0.0874	0.0187
	0.6	0.674	8.80	0.1007	0.0899	0.0066	0.1993	0.0584	0.1575	0.1575	0.0584	0.1993	0.0066	0.0899	0.0191
	1.0	0.723	11.09	0.1049	0.0935	0.0046	0.2193	0.0499	0.1654	0.1654	0.0499	0.2193	0.0046	0.0935	0.0195
	1.5	0.752	12.87	0.1073	0.0961	0.0029	0.2338	0.0420	0.1699	0.1699	0.0420	0.2338	0.0029	0.0961	0.0197
	2.0	0.765	13.87	0.1084	0.0976	0.0018	0.2410	0.0372	0.1720	0.1720	0.0372	0.2410	0.0018	0.0976	0.0198
0.3	0.4	0.642	9.02	0.0977	0.0845	0.0097	0.1763	0.0707	0.1558	0.1558	0.0707	0.1763	0.0097	0.0845	0.0397
	0.6	0.697	12.09	0.1027	0.0861	0.0095	0.1898	0.0700	0.1665	0.1665	0.0700	0.1898	0.0095	0.0861	0.0410
	1.0	0.775	18.68	0.1091	0.0890	0.0084	0.2136	0.0627	0.1803	0.1803	0.0627	0.2136	0.0084	0.0890	0.0426
	1.5	0.828	26.49	0.1132	0.0920	0.0065	0.2376	0.0492	0.1891	0.1891	0.0492	0.2376	0.0065	0.0920	0.0437
	2.0	0.855	32.77	0.1153	0.0943	0.0048	0.2555	0.0366	0.1934	0.1934	0.0366	0.2555	0.0048	0.0943	0.0442
0.4	0.4	0.599	10.15	0.0937	0.0825	0.0101	0.1601	0.0732	0.1509	0.1509	0.0732	0.1601	0.0101	0.0825	0.0642
	0.6	0.652	14.52	0.0986	0.0833	0.0106	0.1668	0.0776	0.1632	0.1632	0.0776	0.1668	0.0106	0.0833	0.0668
	1.0	0.744	26.06	0.1067	0.0847	0.0112	0.1790	0.0835	0.1833	0.1833	0.0835	0.1790	0.0112	0.0847	0.0711
	1.5	0.827	45.95	0.1131	0.0862	0.0113	0.1919	0.0852	0.1995	0.1995	0.0852	0.1919	0.0113	0.0862	0.0746
	2.0	0.878	71.41	0.1169	0.0876	0.0108	0.2033	0.0822	0.2089	0.2089	0.0822	0.2033	0.0108	0.0876	0.0766
0.5	0.0	0.500	4.00	0.0833	0.0810	0.0090	0.1470	0.0630	0.1250	0.1250	0.0630	0.1470	0.0090	0.0810	0.0833

When using these charts for T-beams and box girders $r = \sqrt[3]{(I_{max}/I_{min})} - 1$. Taken from *Beam Factors and Moment Coefficients for Members with Prismatic Haunches*, PCA 1956—ST 81.

Appendix M   Frame Constants and Fixed End Moments for Parabolic Haunched Beams with an Intermediate Hinge

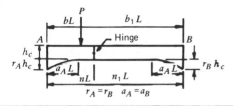

$r_A = r_B$	Carry-Over Factors C'_{AB}	C'_{BA}	Stiffness Factors k'_{AB}	k'_{BA}	Unif. Load F.E.M. Coef. $\times wL^2$ M'_{AB}	M'_{BA}	Concentrated Load F.E.M.—Coef. $\times PL$ b 0.1 M'_{AB}	M'_{BA}	0.15 M'_{AB}	M'_{BA}	0.2 M'_{AB}	M'_{BA}
					$n = 0.15$							
0.0	5.667	0.176	0.11	3.51	0.058	0.093	0.100	0.002	0.149	0.005	0.136	0.030
0.4	5.667	0.176	0.17	5.35	0.055	0.112	0.100	0.001	0.149	0.004	0.135	0.036
0.6	5.667	0.176	0.19	6.22	0.054	0.119	0.100	0.001	0.149	0.003	0.135	0.038
1.0	5.667	0.176	0.25	7.93	0.052	0.132	0.100	0.001	0.150	0.003	0.134	0.041
1.5	5.667	0.176	0.31	9.86	0.049	0.143	0.100	0.001	0.150	0.002	0.133	0.045
2.0	5.667	0.176	0.36	11.66	0.048	0.153	0.100	0.000	0.150	0.002	0.133	0.048
					$n = 0.20$							
0.0	4.000	0.250	0.23	3.69	0.078	0.085	0.099	0.004	0.149	0.007	0.197	0.012
0.4	4.000	0.250	0.36	5.75	0.075	0.101	0.099	0.002	0.149	0.005	0.198	0.010
0.6	4.000	0.250	0.42	6.73	0.073	0.108	0.100	0.002	0.149	0.004	0.198	0.008
1.0	4.000	0.250	0.54	8.71	0.070	0.119	0.100	0.001	0.150	0.003	0.199	0.006
1.5	4.000	0.250	0.69	10.99	0.067	0.130	0.100	0.001	0.150	0.003	0.199	0.005
2.0	4.000	0.250	0.82	13.12	0.065	0.139	0.100	0.001	0.150	0.003	0.199	0.004
					$n = 0.225$							
0.0	3.444	0.290	0.32	3.78	0.088	0.082	0.099	0.004	0.149	0.009	0.197	0.015
0.4	3.444	0.290	0.50	5.96	0.085	0.097	0.099	0.003	0.149	0.008	0.198	0.012
0.6	3.444	0.290	0.59	7.01	0.083	0.103	0.099	0.003	0.149	0.006	0.198	0.009
1.0	3.444	0.290	0.77	9.15	0.080	0.113	0.100	0.002	0.150	0.005	0.199	0.008
1.5	3.444	0.290	0.98	11.64	0.077	0.123	0.100	0.002	0.150	0.004	0.199	0.007
2.0	3.444	0.290	1.18	13.98	0.074	0.132	0.100	0.001	0.150	0.003	0.199	0.006
					$n = 0.25$							
0.0	3.000	0.333	0.43	3.86	0.098	0.081	0.098	0.006	0.149	0.012	0.197	0.019
0.4	3.000	0.333	0.68	6.16	0.094	0.093	0.099	0.004	0.149	0.008	0.198	0.013
0.6	3.000	0.333	0.81	7.30	0.092	0.099	0.099	0.003	0.149	0.006	0.198	0.011
1.0	3.000	0.333	1.07	9.62	0.089	0.108	0.099	0.002	0.150	0.005	0.199	0.009
1.5	3.000	0.333	1.37	12.35	0.086	0.118	0.100	0.002	0.150	0.004	0.199	0.007
2.0	3.000	0.333	1.56	14.95	0.083	0.126	0.100	0.001	0.150	0.003	0.199	0.006
					$n = 0.30$							
0.0	2.333	0.429	0.73	3.97	0.116	0.083	0.097	0.008	0.149	0.017	0.197	0.027
0.4	2.333	0.429	1.20	6.54	0.112	0.090	0.098	0.005	0.149	0.011	0.198	0.019
0.6	2.333	0.429	1.44	7.85	0.110	0.094	0.098	0.004	0.149	0.008	0.198	0.016
1.0	2.333	0.429	1.95	10.59	0.107	0.101	0.099	0.003	0.150	0.007	0.199	0.012
1.5	2.333	0.429	2.56	13.93	0.103	0.108	0.099	0.002	0.150	0.006	0.199	0.010
2.0	2.333	0.429	3.16	17.19	0.100	0.114	0.099	0.002	0.150	0.005	0.199	0.007

When using these tables and charts for T-beams and box girders $r = \sqrt[3]{(I_{max}/I_{min})} - 1$. Taken from *Beam Factors and Moment Coefficients for Members with Intermediate Expansion Hinges*, PCA ST 75.

Note: All carry-over factors and fixed end moment coefficients are negative and all stiffness factors are positive.

Concentrated Load F.E.M.—Coef. $\times PL$

	b										
0.225		0.25		0.3		0.5		0.7		0.9	
M'_{AB}	M'_{BA}	M'_{AB}	M'_{BA}	M'_{AB}	M'_{BA}	M'_{AB}	M'_{BA}	M'_{AB}	M'_{BA}	M'_{AB}	M'_{BA}
					$n = 0.15$						
0.129	0.042	0.123	0.054	0.110	0.076	0.062	0.147	0.025	0.161	0.003	0.083
0.128	0.051	0.122	0.066	0.107	0.096	0.055	0.188	0.019	0.194	0.002	0.089
0.127	0.055	0.120	0.071	0.105	0.104	0.052	0.204	0.016	0.206	0.001	0.092
0.126	0.060	0.118	0.080	0.103	0.118	0.047	0.233	0.013	0.226	0.001	0.094
0.124	0.066	0.116	0.088	0.100	0.132	0.042	0.261	0.010	0.244	0.001	0.096
0.123	0.071	0.115	0.096	0.098	0.142	0.038	0.283	0.008	0.256	0.000	0.097
					$n = 0.20$						
0.189	0.025	0.180	0.036	0.160	0.059	0.091	0.135	0.036	0.155	0.004	0.082
0.188	0.025	0.177	0.041	0.157	0.071	0.082	0.171	0.028	0.187	0.003	0.089
0.188	0.026	0.176	0.043	0.156	0.077	0.078	0.186	0.025	0.200	0.002	0.091
0.187	0.026	0.176	0.047	0.154	0.086	0.072	0.214	0.020	0.220	0.002	0.094
0.187	0.026	0.175	0.051	0.151	0.096	0.065	0.241	0.016	0.238	0.001	0.096
0.185	0.027	0.174	0.054	0.149	0.104	0.059	0.263	0.012	0.251	0.001	0.097
					$n = 0.225$						
0.220	0.019	0.210	0.031	0.188	0.053	0.107	0.129	0.043	0.153	0.004	0.082
0.221	0.015	0.210	0.031	0.186	0.060	0.098	0.163	0.034	0.184	0.004	0.088
0.222	0.012	0.210	0.031	0.185	0.064	0.093	0.177	0.030	0.196	0.003	0.090
0.222	0.010	0.209	0.031	0.183	0.070	0.086	0.203	0.025	0.216	0.002	0.093
0.223	0.008	0.208	0.032	0.181	0.076	0.079	0.230	0.019	0.234	0.002	0.095
0.223	0.006	0.208	0.032	0.179	0.083	0.072	0.251	0.015	0.247	0.001	0.097
					$n = 0.25$						
0.220	0.024	0.241	0.027	0.217	0.049	0.125	0.125	0.050	0.150	0.006	0.082
0.221	0.017	0.243	0.021	0.217	0.050	0.115	0.155	0.040	0.180	0.004	0.088
0.222	0.015	0.244	0.019	0.216	0.051	0.110	0.169	0.036	0.193	0.003	0.090
0.222	0.012	0.245	0.015	0.215	0.054	0.102	0.193	0.029	0.213	0.002	0.093
0.223	0.010	0.246	0.011	0.214	0.057	0.094	0.218	0.023	0.231	0.002	0.095
0.223	0.008	0.247	0.009	0.213	0.061	0.087	0.239	0.019	0.244	0.001	0.096
					$n = 0.30$						
0.220	0.033	0.241	0.038	0.278	0.051	0.162	0.122	0.066	0.147	0.008	0.081
0.221	0.024	0.243	0.028	0.283	0.040	0.153	0.143	0.054	0.175	0.005	0.087
0.222	0.020	0.244	0.024	0.285	0.035	0.148	0.154	0.049	0.186	0.005	0.089
0.222	0.015	0.245	0.019	0.288	0.028	0.140	0.173	0.041	0.205	0.003	0.092
0.223	0.011	0.246	0.014	0.291	0.022	0.131	0.194	0.033	0.224	0.002	0.095
0.223	0.009	0.248	0.011	0.293	0.017	0.124	0.213	0.027	0.237	0.002	0.096

Fixed End Moment Coefficients for Members with Symmetrical Parabolic Soffits and Intermediate Hinges

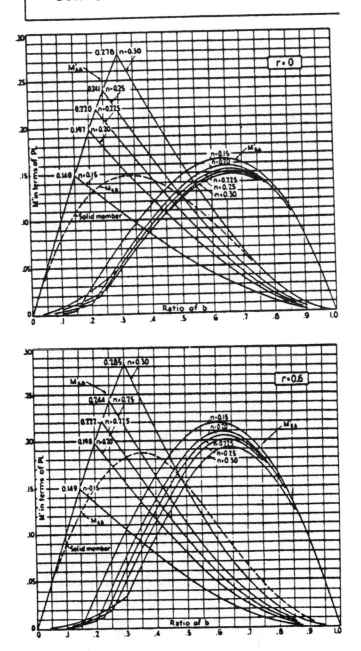

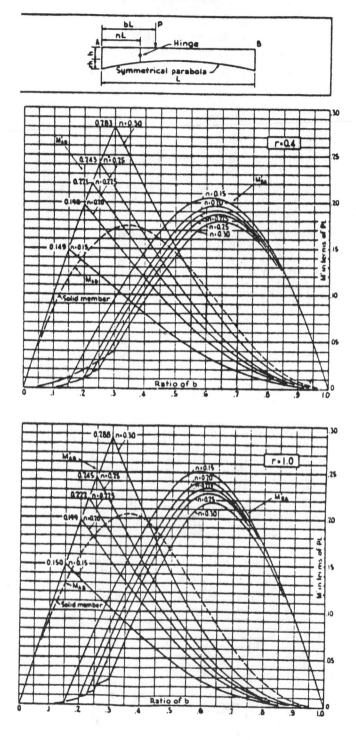

Fixed End Moment Coefficients for Members with Symmetrical Parabolic Soffits and Intermediate Hinges

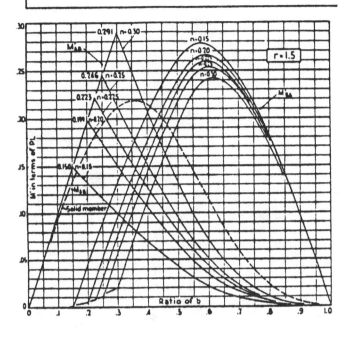

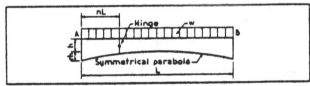

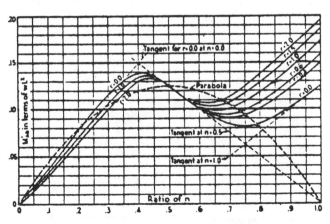

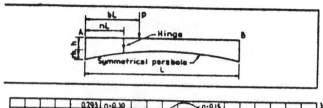

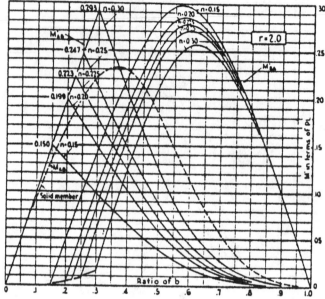

Appendix N   Simple Span Moments at Tenth Points[a]

Load $W = 1$, Span $= 1$	F_0	F_1		0	1	2
	0.01042	0.01042	$wL \div W$	0	0.40	.80
			$V \div W$	0.500	0.480	.420
			$M \div WL$	0	0.04933	.09467
	0.10	0.10	$wL \div W$	0	0.540	.960
			$V \div W$	0.500	0.4720	.3960
			$M \div WL$	0	0.04905	.09280
	0.08333	0.08333	$wL \div W$	1	1	1
			$V \div W$	0.50	0.40	0.30
			$M \div WL$	0.00	0.0450	0.080
	0.06250	0.06250	$wL \div W$	2.00	1.60	1.20
			$V \div W$	0.500	0.320	0.180
			$M \div WL$	0	0.04067	0.06533
	0.050	0.050	$wL \div W$	3.00	1.92	1.080
			$V \div W$	0.500	0.2560	0.1080
			$M \div WL$	0	0.03690	0.05440
	0.0750	0.10	$wL \div W$	0	.2850	0.540
			$V \div W$	0.3750	.36050	0.3190
			$M \div WL$	0	.03701	0.0762
	0.06667	0.10	$wL \div W$	0	0.20	0.40
			$V \div W$	0.33333	0.32333	0.29333
			$M \div WL$	0	0.0330	0.0640
	0.050	0.10	$wL \div W$	0	0.030	0.120
			$V \div W$	0.250	0.2490	0.2420
			$M \div WL$	0	0.02498	0.04960
			$wL \div W$	0	0.500	1.00
			$V \div W$	0.53333	0.50833	0.43333
			$M \div WL$	0	0.05250	0.10000

[a]In this table W = total area or load on the simple span.

3	4	5	6	7	8	9	10
1.20	1.60	2.00	1.60	1.20	0.80	0.40	0
0.320	0.180	0	0.180	0.320	0.420	0.480	0.500
0.1320	0.15733	0.16667	0.15733	0.1320	0.09467	0.04933	0
1.260	1.440	1.500	1.440	1.260	0.960	0.540	0
0.2840	0.1480	0	0.1480	0.2840	0.3960	0.4720	0.500
0.12705	0.14880	0.15625	0.14880	0.12705	0.09280	0.04905	0
1	1	1	1	1	1	1	1
0.20	0.10	0	0.10	0.20	0.30	0.40	0.50
0.1050	0.120	0.1250	0.120	0.1050	0.080	0.0450	0.00
0.80	0.40	0	0.40	0.80	1.20	1.60	2.00
0.080	0.020	0	0.020	0.080	0.180	0.320	0.500
0.0780	0.08267	0.08333	0.08267	0.0780	0.06533	0.04067	0
0.480	0.120	0	0.120	0.480	1.080	1.920	3.00
0.0320	0.0040	0	0.0040	0.0320	0.1080	0.2560	0.500
0.06090	0.06240	0.06250	0.06240	0.06090	0.05440	0.03690	0
0.7650	0.960	1.1250	1.260	1.3650	1.440	1.4850	1.50
0.25350	0.1670	+0.06250	−0.0570	0.18850	0.3290	0.47550	0.6250
0.10001	0.12120	0.13281	0.13320	0.12101	0.09520	.05501	0
0.60	0.80	1.00	1.20	1.40	1.60	1.80	2.00
0.27333	0.17333	+0.08333	−0.02667	0.15667	0.30667	.47667	0.66667
0.0910	0.1120	0.1250	0.1280	0.1190	0.0960	.0570	0
0.270	0.480	0.750	1.080	1.470	1.920	2.430	3.00
0.2230	0.1860	0.1250	+0.0340	−0.0930	0.2620	0.4790	0.750
0.07298	0.09360	0.10938	0.11760	0.11498	0.09760	.06098	0
1.50	2.00	1.6667	1.3333	1.000	0.66667	.33333	0
0.30833	+0.13333	−0.05000	0.20000	0.31667	0.40000	.45000	0.46667
0.13750	0.16000	0.16389	0.15111	0.12500	0.08889	.04611	0

Appendix M Fixed End Moments
Beams with Constant I

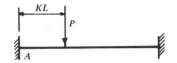 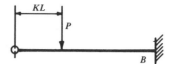

K	M_A	M_B	K	M_A	M_B	K	M_A	M_B
0.01	0.0098	0.0050	0.34	0.1481	0.1504	0.67	0.0730	0.1846
0.02	0.0192	0.0100	0.35	0.1479	0.1536	0.68	0.0696	0.1828
0.03	0.0282	0.0150	0.36	0.1475	0.1567	0.69	0.0663	0.1808
0.04	0.0369	0.0200	0.37	0.1469	0.1597	0.70	0.0630	0.1785
0.05	0.0451	0.0250	0.38	0.1461	0.1626	0.71	0.0597	0.1761
0.06	0.0530	0.0299	0.39	0.1451	0.1654	0.72	0.0564	0.1734
0.07	0.0605	0.0349	0.40	0.1440	0.1680	0.73	0.0532	0.1705
0.08	0.0677	0.0398	0.41	0.1427	0.1706	0.74	0.0500	0.1674
0.09	0.0745	0.0447	0.42	0.1413	0.1730	0.75	0.0469	0.1641
0.10	0.0810	0.0495	0.43	0.1397	0.1753	0.76	0.0438	0.1605
0.11	0.0871	0.0543	0.44	0.1380	0.1774	0.77	0.0407	0.1568
0.12	0.0929	0.0592	0.45	0.1361	0.1795	0.78	0.0378	0.1527
0.13	0.0984	0.0639	0.46	0.1341	0.1814	0.79	0.0348	0.1485
0.14	0.1035	0.0687	0.47	0.1320	0.1831	0.80	0.0320	0.1440
0.15	0.1084	0.0733	0.48	0.1298	0.1847	0.81	0.0292	0.1393
0.16	0.1129	0.0780	0.49	0.1274	0.1862	0.82	0.0266	0.1343
0.17	0.1171	0.0826	0.50	0.1250	0.1875	0.83	0.0240	0.1291
0.18	0.1210	0.0871	0.51	0.1225	0.1887	0.84	0.0215	0.1237
0.19	0.1247	0.0916	0.52	0.1198	0.1897	0.85	0.0191	0.1180
0.20	0.1280	0.0960	0.53	0.1171	0.1906	0.86	0.0169	0.1120
0.21	0.1311	0.1004	0.54	0.1143	0.1913	0.87	0.0147	0.1058
0.22	0.1338	0.1047	0.55	0.1114	0.1918	0.88	0.0127	0.0993
0.23	0.1364	0.1089	0.56	0.1084	0.1922	0.89	0.0108	0.0925
0.24	0.1386	0.1131	0.57	0.1054	0.1924	0.90	0.0090	0.0855
0.25	0.1406	0.1172	0.58	0.1023	0.1925	0.91	0.0074	0.0782
0.26	0.1424	0.1212	0.59	0.0992	0.1923	0.92	0.0059	0.0707
0.27	0.1439	0.1252	0.60	0.0960	0.1920	0.93	0.0046	0.0628
0.28	0.1452	0.1290	0.61	0.0928	0.1915	0.94	0.0034	0.0547
0.29	0.1462	0.1328	0.62	0.0895	0.1909	0.95	0.0024	0.0463
0.30	0.1470	0.1365	0.63	0.0862	0.1900	0.96	0.0015	0.0377
0.31	0.1476	0.1401	0.64	0.0829	0.1890	0.97	0.0009	0.0287
0.32	0.1480	0.1436	0.65	0.0796	0.1877	0.98	0.0004	0.0194
0.33	0.1481	0.1471	0.66	0.0763	0.1863	0.99	0.0001	0.0099

$M = CPL$; C = coefficient from Table: $C = K - 2K^2 + K^3$ for M_A; $C = \frac{1}{2}(K - K^3)$ for M_B.

Appendix P   Fixed End Moments for Beams of Constant Moment of Inertia Using Various Loadings

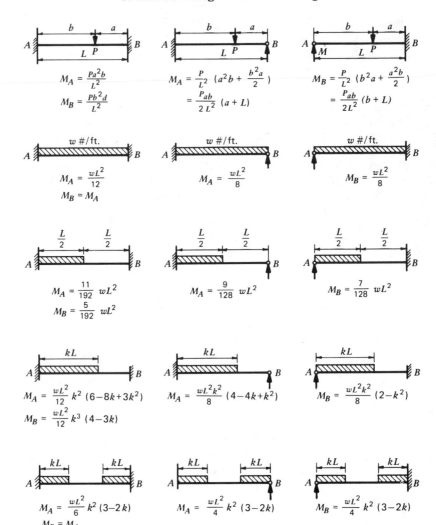

$$M_A = \frac{Pa^2b}{L^2}$$

$$M_B = \frac{Pb^2d}{L^2}$$

$$M_A = \frac{P}{L^2}\left(a^2b + \frac{b^2a}{2}\right)$$
$$= \frac{Pab}{2L^2}(a+L)$$

$$M_B = \frac{P}{L^2}\left(b^2a + \frac{a^2b}{2}\right)$$
$$= \frac{Pab}{2L^2}(b+L)$$

$$M_A = \frac{wL^2}{12}$$
$$M_B = M_A$$

$$M_A = \frac{wL^2}{8}$$

$$M_B = \frac{wL^2}{8}$$

$$M_A = \frac{11}{192}\, wL^2$$
$$M_B = \frac{5}{192}\, wL^2$$

$$M_A = \frac{9}{128}\, wL^2$$

$$M_B = \frac{7}{128}\, wL^2$$

$$M_A = \frac{wL^2}{12} k^2 (6 - 8k + 3k^2)$$
$$M_B = \frac{wL^2}{12} k^3 (4 - 3k)$$

$$M_A = \frac{wL^2 k^2}{8} (4 - 4k + k^2)$$

$$M_B = \frac{wL^2 k^2}{8} (2 - k^2)$$

$$M_A = \frac{wL^2}{6} k^2 (3 - 2k)$$
$$M_B = M_A$$

$$M_A = \frac{wL^2}{4} k^2 (3 - 2k)$$

$$M_B = \frac{wL^2}{4} k^2 (3 - 2k)$$

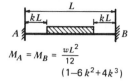

$$M_A = M_B = \frac{wL^2}{12}$$
$$(1-6k^2+4k^3)$$

$$M_A = \frac{wL^2}{8}$$
$$(1-6k^2-4k^3)$$

$$M_B = \frac{wL^2}{8}$$
$$(1-6k^2-4k^3)$$

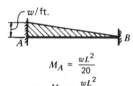

$M_A = M_A$ (for k_1L) −
M_A (for k_2L)
$M_B = M_B$ (for k_1L) −
M_B (for k_2L)

$M_A = M_A$ (for k_1L) −
M_A (for k_2L)

$M_B = M_B$ (for k_1L) −
M_B (for k_2L)

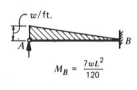

$$M_A = \frac{wL^2}{20}$$
$$M_B = \frac{wL^2}{30}$$

$$M_A = \frac{wL^2}{15}$$

$$M_B = \frac{7wL^2}{120}$$

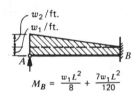
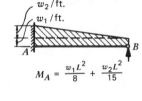

$$M_A = \frac{w_1L^2}{12} + \frac{w_2L^2}{20}$$
$$M_B = \frac{w_1L^2}{12} + \frac{w_2L^2}{30}$$

$$M_A = \frac{w_1L^2}{8} + \frac{w_2L^2}{15}$$

$$M_B = \frac{w_1L^2}{8} + \frac{7w_1L^2}{120}$$

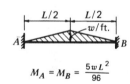
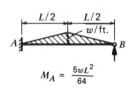
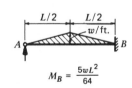

$$M_A = M_B = \frac{5wL^2}{96}$$

$$M_A = \frac{5wL^2}{64}$$

$$M_B = \frac{5wL^2}{64}$$

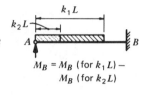

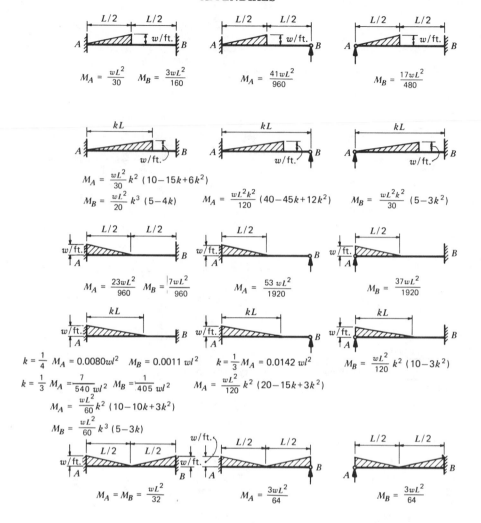

$$M_A = \frac{wL^2}{30} \quad M_B = \frac{3wL^2}{160}$$

$$M_A = \frac{41wL^2}{960}$$

$$M_B = \frac{17wL^2}{480}$$

$$M_A = \frac{wL^2}{30} k^2 (10-15k+6k^2)$$

$$M_B = \frac{wL^2}{20} k^3 (5-4k) \qquad M_A = \frac{wL^2 k^2}{120} (40-45k+12k^2) \qquad M_B = \frac{wL^2 k^2}{30} (5-3k^2)$$

$$M_A = \frac{23wL^2}{960} \quad M_B = \frac{7wL^2}{960}$$

$$M_A = \frac{53 wL^2}{1920}$$

$$M_B = \frac{37wL^2}{1920}$$

$$k = \frac{1}{4} \ M_A = 0.0080wl^2 \quad M_B = 0.0011 \ wl^2 \qquad k = \frac{1}{3} M_A = 0.0142 \ wl^2 \qquad M_B = \frac{wL^2}{120} k^2 (10-3k^2)$$

$$k = \frac{1}{3} \ M_A = \frac{7}{540} wl^2 \quad M_B = \frac{1}{405} wl^2 \qquad M_A = \frac{wL^2}{120} k^2 (20-15k+3k^2)$$

$$M_A = \frac{wL^2}{60} k^2 (10-10k+3k^2)$$

$$M_B = \frac{wL^2}{60} k^3 (5-3k)$$

$$M_A = M_B = \frac{wL^2}{32}$$

$$M_A = \frac{3wL^2}{64}$$

$$M_B = \frac{3wL^2}{64}$$

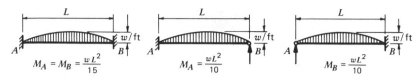

$$M_A = M_B = \frac{wL^2}{15}$$

$$M_A = \frac{wL^2}{10}$$

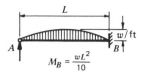

$$M_B = \frac{wL^2}{10}$$

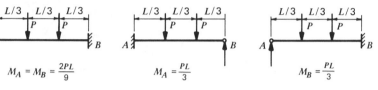

$$M_A = M_B = \frac{PL}{8}$$

$$M_A = \frac{3PL}{16}$$

$$M_B = \frac{3PL}{16}$$

$$M_A = M_B = \frac{2PL}{9}$$

$$M_A = \frac{PL}{3}$$

$$M_B = \frac{PL}{3}$$

$$M_A = M_B = \frac{15PL}{48}$$

$$M_A = \frac{45PL}{96}$$

$$M_B = \frac{45PL}{96}$$

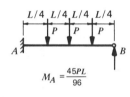

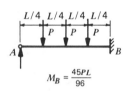

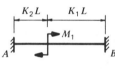

$$M_A = -M_1 k_1 (1 - 3k_2)$$
$$M_B = +M_1 k_2 (1 - 3k_1)$$

$$M_A = -\frac{M_1}{2}(3k_1^{-2} - 1)$$

$$M_B = +\frac{M_1}{2}(3k_2^{-2} - 1)$$

Appendix Q

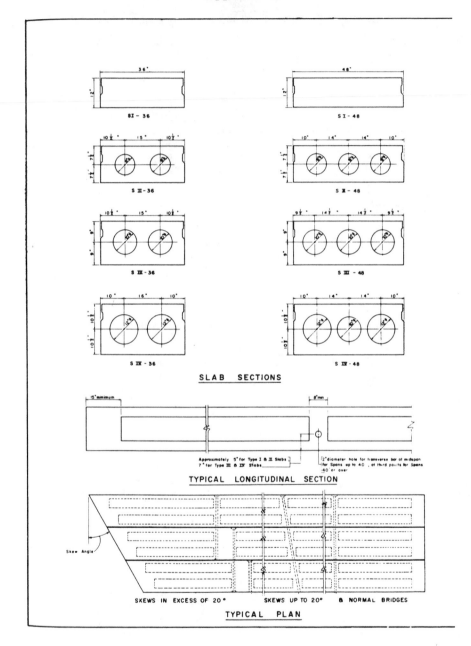

SECTION PROPERTIES

SECTION	AREA(in²)	MOMENT OF 'NERTIA (in⁴)	SECTION MODULUS (in³)	APPROXIMATE SPAN (ft.) H2O-S16-44	H20-44
S I-36 (12"x36")	432	5,184	864	20 - 28	20 - 29
S II-36 (15"x36")	4395	9,725	1,296	27 - 35	28 - 38
S III-36 (18"x36")	491	16,514	1,835	34 - 42	37 - 46
S IV-36 (21"x36")	530	25,747	2,452	41 - 49	45 - 54
S I-48 (12"x48")	576	6,912	1,152	20 - 28	20 - 29
S II-48 (15"x48")	5692	12,897	1,720	27 - 35	28 - 38
S III-48 (18"x48")	628	21,855	2,428	34 - 42	37 - 46
S IV-48 (21"x48")	703	34,517	3,287	41 - 50	45 - 55

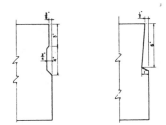

SHEAR KEYS

GENERAL NOTES

SPECIFICATIONS "AASHO Standard Specifications for Highway Bridges" Current edition; "Criteria for Prestressed Concrete Bridges, U.S Bureau of Public Roads",1954 Tentative Recommendations for Prestressed Concrete" by ACI - ASCE Joint Committee 323, and any subsequent revisions approved by the Committee on Bridges and Structures of the AASHO.

LIVE LOAD All highway live loads as specified by AASHO Standard Specifications for Highway Bridges Live loads shall be distributed in accordance with Section 1.3.2 (c) of the "AASHO Standard Specifications for Highway Bridges"

PURPOSE The purpose of the standards shown on this sheet shall be to establish a limited number of simple, practical sections leading to uniformity and simplicity of forming and production methods. These standards shall be applicable to all conditions of highway bridge loading and usage within the approximate span limits indicated in the SUMMARY OF SECTIONS on this sheet. The purpose is specifically not to disrupt or supplant established prestressed concrete slab practice utilizing present plant or forms Similar sections with minor dimensional variations, manufactured with established plant facilities, which meet structural and geometrical requirements of a specific project may be substituted upon submission by the producer of the data necessary to show compliance with the requirements of the job and upon approval of the substitution by the engineer Further, the purpose is not to supercede other standard sections adopted by the AASHO and PCI, but rather to compliment these standards.

SPAN LIMITS Span limits shown are approximate only and are not a rigid limitation of the section. The limits shown are based on f'ci equal to 4000 psi, f'c equal to 5000 psi, 28 foot roadway, an allowance of 30 pounds per square foot for surfacing, the use of straight pretensioning, and allowable stresses as given in the design specifications. The upper span limits may be extended by reduced loading, increased concrete strength, use of light weight concrete, draped tendons, or other approved means within the limits of the design specifications

CONCRETE Recommended minimum strength for concrete in slabs are f'c = 5000 psi at transfer of stressing force, fci = 4000 psi Concrete of greater or less compressive strength, but not less than f'c = 4000 psi, may be used, in which case allowable working stresses and resulting utility of the slabs will be based upon the actual concrete specifications for the particular project

PRESTRESSING REINFORCEMENT Prestressing reinforcement shall generally be designed for particular projects or for prevailing bridge practices and available manufacturing facilities.
Materials for prestressing reinforcement may be any of the materials specified by governing specifications or subsequent developments by manufacturers which have generally been accepted in common prestressed practice.

END BLOCKS The slabs shown utilize end blocks 15" long which have proven satisfactory in many installations The length of end blocks may be increased to accommodate local plant facilities or particular job requirements. Sufficient mild steel reinforcement should be provided in end blocks to resist the tensile forces due to concentrated prestressing loads

DIAPHRAGMS Diaphragms cast within the slab are recommended at midspan for spans up to 40 feet and at third points for spans 40 feet or over

LATERAL TIES Lateral ties shall be provided through the diaphragms in the positions indicated.
Each tie shall be equivalent to a 1½" mild steel bar tensioned to 30,000 pounds or an equal force applied by lateral tensioning of high strength tendons. Tension in 1½" mild steel may be applied by torquing to approximately 600 foot - pounds

SHEAR KEYS After lateral ties have been placed and tightened, shear keys shall be filled with high strength, non - shrinking mortar

FORMS The use of steel forms on concrete founded casting beds is recommended

CHAMFERS AND CORNERS All exposed corners shall be chamfered ¾" or rounded to ¾" radius

FINISH Tops shall be given a broom finish, normal to centerline of roadway

HANDLING In handling, the slabs must be maintained in an upright position at all times and must be picked up only by means of approved devices near the ends of the slabs.

MILD STEEL REINFORCING, BEARING PADS, ANCHORAGES AND MISCELLANEOUS DETAILS All details not shown or specified hereon shall be designed for particular job requirements and shall be in accordance with applicable job specifications

JOINT COMMITTEE
AMERICAN ASSOCIATION OF STATE HIGHWAY OFFICIALS
COMMITTEE ON BRIDGES AND STRUCTURES
AND
PRESTRESSED CONCRETE INSTITUTE

STANDARD PRESTRESSED CONCRETE SLABS FOR HIGHWAY BRIDGE SPANS UP TO 55 FT.

SUBMITTED BY

FOR AASHO   FOR PCI 14887

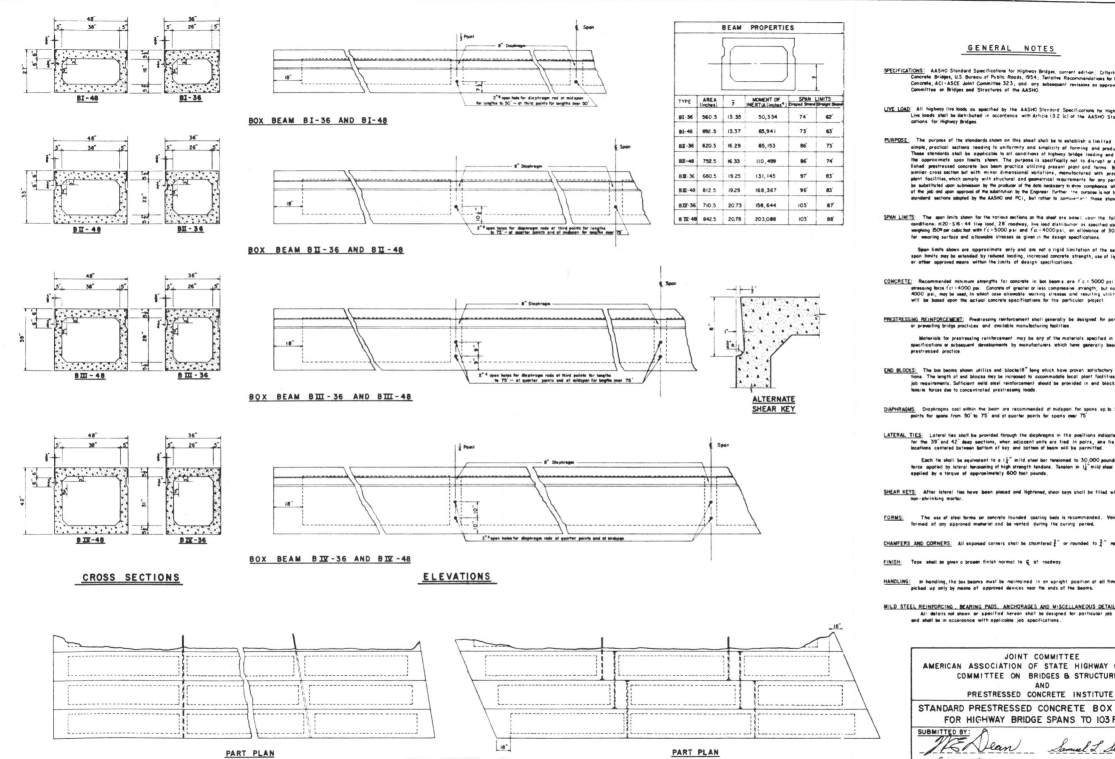

	BEAM PROPERTIES				
TYPE	AREA (inches)	$\bar{y}$	MOMENT OF INERTIA (inches⁴)	SPAN LIMITS Draped Strand	Straight Strand
BI-36	560.5	13.35	50,334	74'	62'
BI-48	692.5	13.37	65,941	73'	63'
BII-36	620.5	16.29	85,153	86'	73'
BII-48	752.5	16.33	110,499	86'	74'
BIII-36	680.5	19.25	131,145	97'	83'
BIII-48	812.5	19.29	168,367	96'	83'
BIV-36	710.5	20.73	158,644	103'	87'
BIV-48	842.5	20.78	203,088	103'	88'

BOX BEAM BI-36 AND BI-48

BOX BEAM BII-36 AND BII-48

BOX BEAM BIII-36 AND BIII-48

BOX BEAM BIV-36 AND BIV-48

ALTERNATE SHEAR KEY

CROSS SECTIONS

ELEVATIONS

PART PLAN
NORMAL SKEWS AND SKEWS UP TO 10°

PART PLAN
SKEWS IN EXCESS OF 10°

GENERAL NOTES

SPECIFICATIONS: AASHO Standard Specifications for Highway Bridges, current edition; Criteria for Prestressed Concrete Bridges, U.S. Bureau of Public Roads, 1954; Tentative Recommendations for Prestressed Concrete, ACI-ASCE Joint Committee 323, and any subsequent revisions as approved by the Committee on Bridges and Structures of the AASHO.

LIVE LOAD: All highway live loads as specified by the AASHO Standard Specifications for Highway Bridges. Live loads shall be distributed in accordance with Article 1.3.2 (c) of the AASHO Standard Specifications for Highway Bridges.

PURPOSE: The purpose of the standards shown on this sheet shall be to establish a limited number of simple, practical sections leading to uniformity and simplicity of forming and production methods. These standards shall be applicable to all conditions of highway bridge loading and usage within the approximate span limits shown. The purpose is specifically not to disrupt or supplant established prestressed concrete box beam practice utilizing present plant and forms. Box beams of similar cross section but with minor dimensional variations, manufactured with presently established plant facilities, which comply with structural and geometrical requirements for any particular project may be substituted upon submission by the producer of the data necessary to show compliance with the requirements of the job and upon approval of the substitution by the Engineer. Further the purpose is not to supersede other standard sections adopted by the AASHO and PCI, but rather to commend those standards.

SPAN LIMITS: The span limits shown for the various sections on this sheet are based upon the following design conditions. H20-S16-44 live load, 28' roadway, live load distribution as specified above, concrete weighing 150# per cubic foot with f'c = 5000 psi and f'ci = 4000 psi, an allowance of 30# per square foot for wearing surface and allowable stresses as given in the design specifications.

Span limits shown are approximate only and are not a rigid limitation of the sections. The upper span limits may be extended by reduced loading, increased concrete strength, use of lightweight concrete, or other approved means within the limits of design specifications.

CONCRETE: Recommended minimum strengths for concrete in box beams are f'c = 5000 psi, at transfer of stressing force f'ci = 4000 psi. Concrete of greater or less compressive strength, but not less than f'c = 4000 psi, may be used, in which case allowable working stresses and resulting utility of the box beams will be based upon the actual concrete specifications for the particular project.

PRESTRESSING REINFORCEMENT: Prestressing reinforcement shall generally be designed for particular projects or prevailing bridge practices and available manufacturing facilities.

Materials for prestressing reinforcement may be any of the materials specified in the governing specifications or subsequent developments by manufacturers which have generally been accepted in prestressed practice.

END BLOCKS: The box beams shown utilize end blocks 18" long which have proven satisfactory in many installations. The length of end blocks may be increased to accommodate local plant facilities or particular job requirements. Sufficient mild steel reinforcement should be provided in end blocks to resist the tensile forces due to concentrated prestressing loads.

DIAPHRAGMS: Diaphragms cast within the beam are recommended at midspan for spans up to 50', at third points for spans from 50' to 75' and at quarter points for spans over 75'.

LATERAL TIES: Lateral ties shall be provided through the diaphragms in the positions indicated except that for the 39" and 42" deep sections, when adjacent units are tied in pairs, one tie at diaphragm locations centered between bottom of key and bottom of beam will be permitted.

Each tie shall be equivalent to a 1¼" mild steel bar tensioned to 30,000 pounds or an equal force applied by lateral tensioning of high strength tendons. Tension in 1¼" mild steel bars may be applied by a torque of approximately 600 foot pounds.

SHEAR KEYS: After lateral ties have been placed and tightened, shear keys shall be filled with high strength, non-shrinking mortar.

FORMS: The use of steel forms on concrete founded casting beds is recommended. Voids may be formed of any approved material and be vented during the curing period.

CHAMFERS AND CORNERS: All exposed corners shall be chamfered ¾" or rounded to ¾" radius.

FINISH: Tops shall be given a broom finish normal to ℄ of roadway.

HANDLING: In handling, the box beams must be maintained in an upright position at all times and must be picked up only by means of approved devices near the ends of the beams.

MILD STEEL REINFORCING, BEARING PADS, ANCHORAGES AND MISCELLANEOUS DETAILS:
All details not shown or specified hereon shall be designed for particular job requirements and shall be in accordance with applicable job specifications.

JOINT COMMITTEE
AMERICAN ASSOCIATION OF STATE HIGHWAY OFFICIALS
COMMITTEE ON BRIDGES & STRUCTURES
AND
PRESTRESSED CONCRETE INSTITUTE

STANDARD PRESTRESSED CONCRETE BOX BEAMS
FOR HIGHWAY BRIDGE SPANS TO 103 FEET

SUBMITTED BY:

FOR AASHO   FOR PCI

AASHO - PCI STANDARD BRIDGE BEAMS

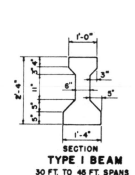
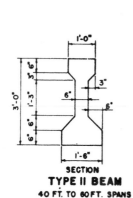
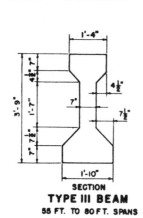
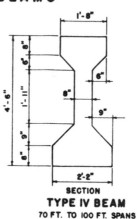

SECTION
TYPE I BEAM
30 FT. TO 45 FT. SPANS

SECTION
TYPE II BEAM
40 FT. TO 60 FT. SPANS

SECTION
TYPE III BEAM
55 FT. TO 80 FT. SPANS

SECTION
TYPE IV BEAM
70 FT. TO 100 FT. SPANS

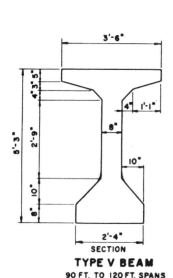
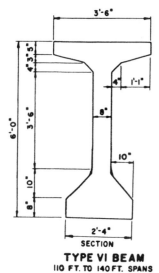

SECTION
TYPE V BEAM
90 FT. TO 120 FT. SPANS

SECTION
TYPE VI BEAM
110 FT. TO 140 FT. SPANS

BEAM PROPERTIES

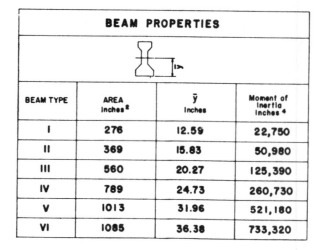

BEAM TYPE	AREA Inches²	ȳ Inches	Moment of Inertia Inches⁴
I	276	12.59	22,750
II	369	15.83	50,980
III	560	20.27	125,390
IV	789	24.73	260,730
V	1013	31.96	521,180
VI	1085	36.38	733,320

END BLOCK DIMENSIONS

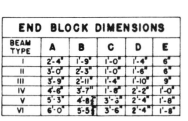
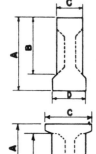

BEAM TYPE	A	B	C	D	E
I	2'-4"	1'-9"	1'-0"	1'-4"	6"
II	3'-0"	2'-3"	1'-0"	1'-6"	6"
III	3'-9"	2'-11"	1'-4"	1'-10"	9"
IV	4'-6"	3'-7"	1'-6"	2'-2"	1'-0"
V	5'-3"	4'-8½"	3'-0"	2'-4"	1'-6"
VI	6'-0"	5'-5"	3'-6"	2'-4"	1'-6"

END BLOCK DETAILS —

FOR POST-TENSIONED BEAMS ONLY

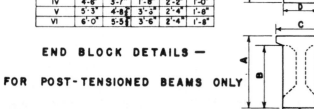

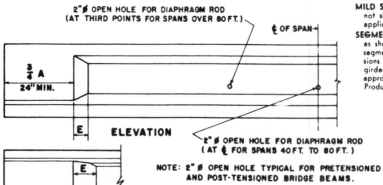

2" Ø OPEN HOLE FOR DIAPHRAGM ROD
(AT THIRD POINTS FOR SPANS OVER 80 FT.)

℄ OF SPAN

¾ A
24" MIN.

ELEVATION

2" Ø OPEN HOLE FOR DIAPHRAGM ROD
(AT ℄ FOR SPANS 40 FT. TO 80 FT.)

NOTE: 2" Ø OPEN HOLE TYPICAL FOR PRETENSIONED
AND POST-TENSIONED BRIDGE BEAMS.

GENERAL NOTES

SPECIFICATIONS: AASHO Standard Specifications for Highway Bridges, latest edition, together with any tentative or supplemental specifications approved by the AASHO Committee on Bridges and Structures.

LIVE LOAD. All Highway Live Loads as specified in AASHO Standard Specifications for Highway Bridges.

PURPOSE: The purpose of the beams shown on this sheet shall be to establish a limited number of simple, practical beam sections leading to uniformity and simplicity of practice, forming, and production methods, and which are applicable to all conditions of highway bridge loading and to all spans within the approximate limits shown. Beams of similar cross sections but with minor dimensional differences, manufactured with presently established plant facilities and which comply with structural and geometric requirements for any particular project may be substituted upon submission by the Producer of data showing compliance with job requirements and approval by the Engineer.

SPAN LIMITS: The span limits shown on this sheet are approximate only and are not mandatory at either limit. Lateral spacing of beams shall be varied in keeping with the requirements of span and loading. The span limits shown contemplate the use of concrete weighing 150 pounds per cubic foot, HS 20-44 live load, simple span construction, cast-in-place deck slabs 6" to 8" thick composed of concrete with f'c not less than 3000 psi and having elastic properties approximately equal to those of the beam concrete. All dead load is assumed to be carried by the beam alone with live load carried by the beam and slab composite section.

By using light weight concrete, continuous construction or live loadings lighter than HS 20-44, span limits may be increased.

CONCRETE: Recommended minimum strengths for concrete in beams are f'c = 5,000 psi; at transfer of stressing force, f'c = 4,000 psi. Concrete of greater compressive strength may be used in which case allowable working stresses and resulting utility of the beams will be based upon the actual concrete specifications for the particular project.

PRESTRESSING REINFORCEMENT: Prestressing reinforcement shall generally be designed for particular projects or for prevailing bridge practices and available manufacturing facilities.

The beams are applicable for use with any acceptable type of prestressing in current practice; namely, pretensioning with straight or deflected strands, post-tensioning or a combination of pretensioning and post-tensioning. Use of draped reinforcement will generally be required for the longer spans in each beam series. Materials for prestressing reinforcement shall be in accordance with the latest applicable designations of ASTM specifications for the particular type of tendons or subsequent developments by manufacturers which have generally been approved and accepted by member departments of AASHO.

Broken wires within individual strands will be permitted up to 2% of the total number of wires in each beam, providing that there is not more than one broken wire per strand. Two or more broken wires per strand will be cause for replacement of the strand even though the two broken wires are within the 2% limitation.

END ZONES: PRETENSIONED: Where all tendons are pretensioned 7-wire strands, the use of end blocks will not be required.

In pretensioned beams, vertical stirrups acting at a unit stress of 20,000 psi. to resist at least 4 per cent of the total prestressing force shall be placed within the distance of d/4 of the end of the beam, the end stirrups to be as close to the end of the beam as practicable. These vertical stirrups are to be provided in addition to those required as shear steel.

POST-TENSIONED: For beams with post-tensioning tendons, end blocks shall be used to distribute the concentrated prestressing forces of the anchorage.

End blocks shall have sufficient area to allow the spacing of the prestressing steel. Preferably, they shall be as wide as the narrow flange of the beam. Their length shall be at least three-fourths of the beam depth or 24 inches minimum. In post-tensioned members, a closely spaced grid of vertical and horizontal bars shall be placed near the face of the end block to resist crushing. Closely spaced horizontal and vertical bars shall be placed through the length of block.

DIAPHRAGMS: Diaphragms of precast or cast-in-place construction using prestressed or non-stressed reinforcement are recommended at span ends. Intermediate diaphragms are not required in spans up to 40 ft.; are recommended at mid-span for spans above 40 ft. and to 80 feet; and are recommended at span third points for spans in excess of 80 feet.

FORMS: The use of steel forms on concrete casting beds is recommended.

CHAMFERS & CORNERS: All exposed corners shall be chamfered not less than ¾" or rounded to ¾" radius. Angles of intersection between webs and flanges shall be rounded to not less than ¾" radius.

FINISH OF TOPS: Tops of all beams shall be left rough. At approximately the time of initial set, all laitance shall be removed with a coarse wire brush.

HANDLING: Beams must be maintained in an upright position at all times and must be picked up only by means of approved devices anchored within the end zones. Disregard of this requirement may result in damage of the member.

MILD STEEL REINFORCING, ELASTOMERIC BEARING PADS, SHOES AND MISCELLANEOUS DETAILS: All details not shown or specified hereon shall be designed for particular job requirements and shall be in accordance with applicable job specifications.

SEGMENTAL CONSTRUCTION: Beams may be built by segmental construction in lieu of full length construction as shown on the plans. In the event segmental construction is elected, the Producer shall, prior to casting any segments, submit to the Engineer for approval complete working drawings showing locations and details and dimensions of construction joints, materials and methods for making closures, and form and casting details. For exterior girders, the closures shall match the color and texture of the adjoining concrete. Working drawings shall be approved by the Engineer before any work involving the drawings is performed. Approval shall not relieve the Producer of responsibility for achieving the desired results.

JOINT COMMITTEE
AMERICAN ASSOCIATION OF STATE HIGHWAY OFFICIALS
COMMITTEE ON BRIDGES AND STRUCTURES AND
PRESTRESSED CONCRETE INSTITUTE

STANDARD PRESTRESSED CONCRETE
BEAMS FOR HIGHWAY BRIDGE SPANS
30 FT. TO 140 FT.

| DATE 1 FEB 1962 | SCALE | REV. NO. △ | DRWG. NO. STD 101-66 |

Index